Frontiers of Nanoscience

Volume 9

Protected Metal Clusters: From Fundamentals to Applications

FRONTIERS OF NANOSCIENCE

Series Editor: Richard E. Palmer
The Nanoscale Physics Research Laboratory,
The School of Physics and Astronomy,
The University of Birmingham, UK

Vol. 1 **Nanostructured Materials** edited by
Gerhard Wilde

Vol. 2 **Atomic and Molecular Manipulation** edited by
Andrew J. Mayne and Gérald Dujardin

Vol. 3 **Metal Nanoparticles and Nanoalloys** edited by
Roy L. Johnston and J.P. Wilcoxon

Vol. 4 **Nanobiotechnology** edited by
Jesus M. de la Fuente and V. Grazu

Vol. 5 **Nanomedicine** edited by
Huw Summers

Vol. 6 **Nanomagnetism: Fundamentals and Applications** edited by
Chris Binns

Vol. 7 **Nanoscience and the Environment** edited by
Jamie R. Lead and Eugenia Valsami-Jones

Vol. 8 **Characterization of Nanomaterials in Complex Environmental and Biological Media** edited by
Mohammed Baalousha and Jamie R. Lead

Vol. 9 **Protected Metal Clusters: From Fundamentals to Applications**
edited by Tatsuya Tsukuda and Hannu Häkkinen

Frontiers of Nanoscience

Volume 9

Protected Metal Clusters: From Fundamentals to Applications

Edited by

Tatsuya Tsukuda
Department of Chemistry,
School of Science,
The University of Tokyo, Tokyo, Japan

Hannu Häkkinen
Departments of Chemistry and Physics,
Nanoscience Center,
University of Jyväskylä,
Jyväskylä, Finland

ELSEVIER

AMSTERDAM • BOSTON • HEIDELBERG • LONDON • NEW YORK • OXFORD
PARIS • SAN DIEGO • SAN FRANCISCO • SINGAPORE • SYDNEY • TOKYO

Elsevier
Radarweg 29, PO Box 211, 1000 AE Amsterdam, Netherlands
The Boulevard, Langford Lane, Kidlington, Oxford OX5 1GB, UK
225 Wyman Street, Waltham, MA 02451, USA

ISBN: 978-0-08-100086-1
ISSN: 1876-2778

For Information on all Elsevier publications
visit our website at http://store.elsevier.com/

Contents

Contributors

Christopher J. Ackerson, Department of Chemistry, Colorado State University, Fort Collins, CO, USA

Christine M. Aikens, Department of Chemistry, Kansas State University, Manhattan, KS, USA

Yuxiang Chen, Department of Chemistry, Carnegie Mellon University, Pittsburgh, PA, USA

Nirmal Goswami, Department of Chemical and Biomolecular Engineering, National University of Singapore, Singapore

Hannu Häkkinen, Department of Physics, and Department of Chemistry, Nanoscience Center, University of Jyväskylä, Jyväskylä, Finland

De-en Jiang, Department of Chemistry, University of California, Riverside, CA, USA

Rongchao Jin, Department of Chemistry, Carnegie Mellon University, Pittsburgh, PA, USA

Wataru Kurashige, Department of Applied Chemistry, Faculty of Science, Tokyo University of Science, Shinjuku-ku, Tokyo, Japan

Jingguo Li, Department of Chemical and Biomolecular Engineering, National University of Singapore, Singapore

Yuichi Negishi, Department of Applied Chemistry, Faculty of Science, Tokyo University of Science, Shinjuku-ku, Tokyo; Photocatalysis International Research Center, Tokyo University of Science, Noda, Chiba; and Department of Materials Molecular Science, Institute for Molecular Science, Okazaki, Aichi, Japan

Thomas W. Ni, Department of Chemistry, Colorado State University, Fort Collins, CO, USA

Yoshiki Niihori, Department of Applied Chemistry, Faculty of Science, Tokyo University of Science, Shinjuku-ku, Tokyo, Japan

Richard E. Palmer, Nanoscale Physics Research Laboratory, School of Physics and Astronomy, University of Birmingham, Birmingham, UK

Shinjiro Takano, Department of Chemistry, School of Science, The University of Tokyo, Bunkyo-ku, Tokyo, Japan

Qing Tang, Department of Chemistry, University of California, Riverside, CA, USA

Marcus A. Tofanelli, Department of Chemistry, Colorado State University, Fort Collins, CO, USA

Tatsuya Tsukuda, Department of Chemistry, School of Science, The University of Tokyo, Bunkyo-ku, Tokyo; Elements Strategy Initiative for Catalysts and Batteries (ESICB), Kyoto University, Katsura, Kyoto, Japan

Yu Wang, Collaborative Innovation Center of Chemistry for Energy Materials, State Key Laboratory for Physical Chemistry of Solid Surfaces, College of Chemistry and Chemical Engineering, Xiamen University, Xiamen, China

Zhi Wei Wang, Beijing Institute of Nanoenergy and Nanosystems, Chinese Academy of Sciences, Beijing, China

Jianping Xie, Department of Chemical and Biomolecular Engineering, National University of Singapore, Singapore

Huayan Yang, Collaborative Innovation Center of Chemistry for Energy Materials, State Key Laboratory for Physical Chemistry of Solid Surfaces, College of Chemistry and Chemical Engineering, Xiamen University, Xiamen, China

Nanfeng Zheng, Collaborative Innovation Center of Chemistry for Energy Materials, State Key Laboratory for Physical Chemistry of Solid Surfaces, College of Chemistry and Chemical Engineering, Xiamen University, Xiamen, China

Acknowledgments

We would like to express our appreciation to Professor Richard Palmer (Birmingham University), the series editor of **Frontiers of Nanoscience**, for giving us an opportunity to edit a book on the currently hot topic of nanomaterials. We are grateful to all our colleagues for their significant efforts in writing the chapters. We thank our Editorial Project Manager, Derek Coleman (Amsterdam), and Acquisitions Editor, Susan Dennis (Oxford), for their guidance and patience throughout this project. H.H. thanks the Wihuri Foundation for supporting a sabbatical leave during which part of this project was completed.

June 2015
Tatsuya Tsukuda/Tokyo, Japan
Hannu Häkkinen/Jyväskylä, Finland

Chapter 1

Introduction

Tatsuya Tsukuda[*,§,1] **and Hannu Häkkinen**[‡,†]
Department of Chemistry, School of Science, The University of Tokyo, Bunkyo-ku, Tokyo, Japan; [§]Elements Strategy Initiative for Catalysts and Batteries (ESICB), Kyoto University, Katsura, Kyoto, Japan; [‡]Department of Physics, Nanoscience Center, University of Jyväskylä, Jyväskylä, Finland; [†]Department of Chemistry, Nanoscience Center, University of Jyväskylä, Jyväskylä, Finland
[1]*Corresponding author: E-mail: tsukuda@chem.s.u-tokyo.ac.jp*

Chapter Outline

1.1 PROTECTED METAL CLUSTERS: A BRIEF HISTORY

Metal clusters composed of less than a few hundred atoms are located between the bulk and atomic states of the corresponding metal and have attracted physicists over the last four decades. The central subject of the early stage of the cluster research was to observe the finite-size effects on physical properties of metal clusters and to understand their microscopic origins. Development of new experimental and theoretical methods has led to a discovery of a variety of remarkable size-specific phenomena and physicochemical properties. For example, the development of versatile methods of cluster production such as laser ablation coupled with mass spectrometry has unveiled magic numbers of clusters due to the closure of electronic and/or geometric structure(s).[1] These observations have led to the establishment of the concepts of electron shell closing based on the jellium model[2] and superatoms.[3] It has been widely recognized that various physicochemical (magnetic, optical, chemical, and thermal) properties of metal clusters deviate significantly from their bulk counterparts and evolve dramatically as a function of size, as exemplified by the metal–insulator transition.[4] During this development, the community has come to be convinced that metal clusters are promising functional units of

Frontiers of Nanoscience, Vol. 9. http://dx.doi.org/10.1016/B978-0-08-100086-1.00001-4

novel materials and has made an effort to develop cluster-based materials under catchphrases: "small is different"[5,6] and "every atom counts."[7,8]

Chemical synthesis has been a challenge to be overcome to initiate, accelerate, and deepen materials science of metal clusters, as evidenced by the explosive growth in materials science of nanocarbons after the large-scale production of C_{60}.[9] In the field of inorganic chemistry, phosphine-protected small Au cluster compounds have been long studied with a special focus on the synthesis and structural determination. One of the most famous examples is Schmid's Au_{55} compound.[10] However, variation of the systems and scope of the application were limited due to the instability and structural and compositional ambiguity. It was in 1994, when the first chemical synthesis of thiolate (RS)-protected Au nanoparticles was reported by Brust and Schiffrin.[11] This simple but inventive method allowed us to treat the metal clusters as conventional chemical compounds. In the late 1990s, these monolayer-protected clusters have been viewed as *nanocrystal gold molecules* by Whetten[12] and *gold nanoelectrode* by Murray.[13,14] The structure models based on hollow-site or bridge-site absorption of thiolates on nanocrystals have been theoretically developed by Landman.[15] Garzón was the first who suggested a strong deformation of the core structure by the thiolate adsorption.[16] Häkkinen proposed a concept of "divide and protect"[17] in which the Au clusters are protected by Au−thiolate oligomers. The first report on the mass spectrometric determination of molecular formula of $Au_n(SR)_m$ in 2005 by Tsukuda has opened a door to the atomically precise synthesis.[18] In 2007, Kornberg made a breakthrough in structure determination of protected metal cluster ($Au_{102}(SR)_{44}$) using single crystal X-ray diffractions.[19] Research interest in basic science and practical applications of the ligand-protected metal clusters has been explosively growing in the last decade, including many other ligand types than thiols and many other metals than gold.[20–44]

1.2 THE AIMS OF THE BOOK

It is an opportune moment after 20 years since the first report on the wet chemical synthesis to write a book concerning ligand-protected clusters in order to provide vivid snapshots of current research trends and innovative applications. This book entitled *Protected Metal Clusters: From Fundamentals to Applications* is included in a series entitled "Frontiers of Nanoscience (Elsevier; series editor, Richard Palmer)" and is aimed to survey development in the last decade in the fundamental concepts and potential applications of atomically precise metal clusters protected by organic ligands. This class of materials is now emerging due to breakthroughs in synthesis and characterization that have taken place during the last few years. This book on these exciting novel nanomaterials has two major aims depending on the audience. It is not trivial for the students and newcomers in this research field to systematically understand the fundamentals from a huge body of literature. Thus the first aim is to provide them the fundamental concepts of the systems explaining how they are formed and characterized. All chapters are written in a tutorial style and in-depth by leading authors so that the reader can catch up the current understanding of the rapidly growing field of protected metal clusters. The second aim is to provide the researchers working in the field of nanoscience with a state-of-the-art information on how and why protected metal clusters will play a potentially important role in the future in areas of molecular electronics, catalysis, sensing, biological imaging, and medical diagnosis and therapy. The book surveys the literature published by the end of 2014 and will serve as a data collection. The readers can share future prospects in the application of protected metal clusters.

1.3 THE OUTLINE OF THE BOOK

This book is organized as follows in order to achieve the above aims. The chapters are composed of three subjects: chemical synthesis, structural characterization, and application (Figure 1).

1.3.1 Synthesis

Current active research on the protected metal clusters relies on the wet chemical synthesis. *Chapters 2–4* focus on the state-of-the-art methods of atomically precise synthesis of metal clusters whose surfaces are protected by monolayers of organic ligands. Key structural parameters for properties and functions of the protected metal clusters, M_nL_m (M = metal, L = ligand), are

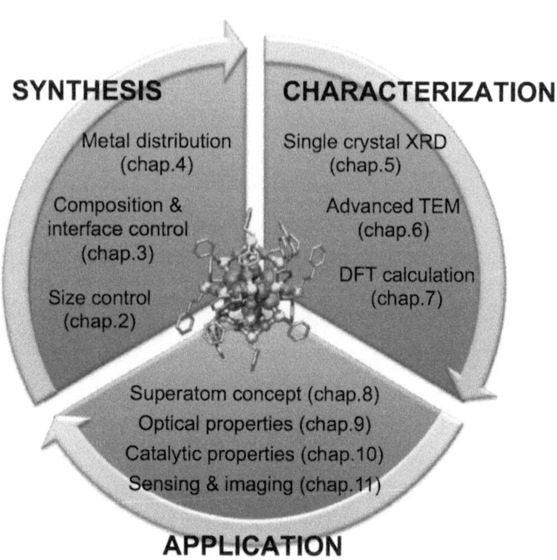

FIGURE 1 Contents of the book.

their chemical compositions (n, m), the nature of the ligands, and mixing mode of metal elements. In this book, M includes Au, Ag, Cu, and their alloys whereas L includes thiolates, selenonate, telluronate, phosphines, alkynes, and their mixture. *Chapter 2*, by Tsukuda, focuses on atomically precise control of size of metal clusters. Size-dependent evolution of electronic and geometric structures is demonstrated as an example. *Chapter 3*, by Negishi, describes the methods for controlling composition of intermetallic clusters and protection of metal clusters by a variety of organic ligands. The influence of the chemical composition of the metal core and the bonding mode at the interface on the fundamental properties will be illustrated. *Chapter 4*, by Zheng, surveys the synthetic strategies how the morphologies and metal distributions of hetero-metallic nanoclusters are manipulated by the proper choice of surface ligands.

1.3.2 Characterization

Structural characterization of the synthesized protected metal clusters is indispensable to establish structure–function correlation and to design new functions, but is a technical challenge. *Chapter 5*, by Ackerson, gives an overview of structural analysis, with emphasis on single crystal X-ray methods, which is the most direct and decisive analytical tool. It presents a comprehensive overview of the structures of thiolate-protected Au, Ag, and bimetallic clusters solved so far. Transmission electron microscopy (TEM) has been conventionally used to estimate the average diameters of the metal nanoparticles. In contrast, it is extremely difficult to determine the atomic scale structures of ultrasmall clusters consisting of less than 100 atoms by TEM owing to fragile nature under electron irradiation and low contrast. *Chapter 6*, by Palmer, describes the advanced TEM approaches recently developed for structural determination of a range of small gold clusters. Theoretical calculation has played an important role to predict structures of the protected metal clusters and to generalize the building-up principle of their structures. The structural prediction of the motif has become highly reliable thanks to the increased ability of computers and development of the algorithm. *Chapter 7*, by Jiang, discusses the recent progress in structure prediction based on the density functional theory (DFT) calculations under the hypothesis of the staple motif (–ligand–metal–ligand–) structure at the metal–ligand interface.

1.3.3 Application

Chapter 8, by Häkkinen, reviews the so-called superatom concept which has become a central guiding principle for considering the stability and physico-chemical properties of the protected metal clusters. It is stressed that the ligands not only protect the metal clusters from aggregation but also regulate the number of valence electrons in the metallic core. The chapter contains an extensive list of most of the structurally known clusters of Au, Ag, Cu, Al, or other main group elements. *Chapter 9*, by Aikens, summarizes optical properties of protected metal clusters which strongly dependent on their size, compositions, and structures. The experimental optical absorption spectra displaying clear onset and multiple peaks often serve as a fingerprint of the structures by comparing with the spectra calculated by time-dependent DFT. The origin of the optical transitions can be assigned with the help of theoretical calculations. Examination of circular dichroism spectra enables elucidation of the factors responsible for the large optical activity observed in these systems. Catalysis is one of the most intriguing properties of small metal clusters. Atomically precise nanoclusters hold great promise in the discovery of unique catalytic processes as well as in advancing the fundamental understanding of catalytic mechanisms at the atomic/molecular level. *Chapter 10*, by Jin, summarizes the catalytic application of atomically precise $Au_n(SR)_m$ nanoclusters. The reactions include catalytic oxidation (e.g., CO oxidation to CO_2, alcohol oxidation to aldehyde), catalytic hydrogenation (e.g., chemoselective hydrogenation), carbon–carbon coupling reactions, etc. The functions of protected metal clusters could be further enriched or enhanced via chemical functionalization, such as modifying the metal core, engineering the ligand surface, and overcoating the cluster surface with other functional materials. *Chapter 11*, by Xie, discusses concepts and strategies for the functionalization of metal clusters, with a specific focus on the principles and effects of functionalization. The chapter highlights some emerging applications of functionalized metal clusters in sensor development, bioimaging, and cancer therapy.

We hope that this book will be beneficial for the readers in giving a snapshot of the diverse and dynamic research area of protected metal clusters and that protected metal clusters will continue to attract significant interest in a future.

REFERENCES

1. Martin TP, Bergmann T, Göhlich H, Lange T. Shell structure of clusters. *J Phys Chem* 1991;**95**:6421–9.
2. Cohen ML, Chou MY, Knight WD, de Heer WA. Physics of metal clusters. *J Phys Chem* 1987;**91**:3141–9.
3. Castleman AW, Khanna SN. Clusters, superatoms, and building blocks of new materials. *J Phys Chem C* 2009;**113**:2664–75.
4. Issendorff B, Cheshnovsky O. Metal to insulator transitions in clusters. *Annu Rev Phys Chem* 2005;**56**:549–80.
5. El-Sayed MA. Small is different: shape-, size-, and composition-dependent properties of some colloidal semiconductor nanocrystals. *Acc Chem Res* 2004;**37**:326–33.
6. Landman U, Luedtke WD. Small is different: energetic, structural, thermal, and mechanical properties of passivated nanocluster assemblies. *Faraday Discuss* 2004;**125**:1–22.
7. Heiz U, Sanchez A, Abbet S, Schneider WD. Catalytic oxidation of carbon monoxide on monodispersed platinum clusters: each atom counts. *J Am Chem Soc* 1999;**121**:3214–7.

8. Landman U. Materials by numbers: computations as tools of discovery. *Proc Natl Acad Sci USA* 2005;**102**:6671–8.
9. a. Kroto HW, Heath JR, O'Brien SC, Curl RF, Smalley RE. C_{60}: Buckminsterfullerene. *Nature* 1985;**318**:162–3;
   b. Krätschmer W, Lamb LD, Fostiropoulos K, Huffman DR. Solid C_{60}: a new form of carbon. *Nature* 1985;**347**:354–8.
10. Schmid G, editor. *Nanoparticles: from theory to application, 2nd, completely revised and updated edition*. New Jersey: Wiley; 2010.
11. Brust M, Walker M, Bethell D, Schiffrin DJ, Whyman R. Synthesis of thiol-derivatised gold nanoparticles in a two-phase liquid-liquid system. *J Chem Soc Chem Commun* 1994:801–2.
12. Whetten RL, Khoury JT, Alvarez MM, Murthy S, Vezmar I, Wang ZL, et al. Nanocrystal gold molecules. *Adv Mater* 1996;**8**:428–33.
13. Chen S, Ingram RS, Hostetler MJ, Pietron JJ, Murray RW, Schaaff TG, et al. Gold nanoelectrodes of varied size: transition to molecule-like charging. *Science* 1998;**280**:2098–101.
14. Templeton AC, Wuelfing WP, Murray RW. Monolayer-protected cluster molecules. *Acc Chem Res* 2000;**33**:27–36.
15. Cleveland CL, Landman U, Schaaff TG, Shafigullin MN, Stephens PW, Whetten RL. Structural evolution of smaller gold nanocrystals: the truncated decahedral motif. *Phys Rev Lett* 1997;**79**:1873–6.
16. Garzón IL, Roviera C, Michaelian K, Beltran MR, Ordejon P, Junquera J, et al. Do thiols merely passivate gold nanoclusters? *Phys Rev Lett* 2000;**85**:5250–1.
17. Häkkinen H, Walter M, Grönbeck H. Divide and protect: capping gold nanoclusters with molecular gold–thiolate rings. *J Phys Chem B* 2006;**110**:9927–31.
18. Negishi Y, Nobusada K, Tsukuda T. Glutathione-protected gold clusters revisited: bridging the gap between gold(I)-thiolate complexes and thiolate-protected gold nanocrystals. *J Am Chem Soc* 2005;**127**:5261–70.
19. Jadzinsky PD, Calero G, Ackerson CJ, Bushnell DA, Kornberg RD. Structure of thiol monolayer-protected gold nanoparticle at 1.1 Å resolution. *Science* 2007;**318**:430–3.
20. a. Häkkinen H. The gold–sulfur interface at the nanoscale. *Nat Chem* 2012;**4**:443–55;
    b. Walter M, Akola J, Lopez-Acevedo O, Jadzinsky PD, Calero G, Ackerson CJ, et al. A unified view of ligand-protected gold clusters as superatom complexes. *Proc Natl Acad Sci USA* 2008;**105**:9157–62.
21. Sardar R, Funston AM, Mulvaney P, Murray RW. Gold nanoparticles: past, present, and future. *Langmuir* 2009;**25**:13840–51.
22. Parker JF, Fields-Zinna CA, Murray RW. The story of a monodisperse gold nanoparticle: $Au_{25}L_{18}$. *Acc Chem Res* 2010;**43**:1289–96.
23. Jin R. Quantum sized, thiolate-protected gold nanoclusters. *Nanoscale* 2010;**2**:343–62.
24. Wang ZW, Toikkanen O, Yin F, Li ZY, Quinn BM, Palmer RE. Counting the atoms in supported, monolayer-protected gold clusters. *J Am Chem Soc* 2010;**132**:2854–5.
25. Aikens CM. Electronic structure of ligand-passivated gold and silver nanoclusters. *J Phys Chem Lett* 2011;**2**:99–104.
26. Tofanelli MA, Ackerson CJ. Superatom electron configuration predicts thermal stability of $Au_{25}(SR)_{18}$ nanoclusters. *J Am Chem Soc* 2012;**134**:16937–40.
27. Qian H, Zhu M, Wu Z, Jin R. Quantum sized gold nanoclusters with atomic precision. *Acc Chem Res* 2012;**45**:1470–9.
28. Tsukuda T. Toward an atomic-level understanding of size-specific properties of protected and stabilized gold clusters. *Bull Chem Soc Jpn* 2012;**85**:151–68.

29. Maity P, Xie S, Yamauchi M, Tsukuda T. Stabilized gold clusters: from isolation toward controlled synthesis. *Nanoscale* 2012;**4**:4027–37.
30. Harkness KM, Tang Y, Dass A, Pan J, Kothalawala N, Reddy VJ, et al. $Ag_{44}(SR)_{30}^{4-}$: a silver-thiolate superatom complex. *Nanoscale* 2012;**4**:4269–74.
31. Yang H, Wang Y, Huang H, Gell L, Lehtovaara L, Malola S, et al. All-thiol-stabilized Ag_{44} and $Au_{12}Ag_{32}$ nanoparticles with single-crystal structures. *Nat Commun* 2013;**4**:2422.
32. Li G, Jin R. Atomically precise gold nanoclusters as new model catalysts. *Acc Chem Res* 2013;**46**:1749–58.
33. Jiang DE. The expanding universe of thiolated gold nanoclusters and beyond. *Nanoscale* 2013;**5**:7149–60.
34. Udayabhaskararao T, Pradeep T. New protocols for the synthesis of stable Ag and Au nanocluster molecules. *J Phys Chem Lett* 2013;**4**:1553–64.
35. Desireddy A, Conn BE, Guo J, Yoon B, Barnett RN, Monahan BM, et al. Ultrastable silver nanoparticles. *Nature* 2013;**501**:399–402.
36. Yuan X, Luo Z, Yu Y, Yao Q, Xie J. Luminescent noble metal nanoclusters as an emerging optical probe for sensor development. *Chem Asian J* 2013;**8**:858–71.
37. Negishi Y. Toward the creation of functionalized metal nanoclusters and highly active photocatalytic materials using thiolate-protected magic gold clusters. *Bull Chem Soc Jpn* 2014;**87**:375–89.
38. Goswami N, Zheng K, Xie J. Bio-NCs – the marriage of ultrasmall metal nanoclusters with biomolecules. *Nanoscale* 2014;**6**:13328–47.
39. Knoppe S, Bürgi T. Chirality in thiolate-protected gold clusters. *Acc Chem Res* 2014;**47**:1318–26.
40. Kumara C, Zuo X, Cullen DA, Dass A. Faradaurate-940: synthesis, mass spectrometry, electron microscopy, high-energy X-ray diffraction, and X-ray scattering study of $Au_{\sim 940\pm 20}(SR)_{\sim 160\pm 4}$ nanocrystals. *ACS Nano* 2014;**8**:6431–9.
41. Konishi K. Phosphine-coordinated pure-gold clusters: diverse geometrical structures and unique optical properties/responses. *Struct Bond* 2014;**161**:49–86.
42. Chen J, Zhang QF, Bonaccorso TA, Williard PG, Wang LS. Controlling gold nanoclusters by diphosphine ligands. *J Am Chem Soc* 2014;**136**:92–5.
43. Azubel M, Koivisto J, Malola S, Bushnell D, Hura GL, Koh AL, et al. Electron microscopy of gold nanoparticles at atomic resolution. *Science* 2014;**345**:909–12.
44. Wan X, Tang Q, Yuan S, Jiang D, Wang Q. Au_{19} nanocluster featuring a V-shaped alkynyl–gold motif. *J Am Chem Soc* 2015;**137**:652–5.

Chapter 2

Controlled Synthesis: Size Control

Shinjiro Takano* and Tatsuya Tsukuda*,§,1
**Department of Chemistry, School of Science, The University of Tokyo, Bunkyo-ku, Tokyo, Japan;*
§Elements Strategy Initiative for Catalysts and Batteries (ESICB), Kyoto University, Katsura, Kyoto, Japan
[1]Corresponding author: E-mail: tsukuda@chem.s.u-tokyo.ac.jp

Chapter Outline

2.1 ATOMICALLY PRECISE SIZE CONTROL: WHY?

Metal nanoparticles and *metal clusters* are states located between the bulk and atomic states of the corresponding metal. Both systems exhibit novel physicochemical properties that are very different from those of the bulk state and strongly dependent on their "size" because of their size-dependent structures. The essential requirement for the study of their basic properties and for practical applications is to synthesize nanoparticles and clusters as stable compounds while controlling their size. Both systems can be stabilized against aggregation by using similar methods including complete passivation by organic

Frontiers of Nanoscience, Vol. 9. http://dx.doi.org/10.1016/B978-0-08-100086-1.00002-6

ligands, partial passivation by polymers, and immobilization on solids. However, the precision required for size control and size evaluation of nanoparticles and clusters is significantly different. Evolution of physicochemical properties of nanoparticles is not noticeable by just removing or adding a single atom but is recognizable when one layer of an atomic shell is removed or added (Figure 1). Thus, the "diameter" of a nanoparticle should be controlled with an accuracy of ~0.5 nm, which can be determined by conventional transmission electron microscopy (TEM) or scanning electron microscopy (SEM). A large variety of "monodisperse" metal nanoparticles with a standard deviation in diameter of less than 10% have been synthesized by controlling the kinetics of the particle growth. Under optimized conditions, nucleation of atoms occurs abruptly in a supersaturated solution without the formation of new nuclei, and the nuclei grow uniformly thereafter to form monodisperse nanoparticles.[1]

In contrast, it has been elucidated by experimental and theoretical studies of bare metal clusters that their intrinsic properties dramatically change even by the addition or removal of a single atom (Figure 1).[2–4] A typical example of such striking size dependence manifests as the "magic numbers" in the mass spectra of metal clusters. Observation of the magic numbers indicates that clusters composed of these specific numbers of atoms are extraordinarily stable compared with neighboring clusters. Therefore, in order to synthesize metal clusters as chemical compounds in the real world, "the number of constituent atoms" must be controlled with atomic precision, which is a nontrivial task. In addition, the cluster size must be determined precisely by mass spectrometry (MS) and not conventional microscopic analysis techniques such as TEM and SEM.

This chapter summarizes the state-of-the-art methods of size-controlled synthesis of metal clusters with a focus on Au and Ag clusters whose surfaces are passivated by monolayers of organic ligands such as thiols (RSH), phosphines (R_3P), and terminal alkynes (RC≡CH).[5–7] In this chapter, these systems are referred to as Au(Ag):SR, Au:PR_3, and Au:C≡CR, respectively.

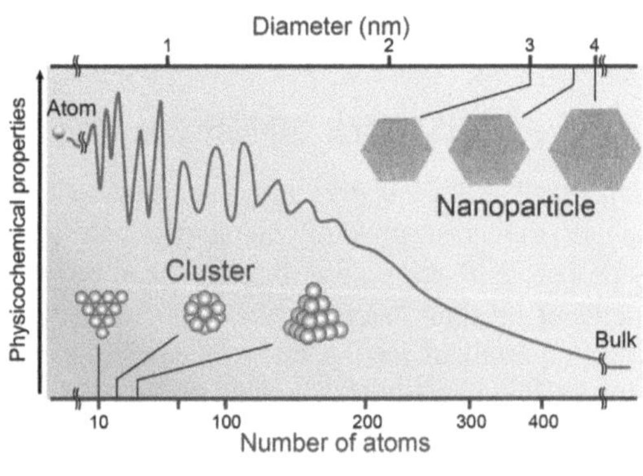

FIGURE 1 Metal nanoparticles and clusters.

The chemical compositions of the protected clusters that had been isolated by the end of 2014 are extensively surveyed. The origin of the high stabilities of these isolated clusters will be discussed in terms of the closure of electronic and geometric shells. Size-dependent evolution of electronic and geometric structures will be demonstrated by taking Au:SR as an example.

2.2 ATOMICALLY PRECISE SIZE CONTROL: HOW?

This section describes experimental techniques for chemical synthesis of metal clusters with atomically defined sizes. The first step in the synthesis of metal clusters is to reduce the precursor metal ions in the presence of protective ligands as in the case of that of protected metal nanoparticles. The yield of small (<2 nm) metal clusters can be enhanced by suppressing the growth of the nuclei. To retard the growth, the surfaces of the nuclei should be rapidly passivated with ligands and all the precursor metal ions should be transformed immediately into the nuclei. Although careful optimization of the synthetic conditions allows us to prepare monodisperse metal clusters, there is a distribution in size in terms of the number of constituent atoms, and the reproducibility of the cluster sizes obtained is not sufficiently high. In what follows, three representative approaches are introduced to overcome these difficulties: template-mediated synthesis, fractionation, and size-focusing synthesis via postsynthetic core etching or slow core growth (Figure 2).

Precise determination of the cluster size is a key for the characterization of small clusters regardless of the synthetic method employed. MS is an indispensable tool for determining the number of not only the metal atoms but also the protecting ligands (Figure 2). Information on molecular formulas is essential for theoretical studies of the structures of ligand-protected metal clusters whose structures cannot always be determined by single crystal X-ray diffraction (XRD). For such a determination, it is crucial to ionize the ligand-protected metal clusters intact (without fragmentation) by using a soft ionization method such as electrospray ionization (ESI)[8–11] or matrix-assisted laser desorption ionization (MALDI).[12] Water-dispersible clusters protected by hydrophilic ligands containing carboxylic, amino, or hydroxyl groups are ionized into multiply charged ions via protonation or deprotonation of the functional groups under given pH conditions.[8,9] To facilitate the dehydration of the clusters, a volatile solvent such as acetonitrile or methanol is added to the aqueous media.[8,9] In contrast, organic dispersible clusters protected by hydrophobic ligands are charged by electrochemical ionization of the metal core[10] or by forming noncovalent adducts with ions such as Cs^+.[11] MALDI-MS is applicable for the analysis of both hydrophobic and hydrophilic Au:SRs by using specific matrices such as *trans*-2-[3-(4-*tert*-butylphenyl)-2-propenyldidene]malononitrile[12] or α,β-diphenylfumaronitrile.[13] The clusters can be ionized intact by careful optimization of the matrix-to-cluster molar ratio (typically >100) and the fluence of the laser (usually a N_2 laser).

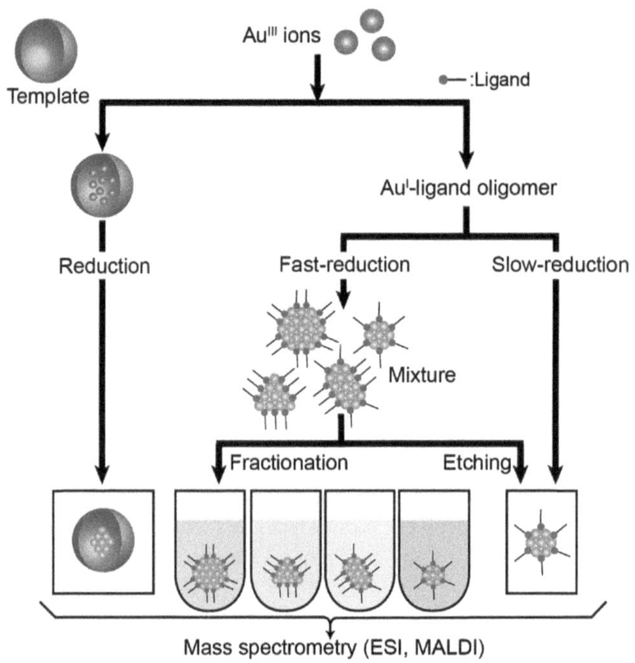

FIGURE 2 Schematic illustration of strategies for size-controlled synthesis of Au clusters.

However, one has to bear in mind that ion intensities in the mass spectra of individual clusters do not directly reflect their relative abundance or populations because the ionization efficiency may depend on the cluster size.

2.2.1 Template-Mediated Synthesis

The first approach for the cluster size control utilizes molecular templates such as dendrimers (DEN) (an arborized polymer with a regularly branched structure from a core),[14,15] ferritin (a self-aggregate constructed by 24-protein subunits with a 7-nm inner diameter),[16] and porphyrins.[17] The basic idea is to regulate the number of precursors ions coordinated to the binding sites located within the cavity or to limit the space for cluster growth by template cages. For example, mass spectrometric studies have demonstrated that a series of Au clusters (Au_5, Au_8, Au_{13}, Au_{23}, and Au_{31}) within poly(amidoamine)-DEN can be synthesized[18] and that highly monodisperse Au clusters ($Au_{\sim 66}$) are confined within a cavity made of six porphyrin molecules.[17]

2.2.2 Fractionation

The second approach is to fractionate size-defined clusters from crude mixtures.[19] The success of the fractionation method lies in the fact that the size distributions of the clusters formed in crude mixtures are not atomically continuous but discrete because of the significant differences in their stabilities. In other words, the crude samples contain only a limited number of stable clusters. In practice, one has to choose a suitable fractionation method depending on the hydrophobic or hydrophilic nature of the clusters, which is determined by the protective ligands. In the following, representative experimental techniques of fractionation or isolation of ligand-protected clusters are presented.

2.2.2.1 Fractional Precipitation or Extraction

Size selection by fractional precipitation or extraction is based on the fact that interactions between clusters in a dispersing medium are dependent on the core size. In general, interactions between clusters become more attractive with an increase in the core size,[20] and as a result, the cluster dispersibility decreases with an increase in cluster size. Thus, by gradually adding a poor solvent to the dispersion of the cluster mixtures, clusters start to precipitate in the order of the core mass. Figure 3 shows laser-desorption

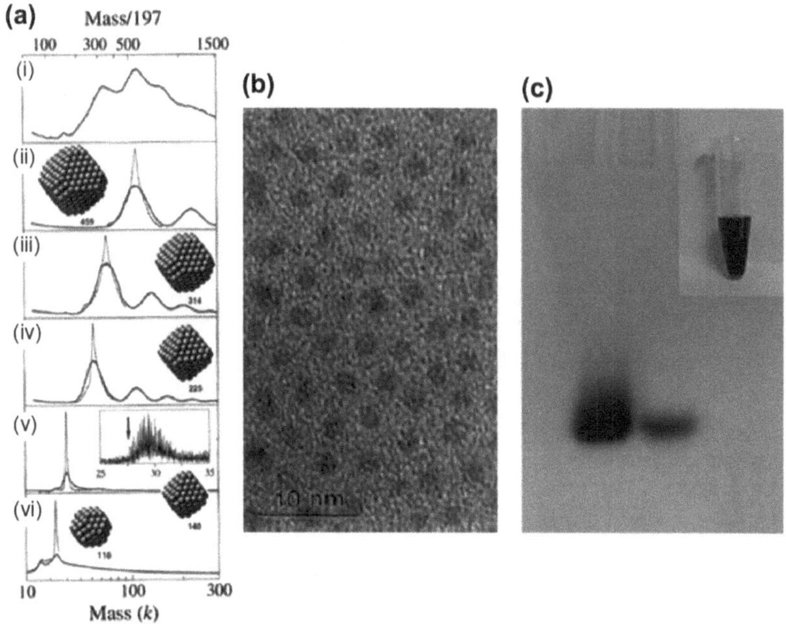

FIGURE 3 Purification of Au:SR by fractional precipitation. (a) MS spectra of the (i) crude mixture and (ii–vi) separated fractions of 92, 63, 45, 28, and 16 kDa, respectively. (b) TEM image of fraction (iii). (c) PAGE result for Au:*p*-MBA before (left) and after (right) purification. *Adapted from Ref. 19 with permission for (a) and (b). Copyright 1996 Wiley-VCH Publishing. Adapted from Ref. 21 with permission for (c). Copyright 2011 American Chemical Society.*

mass spectra of a series of fractions of *n*-alkanethiolate-protected Au clusters (Au:SC*x*, x = number of carbon atoms in the alkyl chain) obtained by the fractional precipitation.[19] Although clusters were detected in the form of $Au_nS_m^+$ due to laser-induced C−S bond breaking, this groundbreaking result clearly demonstrates for the first time that there are a series of highly stable Au:SC*x* clusters with core masses of 16, 28, 45, 63, and 92 kDa. The TEM image shown in Figure 3(b) confirms that fractionated Au:SC12 (core mass = 63 kDa) are indeed monodisperse. Fractional precipitation can also be applied to hydrophilic Au:SR clusters. The size distributions of the as-prepared and fractionated Au:*p*-MBA (*p*-MBA = *p*-mercaptobenzoic acid) clusters were compared by polyacrylamide gel electrophoresis (PAGE, see Section 2.2.2.4).[21] Figure 3(c) shows that the bandwidth was reduced by the fractional precipitation, indicating that the size distribution narrowed remarkably. The fractionated clusters were assigned to molecularly pure $Au_{102}(p\text{-MBA})_{44}$ by ESI and MALDI mass analyses.[21] In reverse, clusters with a given size can be extracted selectively from the solid form of the crude mixtures by optimizing the solvent composition. A series of magic clusters Au_{25}, Au_{38}, and Au_{144} protected by alkanethiolates have been successfully isolated by this method.[22] The benefit of the fractional precipitation or extraction method is that size-defined clusters can be obtained on a large scale by a simple treatment. The drawback is that one has to optimize the conditions for the fractionation of specific clusters based on a trial-and-error approach.

2.2.2.2 Partition Chromatography

Partition chromatography has been successfully applied to the separation of crude mixtures of clusters. Reverse phase chromatography[23–28] utilizes columns comprised of silica or polymer beads whose surfaces are modified with hydrophobic groups such as butyl, octyl, octadecyl, and phenyl groups. The mobile phase is more polar than the stationary phase. More hydrophobic (or less polar) clusters dispersed in the mobile phase have a higher affinity to the stationary phase so that hydrophilic (or more polar) clusters elute faster. Figure 4 exemplifies the isolation of $Au_{102}(SC12)_{44}$, $Au_{130}(SC12)_{50}$, $Au_{144}(SC12)_{60}$, and $Au_{187}(SC12)_{68}$ by reverse phase chromatography.[28] In contrast, normal phase chromatography is applicable to the fractionation of hydrophobic clusters. For example, $Au_{25}(PET)_{18}$ and $Au_{144}(PET)_{60}$ (PET = C_2H_4Ph) have been separated by thin-layer chromatography,[29] and $Au_{30}(EPT)_{13}$, $Au_{35}(EPT)_{18}$, and $Au_{41-43}(EPT)_{21-23}$ (EPT = 9-ethynyl-phenanthrene) have been separated by conventional silica column chromatography.[30]

2.2.2.3 Size Exclusion Chromatography

Size exclusion chromatography (SEC) or gel permeation chromatography separates clusters according to their hydrodynamic volume (diameter).[31]

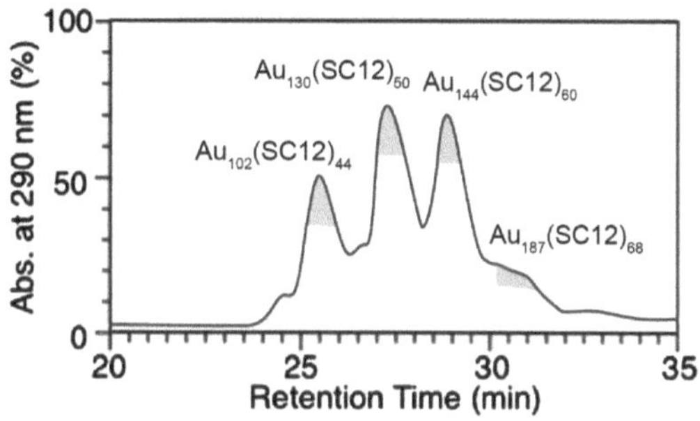

FIGURE 4 Reverse phase chromatogram of a mixture of Au:SC12. The molecular formulas shown at the peaks were determined by positive-ion ESI-FT-ICR-MS. *Adapted from Ref. 28 with permission. Copyright 2012 American Chemical Society.*

In SEC, smaller clusters diffuse further into porous microgels in the column and as a result spend a longer time in the column than larger clusters; that is, larger clusters elute faster than smaller clusters. SEC is applicable to both hydrophobic and hydrophilic clusters by changing the eluent and has been successfully applied to ligand-protected metal clusters.[32–37] Figure 5(a) and (b) shows SEC isolation of $P(PhCONHCH_3)_3$-protected Au_{73-75}[32] and the separation of $Au_{38}(PET)_{24}/Au_{40}(PET)_{24}$,[34] respectively. The resolution of the separation can be improved by operating in recycling mode, in which the clusters are allowed to pass through the column repeatedly.[33,37,38] Figure 5(c) shows SEC separation of $Au_{55}(SC18)_{31}$ (Ref. 35) and $Au_{38}(SC18)_{24}$ in recycling mode.[33] SEC can also be used as an analytical tool to determine the cluster structure. The thickness of a monolayer on a metal cluster can be estimated by subtracting the core diameter determined by TEM from the hydrophobic radius determined by the retention time.[31,39] Size evolution during the synthesis can also be probed by SEC. Figure 5(d) demonstrates that $Ag_{\sim 223}(TBBT)_{\sim 108}$ (TBBT = 4-(*t*-butyl)benzene thiol) is gradually converted to $Ag_{\sim 280}(TBBT)_{\sim 120}$ by incubating with excess thiol.[36]

2.2.2.4 Gel Electrophoresis

Hydrophilic metal clusters can be fractionated by PAGE with a resolution higher than that of SEC.[8,9,40–44] In PAGE, clusters are forced to travel through an acrylamide gel by an electric field applied across the gel. Smaller clusters travel faster through the gel network and so have a higher mobility than larger clusters. To achieve a high separating power for small clusters, the gel must be much more dense than that conventionally used in the separation of proteins. One of the most well-known magic clusters, $Au_{25}(SG)_{18}$ (GS = glutathionate), was isolated and identified for the first time by a combination of PAGE and ESI-MS (Figure 6(a)).[9] One can adjust the mass range at which the high-resolution separation is required by changing the fineness of the polymer

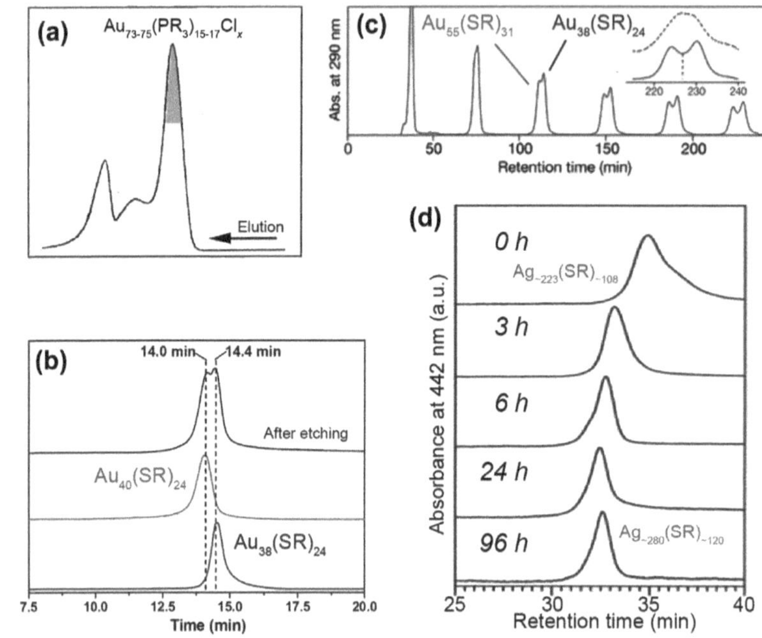

FIGURE 5 (a) SEC isolation of $P(PhCONHCH_3)_3$-protected Au_{73-75}. (b) SEC separation of $Au_{38}(PET)_{24}$ and $Au_{40}(PET)_{24}$. (c) SEC separation of $Au_{55}(SC18)_{31}$ and $Au_{38}(SC18)_{24}$ in recycling mode. (d) SEC assay of the conversion of $Ag_{\sim223}(TBBT)_{\sim108}$ into $Ag_{\sim280}$ $(TBBT)_{\sim120}$ by incubation. *Adapted from Ref. 32 with permission for (a). Copyright Springer-Verlag 1999. Adapted from Ref. 34 with permission for (b). Copyright 2010 American Chemical Society. Adapted from Ref. 33 with permission for (c). Copyright 2006 American Chemical Society. Adapted from Ref. 36 with permission for (d). Copyright The Royal Society of Chemistry 2011.*

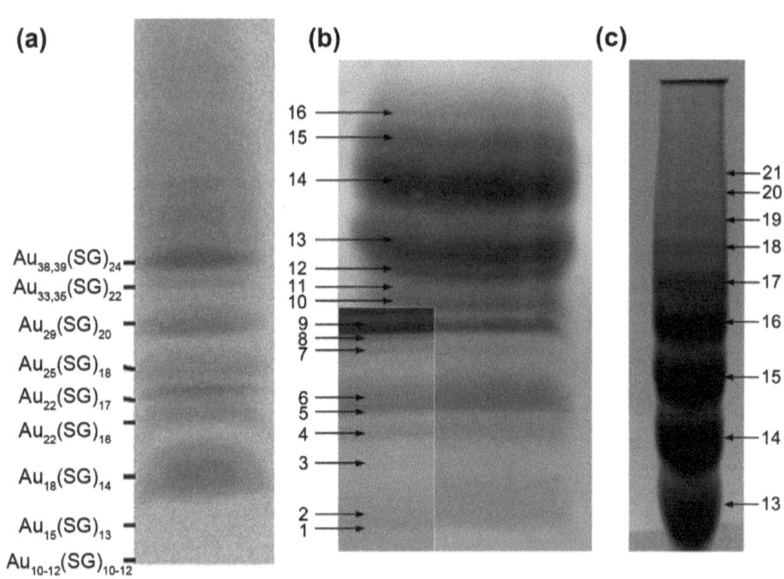

FIGURE 6 PAGE separation of (a) Au:SG and Ag:SG with (b) smaller and (c) larger regions. (see color plate). *Adapted from Ref. 9 with permission for (a). Copyright 2005 American Chemical Society. Adapted from Ref. 44 with permission for (b) and (c). Copyright 2010 American Chemical Society.*

network in the gel. For example, Au:SG clusters were successfully separated into 26 fractions by using two gels with different network densities.[43] PAGE separation of Ag:SG into 21 fractions has also been reported (Figure 6(b) and (c)).[44] PAGE is also used as an analytical tool of the size distribution as in the case of SEC but with a much higher resolution (Figure 3(c)).[21]

2.2.3 Size-Focusing Synthesis

2.2.3.1 Postsynthetic Etching

Use of the fractionation methods described in Section 2.2.2 revealed that there are a series of stable clusters with specific compositions. According to the definition of the magic clusters in the gas phase, removing an atom from such especially stable clusters requires more energy than that from other clusters. Under such circumstances, less stable clusters can be converted into more stable clusters by the sequential removal of Au atoms from the Au core (Figure 7). Namely, the population of stable clusters can be enhanced because the formation of such clusters acts as a kinetic bottleneck in the decomposition processes. How can we etch atoms from the metastable clusters in a controlled manner? It was found that the core mass of Au:SC6 was reduced from ~14 to ~8 kDa by heating in neat dodecanethiol under inert atmosphere.[45] This finding suggests that chemical etching by thiol molecules is a useful top-down method of converting metastable Au clusters into more stable Au clusters. In addition, it was demonstrated that the etching efficiency of Au:SR by free thiol molecules is strongly dependent on cluster size.[46] In the reaction of size-selected $Au_n(SG)_m$ with an excess of glutathione (GSH), only $Au_{25}(SG)_{18}$ remained unetched whereas $Au_n(SG)_m$ with $n \geq 29$ were transformed into stable $Au_{25}(SG)_{18}$ (Figure 8(a)).[46] This remarkable size dependence of the

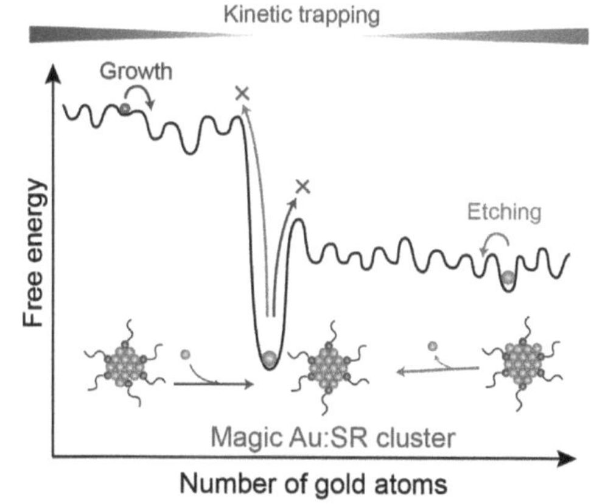

FIGURE 7 Conceptual illustration of focusing synthesis of magic Au:SR clusters.

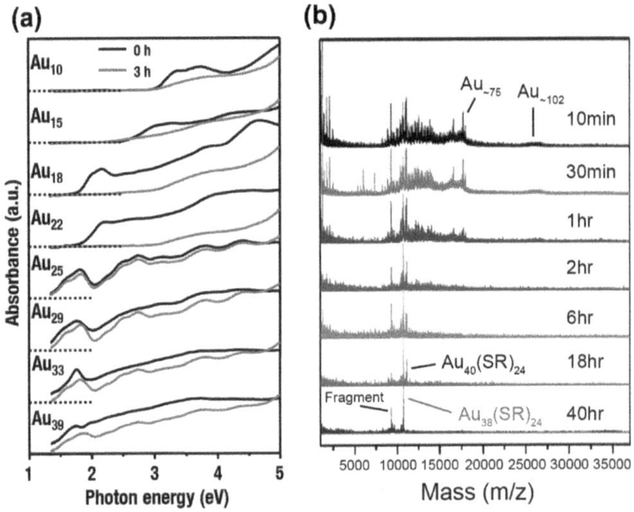

FIGURE 8 (a) Optical spectra of size-selected $Au_n(SG)_m$ before (0 h) and after (3 h) etching reactions with GSH. (b) MALDI mass spectra of Au:SR products as a function of reaction time with PET. *Adapted from Ref. 46 with permission for (a). Copyright 2007 Wiley-VCH publishing. Adapted from Ref. 50 with permission for (b). Copyright 2009 American Chemical Society.*

etching reactivity opens up a promising approach for the selective synthesis of magic clusters on a large scale.[47–57] For example, a broad size distribution of as-prepared Au:SG can be gradually focused into $Au_{38}(PET)_{24}$ during postsynthetic etching with PET, as illustrated in Figure 8(b).[50]

Stable Au:SR clusters can be synthesized in a molecularly pure form and on a large scale by a one-pot method under optimized conditions. Crude Au:SR clusters do not need to be collected during the synthesis if the size distribution of the nascent Au:SR clusters and kinetics of the subsequent core etching are controlled properly. The yield of the one-pot synthesis can be enhanced by starting the etching reactions from a cluster distribution slightly larger than the stable clusters of interest. In situ MS confirmed that metastable Au clusters with a size range of 38–102 formed initially are transformed into $Au_{25}(PET)_{18}$.[51] Sub 100 mg of $Au_{25}(SG)_{18}$ in a pure form was obtained by one-pot mixing of $Au_{11}(PPh_3)_7Cl_3$ and GSH.[47,55]

2.2.3.2 Slow Growth

As described in Section 2.2.3.1, the removal of Au atoms from stable Au:SR clusters requires a large energy as in the case of dissociation of magic metal clusters in the gas phase (Figure 7). However, in sharp contrast to bare metal clusters in the gas phase, a high activation barrier must be surmounted before an Au atom can be attached to a stable Au:SR cluster (cluster growth). The activation barrier arises from the core protection by the ligands and so depends on the coverage. Stable Au:SR clusters are completely protected by thiolates as shown in the Section 2.3. Thus, if cluster growth proceeds slowly enough that passivation is completed at every stage of growth, then there is a chance that stable clusters are kinetically trapped. Selective synthesis of extremely stable $Au_{25}(m\text{-MBA})_{18}$ (*m*-MBA = *m*-mercaptobenzoic acid) has been achieved by this approach.[57] The growth process under slow-reduction conditions was monitored by detecting intermediate species using ESI-MS. Such a mechanistic study confirmed that the growth terminates at the formation of $Au_{25}(m\text{-MBA})_{18}$. However, a larger cluster $[Au_{29}(m\text{-MBA})_{20}]^-$ was observed in the initial stage and gradually disappeared over time. This finding indicates that the etching process from larger clusters also proceeds under slow-reduction conditions.[57]

2.3 ISOLATED GOLD AND SILVER CLUSTERS

By using the methods introduced in Section 2.2, a large variety of ligand-protected Au and Ag clusters have been size-selectively synthesized. This section summarizes the ligand-protected Au and Ag clusters that had been reported by the end of 2014. The origin of the high stability of the isolated clusters is explained in terms of closure of the electronic and geometric shells.

2.3.1 Gold Clusters Protected by Thiolates

2.3.1.1 Stable Chemical Compositions

Since the first report of facile chemical synthesis of Au:SR clusters,[58] extensive studies were initiated in the mid-1990s. Although these early studies indicated the presence of stable Au:SR clusters with specific core masses,[19,59–64] it was not until 2005 that the chemical compositions of stable $Au_n(SR)_m$ (i.e., numbers of Au atoms and thiolates) were determined at a molecular precision for the first time.[9] Since then, isolation of a large number of precisely defined $Au_n(SR)_m$ clusters with n covering the range of 11–940 has been reported.[9,22,28,33–35,37,42,65–101] The chemical compositions of isolated $Au_n(SR)_m$ for various types of thiolates are listed in Table 1 and plotted as open circles in Figure 9. Red circles, blue squares, aqua diamonds, and purple triangles in Figure 9 represent the compositions of Au clusters protected by phenylethanethiolate, alkanethiolates, other thiolates (GSH, thiophenol derivatives, and amino acid derivatives), and thiolates having bulky tail groups (*tert*-butyl, adamanthyl, and cyclohexyl), respectively. Two clusters, $Au_{41}(S\text{-Eind})_{12}$ (Eind = 1,1,3,3,5,5,7,7-octaethyl-s-hydrindacen-4-yl) and $Au_{25}(SDpp)_{11}$ (Dpp = 2,6-diphenylphenyl), exhibit anomalous compositions; the thiol-to-Au ratio is much smaller than that of the other clusters.[37,73] This is ascribed to the low coverage because of the tremendous bulky tail groups. Besides these exceptions, there is a clear correlation between the cluster size and the number of thiolates regardless of the type of

TABLE 1 List of Isolated Au:SR Clusters

$[Au_n(SR)_m]^z$					
n	m	z	n^*	RSH[a]	Single Crystal Structure in Divide-and-Protect Scheme
15	13	0	2	GSH[9]	–
18	14	0	4	GSH[9,65]	–
20	16	0	4	PET,[66] TBBT[67]	$Au_7[Au(SR)_2]_2[Au_3(SR)_4]$ $[Au_8(SR)_8]$[67]
22	16	0	6	GSH[9]	–
22	17	0	5	GSH[9]	–
22	18	0	4	GSH[68]	–
23	16	−1	8	CHT,[69] TBT,[70] AT[71]	$Au_{15}(SR)_4[Au(SR)_2]_2$ $[Au_3(SR)_4]_2$[69]
24	16	0	8	AT[71]	$Au_{15}(SR)_3[Au(SR)_2]$ $[Au_2(SR)_3]$ $[Au_3(SR)_4]_2$[71]
24	20	0	4	PET[72]	–
25	11	0	14	Dpp-SH[73]	–
25	16	0	9	AT[71]	–
25	18	−1	8	GSH,[9] CxSH,[22,74] PET,[75–77] Capt[78]	$Au_{13}[Au_2(SR)_3]_6$[75–77]
		0	7		
		1	6		
28	20	0	8	TBBT[79]	$Au_{20}(SR)_8[Au_2(SR)_3]_4$[79]
29	20	0	9	GSH[9]	–
30	18	0	12	TBT,[80–82] AT[83]	$Au_{22}S(SR)_6[Au(SR)_2]_2$ $[Au_3(SR)_4]_2$[81,82]
33	22	0	11	GSH[9]	–
36	24	0	12	BT,[84] BBT,[85] CxSH[86]	$Au_{28}(SR)_{12}[Au_2(SR)_3]_4$[84–86]
38	24	0	14	GSH,[9] CxSH,[22] PET[34]	$Au_{23}[Au(SR)_2]_3[Au_2(SR)_3]_6$[52]
39	23	0	16	AT[83]	–
39	24	0	15	GSH[9]	–
40	24	0	16	CxSH,[34] PET[34]	–
41	12	0	29	Eind-SH[37]	–
42	24	0	18	GSH[42]	–
43	26	0	17	GSH[43]	–

TABLE 1 List of Isolated Au:SR Clusters—cont'd

$[Au_n(SR)_m]^z$					
n	*m*	*z*	*n**	RSH[a]	Single Crystal Structure in Divide-and-Protect Scheme
44	24	0	20	GSH[42]	–
44	28	0	16	BBT[87]	–
45	25	0	20	GSH[42]	–
54	28	0	26	GSH[42]	–
55	31	0	24	CxSH,[35] PET[88]	–
56	29	0	27	GSH[42]	–
58	29	0	29	GSH[42]	–
64	32	0	32	CHT[89]	–
65	30	0	35	CHT[83]	–
67	30	0	37	CHT[83]	–
67	35	0	32	CxSH,[90] PET[90]	–
99	42	0	57	BT,[91,92] FBT,[92] MOBT,[92] MBT[92]	–
102	44	0	58	MBA[93]	$Au_{79}[Au(SR)_2]_{19}[Au_2(SR)_3]_2$[93]
103	45	0	58	PET[94]	–
104	45	0	59	PET[94]	–
104	46	0	58	PET[94]	–
105	46	0	59	PET[94]	–
130	50	0	80	CxSH,[28] PET[95]	–
137	56	0	81	CxSH,[96] PET[96]	–
144	60	0	84	CxSH,[54] PET[54,97]	–
187	68	0	119	CxSH[28]	–
329	84	0	245	CxSH,[98] PET[98]	–
333	79	0	254	PET[99]	–
~500	~120	0	~380	CxSH,[100] PET[100]	–
~940	~160	0	~780	CxSH,[101] PET[101]	–

[a]*GSH, glutathione; PET, phenylethanthiol; CxSH, n-alkanethiol; TBBT, 4-tert-butylbenzenethiol; AT, 1-adamantanethiol; DPP-SH, 2,6-diphenylphenyl thiol; Capt, Captopril; Eind-SH, 1,1,3,3,5,5,7,7-octaethyl-s-hydrindacen-4-yl thiol; BT, benzenethiol; CHT, cyclohexanethiol; TBT, tert-butylthiol; MBA, 4-mercaptobenzoic acid; MBT, 4-methylbenzenethiol; MOBT, 4-methoxybenzenethiol; FBT, 4-fluorobenzenethiol.*

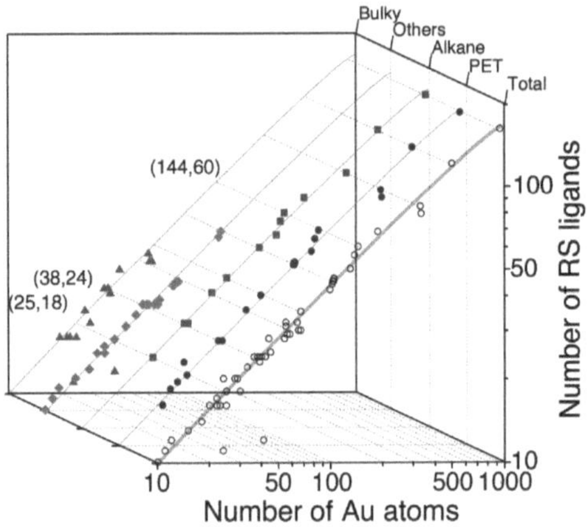

FIGURE 9 Compositional plots of reported Au:SR clusters with various types of thiolates. Red circles (black in print versions), blue squares (dark gray in print versions), aqua diamonds (light gray in print versions), purple triangles (gray in print versions), and open dots correspond to PET, alkanethiolate, other thiolates, bulky thiolate, and all types of thiolates, respectively. Thick gray line represents Eqn (1) whereas other thin lines are guides for the eye.

thiolates.[6,102] A least-squares fit to the compositions revealed that the n and m values follow the relationship

$$m = n^{0.605(9)} \times 10^{0.42(2)} \tag{1}$$

This empirical formula provides a rough estimate of the number of thiolates based on the total number of Au atoms.

Close inspection of the plots reveals that there are commonly observed compositions regardless of the thiolates such as $Au_{25}(SR)_{18}$, $Au_{38}(SR)_{24}$, and $Au_{144}(SR)_{60}$. This observation indicates that these clusters are intrinsically stable and that the tail groups of the thiolates do not significantly affect their stability. However, alkanethiolates do not afford small $Au_n(SR)_m$ with $n < 25$. This implies that the noncovalent interaction between adjacent ligands plays a nontrivial role in the stabilization of smaller clusters. In the case of $Au_{25}(SCx)_{18}$, alkanethiolates are self-organized into bundles by van der Waals interaction only when the chain length x exceeds 10.[74]

2.3.1.2 Origin of Stability

After the breakthrough paper on the single crystal X-ray structure determination of $Au_{102}(SR)_{44}$ published in 2007,[93] geometrical structures have been determined for several Au:SR clusters including $Au_{20}(SR)_{16}$,[67] $[Au_{23}(SR)_{16}]^-$,[69] $Au_{24}(SR)_{16}$,[71] $[Au_{25}(SR)_{18}]^-$,[75,76] $Au_{25}(SR)_{18}$,[77] $Au_{28}(SR)_{20}$,[79] $Au_{30}(SR)_{18}$,[81,82] $Au_{36}(SR)_{24}$,[84–86] and $Au_{38}(SR)_{24}$.[52] Figure 10 shows the crystal structures of $[Au_{25}(SR)_{18}]^-$ and $Au_{102}(SR)_{44}$ as examples. $Au_{25}(SR)_{18}$ has an icosahedral Au_{13} core protected by 6 $-S(R)-[Au-S(R)-]_2$ oligomers

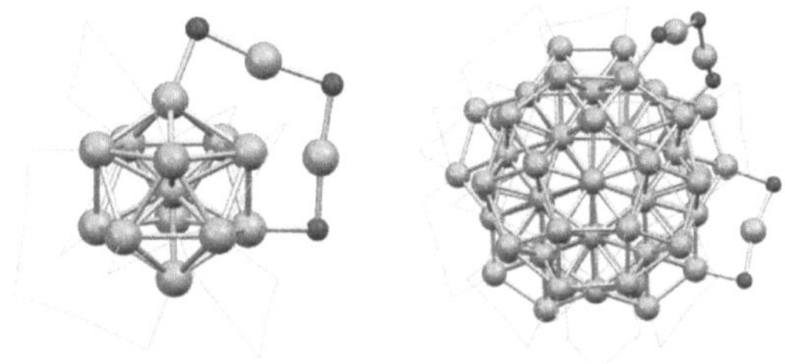

FIGURE 10 Crystal structures of $[Au_{25}(SR)_{18}]^-$ and $Au_{102}(SR)_{44}$.[75,76,93] Yellow (light gray in print versions) and green (dark gray in print versions) balls represent Au and S atoms, respectively. The tail groups of the ligands are omitted for clarity. Gray lines show the framework of the staple motifs. Only one or two staples are highlighted.

called "staples,"[93] while $Au_{102}(SR)_{44}$ has a Mark's decahedral Au_{79} core protected by 19 −S(R)−Au−S(R)− and 2 −S(R)−[Au−S(R)−]$_2$ oligomers. Thus, $Au_{25}(SR)_{18}$ and $Au_{102}(SR)_{44}$ can be formulated as $Au_{13}[Au_2(SR)_3]_6$ and $Au_{79}[Au(SR)_2]_{19}[Au_2(SR)_3]_2$, respectively, according to the "divide-and-protect" scheme.[103] The Au−SR interfacial structure commonly observed in $Au_n(SR)_m$ is in sharp contrast to the classical model in which thiolates are directly bound to the surface of the Au core.[104] Because of this unique interfacial structure, the actual Au core size of $Au_n(SR)_m$ is smaller than that indicated by the value of n. The high stability of $Au_n(SR)_m$ is ascribed to the following geometrical features commonly observed in the structures reported so far.[22] Firstly, the Au core forms a highly symmetric and stable geometric structure. Secondly, all the surface atoms of the Au core are anchored by thiolates at both ends of −S(R)−[Au−S(R)−]$_x$ ($x = 1, 2, 3, 4$). The complete protection of the surface Au atoms by the S anchors of the staples affords thermodynamic and chemical stability against decomposition and etching. Thirdly, the relative population of longer −S(R)−[Au−S(R)−]$_n$ oligomers is higher for smaller clusters (Table 1) because smaller Au cores with larger curvatures prefer protection by longer −S(R)−[Au−S(R)−]$_x$ oligomers, which have bent structures. Mechanical stress imposed on the Au core due to the bonding of the staples is minimized by a proper combination of the chain length.

The electronic structure is also an important determining factor in the stability of $Au_n(SR)_m$ as in the case of bare Au clusters. Magic numbers for bare Au clusters have been explained by the electronic shell model.[105] The small assembly of a positively charged nucleus provides a jellium-like potential field in which valence electrons behave as nearly free electrons. A spherical symmetrical potential affords a series of "superatomic orbitals" with discrete electronic energies, 1S, 1P, 1D, 2S, 1F, 2P, 1G, 2D, 1H, 3S, 1I, etc., in order of energy (Figure 11), where S, P, D, F, G, H, and I represent the value of

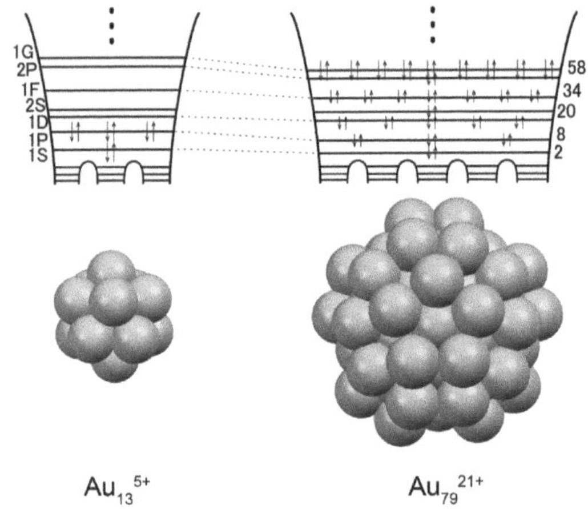

FIGURE 11 Schematic representation of the closed electronic structures of the Au_{13}^{5+} and Au_{79}^{21+} cores of $[Au_{25}(SR)_{18}]^-$ and $Au_{102}(SR)_{44}$, respectively.

the angular momentum $L = 0$, 1, 2, 3, 4, 5, and 6, respectively. Spherical Au clusters gain a high stability when the total number of valence electrons is 2, 8, 18, 20, 34,... because the electronic shell is closed.[106] This concept has been successfully applied to account for the magic stability of $Au_n(SR)_m$ (see Chapter 8 for more details).[107] According to this model, $[Au_n(SR)_m]^z$ clusters (z = net charge) with a total number of valence electrons (n*) of 2, 8, 18, 34, 58, 92,... are stable because their electronic structures are closed, i.e.,

$$n^* = n \times V_A - m - z \quad (2)$$

where V_A (=1) is the number of valence electrons per Au atom. The n* values for other $[Au_n(SR)_m]^z$ calculated using Eqn (2) are listed in Table 1. In several cases, the numbers equal the values expected from electronic shell model. For example, the n* values of $[Au_{25}(SR)_{18}]^-$ and $Au_{102}(SR)_{44}$ are 8 and 58, respectively. The superatomic electronic configurations of the formal cores, Au_{13}^{5+} and Au_{79}^{21+}, are described as $(1S)^2(1P)^6$ and $(1S)^2(1P)^6(1D)^{10}(2S)^2(1F)^{14}(2P)^6(1G)^{18}$, respectively (Figure 11). Deviation from the theoretical values in the small size region is due to nonspherical core structures or stabilization of the open electronic shell structures because the ligand layer acts as a capacitive barrier.[10,62,77,108] Deviation in the large size region suggests that electronic shell closure is less important than geometrical shell closure.

2.3.2 Gold Clusters Protected by Other Ligands

Other ligands have been successfully used to synthesize size-selected Au clusters. Gold clusters protected by mixed ligands of phosphines and halogens (Au:PR_3) have been long known since the 1960s.[1,109] In sharp contrast

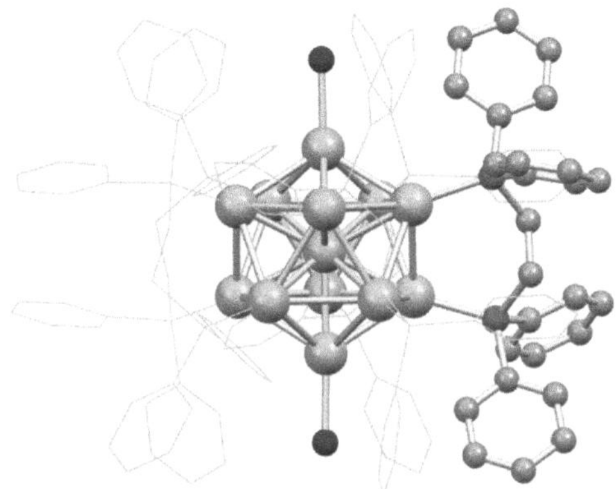

FIGURE 12 Crystal structure of $[Au_{13}(dppe)_{10}Cl_2]^{3+}$.[113] Yellow (light gray in print versions), red (dark gray in print versions), blue (black in print versions), and gray colors represent Au, P, Cl, and C atoms, respectively. H atoms and counteranions are omitted for clarity. One dppe molecule is highlighted by the ball-and-stick model.

to $Au_n(SR)_m$, most of the Au:PR_3 clusters synthesized so far are small in size ($\leq$13) except for a few examples such as Au_{39},[110] Au_{55},[111] and Au_{73-75}.[32] Recently, well-defined and more stable Au clusters in the size range of 7–30 have been synthesized using multidentate phosphines.[112–117] The high stability of isolated Au:PR_3 is also governed by the geometric and electronic structure as in the case of $Au_n(SR)_m$.[107] Extensive X-ray crystallographic studies of small Au_n ($n \leq 13$) clusters have revealed that the phosphine ligands are coordinated directly to the Au core with unique polyhedral structures.[109,110,113–117] As an example, the geometric and electronic structures of $[Au_{13}(dppe)_5Cl_2](PF_6)_3$ (dppe = diphenylphosphinoethane) are considered here.[113] Single crystal X-ray crystallography revealed that the Au_{13} core has an icosahedral motif (Figure 12). The n^* value of the Au_{13} core is calculated to be 8 using Eqn (2) by assuming that phosphine does not remove any charge from the core and a single halogen takes one electron.

Both thiolates and phosphines are ligated to Au clusters such as $[Au_{25}(PPh_3)_{10}(SCx)_5X_2]^{2+}$ and $[Au_{24}(PPh_3)_{10}(SCx)_5X_2]^+$ (X = Cl, Br).[118,119] X-ray crystallography revealed that these clusters contain bi-icosahedral Au_{25} and Au_{24} cores, respectively, whose surfaces are completely passivated; thiolates bridge two Au atoms and phosphine and chlorides are coordinated on the top of the Au atoms. Recently, size-selected clusters such as $Au_{30}(EPT)_{13}$, $Au_{35}(EPT)_{18}$, $Au_{41-43}(EPT)_{21-23}$ (EPT = 9-ethynyl-phenanthrene), $Au_{34}(PA)_{16}$ and $Au_{54}(PA)_{26}$ (PA = phenyl acetylene) have been synthesized by using terminal alkynes as protecting ligands.[30,120] Based on various spectroscopic techniques, it has been proposed that the alkynes in the dehydrogenated form are bound to bridged or hollow sites on the cluster surface.[30] Recent density functional

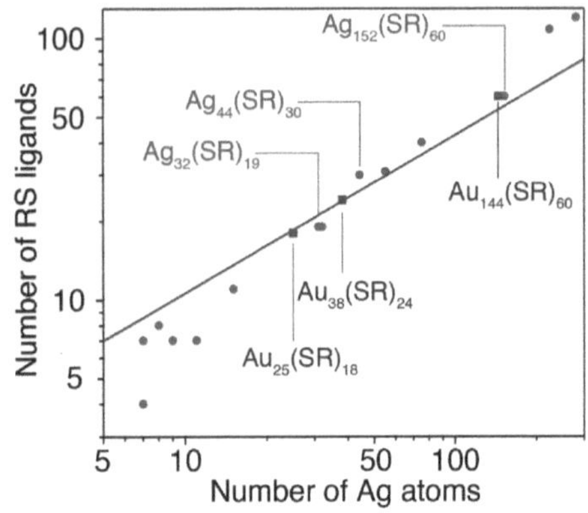

FIGURE 13 Compositional plots of reported Ag:SR clusters (red (black in print versions) circles). For comparison, magic number Au:SR clusters ($Au_{25}(SR)_{18}$, $Au_{38}(SR)_{24}$, and $Au_{144}(SR)_{60}$) are plotted as blue (dark gray in print versions) squares. The solid line represents Eqn (1).

theory calculations suggest that the (RC≡ C−Au−C≡CR) oligomer with the staple motif is formed at the interface of Au clusters as in the case of Au:SR.[121] The oligomeric motif was directly observed by X-ray crystallography in $[Au_{19}(PA)_9(Hdppa)_3]^{2+}$ (Hdppa = *N,N*- bis(diphenylphosphino)amine) at the end of 2014.[122]

2.3.3 Silver Clusters Protected by Thiolates

2.3.3.1 Stable Chemical Compositions

Figure 13 shows a plot of the chemical compositions of $Ag_n(SR)_m$ with sizes ranging from ~10 to ~300.[36,123–135] The number of isolated $Ag_n(SR)_m$ is much smaller than those of $Au_n(SR)_m$ mainly because of the difficulty in the synthesis due to the lower stability. Although the Ag-to-thiol ratios are similar to the Au-to-thiol ratios, the numbers of Ag atoms in $Ag_n(SR)_m$ are different from those of the Au atoms in $Au_n(SR)_m$. This difference suggests the two systems $Ag_n(SR)_m$ and $Au_n(SR)_m$ have different geometrical structure motifs.

2.3.3.2 Origin of Stability

The structure of a representative Ag:SR cluster, $[Ag_{44}(SR)_{30}]^{4-}$, is shown in Figure 14.[131,132] The Ag_{32} core has a Keplerate two-shell $Ag_{12}@Ag_{20}$ structure whose surface is completely passivated by six $Ag_2(SR)_5$ units. Thus, this cluster can be formulated as $\{Ag_{32}[Ag_2(SR)_5]_6\}^{4-}$ according to the divide-and-protect scheme. Although the core and interfacial structures of $[Ag_{44}(SR)_{30}]^{4-}$ are different from those of $Au_n(SR)_m$, all the surface Ag atoms of the core are completely protected by $Ag_2(SR)_5$ units. This complete protection of the Ag_{32}

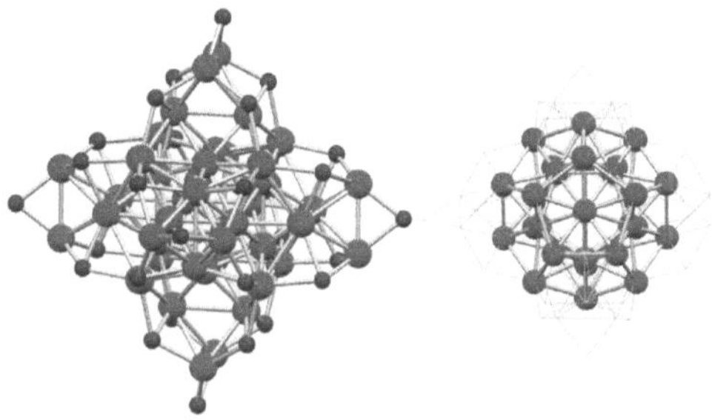

FIGURE 14 Crystal structure of $[Ag_{44}(SR)_{30}]^{4-}$ (left) and the Ag_{32} core moiety (right).[132] Magenta (gray in print versions) and green (dark gray in print versions) balls represent Ag and S atoms, respectively. Organic branches and counteranions are omitted for clarity.

core stabilizes the system both chemically and thermodynamically. The number of valence electrons of the core is calculated to be 18 according to Eqn (2), which indicates that the electronic shell of the Ag_{32} core is closed with a configuration of $(1S)^2(1P)^6(1D)^{10}$.

2.4 SIZE-DEPENDENT EVOLUTION

Size-dependent behavior of structures and properties of metal clusters has been a central issue in basic and applied research since initial studies in the mid-1990s. Powder XRD studies of Au:SR clusters and nanoparticles have demonstrated a structural transition from face-centered cubic (fcc) to decahedral motifs.[19,61] Optical absorption spectroscopy,[59,60] photoluminescence spectroscopy,[136] and electrochemical analysis[62,63] of small Au:SR clusters have revealed quantized electronic structures and molecular-like redox behavior. State-of-the-art synthesis of a series of well-defined $Au_n(SR)_m$ clusters provides us with an opportunity to establish their size-dependent evolution as well as the critical size for a metal-to-molecule transition. This section describes how the electronic and geometric structures of $Au_n(SC12)_m$ evolve with the core size.

2.4.1 Electronic Structures

Optical absorption spectroscopy is the most convenient probe of electronic structures.[137] Figure 15(a) shows the optical absorption spectra of $Au_n(SR)_m$ as a function of size.[138] $Au_n(SR)_m$ with $n \leq 144$ exhibit size-specific profiles with sharp peaks and shoulders that are now used as fingerprints of the composition. Clear optical onset was observed in the spectra of $Au_{25}(SC12)_{18}^{-}$ and $Au_{38}(SC12)_{24}$.[22] The peaks appearing at the lowest energy have been theoretically assigned to optical transitions from the highest occupied molecular orbital to the lowest unoccupied molecular

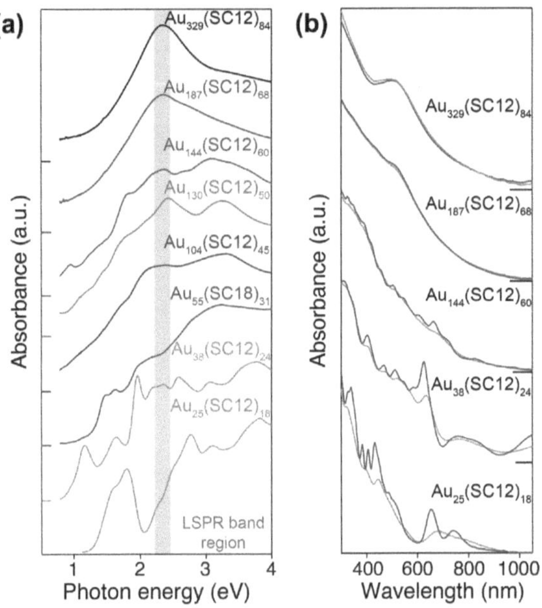

FIGURE 15 (a) Optical absorption spectra of $Au_n(SR)_m$ as a function of size. (b) Temperature dependences of selected $Au_n(SC12)_m$. Red and blue lines correspond to the spectra measured at 290 and 25 K, respectively. The spectra in (a) were corrected by using a Jacobian factor. (see color plate). *Courtesy of Dr Yuichi Negishi.*

orbital localized at the Au cluster core.[99,139,140] In contrast, the spectra of $Au_n(SR)_m$ with $n \geq 187$ exhibit a broadband at ~2.3 eV on otherwise smooth and featureless profiles; this broadband is assignable to localized surface plasmon resonance. These results suggest that the metal-to-molecule transition of electronic structures of $Au_n(SC12)_m$ occurs in the range $187 > n > 144$. This conclusion is further supported by the temperature dependence of the absorption spectra as shown in Figure 15(b). The peaks of $Au_n(SC12)_m$ with $n \leq 144$ become more prominent, and the peak positions are blueshifted at 25 K. Similar phenomena were reported for $Au_{25}(PET)_{18}$, $Au_{38}(PET)_{24}$, and $Au_{144}(PET)_{60}$.[141,142] However, the intensity and position of the broadband at 520−540 nm in the spectra of $Au_n(SC12)_m$ with $n \geq 187$ did not change appreciably even at 25 K. These results indicate that the electronic structure transition occurs between $Au_{144}(SC12)_{60}$ and $Au_{187}(SC12)_{68}$. Electrochemical analysis also provides evidence for the metal-to-molecule transition of the electronic structures.[143] Reflecting the discrete electronic structures, molecular-like redox behavior was observed for Au_{25} $(SR)_{18}$ and $Au_{38}(SR)_{24}$,[22,50,64,74] whereas double layer charging behavior was observed for clusters larger than $Au_{144}(SR)_{60}$,[22,54] indicating that they can be viewed as quantum capacitors.

The effect of size on the electronic structures of $Au_n(SR)_m$ was also studied by other spectroscopic methods including steady-state emission spectroscopy and ultrafast spectroscopy.[144,145] In the bulk metal, photoexcited electrons are quenched very rapidly via electron−electron and electron−phonon coupling processes. In contrast, small $Au_n(SR)_m$ clusters have a much longer relaxation time associated with a large energy gap, and as a result, the quantum yield of the photoluminescence is greatly enhanced.[145] Photoluminescence has been reported for, e.g., $Au_{22}(SR)_{18}$, $Au_{23}(SR)_{16}$, $Au_{25}(SR)_{18}$, $Au_{38}(SR)_{24}$, and $Au_{64}(SR)_{32}$.[9,68,70,78,89] The dynamics of the photoexcited state have been probed by transient absorption spectroscopy and fluorescence upconversion.[144,145] It was demonstrated that the emission lifetimes and two-photon cross sections of $Au_n(SR)_m$ smaller than 2.2 nm are much larger than those of Au nanoparticles.

2.4.2 Geometric Structures

Single crystal XRD analyses of small $Au_n(SR)_m$ revealed that their atomic packing structures are different from the fcc motif of bulk gold. For example, $Au_{25}(SR)_{18}$ (Figure 10) and $Au_{38}(SR)_{24}$ with RS = PET and C*x*SH have icosahedral Au_{13} and bi-icosahedral Au_{23} cores, respectively. These results suggest that the transition between fcc and non-fcc structures occurs in a certain critical size region. However, single crystals have been obtained only for a limited number of clusters (Table 1). Thus, structural information has conventionally been obtained by powder XRD. The powder XRD patterns of a series of $Au_n(SC12)_m$ are presented in Figure 16.[138] The XRD patterns of $Au_{187}(SC12)_{68}$, $Au_{329}(SC12)_{84}$, and $Au_{\sim520}(SC12)_{\sim130}$ are similar to that of Au bulk, suggesting that they have fcc-based cores, although each peak is significantly broadened because of the small size effect. The XRD patterns of $Au_{104}(SC12)_{45}$, $Au_{130}(SC12)_{50}$, and $Au_{144}(SC12)_{60}$ are, however, reproduced well with computational models of $Au_{104}(SH)_{45}$ with a Mark's decahedral (M-Dh) core, $Au_{130}(SH)_{50}$ with an M-Dh core, and $Au_{144}(SH)_{60}$ with an icosahedral core, respectively.[138] These comparisons lead us to conclude that clusters smaller than $Au_{144}(SC_{12}H_{25})_{60}$ have nonbulk-like Au cores. Similar structural transitions have been reported for PET-protected Au clusters; smaller $Au_{103\text{-}105}(PET)_{45\text{-}46}$ and $Au_{144}(PET)_{60}$ have nonbulk-like core structures, whereas larger $Au_{333}(PET)_{79}$ and $Au_{\sim940}(PET)_{\sim160}$ have fcc Au cores.[94,97,99,101]

However, single crystal XRD analysis indicated that some of the small Au clusters such as $Au_{20}(TBBT)_{16}$, $Au_{23}(CHT)_{16}$ (CHT = cyclohexanethiol), $Au_{24}(AT)_{16}$ (AT = 1-adamantanethiol), $Au_{28}(TBBT)_{20}$, and $Au_{36}(TBBT)_{24}$ do have Au cores with an fcc atomic packing.[67,69,71,79,85] Since these thiolates have crowded molecular structures around sulfur, the formation of the small fcc Au core is associated with the structures of the ligands.

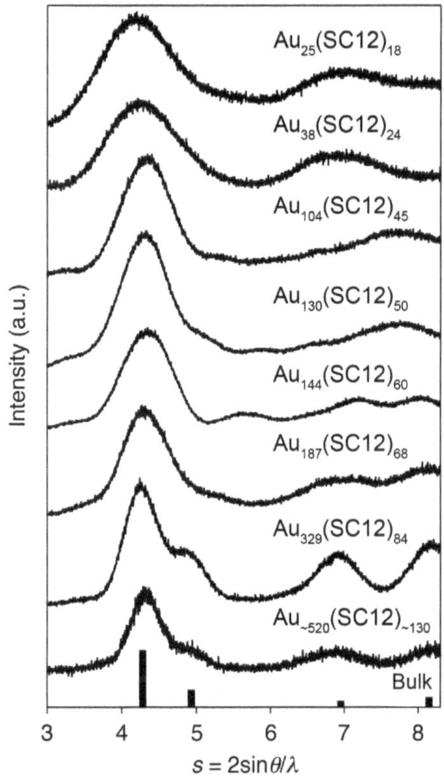

FIGURE 16 XRD patterns of a series of $Au_n(SC12)_m$. *Courtesy of Dr Yuichi Negishi.*

2.5 SUMMARY

This chapter has summarized the state-of-the-art methods of atomically precise synthesis of Au and Ag clusters protected by monolayers of ligands (thiols, phosphines, and terminal alkynes). The representative approaches reviewed include template-mediated synthesis, fractional synthesis, and size-focusing synthesis via postsynthetic core etching or slow core growth. Molecular formulas of the clusters synthesized have been determined by MS coupled with soft ionization (ESI or MALDI). In the last decade, synthesis of a large number of ligand-protected Au and Ag clusters has been reported. In particular, the chemical compositions of those Au:SR that had been isolated by the end of 2014 were extensively surveyed. The origin of the high stabilities of these isolated clusters was discussed in terms of the closure of electronic and geometric shells as in the case of bare Au and Ag clusters. Manifestation of molecular-like optical properties and non-fcc atomic packing below a critical size was demonstrated by taking Au:SR clusters as examples.

REFERENCES

1. Schmid G, editor. *Nanoparticles: from theory to application, 2nd, completely revised and updated edition.* New Jersey: Wiley; 2010.
2. de Heer WA. The physics of simple metal clusters: experimental aspects and simple models. *Rev Mod Phys* 1993;**65**:611–76.
3. Knickelbein MB. Reaction of transition metal clusters with small molecules. *Ann Rev Phys Chem* 1999;**50**:79–115.
4. Castleman AW, Khanna SN. Clusters, superatoms, and building blocks of new materials. *J Phys Chem C* 2009;**113**:2664–75.
5. Jin R. Quantum sized, thiolate-protected gold nanoclusters. *Nanoscale* 2010;**2**:343–62.
6. Maity P, Xie S, Yamauchi M, Tsukuda T. Stabilized gold clusters: from isolation toward controlled synthesis. *Nanoscale* 2012;**4**:4027–37.
7. Lu Y, Chen W. Sub-nanometre sized metal clusters: from synthetic challenges to the unique property discoveries. *Chem Soc Rev* 2012;**41**:3594–623.
8. Schaaff TG, Knight G, Shafigullin MN, Borkman RF, Whetten RL. Isolation and selected properties of a 10.4 kDa gold: glutathione cluster compound. *J Phys Chem B* 1997;**102**: 10643–6.
9. Negishi Y, Nobusada K, Tsukuda T. Glutathione-protected gold clusters revisited: bridging the gap between gold(I)-thiolate complexes and thiolate-protected gold nanocrystals. *J Am Chem Soc* 2005;**127**:5261–70.
10. Negishi Y, Chaki NK, Shichibu Y, Whetten RL, Tsukuda T. Origin of magic stability of thiolated gold clusters: a case study on $Au_{25}(SC_6H_{13})_{18}$. *J Am Chem Soc* 2007;**129**: 11322–3.
11. Tracy JB, Crowe MC, Parker JF, Hampe O, Fields-Zinna CA, Dass A, et al. Electrospray ionization mass spectrometry of uniform and mixed monolayer nanoparticles: $Au_{25}[S(CH_2)_2Ph]_{18}$ and $Au_{25}[S(CH_2)_2Ph]_{18-x}(SR)_x$. *J Am Chem Soc* 2007;**129**: 16209–15.
12. Dass A, Stevenson A, Dubay GR, Tracy JB, Murray RW. Nanoparticle MALDI-TOF mass spectrometry without fragmentation: $Au_{25}(SCH_2CH_2Ph)_{18}$ and mixed monolayer $Au_{25}(SCH_2CH_2Ph)_{18-x}(L)_x$. *J Am Chem Soc* 2008;**130**:5940–6.
13. Kouchi H, Kawasaki H, Arakawa R. A new matrix of MALDI-TOF MS for the analysis of thiolate-protected gold clusters. *Anal Methods* 2012;**4**:3600–3.
14. Crooks RM, Zhao M, Sun L, Chechik V, Yeung LK. Dendrimer-encapsulated metal nanoparticles: synthesis, characterization, and applications to catalysis. *Acc Chem Res* 2001;**34**:181–90.
15. Yamamoto K, Imaoka T. Precision synthesis of subnanoparticles using dendrimers as a superatom synthesizer. *Acc Chem Res* 2014;**47**:1127–36.
16. Ueno T, Suzuki M, Goto T, Matsumoto T, Nagayama K, Watanabe Y. Size-selective olefin hydrogenation by a Pd nanocluster provided in an apo-ferritin cage. *Angew Chem Int Ed* 2004;**43**:2527–30.
17. Sakamoto M, Tanaka D, Tsunoyama H, Tsukuda T, Minagawa Y, Majima Y, et al. Platonic hexahedron composed of six organic faces with an inscribed Au cluster. *J Am Chem Soc* 2012;**134**:816–9.
18. Zheng J, Zhang C, Dickson RM. Highly fluorescent, water-soluble, size-tunable gold quantum dots. *Phys Rev Lett* 2004;**93**:077402–77411.
19. Whetten RL, Khoury JT, Alvarez MM, Murthy S, Vezmar I, Wang ZL, et al. Nanocrystal gold molecules. *Adv Mater* 1996;**8**:428–33.

20. Thomas PJ, Kulkarni GU, Rao CNR. An investigation of two-dimensional arrays of thiolized Pd nanocrystals. *J Phys Chem B* 2000;**104**:8138–44.
21. Levi-Kalisman Y, Jadzinsky PD, Kalisman N, Tsunoyama H, Tsukuda T, Bushnell DA, et al. Synthesis and characterization of $Au_{102}(p\text{-MBA})_{44}$ nanoparticles. *J Am Chem Soc* 2011;**133**:2976–82.
22. Chaki NK, Negishi Y, Tsunoyama H, Shichibu Y, Tsukuda T. Ubiquitous 8 and 29 kDa gold: alkanethiolate cluster compounds: mass-spectrometric determination of molecular formulas and structural implications. *J Am Chem Soc* 2008;**130**:8608–10.
23. Jimenez VL, Leopold MC, Mazzitelli C, Jorgensen JW, Murray RW. HPLC of monolayer-protected gold nanoclusters. *Anal Chem* 2003;**75**:199–206.
24. Song Y, Jimenez V, McKinney C, Donkers R, Murray RW. Estimation of size for 1–2 nm nanoparticles using an HPLC electrochemical detector of double layer charging. *Anal Chem* 2003;**75**:5088–96.
25. Wolfe RL, Murray RW. Analytical evidence for the monolayer-protected cluster $Au_{225}[(S(CH_2)_5CH_3)]_{75}$. *Anal Chem* 2006;**78**:1167–73.
26. Choi MMF, Douglas AD, Murray RW. Ion-pair chromatographic separation of water-soluble gold monolayer-protected clusters. *Anal Chem* 2006;**78**:2779–85.
27. Zhang Y, Shuang S, Dong C, Lo CK, Paau MC, Choi MMF. Application of HPLC and MALDI-TOF MS for studying as-synthesized ligand protected gold nanoclusters products. *Anal Chem* 2009;**81**:1676–85.
28. Negishi Y, Sakamoto C, Ohyama T, Tsukuda T. Synthesis and the origin of the stability of thiolate-protected Au_{130} and Au_{187} clusters. *J Phys Chem Lett* 2012;**3**:1624–8.
29. Ghosh A, Hassinen J, Pulkkinen P, Tenhu H, Ras RHA, Pradeep T. Simple and efficient separation of atomically precise noble metal clusters. *Anal Chem* 2014;**86**:12185–90.
30. Maity P, Takano S, Yamazoe S, Wakabayashi T, Tsukuda T. Binding motif of terminal alkynes on gold clusters. *J Am Chem Soc* 2013;**135**:9450–7.
31. Wilcoxon JP, Martin JE, Provencio P. Size distribution of gold nanoclusters studied by liquid chromatography. *Langmuir* 2000;**16**:9912–20.
32. Gutiérrez E, Powell RD, Furuya FR, Hainfeld JF, Schaaff TG, Shafigullin MN, et al. Greengold, a giant cluster compound of unusual electronic structures. *Eur Phys J D* 1999;**9**:647–51.
33. Tsunoyama H, Negishi Y, Tsukuda T. Chromatographic isolation of "Missing" Au_{55} clusters protected by alkanethiolates. *J Am Chem Soc* 2006;**128**:6036–7.
34. Qian H, Zhu Y, Jin R. Isolation of ubiquitous $Au_{40}(SR)_{24}$ clusters from the 8 kDa gold clusters. *J Am Chem Soc* 2010;**132**:4583–5.
35. Tsunoyama R, Tsunoyama H, Pannopard P, Limtrakul J, Tsukuda T. MALDI mass analysis of 11 kDa gold clusters protected by octadecanethiolate ligands. *J Phys Chem C* 2010;**114**:16004–9.
36. Negishi Y, Arai R, Niihori Y, Tsukuda T. Isolation and structural characterization of magic silver clusters protected by 4-(tert-Butyl)benzyl mercaptan. *Chem Commun* 2011;**47**:5693–5.
37. Nishigaki J, Tsunoyama R, Tsunoyama H, Ichikuni N, Yamazoe S, Negishi Y, et al. A new binding motif of sterically demanding thiolates on a gold cluster. *J Am Chem Soc* 2012;**134**:14295–7.
38. Al-Somali AM, Krueger KM, Falkner JC, Colvin VL. Recycling size exclusion chromatography for the analysis and separation of nanocrystalline gold. *Anal Chem* 2004;**76**:5903–10.

39. Murayama H, Narushima T, Negishi Y, Tsukuda T. Structures and stabilities of alkanethiolate monolayers on palladium clusters as studied by gel permeation chromatography. *J Phys Chem B* 2004;**108**:3496–503.
40. Schaaff TG, Whetten RL. Giant gold-glutathione cluster compounds: intense optical activity in metal-based transitions. *J Phys Chem B* 2000;**104**:2630–41.
41. Negishi Y, Takasugi Y, Sato S, Yao H, Kimura K, Tsukuda T. Magic-numbers Au_n clusters protected by glutathione monolayers ($n = 18, 21, 25, 28, 32, 39$): isolation and spectroscopic characterization. *J Am Chem Soc* 2004;**126**:6518–9.
42. Kimura K, Sugimoto N, Sato S, Yao H, Negishi Y, Tsukuda T. Size determination of gold clusters by polyacrylamide gel electrophoresis in a large cluster region. *J Phys Chem C* 2009;**113**:14076–82.
43. Kothalawala N, West IV JL, Dass A. Size-dependent molecule-like to plasmonic transition in water-soluble glutathione stabilized gold nanomolecules. *Nanoscale* 2014;**6**:683–7.
44. Kumar S, Bolan MD, Bigioni TP. Glutathione-stabilized magic-number silver cluster compounds. *J Am Chem Soc* 2010;**132**:13141–3.
45. Schaaff TG, Whetten RL. Controlled etching of Au:SR cluster compounds. *J Phys Chem B* 1999;**103**:9394–6.
46. Shichibu Y, Negishi Y, Tsunoyama H, Kanehara M, Teranishi T, Tsukuda T. Extremely high stability of glutathionate-protected Au_{25} clusters against core etching. *Small* 2007;**3**:835–9.
47. Shichibu Y, Negishi Y, Tsukuda T, Teranishi T. Large-scale synthesis of thiolated Au_{25} clusters via ligand exchange reactions of phosphine-stabilized Au_{11} clusters. *J Am Chem Soc* 2005;**127**:13464–5.
48. Zhu M, Lanni E, Garg N, Bier ME, Jin R. Kinetically controlled, high-yield synthesis of Au_{25} clusters. *J Am Chem Soc* 2008;**130**:1138–9.
49. Qian H, Zhu M, Anderson UN, Jin R. Facile, karge-scale synthesis of dodecanethiol-stabilized Au_{38} clusters. *J Phys Chem A* 2009;**113**:4281–4.
50. Qian H, Zhu Y, Jin R. Size-focusing synthesis, optical and electrochemical properties of monodisperse $Au_{38}(SC_2H_4Ph)_{24}$ nanoclusters. *ACS Nano* 2009;**3**:3795–803.
51. Dharmaratne AC, Krick T, Dass A. Nanocluster size evolution studied by mass spectrometry in room temperature $Au_{25}(SR)_{18}$ synthesis. *J Am Chem Soc* 2009;**131**:13604–5.
52. Qian H, Eckenhoff WT, Zhu Y, Pintauer T, Jin R. Total structure determination of thiolate-protected Au_{38} nanoparticles. *J Am Chem Soc* 2010;**132**:8280–1.
53. Parker JF, Weaver JEF, McCallum F, Fields-Zinna CA, Murray RW. Synthesis of monodisperse $[Oct_4N^+][Au_{25}(SR)_{18}^-]$ nanoparticles, with some mechanistic observations. *Langmuir* 2010;**26**:13650–4.
54. Qian H, Jin R. Ambient synthesis of $Au_{144}(SR)_{60}$ nanoclusters in methanol. *Chem Mater* 2011;**23**:2209–17.
55. McKenzie LC, Zaikova TO, Hutchison JE. Structurally similar triphenylphosphine-stabilized undecagolds, $Au_{11}(PPh_3)_7Cl_3$ and $[Au_{11}(PPh_3)_8Cl_2]Cl$, exhibit distinct ligand exchange pathways with glutathione. *J Am Chem Soc* 2014;**136**:13426–35.
56. Yuan X, Zhang B, Luo Z, Yao Q, Leong DT, Yan N, et al. Balancing the rate of cluster growth and etching for gram-scale synthesis of thiolate-protected Au_{25} nanoclusters with atomic precision. *Angew Chem Int Ed* 2014;**53**:4623–7.
57. Luo Z, Nachammai V, Zhang B, Yan N, Leong DT, Jiang D, et al. Toward understanding the growth mechanism: tracing all stable intermediate species from reduction of Au(I)-thiolate complexes to evolution of Au_{25} nanoclusters. *J Am Chem Soc* 2014;**136**:10577–80.

58. Brust M, Walker M, Bethell D, Schiffrin DJ, Whyman R. Synthesis of thiol-derivatised gold nanoparticles in a two-phase liquid–liquid system. *J Chem Soc Chem Commun* 1994: 801–2.
59. Schaaff TG, Shafigullin MN, Khoury JT, Vezmar I, Whetten RL, Cullen WG, et al. Isolation of smaller nanocrystal Au molecules: robust quantum effects in optical spectra. *J Phys Chem B* 1997;**101**:7885–91.
60. Hostetler MJ, Wingate JE, Zhong CJ, Harris JE, Vachet RW, Clark MR, et al. Alkanethiolate gold cluster molecules with core diameter from 1.5 to 5.2 nm: core and monolayer properties as a function of core size. *Langmuir* 1998;**14**:17–30.
61. Cleveland CL, Landman U, Schaaff TG, Shafigullin MN, Stephens PW, Whetten RL. Structural evolution of smaller gold nanocrystals: the truncated decahedral motif. *Phys Rev Lett* 1997;**79**:1873–6.
62. Chen S, Ingram RS, Hostetler MJ, Pietron JJ, Murray RW, Schaaff TG, et al. Gold nanoelectrodes of varied size: transition to molecule-like charging. *Science* 1998;**280**:2098–101.
63. Templeton AC, Wuelfing WP, Murray RW. Monolayer-protected cluster molecules. *Acc Chem Res* 2000;**33**:27–36.
64. Parker JF, Fields-Zinna CA, Murray RW. The story of a monodisperse gold nanoparticle: $Au_{25}L_{18}$. *Acc Chem Res* 2010;**43**:1289–96.
65. Ghosh A, Udayabhaskararao T, Pradeep T. One-step route to luminescent $Au_{18}SG_{14}$ in the condensed phase and its closed shell molecular ions in the Gas phase. *J Phys Chem Lett* 2012;**3**:1997–2002.
66. Zhu M, Qian H, Jin R. Thiolate-protected Au_{20} clusters with a large energy gap of 2.1 eV. *J Am Chem Soc* 2009;**131**:7220–1.
67. Zeng C, Liu C, Chen Y, Rosi NL, Jin R. Gold-thiolate ring as a protecting motif in the $Au_{20}(SR)_{16}$ nanocluster and implications. *J Am Chem Soc* 2014;**136**:11922–5.
68. Yu Y, Luo Z, Chevrier DM, Leong DT, Zhang P, Jiang D, et al. Identification of a highly luminescent $Au_{22}(SG)_{18}$ nanocluster. *J Am Chem Soc* 2014;**136**:1246–9.
69. Das A, Li T, Nobusada K, Zeng C, Rosi NL, Jin R. Nonsuperatomic $[Au_{23}(SC_6H_{11})_{16}]^-$ nanocluster featuring bipyramidal Au_{15} kernel and trimeric $Au_3(SR)_4$ motif. *J Am Chem Soc* 2013;**135**:18264–7.
70. Hesari M, Workentin MS. Facile synthesis of $Au_{23}(SC(CH_3)_3)_{16}$ clusters. *J Mater Chem C* 2014;**2**:3631–8.
71. Crasto D, Barcaro G, Stener M, Sementa L, Fortunelli A, Dass A. $Au_{24}(SAdm)_{16}$ nanomolecules: X-ray crystal structure, theoretical analysis, adaptability of adamantane ligands to form $Au_{23}(SAdm)_{16}$ and $Au_{25}(SAdm)_{16}$, and its relation to $Au_{25}(SR)_{18}$. *J Am Chem Soc* 2014;**136**:14933–40.
72. Zhu M, Qian H, Jin R. Thiolate-protected $Au_{24}(SC_2H_4Ph)_{20}$ nanoclusters: superatoms or not? *J Phys Chem Lett* 2010;**1**:1003–7.
73. Nishigaki J, Yamazoe S, Kohara S, Fujiwara A, Kurashige W, Negishi Y, et al. A twisted bi-icosahedral Au_{25} cluster enclosed by bulky arenethiolates. *Chem Commun* 2014;**50**:839–41.
74. Antonello S, Arrigoni G, Dainese T, Nardi MD, Parisio G, Perotti L, et al. Electron transfer through 3D monolayers on Au_{25} clusters. *ACS Nano* 2014;**8**:2788–95.
75. Heaven MW, Dass A, White PS, Holt KM, Murray RW. Crystal structure of the gold nanoparticle $[N(C_8H_{17})_4][Au_{25}(SCH_2CH_2Ph)_{18}]$. *J Am Chem Soc* 2008;**130**:3754–5.
76. Zhu M, Aikens CM, Hollander FJ, Schatz GC, Jin R. Correlating the crystal structure of a thiol-protected Au_{25} cluster and optical properties. *J Am Chem Soc* 2008;**130**:5883–5.
77. Zhu M, Eckenhoff WT, Pintauer T, Jin R. Conversion of anionic $[Au_{25}(SCH_2CH_2Ph)_{18}]^-$ cluster to charge neutral cluster via air oxidation. *J Phys Chem C* 2008;**112**:14221–4.

78. Kumar S, Jin R. Water-soluble $Au_{25}(Capt)_{18}$ nanoclusters: synthesis, thermal stability, and optical properties. *Nanoscale* 2012;**4**:4222–7.
79. Zeng C, Li T, Das A, Rosi NL, Jin R. Chiral structure of thiolate-protected 28-gold-atom nanocluster determined by X-ray crystallography. *J Am Chem Soc* 2013;**135**:10011–3.
80. Crasto D, Dass A. Green gold: $Au_{30}(S\text{-}t\text{-}C_4H_9)_{18}$ molecules. *J Phys Chem C* 2013;**117**: 22094–7.
81. Yang H, Wang Y, Edwards AJ, Yan J, Zheng N. High-yield synthesis and crystal structure of a green Au_{30} cluster co-capped by thiolate and sulfide. *Chem Commun* 2014;**50**:14325–7.
82. Crasto D, Malola S, Brosofsky G, Dass A, Häkkinen H. Single crystal XRD structure and theoretical analysis of the chiral $Au_{30}S(S\text{-}t\text{-}Bu)_{18}$ cluster. *J Am Chem Soc* 2014;**136**: 5000–5.
83. Krommenhoek PJ, Wang J, Hentz N, Johnston-Peck AC, Kozek KA, Kalyuzhny G, et al. Bulky adamantanethiolate and cyclohexanethiolate ligands favor smaller gold nanoparticles with altered discrete sizes. *ACS Nano* 2012;**6**:4903–11.
84. Nimmala PR, Knoppe S, Jupally VR, Delcamp JH, Aikens CM, Dass A. $Au_{36}(SPh)_{24}$ nanomolecules: X-ray crystal structure, optical spectroscopy, electrochemistry, and theoretical analysis. *J Phys Chem B* 2014;**118**:14157–67.
85. Zeng C, Qian H, Li T, Li G, Rosi NL, Yoon B, et al. Total structure and electronic properties of the gold nanocrystal $Au_{36}(SR)_{24}$. *Angew Chem Int Ed* 2012;**51**:13114–8.
86. Das A, Liu C, Zeng C, Li G, Li T, Rosi NL, et al. Cyclopentanethiolato-protected $Au_{36}(SC_5H_9)_{24}$ nanocluster: crystal structure and implications for the steric and electronic effects of ligand. *J Phys Chem A* 2014;**118**:8264–9.
87. Zeng C, Chen Y, Gao L, Jin R. Synthesis of a $Au_{44}(SR)_{28}$ nanocluster: structure prediction and evolution from $Au_{28}(SR)_{20}$, $Au_{36}(SR)_{24}$ to $Au_{44}(SR)_{28}$. *Chem Commun* 2014;**50**:55–7.
88. Qian H, Jin R. Synthesis and electrospray mass spectrometry determination of thiolate-protected $Au_{55}(SR)_{31}$ nanoclusters. *Chem Commun* 2011;**47**:11462–4.
89. Zeng C, Chen Y, Li G, Jin R. Magic size $Au_{64}(S\text{-}c\text{-}C_6H_{11})_{32}$ nanocluster protected by cyclohexanethiolate. *Chem Mater* 2014;**26**:2635–41.
90. Nimmala PR, Yoon B, Whetten RL, Landman U, Dass A. $Au_{67}(SR)_{35}$ nanomolecules: characteristic size-specific optical, electrochemical, structural properties and first-principles theoretical analysis. *J Phys Chem A* 2013;**117**:504–17.
91. Li G, Zeng C, Jin R. Thermally robust $Au_{99}(SPh)_{42}$ nanoclusters for chemoselective hydrogenation of nitro benzaldehyde derivatives in water. *J Am Chem Soc* 2014;**136**: 3673–9.
92. Nimmala PR, Dass A. $Au_{99}(SPh)_{42}$ nanomolecules: aromatic thiolate ligand induced conversion of $Au_{144}(SCH_2CH_2Ph)_{60}$. *J Am Chem Soc* 2014;**136**:17016–23.
93. Jadzinsky PD, Calero G, Ackerson CJ, Bushnell DA, Kornberg RD. Structure of thiol monolayer-protected gold nanoparticle at 1.1 Å resolution. *Science* 2007;**318**:430–3.
94. Dass A, Nimmala PR, Jupally VR, Kothalawala N. $Au_{103}(SR)_{45}$, $Au_{104}(SR)_{45}$, $Au_{104}(SR)_{46}$ and $Au_{105}(SR)_{46}$ nanoclusters. *Nanoscale* 2013;**5**:12082–5.
95. Jupally VR, Dass A. Synthesis of $Au_{130}(SR)_{50}$ and $Au_{130-x}Ag_x(SR)_{50}$ nanomolecules through core size conversion of larger metal clusters. *Phys Chem Chem Phys* 2014;**16**:10473–9.
96. Jupally VR, Dharmaratne AC, Crasto D, Huckaba AJ, Kumara C, Nimmala PR, et al. $Au_{137}(SR)_{56}$ nanomolecules: composition, optical spectroscopy, electrochemistry and electrocatalytic reduction of CO_2. *Chem Commun* 2014;**50**:9895–8.
97. Qian H, Jin R. Controlling nanoparticles with atomic precision: the case of $Au_{144}(SCH_2CH_2Ph)_{60}$. *Nano Lett* 2009;**9**:4083–7.

98. Kumara C, Dass A. $Au_{329}(SR)_{84}$ nanomolecules: compositional assignment of the 76.3 kDa Plasmonic faradaurates. *Anal Chem* 2014;**86**:4227–32.
99. Qian H, Zhu Y, Jin R. Atomically precise gold nanocrystal molecules with surface plasmon resonance. *Proc Natl Acad Sci USA* 2012;**109**:696–700.
100. Kumara C, Zuo X, Ilavsky J, Chapman KW, Cullen DA, Dass A. Super-stable, highly monodisperse plasmonic Faradaurate-500 nanocrystals with 500 gold atoms: $Au_{\sim 500}(SR)_{\sim 120}$. *J Am Chem Soc* 2014;**136**:7410–7.
101. Kumara C, Zuo X, Cullen DA, Dass A. Faradaurate-940: synthesis, mass spectrometry, electron microscopy, high-energy X-ray diffraction, and X-ray scattering study of $Au_{\sim 940\pm 20}(SR)_{\sim 160\pm 4}$ nanocrystals. *ACS Nano* 2014;**8**:6431–9.
102. Dass A. Nano-scaling law: geometric foundation of thiolated gold nanomolecules. *Nanoscale* 2012;**4**:2260–3.
103. Häkkinen H, Walter M, Grönbeck H. Divide and protect: capping gold nanoclusters with molecular gold-thiolate rings. *J Phys Chem B* 2006;**110**:9927–31.
104. Luedtke WD, Landman U. Structure, dynamics, and thermodynamics of passivated gold nanocrystallites and their assemblies. *J Phys Chem* 1996;**100**:13323–9.
105. Wang LM, Wang LS. Probing the electronic properties and structural evolution of anionic gold clusters in the gas phase. *Nanoscale* 2012;**4**:4038–53.
106. Taylor KJ, Pettiette-Hall CL, Cheshnovsky O, Smally RE. Ultraviolet photoelectron spectra of coinage metal clusters. *J Chem Phys* 1992;**96**:3319–29.
107. Walter M, Akola J, Lopez-Acevedo O, Jadzinsky PD, Calero G, Ackerson CJ, et al. A unified view of ligand-protected gold clusters as superatom complexes. *Proc Natl Acad Sci USA* 2008;**105**:9157–62.
108. Tofanelli MA, Ackerson CJ. Superatom electron configuration predicts thermal stability of $Au_{25}(SR)_{18}$ nanoclusters. *J Am Chem Soc* 2012;**134**:16937–40.
109. Konishi K. Phosphine-coordinated pure-gold clusters: diverse geometrical structures and unique optical Properties/Responses. *Struct Bond* 2014;**161**:49–86.
110. Teo BK, Shi X, Zhang H. Pure gold cluster of 1:9:9:1:9:9:1 layered structure: a novel 39-metal-atom cluster $[(Ph_3P)_{14}Au_{39}Cl_6]Cl_2$ with an interesting gold atom in a hexagonal antiprismatic cage. *J Am Chem Soc* 1992;**114**:2743–5.
111. Schmid G, Pfeil R, Boese R, Bandermann F, Meyer S, Calis GHM, et al. $Au_{55}[P(C_6H_5)_3]_{12}Cl_6$ – a gold cluster of an exceptional size. *Chem Ber* 1981;**114**:3634–42.
112. Yanagimoto Y, Negishi Y, Fujihara H, Tsukuda T. Chiroptical activity of BINAP-stabilized undecagold clusters. *J Phys Chem B* 2006;**110**:11611–4.
113. Shichibu Y, Konishi K. Hcl-induced nuclearity convergence in diphosphine-protected ultrasmall gold clusters: a novel synthetic route to "Magic-Number" Au_{13} clusters. *Small* 2010;**6**:1216–20.
114. Wan XK, Lin ZW, Wang QM. Au_{20} nanocluster protected by hemilable phosphines. *J Am Chem Soc* 2012;**134**:14750–2.
115. Wan XK, Yuan SF, Lin ZW, Wang QM. A chiral gold nanocluster Au_{20} protected by tetradentate phosphine ligands. *Angew Chem Int Ed* 2014;**53**:2923–6.
116. Chen J, Zhang QF, Williard PG, Wang LS. Synthesis and structure determination of a new Au_{20} nanocluster protected by tripodal tetraphosphine ligands. *Inorg Chem* 2014;**53**: 3932–4.
117. Chen J, Zhang QF, Bonaccorso TA, Williard PG, Wang LS. Controlling gold nanoclusters by diphosphine ligands. *J Am Chem Soc* 2014;**136**:92–5.

118. Shichibu Y, Negishi Y, Watanabe T, Chaki NK, Kawaguchi H, Tsukuda T. Biicosahedral gold clusters $[Au_{25}(PPh_3)_{10}(SC_nH_{2n+1})_5Cl_2]^{2+}$ (n = 2–18): a stepping stone to cluster-assembled materials. *J Phys Chem C* 2007;**111**:7845–7.
119. Das A, Li T, Nobusada K, Zeng Q, Rosi NL, Jin R. Total structure and optical properties of a phosphine/thiolate-protected Au_{24} nanocluster. *J Am Chem Soc* 2012;**134**:20286–9.
120. Maity P, Tsunoyama H, Yamauchi M, Xie S, Tsukuda T. Organogold clusters protected by phenylacetylene. *J Am Chem Soc* 2011;**133**:20123–5.
121. Tang Q, Jiang D. Insight into the PhC≡C/Au interface. *J Phys Chem C* 2015;**119**:10804–10.
122. Wan X, Tang Q, Yuan S, Jiang D, Wang Q. Au_{19} nanocluster featuring a V-shaped alkynyl–gold motif. *J Am Chem Soc* 2015;**137**:652–5.
123. Wu Z, Lanni E, Chen W, Bier ME, Ly D, Jin R. High yield, large scale synthesis of thiolate-protected Ag_7 clusters. *J Am Chem Soc* 2009;**131**:16672–4.
124. Rao TUB, Pradeep T. Luminescent Ag_7 and Ag_8 clusters by interfacial synthesis. *Angew Chem Int Ed* 2010;**49**:3925–9.
125. Rao TUB, Nataraju B, Pradeep T. Ag_9 quantum cluster through a solid-state route. *J Am Chem Soc* 2010;**132**:16304–7.
126. Baksi A, Bootharaju MS, Chen X, Häkkinen H, Pradeep T. $Ag_{11}(SG)_7$: a new cluster identified by mass spectrometry and optical spectroscopy. *J Phys Chem C* 2014;**118**:21722–9.
127. Bertorelle F, Hamouda R, Rayane D, Broyer M, Antoine R, Dugourd P, et al. Synthesis, characterization and optical properties of low nuclearity liganded silver clusters: $Ag_{31}(SG)_{19}$ and $AG_{15}(SG)_{11}$. *Nanoscale* 2013;**5**:5637–43.
128. Guo J, Kumar S, Bolan M, Desireddy A, Bigioni TP, Griffith WP. Mass spectrometric identification of silver nanoparticles: the case of $Ag_{32}(SG)_{19}$. *Anal Chem* 2012;**84**:5304–8.
129. Bakr OM, Amendola V, Aikens CM, Wenseleers W, Li R, Negro LD, et al. Silver nanoparticles with broad multiband linear optical absorption. *Angew Chem Int Ed* 2009;**48**: 5921–6.
130. Harkness KM, Tang Y, Dass A, Pan J, Kothalawala N, Reddy VJ, et al. $Ag_{44}(SR)_{30}^{4-}$: a silver-thiolate superatom complex. *Nanoscale* 2012;**4**:4269–74.
131. Yang H, Wang Y, Huang H, Gell L, Lehtovaara L, Malola S, et al. All-thiol-stabilized Ag_{44} and $Au_{12}Ag_{32}$ nanoparticles with single-crystal structures. *Nat Commun* 2013;**4**:2422.
132. Desireddy A, Conn BE, Guo J, Yoon B, Barnett RN, Monahan BM, et al. Ultrastable silver nanoparticles. *Nature* 2013;**501**:399–402.
133. Chakraborty I, Mahata S, Mitra A, De G, Pradeep T. Controlled synthesis and characterization of the elusive thiolated Ag_{55} cluster. *Dalton Trans* 2014;**43**:17904–7.
134. Chakraborty I, Udayabhaskararao T, Pradeep T. High temperature nucleation and growth of glutathione protected ~Ag_{75} clusters. *Chem Commun* 2012;**48**:6788–90.
135. Chakraborty I, Govindarajan A, Erusappan J, Ghosh A, Pradeep T, Yoon B, et al. The superstable 25 kDa monolayer protected silver nanoparticle: measurements and interpretation as an icosahedral $Ag_{152}(SCH_2CH_2Ph)_{60}$ cluster. *Nano Lett* 2012;**12**:5861–6.
136. Link S, Beeby A, FitzGerald S, El-Sayed MA, Schaaff TG, Whetten RL. Visible to infrared luminescence from a 28-atom gold cluster. *J Phys Chem B* 2002;**106**:3410–5.
137. Jin R. Atomically precise metal nanoclusters: stable sizes and optical properties. *Nanoscale* 2015;**7**:1549–65.
138. Negishi Y, Nakazaki T, Malola S, Takano S, Niihori Y, Kurashige W, et al. A critical size for nonbulk electronic and geometric structures in dodecanethiolate-protected Au clusters. *J Am Chem Soc* 2015;**137**:1206–12.

139. Lopez-Acevedo O, Tsunoyama H, Tsukuda T, Häkkinen H, Aikens CM. Chirality and electronic structure of the thiolate-protected Au_{38} nanocluster. *J Am Chem Soc* 2010;**132**:8210–8.
140. Akola J, Walter M, Whetten RL, Häkkinen H, Grönbeck H. On the structure of thiolate-protected Au_{25}. *J Am Chem Soc* 2008;**130**:3756–7.
141. Devadas MS, Bairu S, Qian H, Sinn E, Jin R, Ramakrishna G. Temperature-dependent optical properties of monolayer-protected Au_{25} and Au_{38} clusters. *J Phys Chem Lett* 2011;**2**:2752–8.
142. Weissker HC, Barron Escobar H, Thanthirige VD, Kwak K, Lee D, Ramakrishna G, et al. Information on quantum states pervades the visible spectrum of the ubiquitous $Au_{144}(SR)_{60}$ gold nanocluster. *Nat Commun* 2014;**5**:3785.
143. Sardar R, Funston AM, Mulvaney P, Murray RW. Gold nanoparticles: past, present, and future. *Langmuir* 2009;**25**:13840–51.
144. Hartland GV. Optical studies of dynamics in noble metal nanostructures. *Chem Rev* 2011;**111**:3858–87.
145. Yau SH, Varnavski O, Goodson III T. An ultrafast look at Au nanoclusters. *Acc Chem Res* 2013;**46**:1506–16.

Chapter 3

Controlled Synthesis: Composition and Interface Control

Yuichi Negishi,[*,§,‡,1] **Yoshiki Niihori**[*] **and Wataru Kurashige**[*]
[*]Department of Applied Chemistry, Faculty of Science, Tokyo University of Science, Shinjuku-ku, Tokyo, Japan; [§]Photocatalysis International Research Center, Tokyo University of Science, Noda, Chiba, Japan; [‡]Department of Materials Molecular Science, Institute for Molecular Science, Okazaki, Aichi, Japan
[1]Corresponding author: E-mail: negishi@rs.kagu.tus.ac.jp

Chapter Outline

3.1 COMPOSITION AND INTERFACE CONTROL: WHY?

The physical and chemical properties of metal clusters greatly depend on the chemical composition of the metal core as well as the size of the metal core. For example, the catalytic activity of polymer-stabilized Pd_{147} clusters is remarkably improved when the Pd at the surface is partially substituted by Au.[1] In addition, intermetallic compound nanoclusters comprising Pd and Ru exhibit significantly different catalytic activity compared with their monometallic nanocluster counterparts. The catalytic activity obtained by mixing

Frontiers of Nanoscience, Vol. 9. http://dx.doi.org/10.1016/B978-0-08-100086-1.00003-8

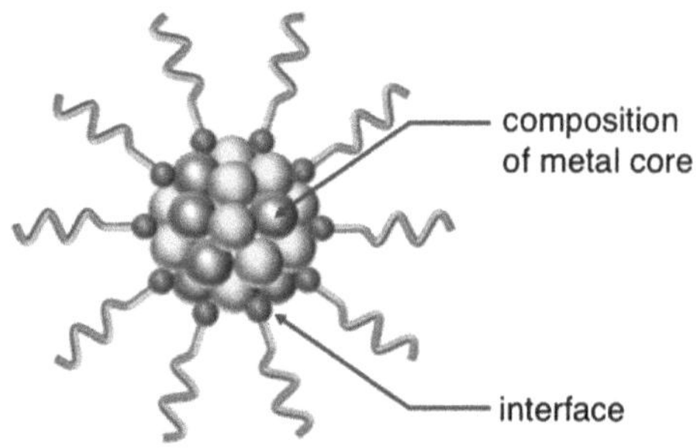

FIGURE 1 Composition and interface of the metal clusters discussed in this chapter.

the two components is higher than that of monometallic nanoclusters of Rh, which is located between these two elements in the periodic table.[2] As illustrated by these examples, synergistic effects caused by mixing different elements generate physical and chemical properties that are different from those of monometallic clusters. Thus, composition control of metal clusters is very interesting from the viewpoint of modification of the physical and chemical properties of the clusters, and results in new applications.

In addition to changes in the metal core, the physical and chemical properties of the metal clusters can also be modified by varying the surrounding ligands. For example, variation of the substituent (R) of thiolate-protected Au clusters ($Au_n(SR)_m$) will alter their solubility, photoluminescence properties,[3,4] and optical activity.[5] In addition, when the functional group (in this case, S) directly bonded to the metal is varied, the bonding mode between Au and the ligand is also modified.[6–9] Such control of the bonding mode between metal and ligand is also expected to result in novel applications.

This chapter focuses on the control of the composition and interface of metal clusters (Figure 1). The first part of this chapter describes methods for synthesizing intermetallic clusters protected by thiolates (SR) and/or phosphines (PR_3), the resultant products, and their synergistic effects. The second part of this chapter concerns gold clusters protected by selenolates (SeR), tellurolates (TeR), or terminal alkynes (C≡CR). Throughout this chapter, we provide an overview of the influence of variations in metal core chemical composition and the bonding mode at the interface on the fundamental properties of small metal clusters.

3.2 COMPOSITION CONTROL

3.2.1 Controlled Synthesis of Intermetallic Clusters

In this section, we describe a method for synthesizing intermetallic compound clusters protected by SR or PR_3. It is well known that SR and PR_3 form strong bonds with Au (Chapter 2). Furthermore, stable clusters (magic clusters) are

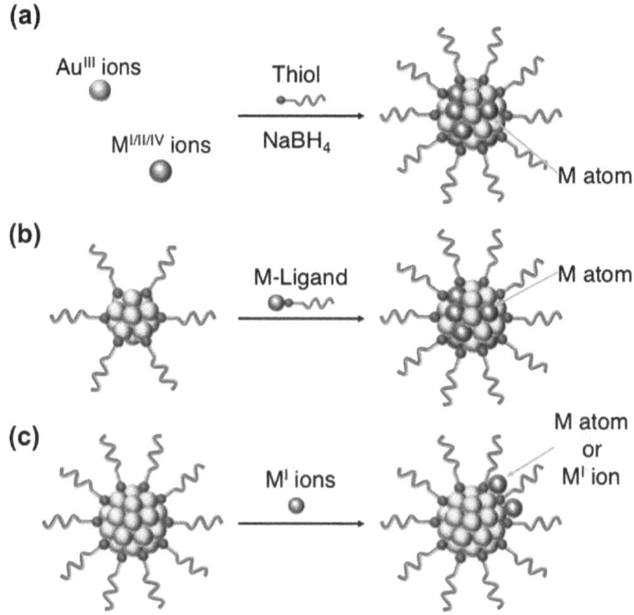

FIGURE 2 Representative synthesis methods of intermetallic clusters; (a) coreduction of multiple types of metal ions, (b) reorganization, and (c) deposition/addition of metal atoms/ions onto existing metal clusters.

present in $Au_n(SR)_m$ clusters and PR_3-protected gold clusters, and a large number of methods have been established for their isolation (Chapter 2). As such, when intermetallic compound clusters are synthesized using SR or PR_3 (or both in some cases) ligands, Au is generally employed as one of the metal elements, and the cluster size is controlled using a method similar to that used for $Au_n(SR)_m$ clusters and PR_3-protected gold clusters (Chapter 2). The priority while synthesizing these intermetallic compound clusters is how to successfully mix the Au with other metallic elements. In the following section, we describe four typical procedures used to address this issue (Figure 2).

3.2.1.1 Coreduction of Multiple Types of Metal Ions

The most common method for mixing Au with other metallic elements is the simultaneous reduction of Au ions with the other metal ions using a reducing agent (Figure 2(a)). This method is called the coreduction method. For example, to synthesize SR-protected intermetallic compound clusters, Au and the other metal ions are mixed in solution, followed by the addition of thiol. The strong reducing agent, $NaBH_4$, is then added to this solution, resulting in the simultaneous reduction of all of the metal ions present. Examples of intermetallic compound clusters synthesized[10–32] using this method include $Au_{25-n}Ag_n(SR)_{18}$ (R = $C_{12}H_{25}$ or C_2H_4Ph; $n = 1-11$; Figure 3(a)),

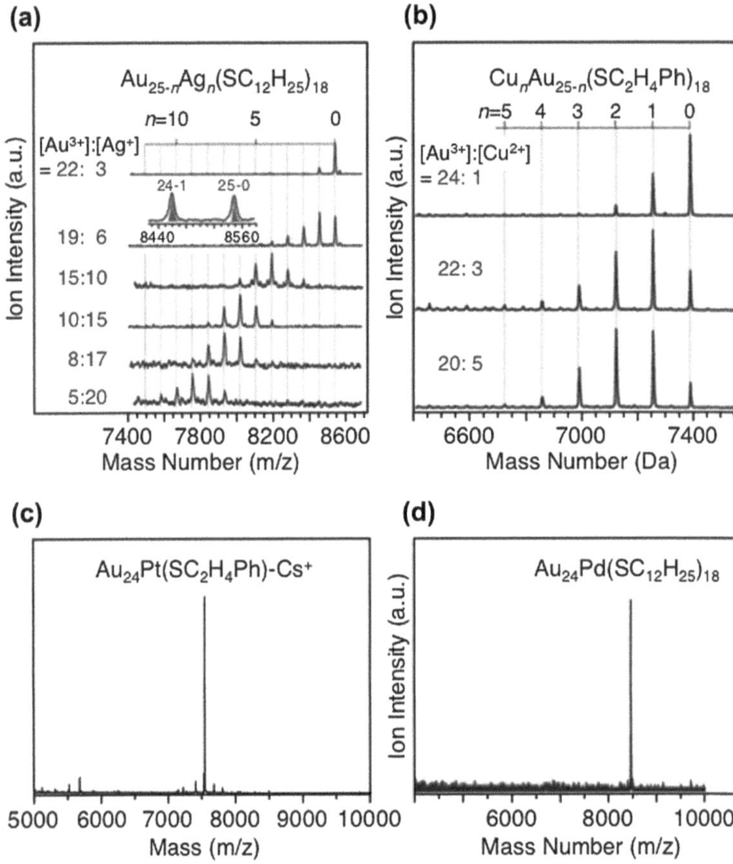

FIGURE 3 Mass spectra of intermetallic clusters protected by thiolates; (a) $Au_{25-n}Ag_n(SC_{12}H_{25})_{18}$, (b) $Au_{25-n}Cu_n(SC_2H_4Ph)_{18}$, (c) $Au_{24}Pt(SC_2H_4Ph)_{18}$, and (d) $Au_{24}Pd(SC_{12}H_{25})_{18}$. In (a) and (b), the initial ion ratios between the two types of metal ions used in the preparation process are shown in red characters. *The figures were adapted from Refs 10,15,19,22 with permission. Copyright 2010 The Royal Society of Chemistry, Copyright 2012 American Chemical Society, Copyright 2012 American Chemical Society, and Copyright 2010 The Royal Society of Chemistry.*

$Au_{25-n}Cu_n(SR)_{18}$ ($R = C_8H_{17}$ or C_2H_4Ph; $n = 1-5$; Figure 3(b)), $Au_{24}Pt(SC_2H_4Ph)_{18}$ (Figure 3(c)), $Au_{24}Pd(SR)_{18}$ ($R = C_{12}H_{25}$ or C_2H_4Ph; Figure 3(d)), $Au_{38-n}Ag_n(SC_2H_4Ph)_{24}$ ($n \sim 1, 3, 4, 6, 8, 9$), $Au_{36}Pd_2(SC_2H_4Ph)_{24}$, $Au_{144-n}Ag_n(SC_2H_4Ph)_{60}$ ($n \sim 30, 34, 52, 53$), $Au_{143/144/145-n}Cu_n(SC_2H_4Ph)_{59/60/61}$ ($n = 1-23$), $Au_{144}Cu(SC_6H_{13})_{60}$, and $Au_{144-n}Pd_n(SC_2H_4Ph)_{60}$ ($n = 1-7$). Using this method, it is also possible to synthesize $Au_{12}Ag_{32}(SR)_{30}$ (R = PhF, PhF_2, or $PhCF_3$) or $Au_{12+n}Cu_{32}(SPhCF_2)_{30+n}$ ($n = 0, 2, 4, 6$) intermetallic compound clusters, where Ag or Cu is bonded to the SR group on the surface.[33,34] Similarly, in the case of PR_3-protected intermetallic compound clusters, a reducing agent ($NaBH_4$ or H_2) is added to a solution containing PR_3. Using this method, intermetallic compound clusters such as $[Au_{10}Pd(PPh_3)_8Cl_2]^0$, $[Au_{13}Cu_8(PPh_2Py)_{12}]^+$ (Py = 2-pyridyl), $[Au_{13}Ag_{12}(P(p\text{-}CH_3C_6H_4)_3)_{10}Br_8]^+$, $[Au_{18}Ag_{19}(P(p\text{-}CH_3C_6H_4)_3)_{12}Br_{11}]^{2+}$, and $Au_{18}Ag_{20}(P(p\text{-}CH_3C_6H_4)_3)_{12}Cl_{14}$ have been synthesized.[35–41] When both SR and PR_3 are present in the solution, an intermetallic compound cluster that contains both ligands is produced; clusters such as $[Au_{13}Cu_2(PPh_3)_6(SPy)_6]^+$ and $[Au_{13}Cu_4(PPh_2Py)_4(SC_6H_4\text{-}tert\text{-}C_4H_9)_8]^+$ have been synthesized in this manner.[36]

Thus, it is possible to use the simultaneous reduction method to synthesize intermetallic compound clusters containing Au, Ag, Cu, Pt, Pd, and Ni regardless of the ligand type. However, two or more types of metal atoms need to be generated simultaneously by the reduction reaction to successfully form intermetallic compound clusters in this manner. Therefore, when there are large differences in redox potential among the precursor metal ions, it is not easy to form intermetallic compound clusters using this method. Hence, intermetallic compound clusters synthesized using this method are presently limited to those comprising Au, Ag, Cu, Pt, Pd, and Ni.

3.2.1.2 Reorganization

Like the simultaneous reduction method, this method is frequently employed for the synthesis of intermetallic compound clusters. In this method, metal clusters are mixed with metal complexes or metal clusters containing different elements to create new stable intermetallic clusters (Figure 2(b)). This method enables the synthesis of intermetallic compound clusters in which the number of the metal atoms is different from those in the precursor metal clusters. For example, mixing PPh_3-protected gold clusters (polydisperse Au cluster or Au_{11} cluster) with Ag-SC_2H_4Ph results in the synthesis of $[Au_{25-x}Ag_x(PPh_3)_{10}(SC_2H_4Ph)_5Cl_2]^{2+}$ ($x \leq 13$), which contains both Au and Ag (Figure 4(a)).[42] This occurs because of the reorganization of the cluster structures during the reaction. Intermetallic compound clusters that contain three different metal elements can be synthesized using this method, following the reaction of PR_3-protected intermetallic compound clusters with monometallic clusters. For example, reaction of $[Au_8Pt(PPh_3)_8]^{2+}$ with $Ag_4(PPh_3)_4Cl_4$ in the presence of $NaBH_4$ results in the formation of $[Au_{12}Ag_{12}Pt(PPh_3)_{10}Cl_7]^+$.[44] $[Au_{12}Ag_{12}Ni(PPh_3)_{10}Cl_7]^+$ has also been synthesized by a similar method.[45] Recently, it has been found that $[Au_{12}Ni(PMe_3)_{11}Cl]^{3+}$, $[Au_{12}Pd(PMe_3)_{11}Cl]^{3+}$, or $[Au_{12}Pt(PMe_3)_{11}Cl]^{3+}$ (Figure 4(b)) can be synthesized by reacting $M(GaCp^*)_4$ (M = Pt, Pd, or Ni; Cp* = 1,2,3,4,5-pentamethyl-cyclopentadienyle) with $Au(PMe_3)Cl_4$.[43] All of these clusters have a core–shell-type geometric structure, where the heteroatom is located at the center of the cluster. As such, this method is suitable for the synthesis of core–shell-type intermetallic compound clusters.

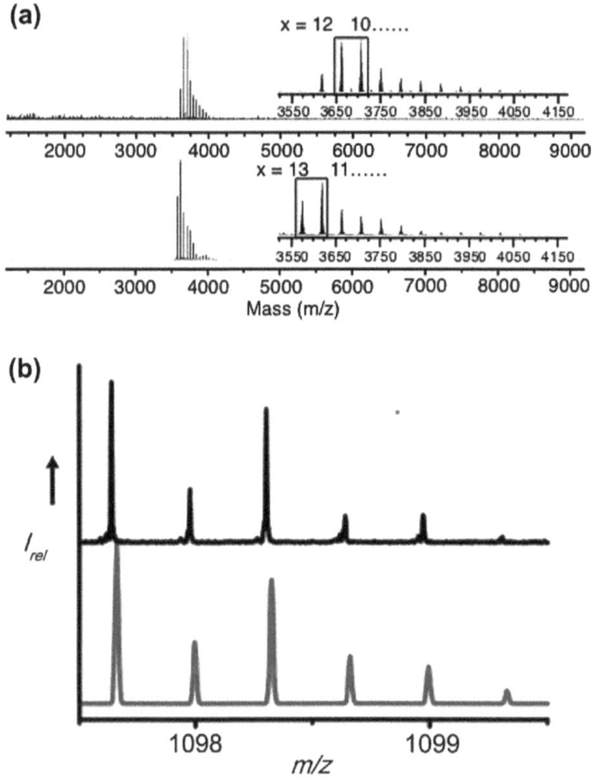

FIGURE 4 Mass spectra of intermetallic clusters protected by both thiolates and phosphines; (a) $[Au_{25-x}Ag_x(PPh_3)_{10}(SC_2H_4Ph)_5Cl_2]^{2+}$ prepared using two different experimental conditions and (b) $[Au_{12}Ni(PMe_3)_{11}Cl]^{3+}$ with calculated isotope pattern. *The figures were adapted from Refs 42,43 with permission. Copyright 2014 Wiley-VCH publishing and Copyright 2014 Wiley-VCH publishing.*

3.2.1.3 Deposition of Metal Atoms onto Metal Clusters by Galvanic or Anti-Galvanic Reaction

On the basis of the Galvanic reaction principle, adding metal nanoparticles with a low redox potential to a solution containing metal ions with a high redox potential will result in the reduction of the metal ions by the metal nanoparticles, leading to the deposition of the metal on the nanoparticle surface.[46–49] For example, when Ag nanoparticles react with chloroauric acid ($HAuCl_4$), an oxidation–reduction reaction proceeds on the surface of the nanoparticles ($3Ag^0 + Au^{3+} \rightarrow Au^0 + 3Ag^+$) because the $AuCl_4^-$/Au (~1 V) redox potential is higher than that of Ag^+/Ag (~0.8 V). As a result, Ag is partially substituted by Au on the surface. Similarly, when Pd nanoparticles react with chloroauric acid ($HAuCl_4$), Pd at the surface of the particles will be

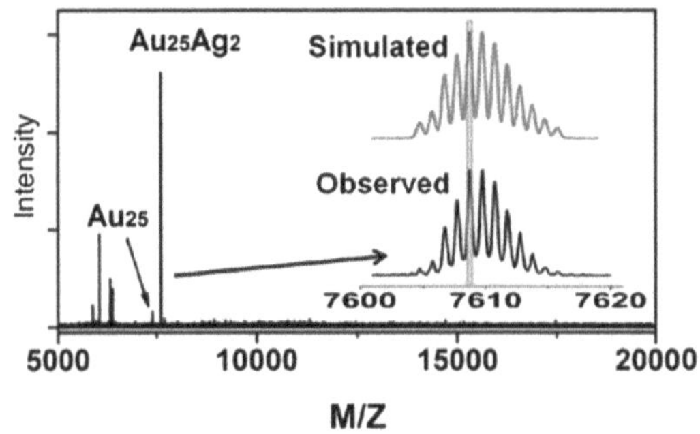

FIGURE 5 Mass spectra of $Au_{25}Ag_2(SC_2H_4Ph)_{18}$ with simulated isotope pattern. *This figure was adapted from Ref. 50 with permission. Copyright 2014 American Chemical Society.*

partially substituted by Au, because the $AuCl_4^-/Au$ (~1 V) redox potential is higher than that of Pd^{2+}/Pd (~0.95 V). The $(Au-Pd)_{147}$ clusters described in the introduction are synthesized using this Galvanic reaction technique.[1]

Conversely, when $Au_n(SR)_m$ clusters having fine metal cores (<2 nm) react with Ag ions, the Ag ions are reduced by the Au, and the Ag is deposited on the cluster surface (Figure 2(c)). Such a reaction exhibits a behavior contrary to the Galvanic reaction, and accordingly, is called an anti-Galvanic reaction (AGR). This type of reaction has been adopted for the synthesis of $Au_{25}Ag_2(SC_2H_4Ph)_{18}$, which is produced by depositing two neutral Ag atoms onto the surface of a $Au_{25}(SC_2H_4Ph)_{18}$ cluster (Figure 5).[50] A particular feature of this synthesis is that the substrate clusters maintain their chemical composition while the other metal atoms are deposited on the cluster surface. Such an AGR is a reaction specific to small metal clusters and is proposed to be induced by the decrease of the redox potential of the Au clusters owing to miniaturization of the metal core and/or thiol coordination.[51]

3.2.1.4 Addition of Metal Ions/Atoms onto Metal Clusters

Examples of the synthesis of intermetallic compound clusters by the addition, as opposed to deposition, of heteroatoms have been reported. For example, mixing $Au_{18}(SG)_{14}$ (SG = glutathione) with Ag^+ in solution results in the formation of intermetallic compound clusters with 1–7 Ag(I) added onto the $Au_{18}(SG)_{14}$ surface.[52] Similarly, intermetallic compound clusters have been synthesized where Ag(I) are added to the surface of $Au_{15}(SG)_{13}$ or $Au_{25}(SG)_{18}$ (Figure 2(c)).[52] For PR_3-protected metal clusters, mixing $[Au_6Pt(PPh_3)_7]^{2+}$ with neutral Hg results in the adsorption of neutral Hg onto the cluster surface to form an intermetallic compound cluster such as $[Au_6PtHg(PPh_3)_7]^{2+}$. The particular feature of this synthesis method is that the chemical composition of the substrate clusters is preserved because the other metal atom simply adds onto the cluster surface.[53]

3.2.2 Synthesized Intermetallic Clusters

Numerous intermetallic compound clusters have been synthesized on the basis of the methods described in Section 3.2.1. In this section, we summarize the chemical compositions and the geometric structures of reported SR-, PR_3-, or both SR- and PR_3-protected intermetallic compound clusters.

3.2.2.1 Intermetallic Clusters Protected by Thiolates

A summary of all SR-protected intermetallic compound clusters that have been synthesized up to end of 2014 is listed in Table 1.[54–56,59] The main metal in the synthesized clusters is generally Au. As mentioned above, this is related to the high binding affinity between SR and Au, and thereby the existence of stable clusters (magic clusters) in $Au_n(SR)_m$ clusters. The elements contained in the isolated intermetallic compound clusters are currently limited to Au, Ag, Cu, Pt, Pd, and Ni regardless of the synthetic method employed. In the case of intermetallic compound clusters comprising Au and Ag, or Au and Cu, a certain distribution of the metal mixing ratio is observed (Figure 3(a) and (b)), except in a limited number of cases.[33,34]

Figure 6 shows examples of determined or predicted geometric structures for SR-protected intermetallic compound clusters. All clusters of $Au_{25-n}Ag_n(SC_2H_4Ph)_{18}$ ($n = 6.7$),[14] $Au_{25-n}Cu_n(SC_2H_4Ph)_{18}$ ($n = 1.4$),[61] $Au_{24}Pt(SC_2H_4Ph)_{18}$ (Refs 19,62), and $Au_{24}Pd(SC_{12}H_{25})_{18}$ (Refs 17,63) have a geometric structure where Au in the gold cluster $Au_{25}(SC_2H_4Ph)_{18}$ is substituted by another metal element (Figure 6(a)). Similarly, the geometric structures obtained when Au in the gold clusters $Au_{38}(SC_2H_4Ph)_{24}$ or $Au_{144}(SC_2H_4Ph)_{60}$ is substituted by a different metal element (Figure 6(b) and (c)) have been predicted for $Au_{38-n}Ag_n(SC_2H_4Ph)_{24}$ ($n = 1-12$), $Au_{36}Pd_2(SC_2H_4Ph)_{24}$, $Au_{144-n}Ag_n(SC_2H_4Ph)_{60}$ ($n \sim 30, 34, 52, 53$), and $Au_{144-n}Cu_n(SC_2H_4Ph)_{60}$ ($n = 1-25$).[25–32,60] It can therefore be assumed that intermetallic compound clusters with the same number of metal atoms as the stable $Au_n(SR)_m$ cluster (Section 3.2) have framework structures similar to those of the stable $Au_n(SR)_m$ cluster. Recent studies on $[Au_{12}Ag_{32}(SR)_{30}]^{4-}$ (R = PhF, PhF_2, or $PhCF_3$) (Figure 6(d)) also have revealed that intermetallic compound clusters with the same number of metal atoms as the stable $[Ag_{44}(SR)_{30}]^{4-}$ cluster (Section 3.2) have a framework structure similar to that of the stable $[Ag_{44}(SR)_{30}]^{4-}$ cluster.[33,34]

In such intermetallic compound clusters, Ag always sits on the surface of the metal core (Figure 6(a) and (c)). Because Ag (78.0 meV Å^{-2}) has a surface energy lower than that of Au (96.8 meV Å^{-2}),[64] Ag is considered to be more energetically stable when it exists on the surface in an intermetallic compound cluster comprising Au and Ag.[63] In the case of Cu, there is no uniform acceptance on the position of substitution. Cu substitution has been reported to occur in the oligomer for $Au_{25-n}Cu_n(SC_2H_4Ph)_{18}$ (Figure 6(a)),[61] but near the center of the metal core for $Au_{144-n}Cu_n(SC_2H_4Ph)_{60}$ (Figure 6(c)).[29] For Pt

TABLE 1 List of Intermetallic Clusters

Number of Metal Atoms	Formula	SR^i	Synthesis Method
6	$Ag_4Pt_2(SR)_4$	DMSA[54]	Coreduction
	$Ag_4Ni_2(SR)_4$	DMSA[55]	Coreduction
13	$Ag_7Au_6(SR)_{10}$	MSA[56]	Galvanic reaction
15	$Au_{12.2}Ag_{2.8}(SR)_{13}$	SG[13]	Coreduction
18	$Au_{14.4}Ag_{3.6}(SR)_{14}$	SG[13]	Coreduction
25	$Au_{25-n}Ag_n(SR)_{18}$, $n=0-11$	$SC_{12}H_{25}$[10]	Coreduction
	$Au_{25-n}Ag_n(SR)_{18}$, $n=1-11$	SC_2H_4Ph[10]	Coreduction
	$Au_{25-n}Ag_n(SR)_{18}$, $n \sim 3$	SC_2H_4Ph[11]	Coreduction
	$Au_{25-n}Ag_n(SR)_{18}$, $n=0-8$	SC_2H_4Ph[16]	Coreduction
	$Au_{18.3}Ag_{6.7}(SR)_{18}$	SC_2H_4Ph[14]	Coreduction
	$Au_{25-n}Ag_n(SR)_{18}$, $n=0-11$	MHA[12]	Coreduction
	$Au_{25-n}Ag_n(SR)_{18}$, $n=0-10$	MOA[12]	Coreduction
	$Au_{25-n}Ag_n(SR)_{18}$, $n=0-9$	MUA[12]	Coreduction
	$Au_{17.6}Ag_{7.4}(SR)_{14}$	SG[13]	Coreduction
	$Au_{25-n}Cu_n(SR)_{18}$, $n=1-5$	SC_2H_4Ph,[15] SC_8H_{17}[15]	Coreduction
	$Au_{25-n}Cu_n(SR)_{18}$, $n=0-4$	SC_2H_4Ph[16]	Coreduction
	$Au_{24}Pt(SR)_{18}$	SC_2H_4Ph[19]	Coreduction
	$Au_{24}Pd(SR)_{18}$	SC_2H_4Ph[21,25]	Coreduction
	$Au_{24}Pd(SR)_{18}$	$SC_{12}H_{25}$[22]	Coreduction
27	$Au_{25}Ag_2(SR)_{18}$	SC_2H_4Ph[50]	Anti-Galvanic reaction
38	$Au_{38-n}Ag_n(SR)_{24}$, $n \sim 1$, 3, 5, 6, 6, 8, 9	SC_2H_4Ph[26]	Coreduction
	$Au_{38-n}Pd_n(SR)_{24}$, $n=1, 2$	SC_2H_4Ph[25,27,57]	Coreduction
44	$[Au_{12}Ag_{32}(SR)_{30}]^{4-}$	SPhF,[33] $SPhF_2$,[33] $SPhCF_3$[33]	Coreduction
	$[Au_{12}Cu_{32}(SR)_{30}]^{4-}$	$SPhCF_3$[34]	Coreduction

Continued

TABLE 1 List of Intermetallic Clusters—cont'd

Number of Metal Atoms	Formula	SR[i]	Synthesis Method
46	$[Au_{14}Cu_{32}(SR)_{32}]^{4-}$	$SPhCF_3$[34]	Coreduction
48	$[Au_{16}Cu_{32}(SR)_{34}]^{4-}$	$SPhCF_3$[34]	Coreduction
50	$[Au_{18}Cu_{32}(SR)_{36}]^{4-}$	$SPhCF_3$[34]	Coreduction
130	$Au_{130-n}Ag_n(SR)_{50}$, $n = 1-14$	SC_2H_4Ph[58]	Coreduction
	$Au_{130-n}Pd_n(SR)_{50}$, $n = 1-4$	SC_2H_4Ph[58]	Coreduction
131	$Au_{130}Ag(SR)_{50}$	SC_2H_4Ph[58]	Coreduction
	$Au_{131-n}Ag_n(SR)_{52}$, $n = 1, 2$	SC_2H_4Ph[58]	Coreduction
137	$Au_{137-n}Ag_n(SR)_{56}$, $n = 1-9$	SC_2H_4Ph[59]	Coreduction
	$Au_{137-n}Pd_n(SR)_{56}$, $n = 1-4$	SC_2H_4Ph[59]	Coreduction
	$Au_{136}Cu(SR)_{57}$	SC_6H_{13}[32]	Coreduction
138	$Au_{138-n}Cu_n(SR)_{56}$, $n = 1, 2$	SC_6H_{13}[32]	Coreduction
143	$Au_{143-n}Cu_n(SR)_{61}$, $n = 3-14$	SC_6H_{13}[29]	Coreduction
144	$Au_{144-n}Ag_n(SR)_{60}$, $n \sim 30, 34, 52, 53$	SC_2H_4Ph[28]	Coreduction
	$Au_{144-n}Ag_n(SR)_{60}$, $n \sim 52$	SC_6H_{13},[28] $SC_{12}H_{25}$[28]	Coreduction
	$Au_{143}Cu(SR)_{61}$	SC_6H_{13}[32]	Coreduction
	$Au_{144-n}Cu_n(SR)_{60}$, $n = 1-16$	SC_6H_{13}[29]	Coreduction
	$Au_{144-n}Pd_n(SR)_{60}$, $n = 0-7$	SC_2H_4Ph[60]	Coreduction
145	$Au_{145-n}Cu_n(SR)_{59}$, $n = 17-23$	SC_2H_4Ph[28]	Coreduction
	$Au_{145-n}Cu_n(SR)_{60}$, $n = 1, 2$	SC_6H_{13}[32]	Coreduction
	$Au_{122}Cu_{23}(SR)_{59}$	SC_6H_{13}[29]	Coreduction

DMSA, *meso*-2,3-dimercaptosuccinic acid; MSA, mercaptosuccinic acid; SG, glutathione; MHA, mercaptohexanoic acid; MOA, meracaptooctanoic acid; MUA, mercaptoundecanoic acid.

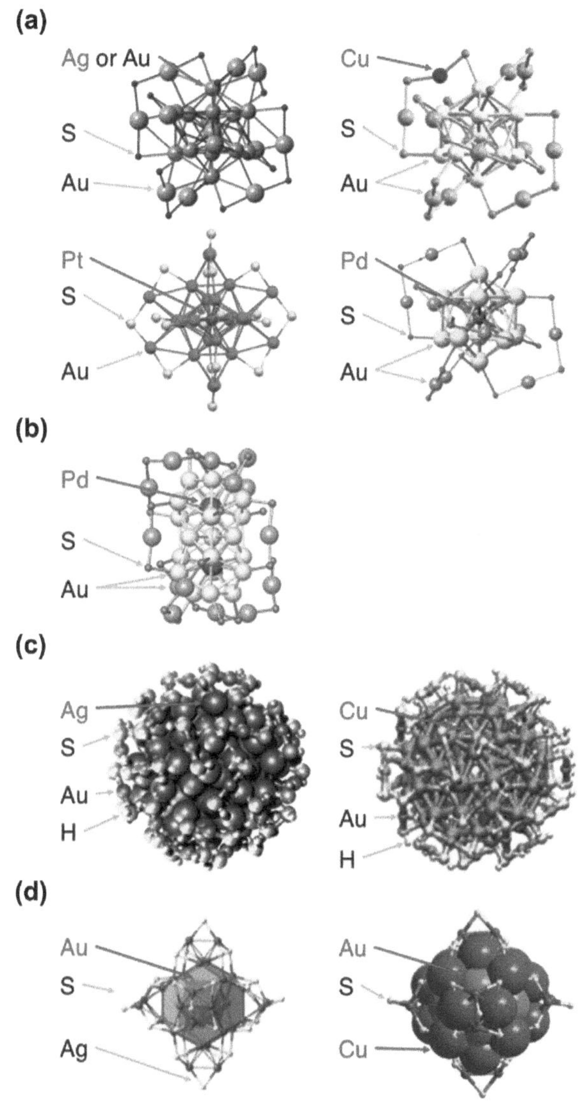

FIGURE 6 Framework structures determined or proposed for intermetallic clusters protected by thiolates; (a) $Au_{25-n}Ag_n(SC_2H_4Ph)_{18}$ ($n = 6.7$), $Au_{25-n}Cu_n(SC_2H_4Ph)_{18}$ ($n = 1.4$), $Au_{24}Pt(SC_2H_4Ph)_{18}$, and $Au_{24}Pd(SC_{12}H_{25})_{18}$, (b) $Au_{36}Pd_2(SCH_3)_{24}$, (c) $Au_{84}Ag_{60}(SH)_{60}$ and $Au_{120}Cu_{24}(SH)_{60}$, and (d) $[Au_{12}Ag_{32}(SPhCF_3)_{30}]^{4-}$ and $[Au_{12}Cu_{32}(SPhCF_3)_{30}]^{4-}$. (see color plate). *The figures were adapted from Refs 14,19,22,27,30,31,33,34,61 with permission. Copyright 2014 American Chemical Society, Copyright 2014 American Chemical Society, Copyright 2012 American Chemical Society, Copyright 2010 The Royal Society of Chemistry, Copyright 2012 The Royal Society of Chemistry, Copyright 2014 American Chemical Society, Copyright 2015 American Chemical Society, Copyright 2013 Macmilian Publishers Ltd, and Copyright 2014 American Chemical Society.*

and Pd, it is assumed that they substitute for Au at the center (or near the center) of the metal core of $Au_n(SR)_m$ (Figure 6(a)),[17,19,60,62,63,65] because Pt (159 meV/Å^2) and Pd (131 meV/Å^2) have higher surface energies than Au (96.8 meV/Å^2).[64]

3.2.2.2 Intermetallic Clusters Protected by Phosphines

PR_3-protected intermetallic compound clusters have been the subject of extensive investigations, even before the study of SR-protected intermetallic compound clusters. The main metal element in the reported clusters is generally Au. Intermetallic compound clusters obtained by mixing Au with Ag, Cu, Pt, Pd, Ni, or Hg have been synthesized (see Section 3.2.1). Figure 7 shows the geometrical structures of PR_3-protected 13 metal atom intermetallic compound clusters. The framework structure of $[Au_9Ag_4(PMePh_2)_8Cl_4]^+$, $[Au_{12}Ni(PMe_3)_{11}Cl]^{3+}$, $[Au_{12}Pd(PMe_3)_{11}Cl]^{3+}$, and $[Au_{12}Pt(PMe_3)_{11}Cl]^{3+}$ (Refs 43,66) is similar to that of the gold cluster $[Au_{13}(PMe_2Ph)_{10}Cl_2]^{3+}$ (Ref. 67), indicating that PR_3-protected intermetallic compound clusters with the same number of metal atoms as the stable PR_3-protected Au clusters have similar framework structures. Regarding substitution with other elements, Ag and Cu appear on the surface of the metal core, while Pt, Pd, and Ni appear at its center, which is similar to that seen for SR-protected intermetallic

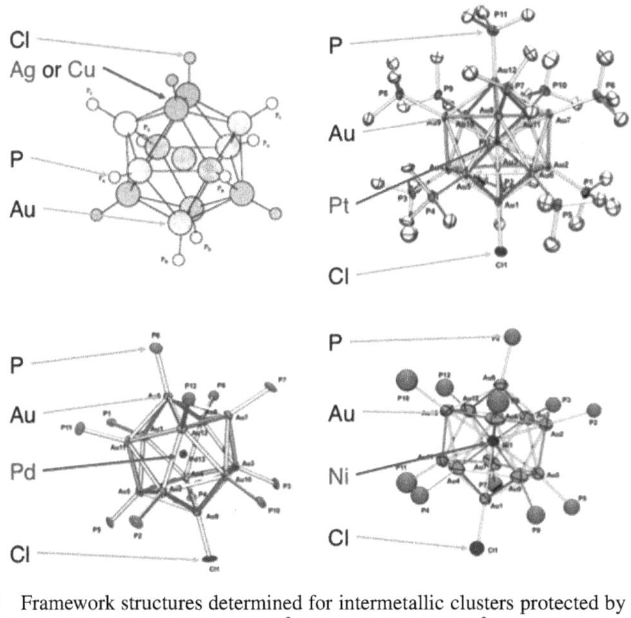

FIGURE 7 Framework structures determined for intermetallic clusters protected by phosphines; $[Au_9Ag_4(PMePh_2)_8Cl_4]^+$, $[Au_{12}Ni(PMe_3)_{11}Cl]^{3+}$, $[Au_{12}Pd(PMe_3)_{11}Cl]^{3+}$, and $[Au_{12}Pt(PMe_3)_{11}Cl]^{3+}$. *The figures were adapted from Refs 43,66 with permission. Copyright 1996 The Royal Society of Chemistry and Copyright 2014 Wiley-VCH publishing.*

compound clusters. In many cases, halogens are also included as a ligand in PR_3-protected metal clusters. When Ag or Cu is present in the metal core, the halogen atoms are preferentially coordinated to these elements.[66]

3.2.2.3 Intermetallic Clusters Protected by Both Thiolates and Phosphines

Clusters including $[Au_{25-x}Ag_x(PPh_3)_{10}(SC_2H_4Ph)_5Cl_2]^{2+}$ ($x \leq 13$), $[Au_{13}Cu_2(PPh_3)_6(SPy)_6]^+$, and $[Au_{13}Cu_4(PPh_2Py)_4(SC_6H_4\text{-}tert\text{-}C_4H_9)_8]^+$ have been reported for intermetallic compound clusters protected both by SR and PPh_3.[36,44] Like SR- or PPh_3-protected structures, these intermetallic compound clusters have a framework structure similar to that of the gold cluster $[Au_{25}(PPh_3)_{10}(SR)_5Cl_2]^{2+}$ ($R = C_2H_5$ or C_2H_4Ph).[68,69] In such a cluster, Ag substitution generally occurs at the Au atoms bonded to Cl, or Au bonding the two Au_{13} cores (Figure 8). When the number of Ag in the cluster is further increased, Ag substitution occurs at the Au bonded to SC_2H_4Ph on the surface (Figure 8).[42]

3.2.3 Effects of the Mixing of Different Elements

A large amount of information has been obtained from studies on the effect of mixing different elements on the physical and chemical properties of intermetallic compound clusters (Table 2). In this section, representative examples will be presented to illustrate the effect of heteroatom mixing.

3.2.3.1 Ag Substitution in Au Clusters

Ag (1.45 Å) is an element belonging to the same group (Group X) in the periodic table as Au (1.44 Å), and has approximately the same atomic radius.[64]

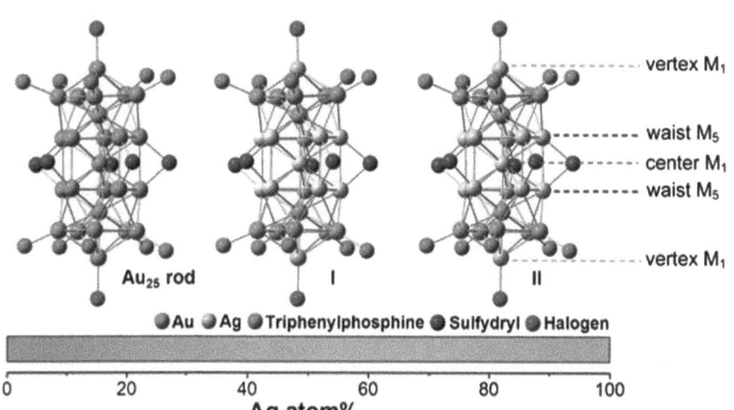

FIGURE 8 Framework structures of $[Au_{25-x}Ag_x(PPh_3)_{10}(SC_2H_4Ph)_5Cl_2]^{2+}$. (see color plate). *This figure was adapted from Ref. 42 with permission. Copyright 2014 Wiley-VCH publishing.*

TABLE 2 List of Properties of Intermetallic Clusters

Cluster	Dopant	Properties Relative to Original Clusters
$Au_{25}(SR)_{18}$	Ag	Increase of HOMO-LUMO gap energy[10,11]
		Increase of LUMO level[11]
		Blue shift of electronic structure in UV–Vis spectrum[11,13]
		Change of photoluminescence property[10,11,13]
		Decrease of optical activity[13]
	Cu	Decrease of HOMO-LUMO gap energy[15]
		Decrease of LUMO level[15]
		Distortion of framework of cluster[15]
		Decrease of stability in solution[15]
	Pt	Change of the excitation pathway[19]
		Change of the charge state[19]
		Increase of the stability against H_2O_2 oxidation[19]
		Increase of the catalytic reactivity of oxidation of styrene[19]
	Pd	Increase of the stability in solution and against laser irradiation[22]
		Change of stable charge state[70]
		Increase of ligand exchange reactivity[21,70]
$Au_{38}(SR)_{24}$	Ag	Change of the UV–Vis spectrum feature at ~750 and ~1050 nm[26]
		Blue shift of electronic structure in UV–Vis spectrum[26]
		Destabilization by increasing of Ag atom doping[26]
	Pd	Increase of the stability in solution[27]
		Change of the UV–Vis spectrum feature[57]
		Change the chiroptical property[57]
		Decrease of the racemization temperature[57]
$Au_{144}(SR)_{60}$	Ag	Change of the UV–Vis spectrum feature at ~310, ~425, and ~560 nm[28]
		Destabilization by high Ag precursor ratio[28]
	Cu	Change of the composition by Cu doping[32]
		Emergence of the LSPR-like feature[29]
	Pd	Diminishment of optical features[60]
$Ag_{44}(SR)_{30}$	Au	Red shift of electronic structure in UV–Vis spectrum[33]
		Increase of the thermal stability[33]

HOMO, highest occupied molecular orbital; LUMO, lowest unoccupied molecular orbital; LSPR, localized surface plasmon resonances.

Therefore, a multitude of substitutions occur between Au and Ag, resulting in continuous changes in the electronic structure of the clusters. Figure 9(a) shows the optical absorption spectra of $Au_{25-n}Ag_n(SC_{12}H_{25})_{18}$ ($n = 1-11$).[10] In the uppermost $Au_{25}(SC_{12}H_{25})_{18}$ spectrum, the peak in the vicinity of 1.5−1.8 eV can be attributed to transition from the highest occupied molecular orbital (HOMO; the occupied Au 6sp orbital in the Au_{13} core) to the lowest unoccupied molecular orbital (LUMO; the unoccupied Au 6sp orbital in the Au_{13} core).[71−73] The position of this peak gradually shifts to higher energy (i.e., shorter wavelength) with increasing number of Ag atoms.[10] This indicates that Ag substitution gradually increases the HOMO-LUMO gap of the

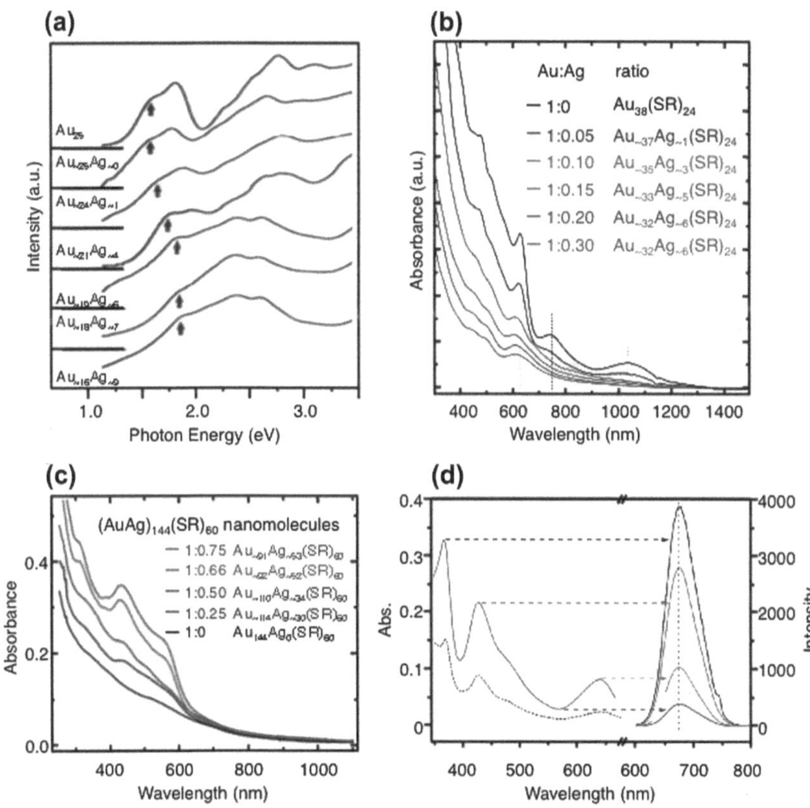

FIGURE 9 Optical absorption spectra of (a) $Au_{25-n}Ag_n(SC_{12}H_{25})_{18}$, (b) $Au_{38-n}Ag_n(SC_2H_4Ph)_{24}$, and (c) $Au_{144-n}Ag_n(SC_2H_4Ph)_{60}$. In (b) and (c), the initial ion ratios between the two types of metal ions used in the preparation process are shown. (d) UV/Vis optical absorption (left) and excitation spectra (left), and emission spectra (right) of $[Au_{25-x}Ag_x(PPh_3)_{10}(SC_2H_4Ph)_5Cl_2]^{2+}$ ($x \leq 13$) at different excitation wavelength. *The figures were adapted from Refs 10,26,28,42 with permission. Copyright 2010 The Royal Society of Chemistry, Copyright 2012 The Royal Society of Chemistry, Copyright 2012 The Royal Society of Chemistry, and Copyright 2014 Wiley-VCH publishing.*

cluster. The photoluminescence energy also gradually increases with the increase in the HOMO-LUMO gap.[10] Ag substitution causes a different influence on the electronic structure of the cluster when the substrate cluster is different. For example, Ag substitution of $Au_{38}(SC_2H_4Ph)_{24}$ gradually extinguishes the peak structure in its optical absorption spectrum (Figure 9(b)),[26] and Ag substitution in $Au_{144}(SC_2H_4Ph)_{60}$ continuously increases the intensity of the absorption peaks around ~560, 425, and 310 nm (Figure 9(c)).[28,32] Ag substitution in $[Au_{25}(PPh_3)_{10}(SC_2H_4Ph)_5Cl_2]^{2+}$ has a significant effect on its quantum yield of photoluminescence. When the number of substitutions reaches 13, the quantum yield increases by up to 200-fold (Figure 9(d)), although there is no substantial change in the emission wavelength.[42]

3.2.3.2 Ag Deposition/Addition onto Au Clusters

When Ag atoms are deposited onto $Au_{25}(SC_2H_4Ph)_{18}$, unlike the case of Ag substitution, no significant change occurs in the electronic structure of the cluster. In contrast, because two Ag atoms act as an active site, the deposition of Ag atoms has a significant effect on the catalytic activity of the cluster. Compared with $Au_{25}(SC_2H_4Ph)_{18}$, $Au_{25}Ag_2(SC_2H_4Ph)_{18}$ exhibits an approximately fivefold increase in catalytic activity toward the hydrolysis of 1,3-Diphenylprop-2-ynyl acetate (Figure 10(a)). The addition of Ag(I) also remarkably influences the photoluminescence of the cluster. For example, $Au_{18}Ag(I)_{1-7}(SG)_{14}$, which is obtained from the addition of 1−7 Ag(I) onto $Au_{18}(SG)_{14}$, exhibits an 18-fold increase in photoluminescence intensity relative to that of the original pure $Au_{18}(SG)_{14}$ (Figure 10(b)).[54]

3.2.3.3 Cu Substitution in Au Clusters

Cu (1.28 Å) also belongs to the same group (Group X) in the periodic table as Au (1.44 Å); however, there is a significant difference in the atomic radius of these two elements.[64] Therefore, Cu substitution generates a strain in the geometric structure of the cluster.[15] Cu substitution in $Au_{25}(SC_2H_4Ph)_{18}$ also causes destabilization of the cluster in solution.[15] Cu substitution also results in different effects on the electronic structure compared with those seen with Ag substitution. For example, Cu substitution in $Au_{25}(SC_2H_4Ph)_{18}$ induces a decrease in the HOMO-LUMO gap of the cluster (Figure 11(a)), in contrast to Ag substitution.[15] Similarly, Cu substitution in $Au_{144}(SC_2H_4Ph)_{60}$ results in the appearance of plasmon-like absorption (Figure 11(b)), which is not caused by Ag substitution (Figure 9(c)).[25,60]

3.2.3.4 Cu Capture in Au Clusters

Recently, clusters mainly comprising $Au_{144}Cu(SC_6H_{13})_{60}$, obtained by the inclusion or capture of a single Cu atom into the $Au_{144}(SC_6H_{13})_{60}$ cluster, have been synthesized. The geometric structure proposed for this cluster suggests that Cu is positioned in the central cavity of $Au_{144}(SR)_{60}$ (Figure 12(a)). The optical

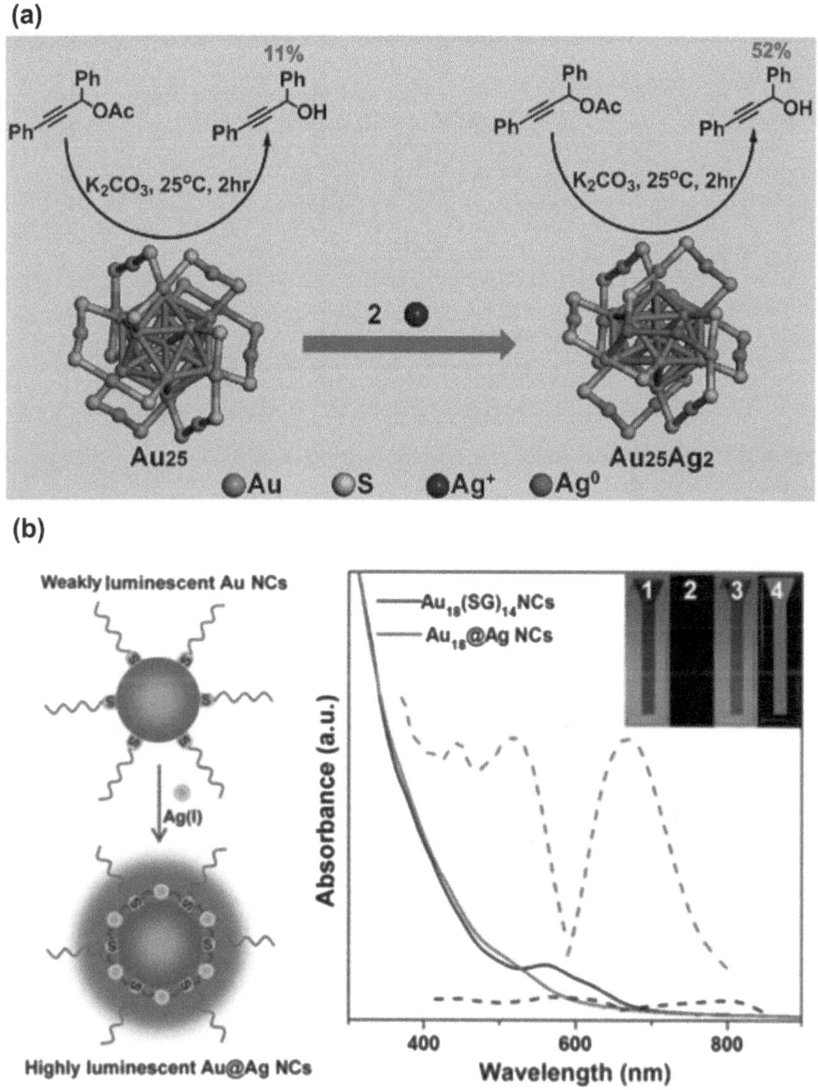

FIGURE 10 (a) Catalytic activity of $Au_{25}Ag_2(SC_2H_4Ph)_{18}$ and (b) schematic and optical properties of $Au_{18}(SG)_{14}$ and $Au_{18}Ag(I)_{1-7}(SG)_{14}$. Insets show digital photos of the parent $Au_{18}(SG)_{14}$ (items 1 and 2) and $Au_{18}Ag(I)_{1-7}(SG)_{14}$ (items 3 and 4), under visible (items 1 and 3) and UV (items 2 and 4) light. (see color plate). *The figures were adapted from Refs 50,52 with permission. Copyright 2014 American Chemical Society and Copyright 2014 The Royal Society of Chemistry.*

absorption spectrum of $Au_{144}Cu(SC_6H_{13})_{60}$, in contrast to that of $Au_{144}(SR)_{60}$, exhibits a peak attributed to plasmon absorption at approximately 540 nm. This indicates that even the capture of a single Cu atom can significantly modify the electronic structure of $Au_{144}Cu(SC_6H_{13})_{60}$ (Figure 12(b)).[32]

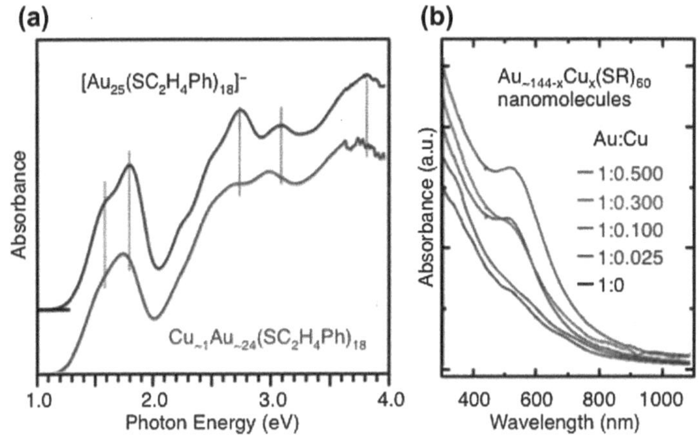

FIGURE 11 Comparison between optical absorption spectra of (a) $Au_{\sim 24}Cu_{\sim 1}(SC_2H_4Ph)_{18}$ and $Au_{25}Cu(SC_2H_4Ph)_{18}$, and (b) $Au_{143/144/145-n}Cu_n(SC_2H_4Ph)_{59/60/61}$ prepared with different [Au]:[Cu] ratios. *The figures were adapted from Refs 15,25 with permission. Copyright 2012 American Chemical Society and Copyright 2014 The Royal Society of Chemistry.*

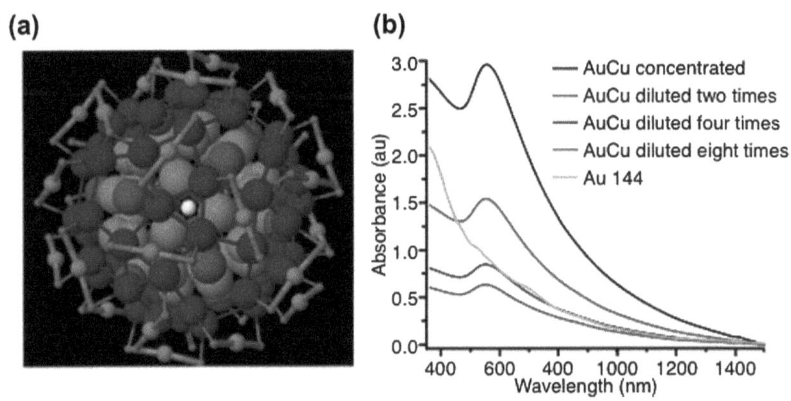

FIGURE 12 (a) Geometric structure proposed for $Au_{144}Cu(SC_6H_{13})_{60}$ and (b) optical absorption spectra of $Au_{144}Cu(SC_6H_{13})_{60}$ measured for different purities and concentrations. *The figures were adapted from Ref. 32 with permission. Copyright 2012 American Chemical Society.*

3.2.3.5 Pt Substitution in Au Clusters

Reports on Pt substitution are limited to the case of $Au_{24}Pt(SC_2H_4Ph)_{18}$ only (Figure 6(a)). Generally, $Au_{25}(SC_2H_4Ph)_{18}$ is obtained as an anion $[Au_{25}(SC_2H_4Ph)_{18}]^-$.[65,74] In contrast, $Au_{24}Pt(SC_2H_4Ph)_{18}$ has been obtained as neutral $[Au_{24}Pt(SC_2H_4Ph)_{18}]^0$.[19] For $[Au_{24}Pt(SC_2H_4Ph)_{18}]^0$ and $[Au_{25}(SC_2H_4Ph)_{18}]^-$, the total number of valence electrons is calculated to be 8 and 6, respectively.[75] This indicates that Pt substitution results in a change in the charge state and total number of valence electrons of the original cluster. Along with these changes, the HOMO-LUMO transition becomes forbidden in

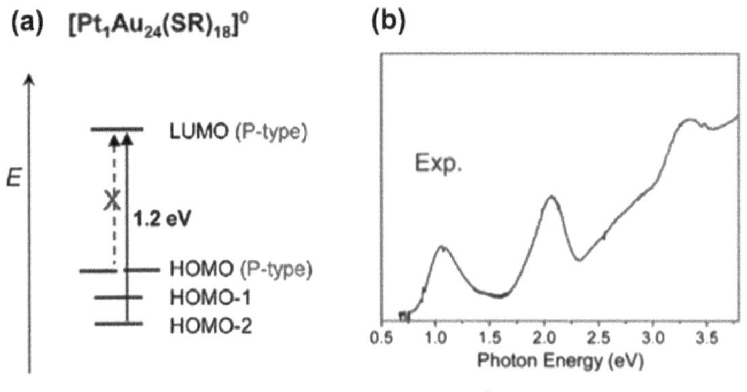

FIGURE 13 (a) Electronic energy levels of $[Au_{24}Pt(SR)_{18}]^0$ and $[Au_{25}(SR)_{18}]^-$, and (b) optical absorption spectrum of $[Au_{24}Pt(SR)_{18}]^0$. *The figures were adapted from Ref. 19 with permission. Copyright 2012 American Chemical Society.*

$[Au_{24}Pt(SC_2H_4Ph)_{18}]^0$, indicating that Pt substitution significantly modifies the optical absorption spectrum of the cluster (Figure 13(a)).[19] In addition, it has been found that Pt substitution enhances the stability of the cluster, and improves its catalytic activity toward the oxidation of styrene (Figure 13(b)).[19]

3.2.3.6 Pd Substitution in Au Clusters

Like Pt, Pd is also an element of group X in the periodic table. Therefore, Pd substitution affects the charge state and electronic structure of a cluster in a manner similar to Pt substitution.[22] Furthermore, Pd substitution in $Au_{25}(SC_{12}H_{25})_{18}$ enhances the stability of the cluster in solution (Figure 14(a)) and increases cluster stability against laser dissociation.[22] Pd substitution in $Au_{25}(SC_{12}H_{25})_{18}$ also improves the reactivity of the cluster in ligand exchange reactions (Figure 14(b)).[70] The effect of Pd substitution has also been studied for $Au_{38}(SC_2H_4Ph)_{24}$ and $Au_{144}(SC_2H_4Ph)_{60}$. These works have revealed that Pd substitution in $Au_{38}(SC_2H_4Ph)_{24}$ improves the stability of the cluster[27] and significantly alters the optical absorption and circular dichroism spectra (Figure 14(c)) of each optical isomer.[57] For $Au_{144}(SC_2H_4Ph)_{60}$, Pd substitution results in a gradual disappearance of peak structure in its optical absorption spectrum (Figure 14(d)).[60]

3.2.3.7 Au Substitution in Ag Clusters

The above parts describe the effects of heteroatom substitution in Au clusters. Conversely, the effect of Au substitution in an Ag cluster has also been clarified for $[Ag_{44}(SR)_{30}]^{4-}$.[33,76] The substitution of 12 central Ag atoms in $[Ag_{44}(SR)_{30}]^{4-}$ (R = PhF, PhF_2, or $PhCF_3$) by Au, forming $[Au_{12}Ag_{32}(SR)_{30}]^{4-}$ (Figure 6(d)), results in a decrease in the intensities of the two peaks at 500–700 nm in its optical absorption spectrum (Figure 15(a)). Moreover, such a substitution improves the stability of the cluster against degradation in solution under harsh conditions (80 °C) (Figure 15(b)). A cluster resulting from the substitution of the remaining 32 Ag atoms in

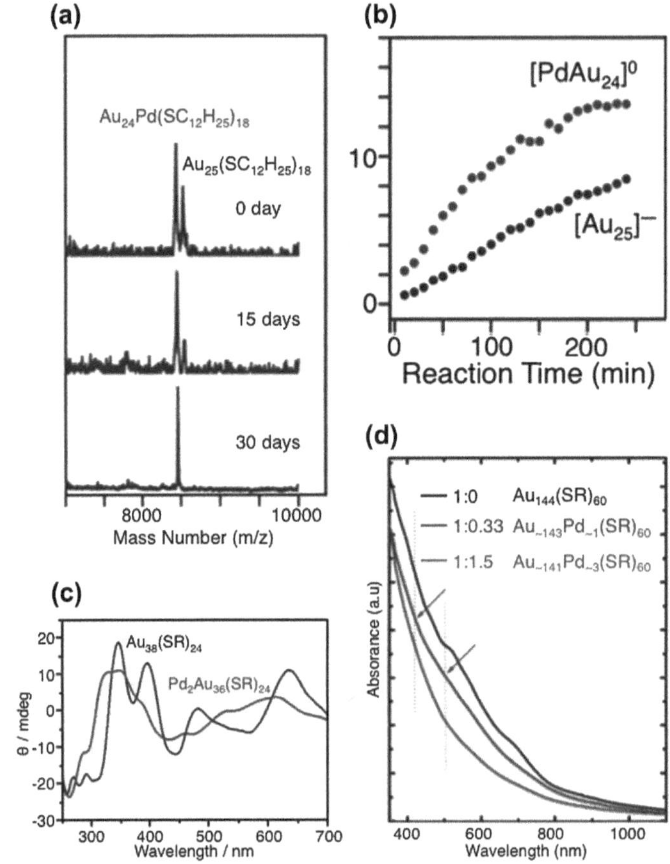

FIGURE 14 Effect of Pd substitution on (a) the stability in solution at 50 °C and (b) the ligand exchange reactivity of $Au_{25}(SC_{12}H_{25})_{18}$, (c) the optical activity of each isomer of $Au_{38}(SC_2H_4Ph)_{24}$, and (d) optical absorption spectra of $Au_{144-n}Pd_n(SC_2H_4Ph)_{60}$. *The figures were adapted from Refs 22,57,60,70 with permission. Copyright 2010 The Royal Society of Chemistry, Copyright 2013 The Royal Society of Chemistry, Copyright 2012 American Chemical Society, and Copyright 2013 The Royal Society of Chemistry.*

$[Au_{12}Ag_{32}(SPhCF_3)_{30}]^{4-}$ by Cu, $[Au_{12}Cu_{32}(SPhCF_3)_{30}]^{4-}$, has also been synthesized.[34] Study of this resulting cluster has revealed that such Cu substitution decreases the stability of the cluster.[34]

3.3 INTERFACIAL CONTROL

3.3.1 Controlled Synthesis of Au Clusters Protected by Other Ligands

This section describes the synthesis of gold clusters protected by SeR, TeR, or C≡CR. Se and Te are elements belonging to the same group (Group XVI) as S

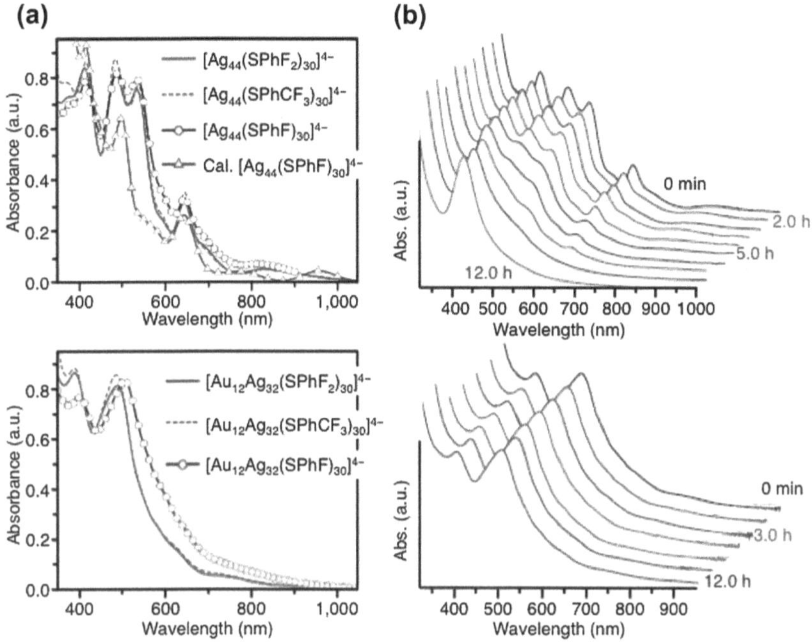

FIGURE 15 Comparison between (a) optical absorption spectra of $[Ag_{44}(SR)_{30}]^{4-}$ and $[Au_{12}Ag_{32}(SR)_{30}]^{4-}$ (R = PhF, PhF_2, or $PhCF_3$), and (b) stability in solution under a harsh condition of $[Ag_{44}(SPhF)_{30}]^{4-}$ (upper) and $[Au_{12}Ag_{32}(SPhF)_{30}]^{4-}$ (bottom). *The figures were adapted from Ref. 33 with permission. Copyright 2013 Macmilian Publishers Ltd.*

in the periodic table. Therefore, SeR and TeR are also classified into a group known as chalcogenates. Gold clusters that are protected by these ligands can be synthesized using the same methods as those employed for $Au_n(SR)_m$ clusters (Chapter 2). The synthetic methods used in the past can be divided into two groups. In the first group, gold ions are reduced with a reducing agent in solution in the presence of a ligand (in this case, RSeH or $(RSe)_2$) (Figure 16(a)).[77–80] The resulting mixture is subjected to separation according to particle size (Section 3.2.2.1). It is also possible to synthesize SeR-protected intermetallic compound clusters using this method.[81,82] In the second group, $Au_n(SR)_m$ clusters are first produced, and are then reacted with RSeH or $(RSe)_2$, resulting in the substitution of SR ligands on the cluster surface by SeR, and in the formation of $Au_n(SeR)_m$ clusters (ligand exchange reaction; Figure 16(a)).[83–85] In the case of the second method, generally, $Au_n(SeR)_m$ clusters with the same number of Au atoms as the precursor $Au_n(SR)_m$ cluster are obtained as a main product. $Au_n(SeR)_m$ clusters such as $Au_{25}(SePh)_{18}$ (Figure 17(a)) and $Au_{38}(SeR)_{24}$ (R = C_8H_{17} or $C_{12}H_{25}$) have been synthesized in high purity using this method. The latter ligand exchange reaction is also used in the synthesis of clusters that contain TeR in a ligand shell.[86] However,

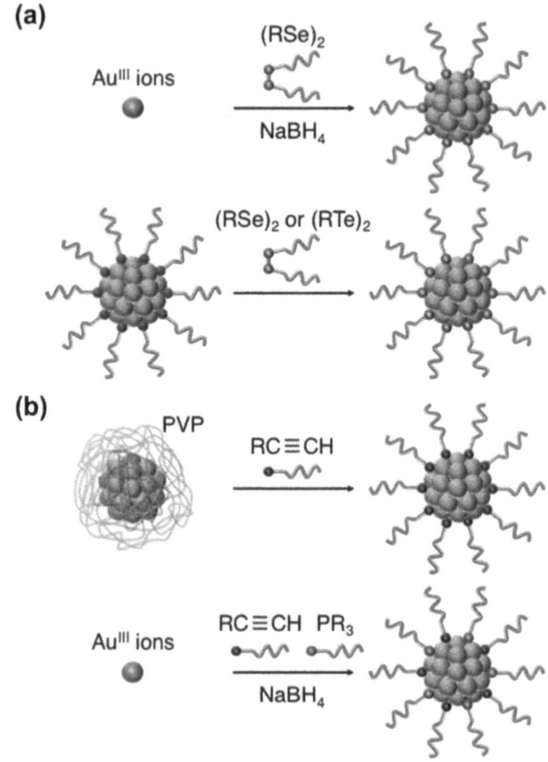

FIGURE 16 Representative synthesis methods of (a) RSe- or RTe-protected and (b) RC≡C-protected Au clusters.

isolation of clusters where all the ligands are substituted by TeR has not yet been achieved for $Au_n(TeR)_m$ clusters (Figure 17(b)).[86]

In contrast, C≡CR-protected gold clusters ($Au_n(C{\equiv}CR)_m$) cannot be synthesized by Au ion reduction in the presence of the ligand.[87] Instead, these clusters can be synthesized by the reaction of polymer-stabilized Au clusters with RC≡CH (Figure 16(b)). For example, $Au_n(C{\equiv}CPh)_m$ between $Au_{42}(C{\equiv}CPh)_{22}$ and $Au_{110}(C{\equiv}CPh)_{40}$ have been synthesized by reacting poly(*N*-vinyl-2-pyrrolidone) (PVP)-stabilized small gold clusters (1.2 ± 0.2, 1.4 ± 0.2, or 1.8 ± 0.3 nm) with PhC≡CH in toluene.[88] Under modified experimental conditions, it is also possible to selectively synthesize $Au_{34}(C{\equiv}CPh)_{16}$ or $Au_{54}(C{\equiv}CPh)_{26}$ using this method (Figure 17(c)).[87,89] With 9-ethynyl-phenanthrene (EPT) as the ligand, it is possible to selectively synthesize clusters such as $Au_{30}(EPT)_{13}$ and $Au_{35}(EPT)_{18}$.[89] However, when $NaBH_4$ is added to a solution containing PhC≡CAu and $HdppaAu_2(SbF_6)_2$ (Hdppa = *N*,*N*-bis(diphenylphosphino)amine), an Au cluster is formed that is protected by both ligands (Figure 16(b)). $[Au_{19}(C{\equiv}CPh)_9(Hdppa)_3]^{2+}$ (Figure 17(d)) has been synthesized by this method.[90]

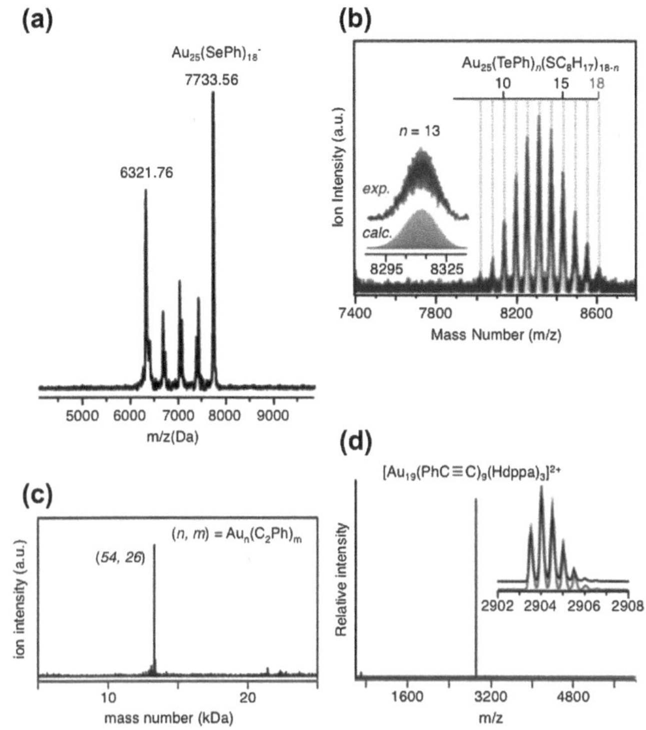

FIGURE 17 Mass spectra of (a) $[Au_{25}(SePh)_{18}]^-$, (b) $Au_{25}(TePh)_n(SC_8H_{17})_{18-n}$, (c) $Au_{54}(C{\equiv}CPh)_{26}$, and (d) $[Au_{19}(C{\equiv}CPh)_9(Hdppa)_3]^{2+}$. *The figures were adapted from Refs 80,86,89,90 with permission. Copyright 2014 The Royal Society of Chemistry, Copyright 2014 American Chemical Society, Copyright 2012 The Royal Society of Chemistry, and Copyright 2015 American Chemical Society.*

3.3.2 Synthesized Au Clusters

Numerous syntheses have been reported for SeR-, TeR-, or C≡CR-protected gold clusters. In this section, we review the chemical composition and geometric structure of the reported clusters.

3.3.2.1 Au Clusters Protected by the Other Calcogenates

Table 3 summarizes the $Au_n(SeR)_m$ clusters synthesized using the methods described in the previous section. As shown in Table 3, $Au_{18}(SePh)_{14}$, $Au_{24}(SePh)_{20}$, $Au_{25}(SeR)_{18}$ ($R = C_8H_{17}$, $C_{12}H_{25}$, C_2H_4Ph, or Ph), and $Au_{38}(SeR)_{24}$ ($R = C_8H_{17}$ or $C_{12}H_{25}$) have been isolated.[77–80,83–85] Formation of clusters such as $Au_{15}(SePh)_{13}$, $Au_{20}(SePh)_{16}$, and $Au_{36}(SeC_8H_{17})_{24}$ has also been observed in the mixtures.[84,85] The number of metal atoms and ligands in these $Au_n(SeR)_m$ are the same as their stable $Au_n(SR)_m$ cluster counterparts (Chapter 2). Among these $Au_n(SeR)_m$ clusters, $Au_{25}(SeR)_{18}$ has been revealed

TABLE 3 List of Clusters Protected by SeR or TeR

Number of Metal Atoms	Formula	Ligand	Synthesis Method
15	$Au_{15}(SeR)_{13}$	SePh[85]	Ligand exchange reaction
18	$Au_{18}(SeR)_{14}$	SePh[85]	Ligand exchange reaction
19	$Au_{19}(SeR)_{15}$	SePh[85]	Ligand exchange reaction
20	$Au_{20}(SeR)_{16}$	SePh[85]	Ligand exchange reaction
24	$Au_{24}(SeR)_{20}$	SePh[78]	Coreduction
25	$Au_{25}(SeR)_{18}$	SeC_8H_{17},[77] $SeC_{12}H_{25}$,[77] SePh,[79,80] SeC_2H_4Ph[80]	Coreduction
	$Au_{25}(SeR)_{18}$	SePh[85]	Ligand exchange reaction
38	$Au_{38}(SeR)_{24}$	SeC_8H_{17},[84] $SeC_{12}H_{25}$[84]	Ligand exchange reaction
25	$Au_{25}(SR)_{18-n}(TeR)_n$	SC_8H_{17} and TePh[86]	Ligand exchange reaction

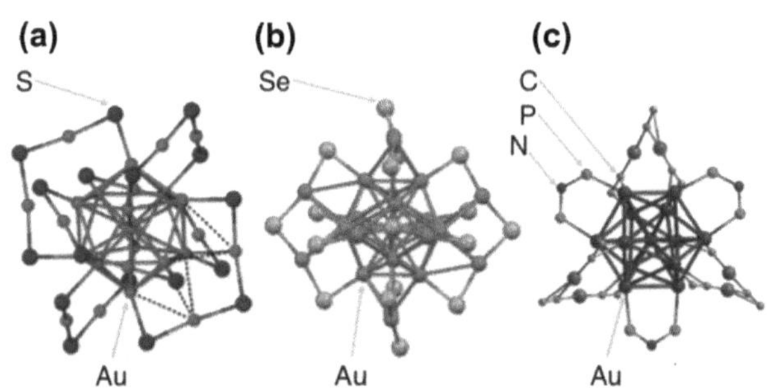

FIGURE 18 Geometric structure of (a) $[Au_{25}(SC_2H_4Ph)_{18}]^-$, (b) $[Au_{25}(SePh)_{18}]^-$, and (c) $[Au_{19}(C{\equiv}CPh)_9(Hdppa)_3]^{2+}$. In this figure, only Au, S, Se, C≡C, and P are depicted. *The figures were adapted from Refs 74,80,90 with permission. Copyright 2008 American Chemical Society, Copyright 2014 The Royal Society of Chemistry, and Copyright 2015 American Chemical Society.*

by single crystal X-ray structural analysis to have the same geometric structure as $Au_{25}(SR)_{18}$ (Figure 18(a) and (b)).[80] It is well known that the stability of $Au_n(SR)_m$ is explained by its total number of valence electrons and geometric structure (Chapter 2). These results indicate that in the smaller size region

($n \leq 38$), the stability of the $Au_n(SeR)_m$ clusters is also governed by the same factors as for $Au_n(SR)_m$ clusters.[84]

In contrast, no $Au_n(TeR)_m$ cluster, where all of the ligands are TeR, has been isolated in sufficient quantities thus far. The ligand exchange method shown in Figure 16(a) is also an efficient way to synthesize $Au_n(TeR)_m$ clusters,[86] and $Au_{25}(TePh)_{18}$ has also been produced by this method (Figure 17(b)). However, the quantity of the product was extremely small, making it difficult to isolate $Au_{25}(TeR)_{18}$.

3.3.2.2 Au Clusters Protected by Terminal Alkynes

In terms of $Au_n(C{\equiv}CR)_m$ clusters where all ligands are terminal alkyne groups, $Au_{34}(C{\equiv}CPh)_{16}$, $Au_{54}(C{\equiv}CPh)_{26}$, $Au_{30}(EPT)_{13}$, and $Au_{35}(EPT)_{18}$ have been synthesized. In these clusters, the ratio of the number of ligands to the number of metal atoms is different from that in $Au_n(SR)_m$ clusters, and the ligand (C≡CR) is presumed to be coordinated to the metal core in a bonding mode that is different to that observed for SR.[87]

Single crystal X-ray structure analysis has revealed the geometric structure for the two-ligand cluster $[Au_{19}(C{\equiv}CPh)_9(Hdppa)_3]^{2+}$ (Figure 18(c)), demonstrating that in this cluster, the Au_{13} core is covered by oligomers comprising (Ph)C≡C–Au–C≡C(Ph)–Au–C≡C(Ph).[90] It is worth noting that theoretical calculations predicted this type of interfacial structure for the bonding between Au and C≡CPh before the experimental findings.[91] This interfacial structure is very similar to that of $Au_{25}(SR)_{18}$, where S(R)–Au–S(R)–Au–S(R) covers the Au_{13} core (Figure 18(a)). Formation of a similar interfacial structure has also been reported for the intermetallic compound cluster Au_6Ag_6(ethisterone)$_{12}$.[92] The total number of valence electrons in $[Au_{19}(C{\equiv}CPh)_9(Hdppa)_3]^{2+}$ can be calculated to be 8, which is same as that of $[Au_{25}(SR)_{18}]^-$. Thus, overall, there are many features in common between $[Au_{19}(C{\equiv}CPh)_9(Hdppa)_3]^{2+}$ and $[Au_{25}(SR)_{18}]^-$ despite their differences in ligand type and number of metal atoms.

3.3.3 Effects of the Use of Each Ligand

3.3.3.1 Other Chalcogenates

As described above, $Au_n(SeR)_m$ clusters have a framework structure that is basically similar to that of $Au_n(SR)_m$ clusters (Figure 18(a) and (b)). That is, change of the ligand from SR to SeR does not significantly affect the framework structure of Au clusters. However, closer examination makes it clear that the $Au_n(SeR)_m$ framework structure is slightly expanded in comparison with the $Au_n(SR)_m$ cluster because Se (1.17 Å) has an atomic radius slightly larger than that of S (1.02 Å).[80,84] Similar expansion of the framework structure has been also observed for clusters containing TeR in the ligand shell, because Te (1.44 Å) has an atomic radius larger than that of S (1.02 Å).[86] As for electronic structure, $Au_n(SeR)_m$ clusters has a slightly smaller

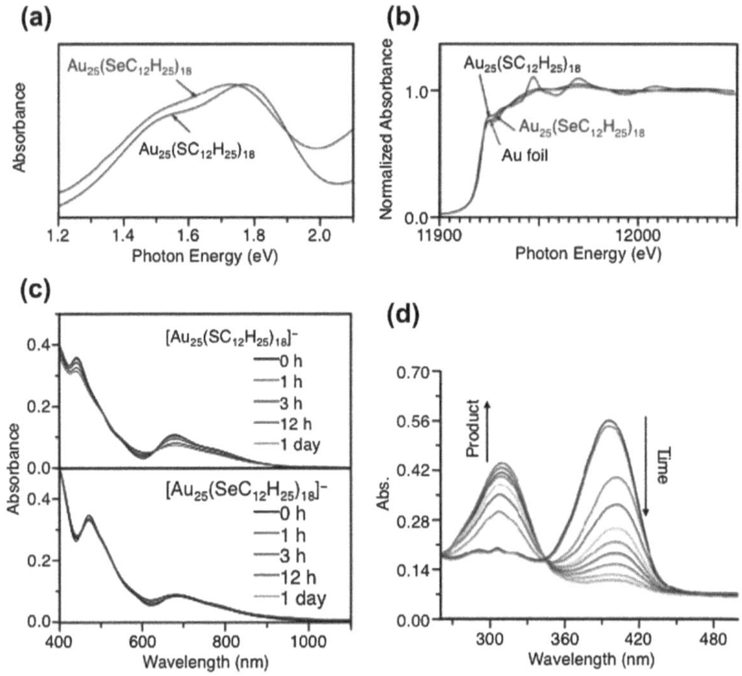

FIGURE 19 Comparison of (a) optical absorption spectra of $Au_{25}(SeC_{12}H_{25})_{18}$ and $Au_{25}(SC_{12}H_{25})_{18}$, (b) Au L_3-edge XANES spectra of $Au_{25}(SeC_{12}H_{25})_{18}$ and $Au_{25}(SC_{12}H_{25})_{18}$, and (c) stability in solution under harsh conditions of $Au_{25}(SeC_{12}H_{25})_{18}$ and $Au_{25}(SC_{12}H_{25})_{18}$. (d) Representative UV−Vis optical absorption spectra of the reduction of 4-nitrophenolate to 4-aminophenol over $Au_{25}(SePh)_{18}$. *The figures were adapted from Refs 80,84,93 with permission. Copyright 2013 American Chemical Society, Copyright 2012 American Chemical Society, and Copyright 2014 The Royal Society of Chemistry.*

HOMO-LUMO gap than that of $Au_n(SR)_m$ clusters (Figure 19(a)), indicating that the use of SeR as a ligand causes a decrease in the HOMO-LUMO gap of the clusters.[80,84] A more remarkable decrease in the HOMO-LUMO gap has been observed for clusters containing TeR in the ligand shell.[86]

The most interesting aspect in the use of these ligands is probably the modification of the Au−ligand bonding mode. According to studies related to two-dimensional self-assembled monolayers, the charge transfer from Au to the ligand is less for Au−SeR than Au−SR.[6] Furthermore, changing the nature of the bonding mode in this way resulted in a stronger Au−SeR bond compared with the Au−SR bond, as well as a higher conductivity in the Au−SeR bond.[7−9] Similar changes in bonding mode also occur in gold clusters. For example, in $Au_n(SeC_{12}H_{25})_m$ clusters, charge transfer from Au to the ligand is reduced (Figure 19(b)), and the Au−ligand bond becomes much more covalent than that in $Au_n(SC_{12}H_{25})_m$ clusters.[77,84] Owing to these changes in the bonding mode, $Au_{25}(SeR)_{18}$ ($R = C_{12}H_{25}$ or C_8H_{17}) exhibits higher stability against degradation in solution than $Au_{25}(SR)_{18}$ ($R = C_{12}H_{25}$ or C_8H_{17}) (Figure 19(c)).[93] In addition to improved stability, it is expected that using SeR as a ligand improves the conductivity between the gold core and the ligand.[8,94,95] Therefore, new findings are anticipated for the conductivity of $Au_n(SeR)_m$ clusters in the future. Furthermore, recent studies have demonstrated that $Au_{25}(SePh)_{18}$ show the catalytic activity toward the reduction of 4-nitrophenolate (Figure 19(d)).[80]

3.3.3.2 C≡CR

$Au_n(C{\equiv}CPh)_m$ and $Au_n(EPT)_m$ clusters also exhibit photoluminescence similar to that observed for $Au_n(SR)_m$ clusters. However, the mechanism of photoluminescence from $Au_n(C{\equiv}CR)_m$ clusters is different from that of $Au_n(SR)_m$ clusters.[87] The origin of the photoluminescence observed for $Au_n(C{\equiv}CPh)_m$ and $Au_n(EPT)_m$ clusters is mainly caused by the photoluminescence properties of their ligands (Figure 20(a)). In particular, photoluminescence in the $Au_n(EPT)_m$ cluster is attributed to excimer emissions from the ligands (Figure 20(a)).[87] For $[Au_{19}(C{\equiv}CPh)_9(Hdppa)_3]^{2+}$, optical absorption spectroscopy shows that $[Au_{19}(C{\equiv}CPh)_9(Hdppa)_3]^{2+}$ has smaller HOMO-LUMO gap than $[Au_{25}(SR)_{18}]^-$ (Figure 20(b)).[90]

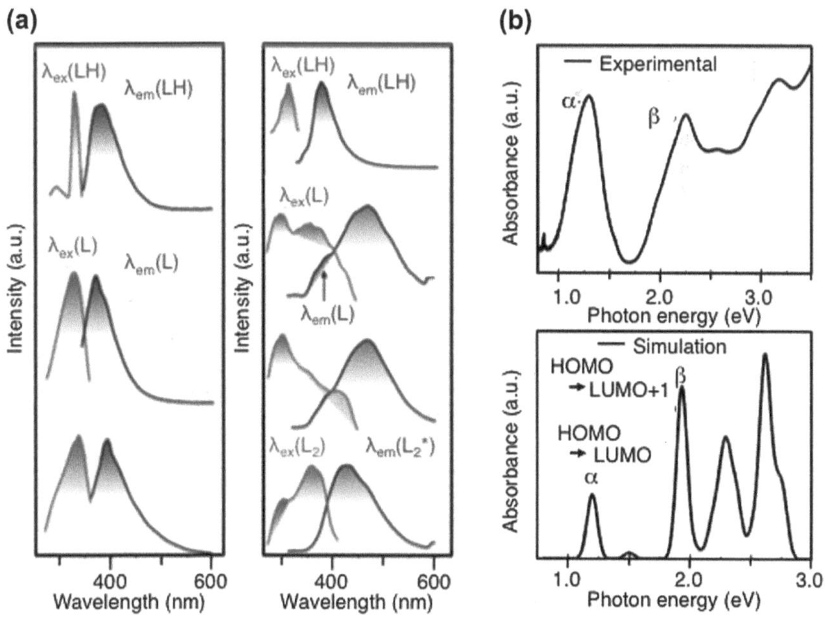

FIGURE 20 (a) Photoluminescence (black) and excitation (red) spectra of (left) PhC≡CH, $Au_{34}(C{\equiv}CPh)_{16}$, and $Au_{54}(C{\equiv}CPh)_{26}$, and (right) EPT, $Au_{30}(EPT)_{13}$, $Au_{35}(EPT)_{18}$, and $Au_{41-43}(EPT)_{21-23}$. (b) Experimental and simulated optical absorption spectrum of $[Au_{19}(C{\equiv}CPh)_9(Hdppa)_3]^{2+}$. *The figures were adapted from Refs 87,90 with permission. Copyright 2013 American Chemical Society and Copyright 2015 American Chemical Society.*

3.4 SUMMARY AND PERSPECTIVE

In this chapter, we have conducted a survey of SR- or PR_3-protected intermetallic compound clusters, as well as SeR-, TeR-, or terminal alkyne (C≡CR)-protected gold clusters, with a particular focus on control of the metal cluster composition and interface. Except PR_3-protected compound clusters, the first synthesis of these clusters was reported only 5 years ago (in 2009). However, an extensive range of studies has been conducted since with rapid development of their synthetic methods and an improved understanding of their physical and chemical characteristics. The studies have shown that the control of the composition and interface of these clusters leads to changes in their physical and chemical properties, and thereby facilitates their use in a wider range of applications. In the future, methods of synthesizing these clusters will most likely be tailored to meet the requirements of specific applications.

REFERENCES

1. Zhang H, Watanabe T, Okumura M, Haruta M, Toshima N. Catalytically highly active top gold atom on palladium nanocluster. *Nat Mater* 2012;**11**:49–52.
2. Kusada K, Kobayashi H, Ikeda R, Kubota Y, Takata M, Toh S, et al. Solid solution alloy nanoparticles of immiscible Pd and Ru elements neighboring on Rh: changeover of the thermodynamic behavior for hydrogen storage and enhanced CO-oxidizing ability. *J Am Chem Soc* 2014;**136**:1864–71.
3. Shibu ES, Muhammed MAH, Tsukuda T, Pradeep T. Ligand exchange of $Au_{25}SG_{18}$ leading to functionalized gold clusters: spectroscopy, kinetics, and luminescence. *J Phys Chem C* 2008;**112**:12168–76.
4. Wu Z, Jin R. On the ligand's role in the fluorescence of gold nanoclusters. *Nano Lett* 2010;**10**:2568–73.
5. Knoppe S, Bürgi T. Chirality in thiolate-protected gold clusters. *Acc Chem Res* 2014;**47**:1318–26.
6. de la Llave E, Scherlis DA. Selenium-based self-assembled monolayers: the nature of adsorbate–surface interactions. *Langmuir* 2010;**26**:173–8.
7. Szelagowska-Kunstman K, Cyganik P, Schüpbach B, Terfort A. Relative stability of thiol and selenol based SAMs on Au(111) – exchange experiments. *Phys Chem Chem Phys* 2010;**12**:4400–6.
8. Romashov LV, Ananikov VP. Self-assembled selenium monolayers: from nanotechnology to materials science and adaptive catalysis. *Chem Eur J* 2013;**19**:17640–60.
9. Yokota K, Taniguchi M, Kawai T. Control of the electrode-molecule interface for molecular devices. *J Am Chem Soc* 2007;**129**:5818–9.
10. Negishi Y, Iwai T, Ide M. Continuous modulation of electronic structure of stable thiolate-protected Au_{25} cluster by Ag doping. *Chem Commun* 2010;**46**:4713–5.
11. Kauffman DR, Alfonso D, Matranga C, Qian H, Jin R. A quantum alloy: the ligand-protected $Au_{25-n}Ag_n(SR)_{18}$ cluster. *J Phys Chem C* 2013;**117**:7914–23.
12. Dou X, Yuan X, Yao Q, Luo Z, Zheng K, Xie J. Facile synthesis of water-soluble $Au_{25-x}Ag_x$ nanoclusters protected by mono- and bi-thiolate ligands. *Chem Commun* 2014;**50**:7459–62.

13. Kobayashi R, Nonoguchi Y, Sasaki A, Yao H. Chiral monolayer-protected bimetallic Au—Ag nanoclusters: alloying effect on their electronic structure and chiroptical activity. *J Phys Chem C* 2014;**118**:15506—15.
14. Kumara C, Aikens CM, Dass A. X-ray crystal structure and theoretical analysis of $Au_{25-x}Ag_x(SCH_2CH_2Ph)_{18}^-$ Alloy. *J Phys Chem Lett* 2014;**5**:461—6.
15. Negishi Y, Munakata K, Ohgake W, Nobusada K. Effect of copper doping on electronic structure, geometric structure, and stability of thiolate-protected Au_{25} nanoclusters. *J Phys Chem Lett* 2012;**3**:2209—14.
16. Gottlieb E, Qian H, Jin R. Atomic-level alloying and de-alloying in doped gold nanoparticles. *Chem Eur J* 2013;**19**:4238—43.
17. Negishi Y, Kurashige W, Niihori Y, Nobusada K. Toward the creation of stable, functionalized metal clusters. *Phys Chem Chem Phys* 2013;**15**:18736—51.
18. Jin R, Nobusada K. Doping and alloying in atomically precise gold nanoparticles. *Nano Res* 2014;**7**:285—300.
19. Qian H, Jiang D-e, Li G, Gayathri C, Das A, Gil RR, et al. Monoplatinum doping of gold nanoclusters and catalytic application. *J Am Chem Soc* 2012;**134**:16159—62.
20. Negishi Y. Toward the creation of functionalized metal nanoclusters and highly active photocatalytic materials using thiolate-protected magic gold clusters. *Bull Chem Soc Jpn* 2014;**87**:375—89.
21. Fields-Zinna CA, Crowe MC, Dass A, Weaver JEF, Murray RW. Mass spectrometry of small bimetal monolayer-protected clusters. *Langmuir* 2009;**25**:7704—10.
22. Negishi Y, Kurashige W, Niihori Y, Iwasa T, Nobusada K. Isolation, structure, and stability of a dodecanethiolate-protected Pd_1Au_{24} cluster. *Phys Chem Chem Phys* 2010;**12**:6219—25.
23. Niihori Y, Matsuzaki M, Pradeep T, Negishi Y. Separation of precise compositions of noble metal clusters protected with mixed ligands. *J Am Chem Soc* 2013;**135**:4946—9.
24. Niihori Y, Matsuzaki M, Uchida C, Negishi Y. Advanced use of high-performance liquid chromatography for synthesis of controlled metal clusters. *Nanoscale* 2014;**6**:7889—96.
25. Qian H, Barry E, Zhu Y, Jin R. Doping 25-atom and 38-atom gold nanoclusters with palladium. *Acta Phys Chim Si* 2011;**27**:513—9.
26. Kumara C, Dass A. AuAg alloy nanomolecules with 38 metal atoms. *Nanoscale* 2012;**4**:4084—6.
27. Negishi Y, Igarashi K, Munakata K, Ohgake W, Nobusada K. Palladium doping of magic gold cluster $Au_{38}(SC_2H_4Ph)_{24}$: formation of $Pd_2Au_{36}(SC_2H_4Ph)_{24}$ with higher stability than $Au_{38}(SC_2H_4Ph)_{24}$. *Chem Commun* 2012;**48**:660—2.
28. Kumara C, Dass A. $(AuAg)_{144}(SR)_{60}$ alloy nanomolecules. *Nanoscale* 2011;**3**:3064—7.
29. Dharmaratne AC, Dass A. $Au_{144-x}Cu_x(SC_6H_{13})_{60}$ nanomolecules: effect of Cu incorporation on composition and plasmon-like peak emergence in optical spectra. *Chem Commun* 2014;**50**:1722—4.
30. Malola S, Lehtovaara L, Häkkinen H. TDDFT analysis of optical properties of thiol monolayer-protected gold and intermetallic silver—gold $Au_{144}(SR)_{60}$ and $Au_{84}Ag_{60}(SR)_{60}$ clusters. *J Phys Chem C* 2014;**118**:20002—8.
31. Malola S, Hartmann MJ, Häkkinen H. Copper induces a core plasmon in intermetallic $Au_{(144,145)-x}Cu_x(SR)_{60}$ nanoclusters. *J Phys Chem Lett* 2015;**6**:515—20.
32. Bhattarai N, Black DM, Boppidi S, Khanal S, Uribe DB, Tlahuice-Flores A, et al. ESI-MS identification of abundant copper-gold clusters exhibiting high plasmonic character. *J Phys Chem C* 2015;**119**:10935—42.
33. Yang H, Wang Y, Huang H, Gell L, Lehtovaara L, Malola S, et al. All-thiol-stabilized Ag_{44} and $Au_{12}Ag_{32}$ nanoparticles with single-crystal structures. *Nat Commun* 2013;**4**:2422.

34. Yang H, Wang Y, Yan J, Chen X, Zhang X, Häkkinen H, et al. Structural evolution of atomically precise thiolated bimetallic $[Au_{12+n}Cu_{32}(SR)_{30+n}]^{4-}$ ($n = 0, 2, 4, 6$) nanoclusters. *J Am Chem Soc* 2014;**136**:7197–200.
35. Kurashige W, Negishi Y. Synthesis, stability, and photoluminescence properties of $PdAu_{10}(PPh_3)_8Cl_2$ clusters. *J Clust Sci* 2012;**23**:365–74.
36. Yang H, Wang Y, Lei J, Shi L, Wu X, Mäkinen V, et al. Ligand-stabilized $Au_{13}Cu_x$ ($x = 2, 4, 8$) bimetallic nanoclusters: ligand engineering to control the exposure of metal sites. *J Am Chem Soc* 2013;**135**:9568–71.
37. Teo BK, Zhang H, Shi X. Molecular architecture of a novel vertex-sharing biicosahedral cluster $[(p\text{-}Tol_3P)_{10}Au_{13}Ag_{12}Br_8](PF_6)$ containing a staggered-staggered-staggered configuration for the 25-atom metal framework. *Inorg Chem* 1990;**29**:2083–91.
38. Teo BK, Hong MC, Zhang H, Huang DB. Cluster of clusters: structure of the 37-atom cluster $[(p\text{-}Tol_3P)_{12}Au_{18}Ag_{19}Br_{11}]^{2+}$ and a novel series of supraclusters based on vertex-sharing icosahedra. *Angew Chem Int Ed* 1987;**26**:897–900.
39. Teo BK, Zhang H, Shi X. Cluster of clusters: a modular approach to large metal clusters. structural characterization of a 38-atom cluster $[(p\text{-}Tol_3P)_{12}Au_{18}Ag_{20}Cl_{14}]$ based on vertex-sharing triicosahedra. *J Am Chem Soc* 1990;**112**:8552–62.
40. Teo BK, Zhang H. Polyicosahedricity: Icosahedron to icosahedron of icosahedra growth pathway for bimetallic (Au–Ag) and trimetallic (Au–Ag–M; m = Pt, Pd, Ni) supraclusters; synthetic strategies, site preference, and stereochemical principles. *Coord Chem Rev* 1995;**143**:611–36.
41. Pignolet LH, Aubart MA, Craighead KL, Gould RAT, Krogstad DA, Wiley JS. Phosphine-stabilized, platinum-gold and palladium-gold cluster compounds and applications in catalysis. *Coord Chem Rev* 1995;**143**:219–63.
42. Wang S, Meng X, Das A, Li T, Song Y, Cao T, et al. 200-fold quantum yield boost in the photoluminescence of silver-doped Ag_xAu_{25-x} nanoclusters: the 13th silver atom matters. *Angew Chem Int Ed* 2014;**53**:2376–80.
43. Puls A, Jerabek P, Kurashige W, Förster M, Molon M, Bollermann T, et al. A novel concept for the synthesis of multiply doped gold clusters $[(M@Au_nM'_m)L_k]^{q+}$. *Angew Chem Int Ed* 2014;**53**:4327–31.
44. Teo BK, Zhang H, Shi X. Design, synthesis, and structure of the largest trimetallic cluster, $[(Ph_3P)_{10}Au_{12}Ag_{12}PtCl_7]Cl$: the first example of a trimetallic biicosahedral supracluster and its implication for the vertex-sharing polyicosahedral growth of the gold/silver/platinum ternary cluster system. *J Am Chem Soc* 1993;**115**:8489–90.
45. Teo BK, Zhang H, Shi X. Site preference in vertex-sharing polyicosahedral supraclusters containing groups 10 and 11 metals and their bonding implications: syntheses and structures of the first Au–Ag–M (M = Pt, Ni) biicosahedral clusters $[(Ph_3P)_{10}Au_{12}Ag_{12}PtCl_7]Cl$ and $[(Ph_3P)_{10}Au_{12}Ag_{12}NiCl_7](SbF_6)$. *Inorg Chem* 1994;**33**:4086–97.
46. Wang C, Wang Y, Xu L, Shi X, Li X, Xu X, et al. A galvanic replacement route to prepare strongly fluorescent and highly stable gold nanodots for cellular imaging. *Small* 2013;**9**:413–20.
47. Xia X, Wang Y, Ruditskiy A, Xia Y. 25th anniversary article: galvanic replacement: a simple and versatile route to hollow nanostructures with tunable and well-controlled properties. *Adv Mater* 2013;**25**:6313–33.
48. Wan D, Xia X, Wang Y, Xia Y. Robust synthesis of gold cubic nanoframes through a combination of galvanic replacement, gold deposition, and silver dealloying. *Small* 2013;**9**:3111–7.
49. Zhang L, Wang Y, Tong L, Xia Y. Synthesis of colloidal metal nanocrystals in droplet reactors: the pros and cons of interfacial adsorption. *Nano Lett* 2014;**14**:4189–94.

50. Yao C, Chen J, Li MB, Liu L, Yang J, Wu Z. Adding two active silver atoms on Au_{25} nanoparticle. *Nano Lett* 2015;**15**:1281–7.
51. Sun J, Wu H, Jin Y. Synthesis of thiolated Ag/Au bimetallic nanoclusters exhibiting an anti-galvanic reduction mechanism and composition-dependent fluorescence. *Nanoscale* 2014;**6**:5449–57.
52. Dou X, Yuan X, Yu Y, Luo Z, Yao Q, Leong DT, et al. Lighting up thiolated Au@Ag nanoclusters via aggregation-induced emission. *Nanoscale* 2014;**6**:157–61.
53. Could RAT, Pignolet LH. Heterometallic Au–Pt–Hg and Au–Pt–Hg–Co phosphine–stabilized cluster compounds. X-ray crystal and molecular structure of $[(PPh_3)Pt(AuPPh_3)_6(HgNO_3)]NO_3$. *Inorg Chem* 1994;**33**:40–6.
54. Biltek SR, Sen A, Pedicini AF, Reber AC, Khanna SN. Isolation structural characterization of a silver–platinum nanocluster, $Ag_4Pt_2(DMSA)_4$. *J Phys Chem A* 2014;**118**:8314–9.
55. Biltek SR, Mandal S, Sen A, Reber AC, Pedicini AF, Khanna SN. Synthesis and structural characterization of an atom-precise bimetallic nanocluster, $Ag_4Ni_2(DMSA)_4$. *J Am Chem Soc* 2013;**135**:26–9.
56. Udayabhaskararao T, Sun Y, Goswami N, Pal SK, Balasubramanian K, Pradeep T. Ag_7Au_6: a 13-atom alloy quantum cluster. *Angew Chem Int Ed* 2012;**51**:2155–9.
57. Barrabés N, Zhang B, Bürgi T. Racemization of chiral $Pd_2Au_{36}(SC_2H_4Ph)_{24}$: doping increases the flexibility of the cluster surface. *J Am Chem Soc* 2014;**136**:14361–4.
58. Jupally VR, Dass A. Synthesis of $Au_{130}(SR)_{50}$ and $Au_{130-x}Ag_x(SR)_{50}$ nanomolecules through core size conversion of larger metal clusters. *Phys Chem Chem Phys* 2014;**16**:10473–9.
59. Jupally VR, Dharmaratne AC, Crasto D, Huckaba AJ, Kumara C, Nimmala PR, et al. $Au_{137}(SR)_{56}$ nanomolecules: composition, optical spectroscopy, electrochemistry and electrocatalytic reduction of CO_2. *Chem Commun* 2014;**50**:9895–8.
60. Kothalawala N, Kumara C, Ferrando R, Dass A. $Au_{144-x}Pd_x(SR)_{60}$ nanomolecules. *Chem Commun* 2013;**49**:10850–2.
61. Yamazoe S, Kurashige W, Nobusada K, Negishi Y, Tsukuda T. Preferential location of coinage metal dopants (M = Ag or Cu) in $[Au_{25-x}M_x(SC_2H_4Ph)_{18}]^-$ ($x \sim 1$) as determined by extended X-ray absorption fine structure and density functional theory calculations. *J Phys Chem C* 2014;**118**:25284–90.
62. Christensen SL, MacDonald MA, Chatt A, Zhang P, Qian H, Jin R. Dopant location, local structure, and electronic properties of $Au_{24}Pt(SR)_{18}$ nanoclusters. *J Phys Chem C* 2012;**116**:26932–7.
63. Negishi Y, Kurashige W, Kobayashi Y, Yamazoe S, Kojima N, Seto M, et al. Formation of a $Pd@Au_{12}$ superatomic core in $Au_{24}Pd_1(SC_{12}H_{25})_{18}$ probed by ^{197}Au Mössbauer and Pd K-edge EXAFS spectroscopy. *J Phys Chem Lett* 2013;**4**:3579–83.
64. Ferrando R, Jellinek J, Johnston RL. Nanoalloys: from theory to applications of alloy clusters and nanoparticles. *Chem Rev* 2008;**108**:845–910.
65. Kurashige W, Niihori Y, Sharma S, Negishi Y. Recent progress in the functionalization methods of thiolate-protected gold clusters. *J Phys Chem Lett* 2014;**5**:4134–42.
66. Copley RCB, Mingos DMP. Synthesis and characterization of the centred icosahedral cluster series $[Au_9M^{IB}{}_4Cl_4(PMePh_2)_8][C_2B_9H_{12}]$, where M^{IB} = Au, Ag or Cu. *J Chem Soc Dalton Trans* 1996:491–500.
67. Briant CE, Theobald BRC, White JW, Bell LK, Mingos DMP, Welch AJ. Synthesis and X-ray structural characterization of the centred icosahedral gold cluster compound $[Au_{13}(PMe_2Ph)_{10}C1_2](PF_6)_3$; the realization of a theoretical prediction. *J Chem Soc Chem Commun* 1981;**5**:201–2.

68. Shichibu Y, Negishi Y, Watanabe T, Chaki NK, Kawaguchi H, Tsukuda T. Biicosahedral gold clusters $[Au_{25}(PPh_3)_{10}(SC_nH_{2n+1})_5Cl_2]^{2+}$ ($n = 2-18$): a stepping stone to cluster-assembled materials. *J Phys Chem C* 2007;**111**:7845–7.
69. Qian H, Eckenhoff WT, Bier ME, Pintauer T, Jin R. Crystal structures of Au_2 complex and Au_{25} nanocluster and mechanistic insight into the conversion of polydisperse nanoparticles into monodisperse Au_{25} nanoclusters. *Inorg Chem* 2011;**50**:10735–9.
70. Niihori Y, Kurashige W, Matsuzaki M, Negishi Y. Remarkable enhancement in ligand-exchange reactivity of thiolate-protected Au_{25} nanoclusters by single Pd atom doping. *Nanoscale* 2013;**5**:508–12.
71. Zhu M, Aikens CM, Hollander FJ, Schatz GC, Jin R. Correlating the crystal structure of a thiol-protected Au_{25} cluster and optical properties. *J Am Chem Soc* 2008;**130**:5883–5.
72. Akola J, Walter M, Whetten RL, Häkkinen H, Grönbeck H. On the structure of thiolate-protected Au_{25}. *J Am Chem Soc* 2008;**130**:3756–7.
73. Jiang D-e, Kühn M, Tang Q, Weigend F. Superatomic orbitals under spin-orbit coupling. *J Phys Chem Lett* 2014;**5**:3286–9.
74. Heaven MW, Dass A, White PS, Holt KM, Murray RW. Crystal structure of the gold nanoparticle $[N(C_8H_{17})_4][Au_{25}(SCH_2CH_2Ph)_{18}]$. *J Am Chem Soc* 2008;**130**:3754–5.
75. Walter M, Akola J, Lopez-Acevedo O, Jadzinsky PD, Calero G, Ackerson CJ, et al. A unified view of ligand-protected gold clusters as superatom complexes. *Proc Natl Acad Sci USA* 2008;**105**:9157–62.
76. Gell L, Häkkinen H. Theoretical analysis of the $M_{12}Ag_{32}(SR)_{40}^{4-}$ and $X@M_{12}Ag_{32}(SR)_{30}^{4-}$ nanoclusters (M = Au, Ag; X = H, Mn). *J Phys Chem C* 2015;**119**:10943–8.
77. Negishi Y, Kurashige W, Kamimura U. Isolation and structural characterization of an octaneselenolate-protected Au_{25} cluster. *Langmuir* 2011;**27**:12289–92.
78. Song Y, Wang S, Zhang J, Kang X, Chen S, Li P, et al. Crystal structure of selenolate-protected $Au_{24}(SeR)_{20}$ nanocluster. *J Am Chem Soc* 2014;**136**:2963–5.
79. Song Y, Cao T, Deng H, Zhu X, Li P, Zhu M. Kinetically controlled, high-yield, direct synthesis of $[Au_{25}(SePh)_{18}]^-TOA^+$. *Sci Chin Chem* 2014;**57**:1218–24.
80. Song Y, Zhong J, Yang S, Wang S, Cao T, Zhang J, et al. Crystal structure of $Au_{25}(SePh)_{18}$ nanoclusters and insights into their electronic, optical and catalytic properties. *Nanoscale* 2014;**6**:13977–85.
81. Kurashige W, Munakata K, Nobusada K, Negishi Y. Synthesis of stable Cu_nAu_{25-n} nanoclusters ($n = 1-9$) using selenolate ligands. *Chem Commun* 2013;**49**:5447–9.
82. Sarkar S, Chakraborty I, Panwar MK, Pradeep T. Isolation and tandem mass spectrometric identification of a stable monolayer protected silver–palladium alloy cluster. *J Phys Chem Lett* 2014;**5**:3757–62.
83. Meng X, Xu Q, Wang S, Zhu M. Ligand-exchange synthesis of selenophenolate-capped Au_{25} nanoclusters. *Nanoscale* 2012;**4**:4161–5.
84. Kurashige W, Yamazoe S, Kanehira K, Tsukuda T, Negishi Y. Selenolate-protected Au_{38} nanoclusters: isolation and structural characterization. *J Phys Chem Lett* 2013;**4**:3181–5.
85. Xu Q, Wang S, Liu Z, Xu G, Meng X, Zhu M. Synthesis of selenolate-protected $Au_{18}(SeC_6H_5)_{14}$ nanoclusters. *Nanoscale* 2013;**5**:1176–82.
86. Kurashige W, Yamazoe S, Yamaguchi M, Nishido K, Nobusada K, Tsukuda T, et al. Au_{25} clusters containing unoxidized tellurolates in the ligand shell. *J Phys Chem Lett* 2014;**5**:2072–6.
87. Maity P, Takano S, Yamazoe S, Wakabayashi T, Tsukuda T. Binding motif of terminal alkynes on gold clusters. *J Am Chem Soc* 2013;**135**:9450–7.
88. Maity P, Tsunoyama H, Yamauchi M, Xie S, Tsukuda T. Organogold clusters protected by phenylacetylene. *J Am Chem Soc* 2011;**133**:20123–5.

89. Maity P, Wakabayashi T, Ichikuni N, Tsunoyama H, Xie S, Yamauchi M, et al. Selective synthesis of organogold magic clusters $Au_{54}(C{\equiv}CPh)_{26}$. *Chem Commun* 2012;**48**:6085–7.
90. Wan XK, Tang Q, Yuan SF, Jiang D-e, Wang QM. Au_{19} nanocluster featuring a V-shaped alkynyl–gold motif. *J Am Chem Soc* 2015;**137**:652–5.
91. Tang Q, Jiang D-e. Insights into the $PhC{\equiv}C/Au$ interface. *J Phys Chem C* 2015;**119**: 10804–10.
92. Guo WJ, Wang YT, Kong DX, Wang JY, Wei QH, Chen GN. Luminescent Ag_6Au_6 heterometallic ethisterone cluster and probe for estrogen receptor α. *Chem Eur J* 2015;**21**:4205–8.
93. Kurashige W, Yamaguchi M, Nobusada K, Negishi Y. Ligand-induced stability of gold nanoclusters: thiolate versus selenolate. *J Phys Chem Lett* 2012;**3**:2649–52.
94. Li Y, Zaluzhna O, Xu B, Gao Y, Modest JM, Tong YYJ. Mechanistic insight into the Brust–Schiffrin two–phase synthesis of organo–chalcogenate–protected metal nanoparticles. *J Am Chem Soc* 2011;**133**:2092–5.
95. Ventra MD, Lang ND. Transport in nanoscale conductors from first principles. *Phys Rev B* 2001;**65**:045402.

Chapter 4

Structural Engineering of Heterometallic Nanoclusters

Yu Wang, Huayan Yang and Nanfeng Zheng[1]

Collaborative Innovation Center of Chemistry for Energy Materials, State Key Laboratory for Physical Chemistry of Solid Surfaces, College of Chemistry and Chemical Engineering, Xiamen University, Xiamen, China

[1]*Corresponding author: E-mail: nfzheng@xmu.edu.cn*

Chapter Outline

4.1 INTRODUCTION

By integrating different metals together, heterometallic nanoparticles have been emerging as a type of important nanomaterials that often exhibit unique chemical and physical properties that are not found in their homometallic counterparts. Compared to homometallic nanoparticles, heterometallic nanoparticles display more structural diversity in terms of the metal

Frontiers of Nanoscience, Vol. 9. http://dx.doi.org/10.1016/B978-0-08-100086-1.00004-X

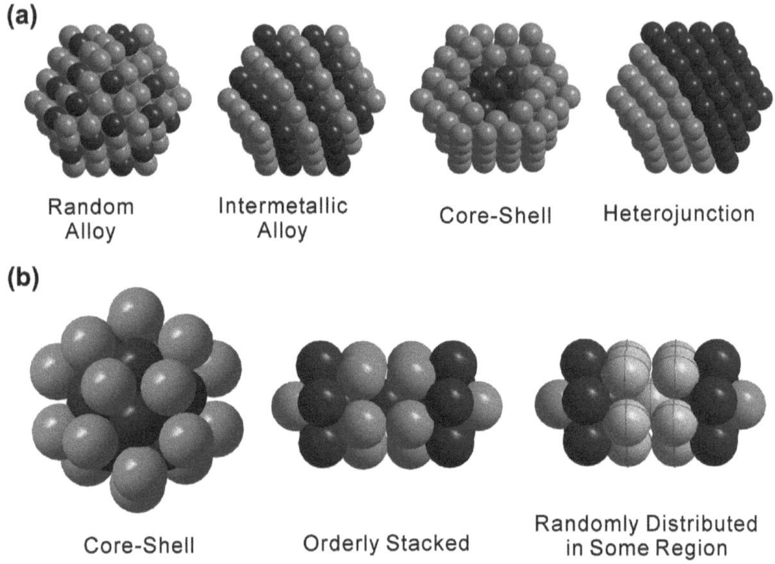

FIGURE 1 Metal distributions in (a) organic-capped bimetallic nanoparticles and (b) organic-capped atomically precise bimetallic nanoclusters. Dark- and light-colored solid spheres represent two different metals. Spheres with orbitals represent mixed metal sites.

distributions.[1–3] As illustrated in Figure 1(a), when two different metals are brought together to form bimetallic nanostructures, the following structures are possible: (1) alloyed structure, (2) core–shell structure with only one metal exposed, (3) heterojunction structure with both metals exposed. Even in alloyed structures, metal atoms can be distributed randomly as in solid solutions or arranged orderly to form intermetallic nanoparticles. The distribution of metal atoms in heterometallic nanoparticles is critical to many of their overall properties. Heterometallic nanoparticles with the same size, shape, and composition could have significant differences in catalytic activity.[4–10] For instance, Ag nanoparticles coated with a thin layer of Pd atoms significantly enhance the production of H_2 from formic acid at room temperature.[4] The catalytic property of Ru-Pt core–shell nanoparticles for preferential CO oxidation in hydrogen-rich environments is significantly improved compared to their alloy counterparts. Au-Pd alloy nanoparticles exhibit higher electrocatalytic activity and stability for formic acid oxidation than their core–shell counterparts.[6,10] The multiple metal components in a heterometallic nanostructure often interplay with each other to induce synergetic effect for catalysis. Synergetic effects in heterometallic nanocatalysts are mainly induced by altering the electronic structure or creating collaborative sites due to the presence of multiple metal components.[2] Currently, there are quite several techniques for the characterization of metal distributions in heterometallic nanoparticles such as atom probe tomography, low-energy ion scattering spectroscopy, and so on.[1,4] However, most techniques do not allow us to characterize the metal distribution at the atomic scale, preventing us from deeply understanding the nature of synergetic effects in heterometallic nanoparticles.

Atomically precise metal nanoclusters are a special family of metal nanomaterials that are readily crystallized into single crystals for determining their molecular structures, representing an ideal system to study the surface and interface chemistry of metal nanomaterials including heterometallic nanoparticles. The research on metal nanoclusters never stops in the past several decades because they are an ideal system to investigate the fine structure of solid surfaces. Early studies of heterometallic nanoclusters mainly focused on carbonyl clusters,[1,11–13] and phosphine-stabilized Au-based heterometallic nanoclusters.[14,15] With the development of a two-phase method to prepare organic-protected metal nanoparticles in 1994 by Brust and coworkers,[16] organic-capped metal nanoparticles have been quickly applied to different fields. In the field of organic-protected metal nanoparticles, there is a strong desire among scientists to know their detailed surface structures because the binding structure of capping ligand is vital to the overall properties of organic-capped metallic nanoparticles. In 2007, Kornberg's group determined the first crystal structure of thiolated nanocluster (Au_{102}).[17] Unexpectedly, "staple" −SR−Au−SR− structural units were revealed on the surface of thiolated Au nanoclusters. With the discovery of a series of thiolated Au nanoclusters with crystallographically determined structures, atomically precise metal nanoclusters have been attracting increasing research attention in recent years. The preparation and structural characterizations of more and more clusters with various sizes and structures have been achieved.

Compared with the significant progress in identification of pure Au nanoclusters stabilized, less success has been achieved in the synthesis of heterometallic nanoclusters beyond the early development of phosphine-capped AuAg nanoclusters.[14,15,18] However, the syntheses and structure characterization of structurally more diverse heterometallic nanoclusters are important to create model systems to better understand the surface and interface chemistry of functional heterometallic nanoparticles having larger size but hardly identified their surface structures. There are common structure features among the reported heterometallic nanoclusters: (1) the surface metal atoms have to be bound by ligands, (2) the metal cores are generally not a fragment of face-centered cubic metals, (3) the arrangement of surface metal atoms is highly related to the surface ligands.[19,20] Due to the small size of nanoclusters and the involvement of a large number of surface ligands, the globally random distribution of metals is not commonly observed in organic-protected heterometallic nanoclusters. As shown in Figure 1(b), the metal atoms in heterometallic nanoclusters are mainly geometrically arranged in the following three situations: (1) distributed in a core−shell fashion, (2) orderly stacked, (3) randomly distributed in some region. The metal distributions are highly determined by the ligands used in the clusters' syntheses.

While important to many of their properties, the structure engineering of organic-capped heterometallic nanoclusters, complicated by the different properties of metals and surface-capping ligands, is challenging. We feel it is necessary to summarize what have been done on the surface controlling of organic-capped heterometallic nanoclusters. Considering general synthesis and characterization techniques are provided in other chapters of this book, this chapter will focus our discussions on the synthetic and structure controlling strategies for organic-capped heterometallic nanoclusters with atomically precise structures, particularly phosphine- and thiol-protected heterometallic nanoclusters.

4.2 SYNTHETIC STRATEGIES TOWARD HETEROMETALLIC NANOCLUSTERS

To bring two or more different metal atoms together to form heterometallic nanoclusters (Figure 2), two major synthetic strategies have been developed during the past decades: (1) coreduction of different metal precursors by a strong reducing agent in solutions, (2) rearrangements of premade monometallic or multimetallic nanoclusters. Metal precursors used in the synthesis of metal nanoclusters could be simple metal salts (e.g., $HAuCl_4$, $AgNO_3$) or complexes (e.g., $ClAuPPh_3$). Different metals have different redox potentials. During the reduction process, the metal species with the higher redox potentials are generally reduced first. In the coreduction method, the most challenging issue to prepare heterometallic clusters is thus to avoid the separation of reduction processes of two different metal precursors. The strong reducing agent, $NaBH_4$, has been widely demonstrated as an excellent reducing agent

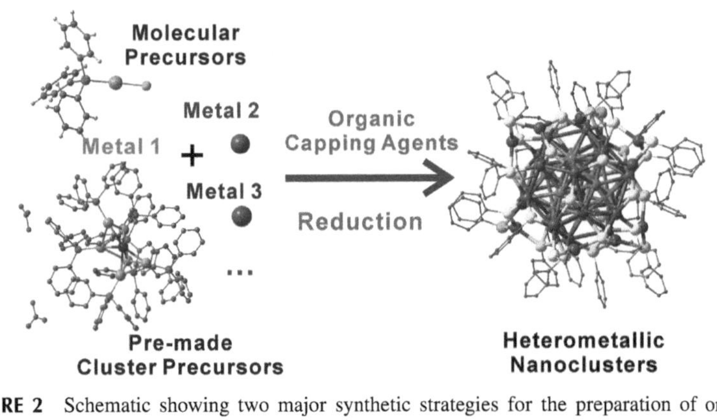

FIGURE 2 Schematic showing two major synthetic strategies for the preparation of organic-capped heterometallic nanoclusters using molecular or premade cluster precursors. (see color plate).

for the synthesis of heterometallic nanoclusters. The major driving force to bring different metal atoms together is the cohesive energy due to the formation of metal–metal bonds. The use of strong reducing agents helps to simultaneously reduce different metal precursors for the formation of metal–metal bonds. However, to stabilize the reduced metal species as metallic nanoclusters instead of poorly defined nanoparticles, the involvement of strong capping ligands, such as phosphines and thiolates, are typically required during the synthesis of metal nanoclusters.

4.2.1 Coreduction of Metal Precursors to Heterometallic Nanoclusters

Teo et al. have successfully prepared a series of phosphine-protected Au–Ag heterometallic nanoclusters by reducing a mixture of gold and silver complexes (e.g. $Au(PR_3)X$, $Ag_2(PR_3)_2X_2$) with $NaBH_4$ in ethanol.[21–38] One of the latest phosphine-protected Au–Ag nanoclusters reported is $(Me_2PhP)_{10}Au_{12}Ag_7(NO_3)_9$, which was prepared by coreduction of $Au(NO_3)(PMe_2Ph)$ and $AgNO_3$ with $NaBH_4$ in ethanol.[39] To have Pd doped into the bimetallic clusters, zero-valent metal precursors, such as $Pd(PPh_3)_4$, were also used in the synthesis.[15,40] For example, Pignolet et al. reported the synthesis of $[Au_8Pd(PPh_3)_8]^{2+}$ by reacting $Pd(PPh_3)_4$ and 3 equiv of $Au(PPh_3)NO_3$ in CH_2Cl_2 with 2 equiv of $NaBH_4$ in MeOH at room temperature.[40] $NaBH_4$ is soluble in alcohols and water, but insoluble in organic solvents such as benzene, toluene, and chloroform. Large bimetallic nanoclusters do not have good solubilities in ethanol and are typically obtained as precipitates or microcrystallites from the reaction mixtures. In order to obtain big single crystals for structure determination, typically careful recrystallization processes are still required.

The one-phase reduction method has been extremely successful in the synthesis of phosphine-stabilized bimetallic nanoclusters. However, less success has been achieved in preparing thiolated metal nanoparticles or nanoclusters using the one-phase method. One very important contribution of the two-phase Brust reduction method developed in 1994 is that the method has nicely resolved the solubility issue.[16] In the developed two-phase method, inorganic metal precursors are transferred into an organic phase with the help of a phase-transferring agent, for example, tetraoctylammonium bromide. $NaBH_4$ is dissolved in iced water and reduces metal precursors in the organic phase through the two-phase interface. The produced organic-capped nanoparticles/nanoclusters are dispersed in the organic phase while the by-products are left in the water phase. During the past several years, the two-phase method has been demonstrated as a very effective strategy for the preparation of monodisperse thiolate-protected monometallic and multimetallic nanoclusters as well. For example, the two-phase method led to the successful synthesis and structure determination of the hydrophobic thiolated Au nanocluster,

$[Au_{25}(SR)_{18}]^-$.[41,42] A series of hydrophobic $[Ag_{44}(SR)_{30}]^{4-}$ clusters were also prepared by using a modified two-phase method.[43] Several groups have been able to obtain $Pd_1Au_{24}(SR)_{18}$/$Pt_1Au_{24}(SR)_{18}$ from the coreduction of Au and Pd/Pt salt precursors although the doped cluster was mixed together with many other species.[44–47] Interestingly, the $Pd_1Au_{24}(SR)_{18}$ composition was obtained regardless of the starting ratio of Au(III) to Pd(II) salt in the synthesis, showing that only one Pd atom can be doped into the 25-atom Au nanocluster. More recently, the two-phase method has been improved by excluding the use of phase-transferring agents for the preparation of bimetallic nanoclusters.[48–50] Two series of thiolated AuAg and AuCu clusters, $[Au_{12}Ag_{32}(SR)_{30}]^{4-}$ and $[Au_{12+n}Cu_{32}(SR)_{30+n}]^{4-}$ ($n = 0, 2, 4, 6$), have been successfully prepared in two-phase systems by reducing the mixture of $AuPPh_3Cl$ and $AgNO_3$/$Cu(ClO_4)_2$ in $CHCl_3$/MeOH with $NaBH_4$ in iced water.[48–50]

While $NaBH_4$ is a strong reducing agent, it easily hydrolyzes in water particularly in the presence of metals. This means that $NaBH_4$ is not a stable reducing agent that allows long-time running of reduction process involving the use of water. In two-phase systems, however, the long running time is important to fully reduce metal precursors and thus produces nanoparticles/nanoclusters in high yields. Two strategies have been developed to prevent the fast decomposition of $NaBH_4$ during synthesis: (1) carry out the synthesis in ice bath, (2) introduce amine (e.g., triethylamine) to stabilize the reducing agent as a more stable borane–amine complex,[51] thus providing a more sustainable reducing atmosphere. The successful high-yield one-pot synthesis of $[Au_{12}Ag_{32}(SR)_{30}]^{4-}$ nanoclusters on a scale of tens of grams has nicely demonstrated the importance of introducing amine in the synthesis of metal nanoclusters.[48]

Utilizing the reduction chemistry to bring different metal atoms together is an effective strategy to create bimetallic nanoclusters. Besides $NaBH_4$, other reducing agents have also been utilized in the synthesis of bimetallic nanoclusters, for example, $[(PPh_3)Pt(AuPPh_3)_6]^{2+}$ was made by the reaction of $[(PPh_3)_2(NO_3)Pt(AuPPh_3)_2]^+$ and H_2 with CH_2Cl_2 as the solvent.[52] $[Pt(H)(PPh_3)(AuPPh_3)_7]^{2+}$ was prepared by the reaction of H_2 with $Au(PPh_3)NO_3$ and $Pt(PPh_3)_3$.[53]

4.2.2 Preparation of Heterometallic Nanoclusters from Premade Clusters

With the help of reducing agents, atoms of one metal can be deposited onto a premade cluster of another metal as well.[15] For example, $(Ph_3P)_{10}Au_{11}Ag_{12}Pt_2Cl_7$ was prepared by reducing a mixture of $[(Ph_3P)_7Au_6Pt](NO_3)_2$ and $(Ph_3P)_4Ag_4Cl_4$ with $NaBH_4$.[29] The reduction of a mixture of $[(Ph_3P)_8Au_8Pt](NO_3)_2$ and $(Ph_3P)_4Ag_4Cl_4$ with $NaBH_4$ in ethanol led to the formation of $[(Ph_3P)_{10}Au_{12}Ag_{12}PtCl_7]Cl$.[30] Using a similar method, $[(Ph_3P)_{10}Au_{12}Ag_{12}NiCl_7]SbF_6$ was synthesized from $[(Ph_3P)_8Au_8Ni](NO_3)_2$

and $(Ph_3P)_4Ag_4Cl_4$ in the presence of $NaBH_4$.[35] $(Ph_3P)_6Au_6Ag_8PtI_6$ was synthesized by reducing a mixture of $[(Ph_3P)_8Au_7PtH](NO_3)_2$ and $[Ph_3PAgI]_4$ in ethanol using $NaBH_4$ over a period of 30 h.[38]

Moreover, the difference in electronegativity between the two metals allows one metal to serve as an effective reducing agent to reduce the other metal salts, forming intermetallic clusters. Pradeep et al. reported the synthesis of $Ag_7Au_6(SR)_{10}$ by galvanic exchange reaction of $Ag_{7,8}$ with $HAuCl_4$.[54] Zhu and coworkers have recently demonstrated that the reactions of phosphine-capped polydisperse Au nanoparticles or atomically precise Au nanoclusters with the $Ag(I)SC_2H_4Ph$ complex led to the formation of rod-like $[Ag_xAu_{25-x}(PPh_3)_{10}(SC_2H_4Ph)_5Cl_2]^{2+}$ clusters.[55]

Starting from premade metal nanoclusters, heterometallic nanoclusters with more diverse structures and compositions were readily prepared by simple electrophilic or nucleophilic addition reactions as well.[15] The addition of an electron acceptor (e.g., Ag^+, Cu^+) to centered Au or mixed Pt–Au nanoclusters can lead to the formation of electrophilic addition products. $Pt(AuPPh_3)_8^{2+}$ and $Pt(CO)(AuPPh_3)_8^{2+}$ reacted with Ag^+ to form the addition products $PtAg(AuPPh_3)_8^+$ and $Pt(CO)Ag(AuPPh_3)_8^+$, respectively.[56] Similarly, the reaction of $[(PPh_3)Pt(AuPPh_3)_6]^{2+}$ and an electron donor (CO) in CH_2Cl_2 yielded $[(PPh_3)(CO)Pt(AuPPh_3)_6]^{2+}$ as the electrophilic addition product.[52]

4.3 LIGAND-INDUCED STRUCTURAL ENGINEERING OF HETEROMETALLIC NANOCLUSTERS

Compared to heterometallic nanoparticles having larger particle size, a large portion of metal atoms in heterometallic nanoclusters are located on their surfaces. Without the appropriate protection by ligands, these small-sized clusters have high surface energy and are easily aggregated or even sintered into larger particles. Thus, protecting ligands need to be introduced during the preparation of stable heterometallic nanoclusters. The ligands introduced in the systems often exhibit quite different coordination behaviors to different metals. Therefore, the use of different types of surface-capping ligands creates great opportunities for us to engineer the structures of heterometallic nanoclusters. In this section, the influence of two different ligands (i.e., phosphines and thiolates) on the structures of heterometallic nanoclusters will be mainly discussed. As for early work on phosphine-capped heterometallic nanoclusters, since there have been several excellent reviews,[14,15,18] we will not discuss too much the detailed structures of those clusters but mainly focus the discussions on how surface ligands help to engineer their surface structures.

4.3.1 Phosphine-Stabilized Heterometallic Nanoclusters

Metallic clusters are typically prepared by chemically reducing metal precursors with high valence under the protection of appreciate ligands which could be simply the ligands on the metal precursors. For instance, $ClAuPPh_3$ has been well demonstrated as an effective Au precursor to prepare Au-containing heterometallic nanoclusters, particularly in the early work starting from 1980s.[14,15] Phosphine-stabilized nanoclusters have thus developed as an important class of heterometallic nanoclusters. Phosphine has two valence electrons and empty 3d orbitals, making it an excellent ligand to various metals. In the case of Au, phosphine coordinates to Au with atop mode. One phosphine coordinates to only one Au atom. In chemically prepared organic ligand-stabilized clusters, the average valence of metal atoms is typically between 0 and 1, which is also true for phosphine-capped heterometallic nanoclusters. Since phosphines are normally neutral ligands, negatively charged components have to accompany with phosphine-stabilized metallic clusters to balance the positive charge from the metal components. The negatively charged components may be in the forms of anionic ligands coordinating to metal sites or counteranions of the clusters. As a result, halides (i.e., Cl^-, Br^-, I^-), NO_3^-, PF_6^-, or SbF_6^-, some of which are derived from the metal precursors, are commonly associated with phosphine-stabilized metallic clusters.

Different ligands have different coordination structures and strengths with different metal atoms. The copresence of both phosphine and anionic ligands helps to control the surface structure of metal clusters. So far, monophosphine, such as triphenylphosphine, tris-*p*tolylphosphine, dimethylphenylphosphine, and triethylphosphine; and diphosphine ligands have been used in the synthesis of heterometallic nanoclusters.[23–25,57–59] A large number of phosphine-stabilized heterometallic nanoclusters have been prepared and structurally identified particularly in 1980s and 1990s.[15,21–40,52,53,56–80] The clusters can be classified into the following two different types: (1) clusters based on icosahedral units, (2) clusters containing less than 13 metal atoms. When containing more than 13 metal atoms, clusters adopt icosahedra as their main building units. The icosahedral building blocks are readily connected together by vertex-sharing, edge-sharing, face-sharing, and even interpenetrating to form large clusters. Teo and coworkers have put forward the concept of "cluster of clusters" for the growth of a series of phosphine-capped heterometallic nanoclusters.[14,31]

By sharing a common vertex, two centered icosahedra are readily joined together to form rod-like clusters containing 25 metal atoms (Figure 3). Several examples are $[(R_3P)_{10}Au_{11}Ag_{12}M_2X_{9-m}]^{m+}$ (R = *p*-Tol_3P, Ph_3P, Me_2PhP; X = Br, Cl; M = Au, Pt, Ni), $(Ph_3P)_{10}Au_{10}Ag_{13}Pt_2Cl_7$, and $[(Ph_3P)_{12}Au_{13}Ag_{12}Cl_6]^+$.[14,21–30] As shown in Figure 3(a) and (b), the 25 metal atoms in the clusters are classified into 3 different positions: (1) 3 atoms embedded at the center, (2) 20 atoms on the side faces of the rod, (3) 2 atoms at the 2 tips. Except the three 12-coordinated atoms at the centers, all the other 22 metal atoms are capped either by phosphine or halide ligands. While the 3

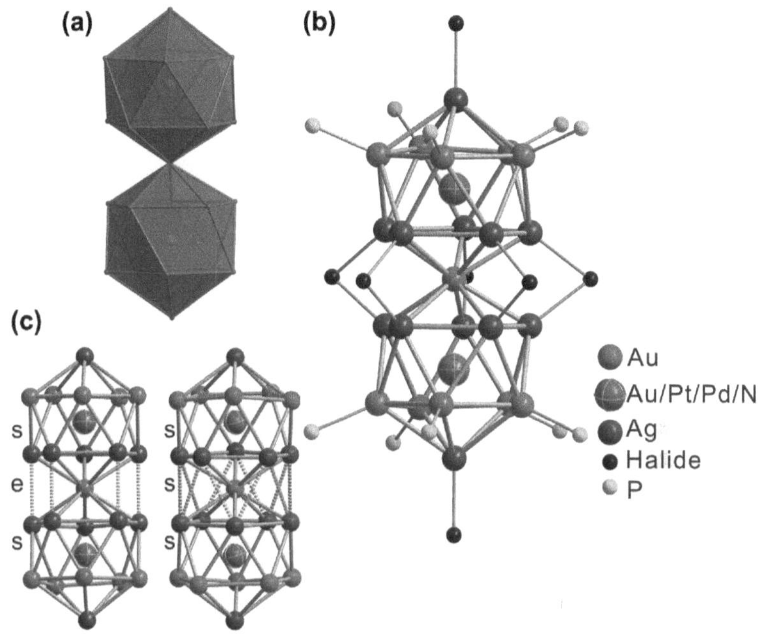

FIGURE 3 The structural anatomy of phosphine-capped Au-based heterometallic nanoclusters containing 25 metal atoms. (a) The structural description of the clusters as vertex-sharing bi-icosahedra. (b) The preference sites of different metal atoms and ligands in the clusters. (c) The two commonly observed arrangements of two icosahedral units in the clusters.

central atoms can be Au, Pt, Pd, or Ni, the other 22 atoms are arranged in an ordered fashion. In all of the reported phosphine-capped 25-atom clusters, the 2 tips are Ag atoms. Four-stacked Au_5–Ag_5–Ag_5–Au_5 pentagonal rings are sandwiched by the two Ag tips. The involvement of two different types of ligands (i.e., phosphine and halide) significantly plays an important role in determining the distribution of surface metal atoms in the clusters. Each surface Au atom is protected by one phosphine ligand. Halide ligands are binding to the surface Ag atoms in two different structures: (1) two terminal halides are located at the two Ag tips, (2) two-, three-, or four-coordinated halides are found between the two Ag_5 pentagonal rings.

Besides the effect on surface metal distribution, the properties of surface ligands also affect the overall framework structure and even detailed local structure of the clusters. The use of different phosphines in the synthesis of phosphine-capped 25-metal-atom heterometallic nanoclusters resulted in the formation of clusters with different relative configurations of the two centered icosahedra (Figure 3(c)). In $[(p\text{-Tol}_3P)_{10}Au_{13}Ag_{12}Br_8]^+$, the four adjacent metal pentagons are staggered (hereafter designated as s) with respect to one another, giving rise to the staggered–staggered–staggered (sss) configuration.[21] When $(p\text{-Tol}_3P)AuBr$ and $[(p\text{-Tol}_3P)AgBr]_4$ in the synthesis were replaced by $(Ph_3P)_2Ag_2Br_2$ and $(Ph_3P)AuBr$, respectively, $[(Ph_3P)_{10}Au_{13}Ag_{12}Br_8]^+$ was obtained.[23] The two inner Ag_5 pentagons of the cluster are eclipsed, giving a staggered–eclipsed–staggered (ses) configuration. While the sss configuration in the cationic $[(p\text{-}Tol_3P)_{10}Au_{13}Ag_{12}Br_8]^+$ gives rise to icosahedral (or bicapped pentagonal antiprismatic) cages, the ses configuration gives rise to bicapped pentagonal prismatic cages.[23] The rotation of two Au_7Ag_6 icosahedra leads to the variation of the distances between adjacent Ag atoms in the two inner Ag_5 pentagons. The surface of the two inner Ag_5 pentagons can accommodate five, six, and even seven halide ligands. Doubly (d), triply (t), and quadruply bridging halides have been observed in the reported phosphine-capped 25-atom heterometallic nanoclusters. In the cases of clusters containing six halide bridging ligands, the halide arrangements observed so far are qd_5, *ortho*-q_2d_4, *ortho*-t_2d_4, *para*-d_2t_4, and *para*-t_2d_4.[21–23,27,28] Moreover, with the less bulky $MePh_2P$ (in comparison with Tol_3P on $[(p\text{-}Tol_3P)_{10}Au_{13}Ag_{12}C1_7]^{2+}$ or Ph_3P on $[(Ph_3P)_{10}Au_{13}Ag_{12}Br_8]^+$), the two inner Ag_5 pentagons in $(MePh_2P)_{10}Au_{13}Ag_{12}Br_9$ readily accommodate seven halide ligands.[23–25] Interestingly, the successful synthesis of $[(Ph_3P)_{12}Au_{13}Ag_{12}Cl_6]^{m+}$ having 12 phosphine ligands on its surface was also reported. The cluster is unique with the binding of one phosphine (instead of halide) to each of the two apical Ag atoms.[26]

Phosphine and halide prefer to bind on Au and Ag, respectively. Similar ligand effects are also well documented in the phosphine-stabilized clusters containing 37 or 38 metal atoms.[31,32] Those 37/38-metal-atom clusters are derivatives of a 36-metal-atom unit featuring three vertices-sharing, centered icosahedra. The overall structure of a 37/38-metal-atom cluster can be viewed as three centered icosahedra sharing three vertices plus one/two capping metal atoms located on the idealized threefold axis (Figure 4). In the 37/38 atoms bimetallic clusters, there are the following four different metal sites: (1) 3 centered atoms of the three icosahedra, (2) 3 shared vertices by the three icosahedra, (3) 30 atoms on the surface, (4) 1 or 2 atoms at the capping sites. While both the 3 centered atoms and the 3 shared vertices are 12-coordinated, all the other metal atoms are capped either by phosphine or halide ligands. In all of the reported phosphine-capped 37/38-atom clusters, the capping sites are Ag atoms. The 36 metal atoms in the three vertices-sharing, centered icosahedra are arranged in an ordered fashion. Similar to heterometallic nanoclusters containing 25 metal atoms, the three centered atoms can be Au, Pd, Pt, or Ni. The three shared vertices are Au atoms. Except the centered atoms, all the other atoms bound to the three shared vertices are Ag atoms. In the core 36-metal-atom unit, there are 18 Ag atoms. The remained 12 sites are occupied by Au. The coordination chemistry of phosphine-capped 37/38-metal-atom clusters is quite similar to that of phosphine-capped heterometallic nanoclusters containing 25 metal atoms. While the surface Au atoms are coordinated by phosphines, the Ag sites are mainly bound by halide ligands. In some

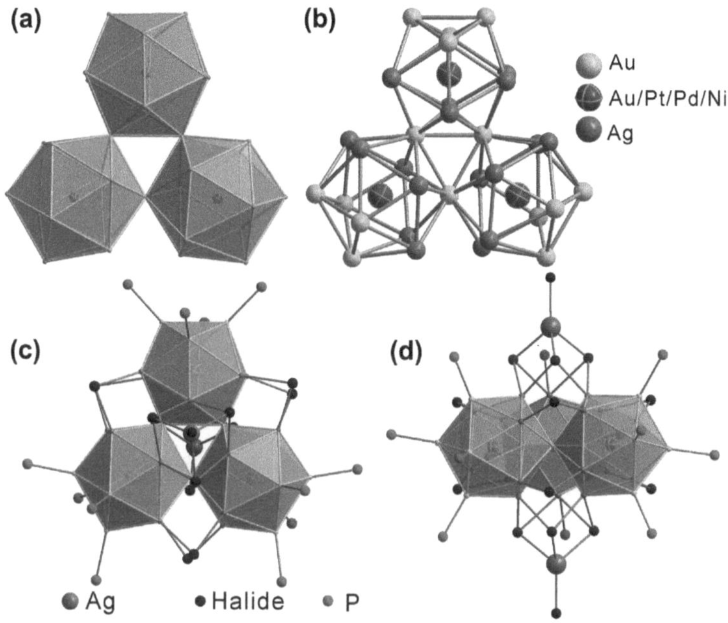

FIGURE 4 The structural anatomy of phosphine-capped Au-based heterometallic nanoclusters containing 37/38 metal atoms. (a) The 36-metal-atom unit featuring three vertices-sharing, centered icosahedra. (b) The preference metal distribution in the clusters. (c and d) The top and side view of the $[(Ph_3P)_{14}Au_{18}Ag_{20}Cl_{12}]^{2+}$ cluster. All carbon atoms are omitted for clarity. (see color plate).

structures, the binding of phosphine on the capping Ag sites was also found possible, for example, in the case of $[(R_3P)_{14}Au_{18}Ag_{20}Cl_{12}]Cl_2$.[32–34]

In the structures of vertex-sharing 25-metal-atom clusters, halides bridge two icosahedra together through multiply coordinating to silver atoms. When halides were excluded in the synthesis of phosphine-capped heterometallic nanoclusters, 19-metal-atom clusters featuring an interpenetration of two centered icosahedra were also prepared.[39] In the synthesis of $(Me_2PhP)_{10}Au_{12}Ag_7(NO_3)_9$ and $Au_{17}Ag_2(PMe_2Ph)_{10}(NO_3)_9$, $Au(NO_3)(PMe_2Ph)$ and $AgNO_3$ are used instead of halide-containing metal precursors (e.g., $HAuCl_4$, R_3PAuX).[39] The 19 metal atoms in the clusters are classified into three different positions (Figure 5(a)): (1) 2 atoms embedded at the center, (2) 15 atoms on the side faces of the rod, (3) 2 atoms at the 2 tips. Except the two 12-coordinated atoms at the centers, all the other 17 metal atoms are capped either by phosphine or nitrate ligands. As in the structure of these two clusters, three metal pentagonal rings with a staggered–staggered (ss) configuration are sandwiched between the two Ag tips. While the inner pentagon can be Au_5 or Ag_5, the exterior pentagons and the two center atoms are Au atoms. As shown in Figure 5(b), while the 10 phosphine ligands

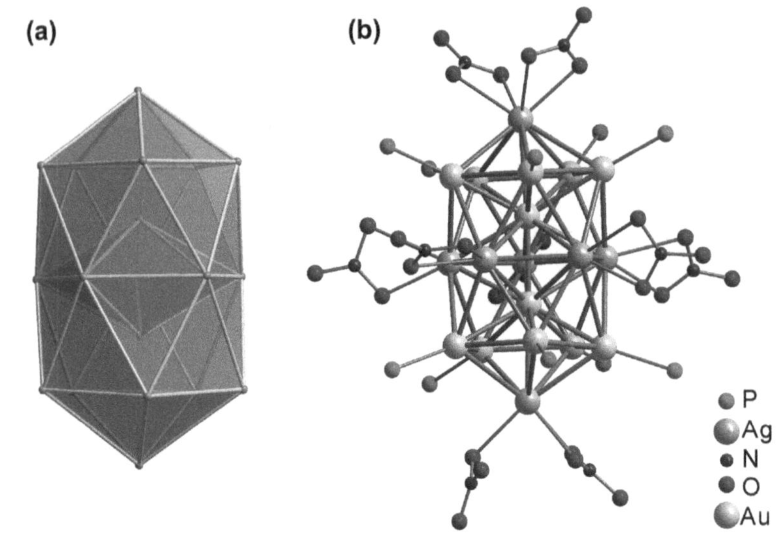

FIGURE 5 The structure of the $(Me_2PhP)_{10}Au_{12}Ag_7(NO_3)_9$ cluster. (a) The two interpenetrating icosahedral units of the metal framework. (b) The detailed molecular structure of the cluster. All carbon atoms are omitted for clarity. (see color plate).

coordinate to 10 peripheral Au atoms, the 9 nitrate ions coordinate to 7 Ag atoms in $(Me_2PhP)_{10}Au_{12}Ag_7(NO_3)_9$. Each of the two tip Ag atoms is coordinated by two nitrate ions, and each Ag atom in the inner Ag_5 pentagon is bound by one nitrate ion. It is also possible to have nitrate coordinated to surface Au sites if surface Ag atoms are substituted with Au. For example, the binding of NO_3^- to Au atoms was observed in $Au_{17}Ag_2(PMe_2Ph)_{10}(NO_3)_9$.[39]

Electronegativity and cohesive energy have been considered as two of important factors that determine the preferential sites of different metals in organic-capped heterometallic nanoclusters.[14] Metals with high cohesive energy prefer to occupy high-coordinated sites such as centers of icosahedra. The more electronegativity of Au than Ag results in the preferential binding of phosphines and halides on Au and Ag, respectively. Together with various coordination modes between different ligands and metals, the ligand−metal interactions play important roles in tuning the metal distribution and the overall structure of organic-capped heterometallic nanoclusters. In the structures of phosphine and halide costabilized Au−Ag−Pt clusters, Pt-rich core and Au/Ag-rich surface were often observed,[29,30,35,67] which can be explained by a higher cohesive energy of Pt than Au and Ag. However, when CO is introduced in the synthesis, the exposure of Pt on the surface of heterometallic nanoclusters is possible due to the much stronger interaction between CO and Pt. Such a ligand effect in controlling the metal distribution has also well documented in phosphine-capped heterometallic nanoclusters containing less than 13 metal atoms.[52,79]

4.3.2 Thiolate-Stabilized Heterometallic Nanoclusters

Different from phosphine ligands, thiolates are anionic ligands and can thus help to balance the positive charges on metal components without the need to further introduce other anionic ligands. More importantly, thiolate ligands have a much richer coordination chemistry to various noble metals as compared with phosphines. As widely reported in the literature, thiolates (SR) on Au clusters mainly serve as two-coordinated surface ligands, which is an important structure feature for creating $-Au_n(SR)_{n+1}-$ staple surface motifs in thiolated Au nanoclusters.[41,42,81] When coordinated to other metals such as Ag and Cu, the coordination modes of thiolates become very rich. Thiolates readily serve as two-coordinated (μ_2-), three-coordinated (μ_3-), and four-coordinated (μ_4-) ligands, which has been well documented in the systems of thiolated silver and copper chalcogenide nanoclusters.[82–84] More recently, Zheng's group has also nicely demonstrated the similar rich coordination chemistry in Ag(0)-containing nanoclusters.[85,86] A series of Ag clusters coprotected by phosphine/diphosphine were successfully prepared and structurally determined by X-ray single crystal diffraction. The crystal structure analysis of those clusters confirmed that thiolates also display a rich coordination chemistry in Ag(0)-containing nanoclusters. μ_3 thiolates are found in the luminescent $Ag_{14}(SR_{14})(PPh_3)_8$ cluster.[85] Thiolates with three different coordination modes (e.g., μ_2-, μ_3-, and μ_4-) are revealed in $Ag_{16}(DPPE)_4(SR)_{14}$ (DPPE = bis(diphenylphosphino)ethane) and $[Ag_{32}(DPPE)_5(SR)_{24}]^{2-}$ clusters.[86]

The abundant coordination modes of the thiolate ligands are expected to enrich the structures of all-thiol-protected metal clusters as well. In the past decade, much research effort has been paid to prepare thiolated silver clusters. Although all-thiolate-stabilized Ag or Au/Ag nanoclusters with uniform size and even with precise number of atoms have been widely achieved by several research groups,[43,54,87–92] no crystal structure of all-thiolate-protected silver nanoparticles was resolved before the solution of crystal structures of $[Ag_{44}(SR)_{30}]^{4-}$ and $[Au_{12}Ag_{32}(SR)_{30}]^{4-}$ independently by two groups, Zheng and Bigioni.[48,93] The crystallization of Ag(0)-containing clusters into reasonably good single crystals for structure analysis is challenging. In Zheng's studies, bulky countercations (e.g., PPh_4^+, $(PPh_3)_2N^+$) were introduced to promote the crystallization of various hydrophobic nanoclusters into single crystals, representing an important progress toward the resolution of thiolated Ag(0)/Cu(0)-containing nanoclusters.[48,49,85,86,94]

Due to the richer coordination chemistry of thiolates to Ag/Cu than to Au, introducing Ag into the synthetic systems of thiolated Au nanoclusters should significantly alter the structures of AuAg clusters to be obtained. In fact, such an expectation was nicely fulfilled with the successful synthesis and structure characterizations of series of $[Au_{12}Ag_{32}(SR)_{30}]^{4-}$ and $[Au_{12+n}Cu_{32}(SR)_{30+n}]^{4-}$ ($n = 0, 2, 4, 6$) nanoclusters.[48–50] Both families of nanoclusters were synthesized by a modified two-phase method. In a typical synthesis of $[Au_{12}Ag_{32}(SR)_{30}]^{4-}$, $ClAuPPh_3$ or $ClAuPPh_2Py$ (PPh_2Py = diphenyl 2-pyridylphoshpine) that is soluble in organic solvents was used as the Au precursor. The silver precursor could be $AgNO_3$, $AgBF_4$, $AgCF_3COO$, or any other soluble Ag salt. The metal precursors were chemically reduced by $NaBH_4$ in the presence of thiol (e.g., 3,4-difluorobenzenethiol, 4-fluorobenzenethiol, 4-trifluoromethylbenzenethiol), $PPh_4Br/(PPh_3)_2NCl$, and triethylamine in a mixed solvent of dichloromethane and methanol at 0 °C in an ice bath.

With the combination of Au and Ag, as revealed by X-ray crystal structure analysis (Figure 6), $[Au_{12}Ag_{32}(SR)_{30}]^{4-}$ exhibits two unique structural features distinguishing from thiolated pure Au clusters: (1) no staple units are

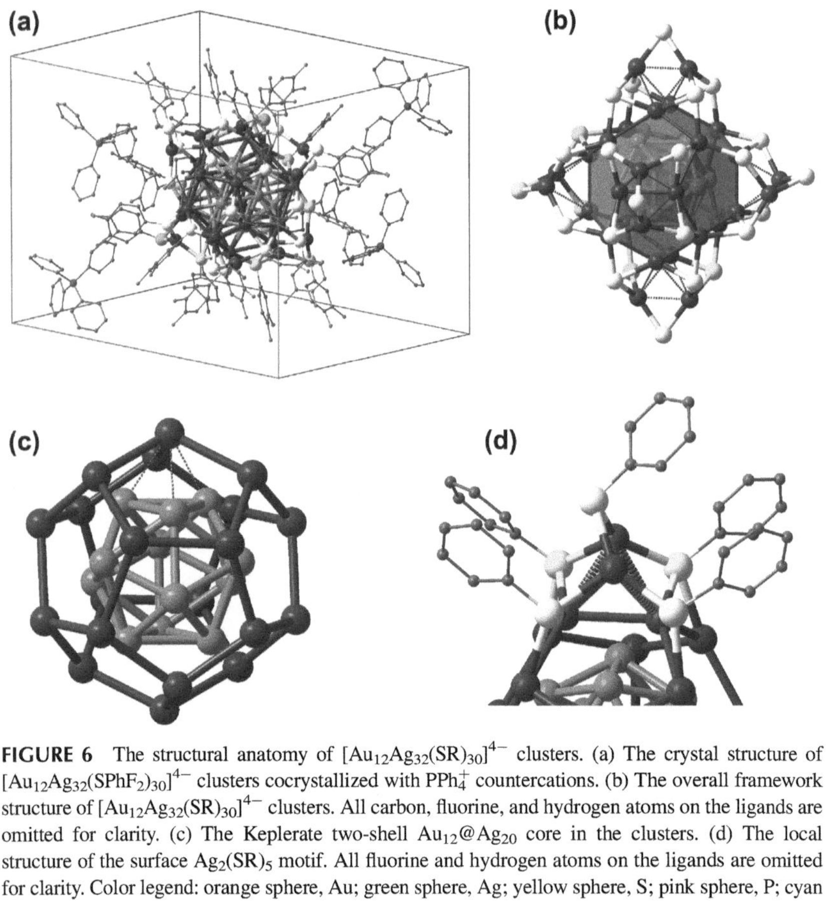

FIGURE 6 The structural anatomy of $[Au_{12}Ag_{32}(SR)_{30}]^{4-}$ clusters. (a) The crystal structure of $[Au_{12}Ag_{32}(SPhF_2)_{30}]^{4-}$ clusters cocrystallized with PPh_4^+ countercations. (b) The overall framework structure of $[Au_{12}Ag_{32}(SR)_{30}]^{4-}$ clusters. All carbon, fluorine, and hydrogen atoms on the ligands are omitted for clarity. (c) The Keplerate two-shell $Au_{12}@Ag_{20}$ core in the clusters. (d) The local structure of the surface $Ag_2(SR)_5$ motif. All fluorine and hydrogen atoms on the ligands are omitted for clarity. Color legend: orange sphere, Au; green sphere, Ag; yellow sphere, S; pink sphere, P; cyan sphere, F; gray sphere, C. (see color plate).

present on their surface, (2) a hollow icosahedral Au_{12} cage is sitting at the center of the clusters. The $[Au_{12}Ag_{32}(SR)_{30}]^{4-}$ cluster can be structurally described as a Keplerate two-shell $Au_{12}@Ag_{20}$ core protected by six $Ag_2(SR)_5$ units on its surface. For a hollow icosahedral Au_{12} cage, there are 20 Au_3 faces. According to previous experimental and theoretical studies, pure hollow icosahedral Au_{12} cages should be unstable and thus need to be stabilized by being encapsulation of transition metal atoms. In $[Au_{12}Ag_{32}(SR)_{30}]^{4-}$, the central hollow icosahedral Au_{12} cage is stabilized with all of its 20 Au_3 faces fully face capped by Ag atoms to form the two-shell $Au_{12}@Ag_{20}$ core which has a nearly perfect I_h point symmetry. The two-shell $Au_{12}@Ag_{20}$ is further protected by six $Ag_2(SR)_5$ units. The structure of surface $Ag_2(SR)_5$ units has not been found in thiolated Au clusters. Both two- and three-coordinated thiolates are present on the surface of $[Au_{12}Ag_{32}(SR)_{30}]^{4-}$ and their pure silver counterparts $[Ag_{44}(SR)_{30}]^{4-}$ as well. The rich coordination chemistry of thiolate to Ag rules out that staple units are the essential surface motifs to stabilize thiolated Au-containing nanoclusters. Moreover, there are no 12-coordinated metal atoms in $[Au_{12}Ag_{32}(SR)_{30}]^{4-}$ clusters due to the presence of the hollow Au_{12} cage at their centers.

Considering Au and Ag have almost identical atomic radii, it is also somewhat surprising that the combination of Au and Ag did not lead to the formation of nanoclusters with randomly distribution of Au and Ag sites. In phosphine-stabilized AuAg clusters, Au and Ag are orderly distributed on their surfaces mainly because of preferential binding of phosphines on Au and halides on Ag. In $[Au_{12}Ag_{32}(SR)_{30}]^{4-}$, Au and Ag are distributed in a core–shell fashion rather than a random fashion. The presence of only Ag atoms on the surface of thiolated AuAg clusters clearly suggests the preferential binding of thiolate on Ag over Au during the chemical reduction process, which is different from the situation in phosphine-stabilized AuAg clusters. This phenomenon further confirms the critical role of ligands in determining the metal distributions in heterometallic nanoclusters and thus their overall structures.

In the periodic table, Cu belongs to the same group as Au and Ag. Cu has a smaller atomic radius (1.28 Å) than Au and Ag (1.44 Å). A series of $Cu_xAu_{25-x}(SC_2H_4Ph)_{18}$ ($x = 1-5$) were synthesized by reacting $HAuCl_4$ and $CuCl_2$ with PhC_2H_4SH in methanol followed by reducing the resulting complexes with $NaBH_4$.[95] When $C_8H_{17}SeH$ was used rather than PhC_2H_4SH, $Cu_nAu_{25-n}(SeC_8H_{17})_{18}$ with up to nine Cu atoms were obtained.[96] No crystal structure was actually resolved for the Cu_nAu_{25-n} clusters, making it impossible to evaluate their structure difference with their $Au_{25}(SR)_{18}$ counterparts. Fortunately, when a similar synthetic strategy of $[Au_{12}Ag_{32}(SR)_{30}]^{4-}$ was adopted to synthesize all-thiol-stabilized AuCu bimetallic nanoclusters, a series of $[Au_{12+n}Cu_{32}(SR)_{30+n}]^{4-}$ ($n = 0, 2, 4, 6$) nanoclusters were successfully prepared and recrystallized into single crystals good enough for X-ray analysis (Figure 7).[48,49] This series of

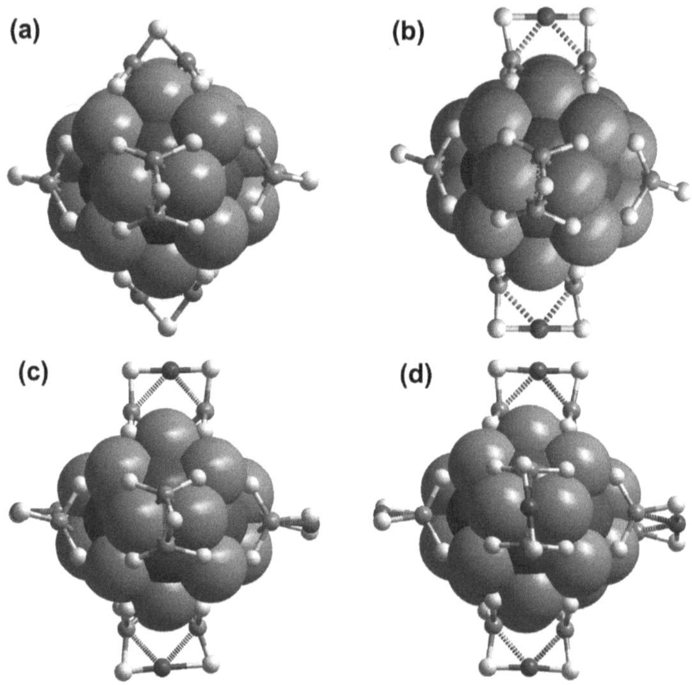

FIGURE 7 The structures of $[Au_{12+n}Cu_{32}(SR)_{30+n}]^{4-}$ clusters, $n = 0$ (a), 2 (b), 4 (c), and 6 (d). The Keplerate two-shell $Au_{12}@Cu_{20}$ cores are depicted in the space-filling model. All carbon, fluorine, and hydrogen atoms on the ligands are omitted for clarity. Color legend: orange sphere, Au; green sphere, Cu; yellow sphere. (see color plate).

$[Au_{12+n}Cu_{32}(SR)_{30+n}]^{4-}$ clusters share both structural similarity and structural difference with $[Au_{12}Ag_{32}(SR)_{30}]^{4-}$ clusters. When $n = 0$, the $[Au_{12}Cu_{32}(SR)_{30}]^{4-}$ cluster has an almost identical framework structure to that of $[Au_{12}Ag_{32}(SR)_{30}]^{4-}$ by simply replacing Ag sites with Cu. However, compared to Ag, Cu has a much smaller atomic radius. Such a difference makes a significant impact in the detailed surface structure of $[Au_{12}Cu_{32}(SR)_{30}]^{4-}$. Similar to $[Au_{12}Ag_{32}(SR)_{30}]^{4-}$, there are six $Cu_2(SR)_5$ units on the surface of $[Au_{12}Cu_{32}(SR)_{30}]^{4-}$. The average Cu−Cu distance in six $Cu_2(SR)_5$ units on the surface is 2.763 Å (2.761−2.781 Å), 8% longer than the sum of Cu atomic radii (1.28 Å), indicating that there is no strong metal−metal interaction within the two Cu atoms of the $Cu_2(SR)_5$ units. The absence of a strong Cu−Cu interaction in the $Cu_2(SR)_5$ units makes it possible to have one Au atom insert between the two Cu atoms during synthesis. Indeed, the structures of $[Au_{12+n}Cu_{32}(SR)_{30+n}]^{4-}$ clusters were found highly sensitive to their synthesis although the protocol was quite similar. A subtle change in the synthesis condition of $[Au_{12}Cu_{32}(SR)_{30}]^{4-}$ could lead to the formation of $[Au_{12+n}Cu_{32}(SR)_{30+n}]^{4-}$ with different n of 2, 4, and 6. These clusters can be better described as a core $Au_{12}@Cu_{20}$ Keplerate cage surface-capped by $6 - n$ $Cu_2(SR)_5$ units and n $Cu_2Au(SR)_6$ units ($n = 2, 4, 6$). There are three pairs of $Cu_2(SR)_5$ units on $[Au_{12}Cu_{32}(SR)_{30}]^{4-}$. Replacing one pair of the surface $Cu_2(SR)_5$ units on $[Au_{12}Cu_{32}(SR)_{30}]^{4-}$ by $Cu_2Au(SR)_6$ units leads to $[Au_{14}Cu_{32}(SR)_{32}]^{4-}$. Further replacing the surface $Cu_2(SR)_5$ units by two or three pairs of $Cu_2Au(SR)_6$ units yields $[Au_{16}Cu_{32}(SR)_{34}]^{4-}$ or $[Au_{18}Cu_{32}(SR)_{36}]^{4-}$ clusters, respectively. Compared to the $Cu_2(SR)_5$ surface motif, the $Cu_2Au(SR)_6$ unit has one Au atom sitting between the two Cu atoms and involves the insertion of an extra thiolate ligand as well. Importantly, the insertion of Au—SR in the surface $Cu_2(SR)_5$ units does not alter the count of valence electrons in the clusters. The different surface structures in the $[Au_{12+n}Cu_{32}(SR)_{30+n}]^{4-}$ clusters could be induced by changing metal precursors and countercations.

For bimetallic nanoalloy systems, the distribution of two metals is important for many of their applications such as catalysis. The atomic radii and surface energy are the major factors influencing the metal arrangements in nanoalloys. For example, theoretical calculations predicted that smaller Cu atoms in AuCu nanoalloys should be located at the sterically confined icosahedral core.[1,97] However, there was so far a lack of direct experimental evidences to support the calculations. Very recently, the crystal structure of $[Au_{25-x}Ag_x(SCH_2CH_2Ph)_{18}]^-$ indicates the Ag atoms are likely located at the 12 vertices of the icosahedral core, and SR—Au—SR staple units are present on the surface of the cluster.[98] It is obvious that more systematic studies are still needed to fully understand the metal distribution in thiolated bimetallic nanoclusters.

Another important phenomenon that needs to be better understood in future studies is regarding the stability of clusters.[50] Based on the crystal analysis, the series of $[Ag_{44}(SR)_{30}]^{4-}$, $[Au_{12}Ag_{32}(SR)_{30}]^{4-}$, and $[Au_{12+n}Cu_{32}(SR)_{30+n}]^{4-}$ ($n = 0, 2, 4, 6$) clusters all have a charge of 4^- which can be balanced by four monovalent countercations (e.g., PPh_4^+, $(PPh_3)_2N^+$, NBu_4^+). According to the superatom electron-counting rules, all of these three series of clusters have 18-electron superatom shell closure in their metal cores. Regardless the difference in their surface-capping thiolates and countercations, these clusters are expected to have similar stabilities. Experimental studies have demonstrated that the thermal stabilities of $[Ag_{44}(SR)_{30}]^{4-}$ and $[Au_{12}Ag_{32}(SR)_{30}]^{4-}$ clusters highly depend on their surface thiolate ligands and countercations as well. 3,4-difluorobenzenethiol was a better surface ligand than 4-(trifluoromethyl) benzenethiol or 4-fluorobenzenethiol to stabilize the clusters.[50] The clusters having more bulky $(PPh_3)_2N^+$ were found less stable than those with PPh_4^+ as countercations.[50] $[Au_{12}Ag_{32}(SR)_{30}]^{4-}$ was more stable than $[Ag_{44}(SR)_{30}]^{4-}$ even with the same surface ligand and countercations. The superatom theory could not provide insights on the stability differences among these clusters. More research efforts are still needed to understand why a small structural change in the structure could result in a significant change in the stability.

4.3.3 Heterometallic Nanoclusters Costabilized by Phosphines and Thiolates

In ligand-stabilized metal clusters, all surface metal sites are typically bound by ligands, such as phosphine, thiolate, halide, CO, and alkyne. As discussed above, the coordination chemistry of ligands on surface metal atoms of metal nanoclusters varies with the types of ligands and their coordinating metals as well, providing us great opportunities to finely tune the structures of heterometallic nanoclusters. Most reported ligand-stabilized heterometallic nanoclusters are protected mainly by simple ligands that do not have multiple coordinating sites. Although ligands, such as halides and thiolates, have different coordination modes, they do not bind to multiple metal sites with different properties. For example, in the case of AuAg nanoclusters costabilized by halide and phosphine, phosphine and halide selectively binds to Au and Ag, respectively. Introducing multiple different binding sites on surface ligands is expected to serve as an effective method to precisely control the binding number and thus surface metal distribution in heterometallic nanoclusters.

Taking clusters based on an icosahedral Au core (hollow Au_{12} or centered Au_{13}) as an example, one should realize that there are 20 Au_3 faces on the surface of the icosahedron. The number of face-capping metal atoms can be varied. In the cases of $[Au_{12}Ag_{32}(SR)_{30}]^{4-}$ and $[Au_{12+n}Cu_{32}(SR)_{30+n}]^{4-}$ clusters, all of the 20 Au_3 faces are face-capped by Ag and Cu, respectively.[48–50] The face-capping of less than 20 Au_3 faces is structurally possible. In $[Au_{25}(SR)_{18}]^-$, 12 of 20 Au_3 faces in the Au_{13} core are capped by Au atoms.[41,42] But the question remains unanswered whether the number of face-capping metal atoms can be precisely controlled. In order to stabilize metallic clusters based on an icosahedral Au core, the 12 vertex atoms of the icosahedral core have to be well protected by ligands or metals due to the low coordination number of these vertex sites. In $[Au_{25}(SR)_{18}]^-$, there is actually one thiolate binding to each of 12 vertex atoms on the icosahedral core. The binding of 12 organic ligands on the 12 vertices could however create different numbers of extra coordination sites if the 12 surface ligands are appropriately selected.

To achieve the incorporation of different numbers of metal atoms onto the surface of centered Au_{13} icosahedron, one has to control the number of coordination sites by introducing multidentate ligands. Such a structure-controlling strategy has been recently developed by Zheng et al.[94] Pyridyl-bearing phosphine or thiolate ligands (e.g., 2-pyridinethiolate (2-SPy) and diphenyl-2-pyridylphosphine (PPh_2Py)) were introduced in their systems to precisely tune the binding sites of 12 surface ligands. As shown in Figure 8(a), when coordinating Au using their sulfur and phosphorous site, 2-SPy and PPh_2Py readily bind to additional metal sites using their pyridyl groups. The number of extra binding sites created on the surface of Au_{13} icosahedron was

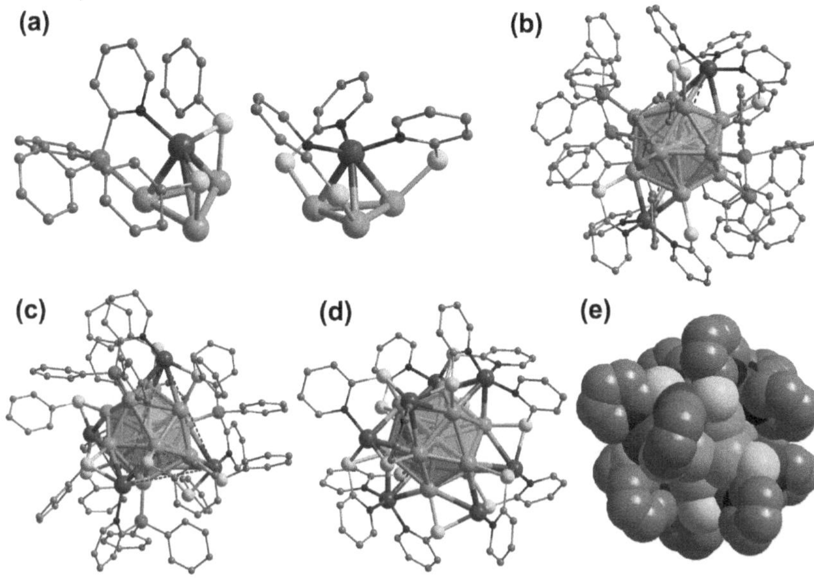

FIGURE 8 The structures of $Au_{13}Cu_n$ ($n = 2, 4, 8$) clusters with different number of surface Cu atoms on centered icosahedral Au_{13}. (a) The local coordination structures of pyridyl-bearing ligands, diphenyl-2-pyridylphosphine (PPh_2Py) and 2-pyridinethiolate (SPy). (b) The structure of $[Au_{13}Cu_2(PPh_3)_6(SPy)_6]^+$ stabilized by phosphine and pyridyl-bearing thiolate. (c) The structure of $[Au_{13}Cu_4(PPh_2Py)_4(SC_6H_4\text{-}tert\text{-}C_4H_9)_8]^+$ capped by thiolate and pyridylphosphine. (d, e) The stick-and-ball (d) and space-filling (e) structure models of $[Au_{13}Cu_8(SPy)_{12}]^+$. All hydrogen atoms are omitted for clarity. Color legend: orange sphere, Au; green sphere, Ag; yellow sphere, S; pink sphere, P; gray sphere, C. (see color plate).

systematically tunable by the selective use of surface phosphine and thiolate ligands with or without pyridyl groups.

In the synthetic experiments, the Au and Cu precursors were mixed together with desired ligands and coreduced by $NaBH_4$ to prepare AuCu bimetallic nanoclusters. When both PPh_3 and 2-pyridinethiol were introduced in the synthesis as the capping agents, the obtained cluster was $[Au_{13}Cu_2(PPh_3)_6(SPy)_6]^+$.[94] As illustrated in Figure 8(b), the surface of the cluster is passivated by six PPh_3 and six 2-SPy. Every PPh_3 or 2-SPy ligand binds to a surface Au atom on 12 vertices of the Au_{13} icosahedron using P or S atom, respectively. The six PPh_3 ligands coordinate (one each) to the six gold atoms at the "equatorial" positions of the Au_{13} core in a radial fashion. The other six vertex Au atoms are bound by the six 2-SPy ligands (one each). The distribution of the six 2-SPy ligands leads to the presence of three pyridyl motifs in each axial direction. The two sets of three pyridyl units hold two Cu atoms to face cap two Au_3 faces at the opposite sides of the cluster. In $[Au_{13}Cu_2(PPh_3)_6(SPy)_6]^+$, each 2-SPy serves as a bidentate ligand with S binding to Au and pyridyl coordinating to Cu. The two face-capping Cu atoms on the cluster are three-coordinated by three pyridyl groups.

When PPh_2Py and SC_6H_4-*tert*-C_4H_9 were used as the capping agents, the synthesis led to the formation of $[Au_{13}Cu_4(PPh_2Py)_4(SC_6H_4\text{-}tert\text{-}C_4H_9)_8]^+$ (Figure 8(c)) with four Cu atoms face-capping four Au_3 faces of the icosahedral Au_{13} core.[94] In $[Au_{13}Cu_4(PPh_2Py)_4(SC_6H_4\text{-}tert\text{-}C_4H_9)_8]^+$, both PPh_2Py and $SC_6H_4\text{-}tert\text{-}C_4H_9^-$ serve as bidentate ligands. While each PPh_2Py binds to Au using its P site and Cu using its pyridyl motif, each $SC_6H_4\text{-}tert\text{-}C_4H_9^-$ coordinates to one Au and one Cu. Therefore, Cu atoms in $[Au_{13}Cu_4(PPh_2Py)_4(SC_6H_4\text{-}tert\text{-}C_4H_9)_8]^+$ are three-coordinated by two S and one pyridyl group, different from the situation in $[Au_{13}Cu_2(PPh_3)_6(SPy)_6]^+$.

Structurally, the number of metal atoms face-capping the triangular Au_3 faces could be varied from 0 to 20. However, the binding of 12 ligands on Au_{13} might result in significant steric hindrance to limit the number of surface face-capping metal atoms that can be incorporated. To increase the number of face-capping metal atoms, less bulky ligands with multiple binding sites are highly desirable. In Zheng's studies, when only SPy^- was introduced in the synthesis of bimetallic AuCu nanoclusters, $[Au_{13}Cu_8(SPy)_{12}]^+$ was prepared successfully and structurally identified.[94] As shown in Figure 8(d), Each SPy^- in the $Au_{13}Cu_8$ cluster binds to one Au through Au–S and two Cu atoms through Cu–S and Cu–N. Together with the 12 binding sites to Au, the 12 SPy^- surface ligands create 24 binding sites (12 S and 12 pyridyls) available for holding 8 Cu atoms face-capping the Au_{13} core. The eight face-capping Cu atoms in $[Au_{13}Cu_8(SPy)_{12}]^+$ adopt both Cu coordination modes revealed in $[Au_{13}Cu_2(PPh_3)_6(SPy)_6]^+$ and $[Au_{13}Cu_4(PPh_2Py)_4(SC_6H_4\text{-}tert\text{-}C_4H_9)_8]^+$. Each of the six Cu atoms that are equatorially distributed is coordinated by two S and one 2-pyridyl. And each of the two oppositely oriented Cu atoms is coordinated by three pyridyl groups. As discussed above, the available binding sites of the surface ligands control the number of metal atoms to be incorporated onto Au_{13}. While leaving more space for holding more face-capping metal atoms, the presence of multidentate ligands with less bulkiness on Au_{13} also makes it possible to access the Au atoms. Compared with $[Au_{13}Cu_2(PPh_3)_6(SPy)_6]^+$ and $[Au_{13}Cu_4(PPh_2Py)_4(SC_6H_4\text{-}tert\text{-}C_4H_9)_8]^+$ that have bulky ligands, the presence of 12 SPy^- ligands on the surface of $[Au_{13}Cu_8(SPy)_{12}]^+$ leaves open space for small molecules to access its Au sites (Figure 8(e)).

Phosphine and thiolate have comparable binding capabilities to Au. The specific method applied in the synthesis of bimetallic clusters of Au is sometimes very critical in determining the metal distribution and thus their properties of the prepared bimetallic clusters. Zhu and coworkers have recently demonstrated that the use of different Au precursors in the synthesis of $[Ag_xAu_{25-x}(PPh_3)_{10}(SC_2H_4Ph)_5Cl_2]^{2+}$ clusters (Figure 9) made a significant impact on the number of Ag that could be incorporated into the clusters.[55] The clusters were prepared by reacting premade PPh_3-protected Au

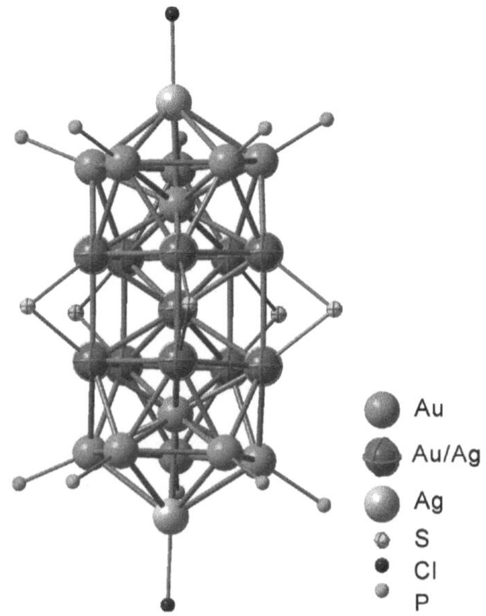

FIGURE 9 The structure of the $[Ag_xAu_{25-x}(PPh_3)_{10}(SC_2H_4Ph)_5Cl_2]^{2+}$ cluster. All carbon and hydrogen atoms on the surface ligands are omitted for clarity.

nanoparticles or nanoclusters with Ag(I) thiolate. When polydisperse PPh_3-capped Au nanoparticles were used as the precursor, the obtained Ag_xAu_{25-x} clusters contained Ag atoms less than 12. In comparison, when $[Au_{11}(PPh_3)_8Cl_2]^+$ was used as the Au precursor, the obtained Ag_xAu_{25-x} clusters could have more than 12 Ag atoms.

4.4 PROPERTIES OF ORGANIC-PROTECTED HETEROMETALLIC NANOCLUSTERS

The use of appropriate ligands controls the detailed structure of heterometallic nanoclusters at different aspects: (1) the distribution of metal atoms, (2) the exposure of metal sites on clusters, (3) the number of secondary metal atoms that can be incorporated in structurally well-defined clusters. Such structural tunings induced by surface ligands make influences on the overall properties of heterometallic nanoclusters. Heterometallic nanoclusters usually exhibit optical properties that are different from their homometallic counterparts. As shown in Figure 10, the $[Au_{12}Ag_{32}(SR)_{30}]^{4-}$ clusters display UV−Vis absorption spectra that are quite different from the $[Ag_{44}(SR)_{30}]^{4-}$ clusters. While the $[Ag_{44}(SR)_{30}]^{4-}$ clusters display six apparent peaks (at 374, 411, 483, 535, 641, and 833 nm) and two shoulder peaks (at 590 and 689 nm), the $[Au_{12}Ag_{32}(SPhF_2)_{30}]^{4-}$ clusters display two strong but relatively broad peaks at 390 and 490 nm with three weak shoulder peaks at 576, 620, and 728 nm.[48]

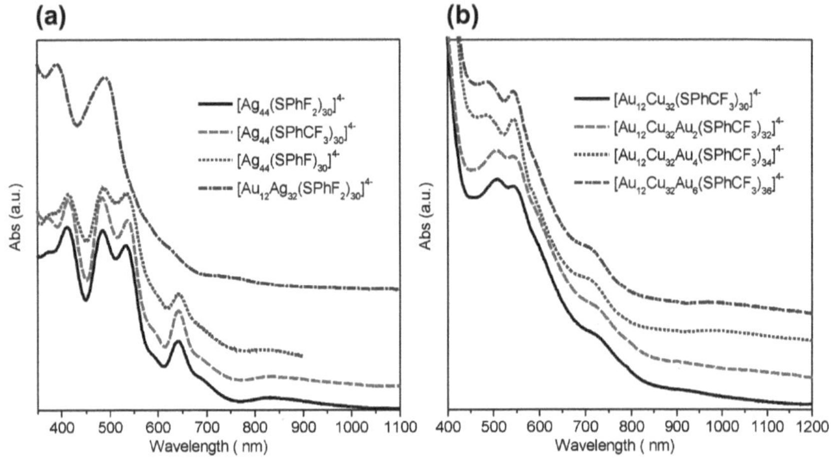

FIGURE 10 The UV–Vis absorption spectra of (a) $[Ag_{44}(SR)_{30}]^{4-}$ and $[Au_{12}Ag_{32}(SR)_{30}]^{4-}$ with different surface thiolates, and (b) $[Au_{12+n}Cu_{32}(SPhCF_3)_{32+n}]^{4-}$ clusters (i.e., $[Au_{12}Cu_{32}(SPhCF_3)_{30}]^{4-}$, $[Au_{14}Cu_{32}(SPhCF_3)_{32}]^{4-}$, $[Au_{16}Cu_{32}(SPhCF_3)_{34}]^{4-}$, and $[Au_{18}Cu_{32}(SPhCF_3)_{36}]^{4-}$).

In the cases of $[Au_{12+n}Cu_{32}(SR)_{30+n}]^{4-}$ clusters, the different number of n did not lead to difference in their optical absorption.[49] These results clearly indicate that the optical properties of metallic nanoclusters are mainly determined by their metal core structure rather than their surface structure. Sometimes, a subtle change in the core structure of heterometallic nanoclusters can result in a dramatic change in their optical properties. For example, $[Ag_xAu_{25-x}(PPh_3)_{10}(SC_2H_4Ph)_5Cl_2]^{2+}$ clusters prepared from different Au precursors led to the formation of clusters containing different number of Ag atoms in their cores.[55] While the Ag_xAu_{25-x} clusters containing Ag atoms less than 12 were found only weakly fluorescent, the Ag_xAu_{25-x} clusters having more than 12 Ag atoms were highly luminescent with a quantum yield up to 40%. Such a significant difference could be induced by only one atom's difference, which is hardly studied in heterometallic nanoparticulate systems.

Heterometallic nanoclusters would be an excellent system to correlate the catalysis with their detailed surface and electronic structures. Doping Pt into $Au_{25}(SR)_{18}$ clusters has been demonstrated as an effective method to tune the electronic structure for optimizing catalysis. $PtAu_{24}(SR)_{18}$ was found much active than $Au_{25}(SR)_{18}$ in the catalytic performance for the styrene oxidation.[47] However, the catalytic reactions on organic-protected metal nanoclusters demonstrated so far are still limited, which is mainly due to the blockage of catalytically active sites by organic capping ligands and instability of clusters under harsh reaction conditions. Thus, there is a huge demand in making metal nanoclusters with exposed metal sites while protected by organic ligands. In the series of $Au_{13}Cu_n$ clusters ($n = 2, 4, 8$), Zheng and coworkers have demonstrated the importance to utilize strong but less bulky ligands to create accessible metal sites for catalysis. In the cases of $Au_{13}Cu_2$ and $Au_{13}Cu_4$ clusters, the presence of bulky ligands (e.g., PPh_3, PPh_2Py, 4-*tert*-butylbenzylthiolate) on their surface makes the metal sites inaccessible for catalysis. With less bulky ligands, SPy^-, on its surface, $[Au_{13}Cu_8(SPy)_{12}]^+$ was highly active in the hydrogenation of 4-nitrophenol by sodium borohydride ($NaBH_4$) to 4-aminophenol, a model reaction to evaluate the accessibility of Au sites in the three clusters.[94] This result indicates that it is still possible to create open metal sites for catalysis on organic-capped heterometallic nanoclusters when appropriate ligands are chosen.

4.5 SUMMARY

In the past three decades, increasing research attention has been directed to the synthesis of organic-capped heterometallic nanoclusters. Phosphines and thiolates have been demonstrated as effective surface ligands to stabilize various heterometallic nanoclusters containing tens of metal atoms. In this chapter, we have mainly summarized the work that has been done on phosphine- and thiolate-protected heterometallic nanoclusters, particularly Au-based nanoclusters. The binding structure of ligands varies from one metal to another even among the metals in the same group of the periodic table. The use of different ligands in the synthesis of heterometallic nanoclusters provides us an effective strategy to engineer the structures of organic-capped heterometallic nanoclusters. When phosphines were involved, the structures of heterometallic nanoclusters obtained in the early studies were mainly based on icosahedral units. Two important structures are 25- and 37/38-metal atom clusters, which can be structurally described as cluster-of-clusters by jointing small-centered icosahedral M_{13} units together into the large clusters. Since phosphines are neutral terminal ligands, anionic ligands (e.g., halides) or counteranions have to be involved in the structures of phosphine-protected heterometallic nanoclusters. In the reported phosphine-protected heterometallic nanoclusters, the preferential binding of halides to Ag and phosphines to Au defines the ordering of metal atoms, particularly the surface atoms, on the clusters. Compared with phosphines, thiolates are a class of anionic ligands exhibiting a richer coordination chemistry to various metals. During the past 7 years, the use of thiolates as surface-capping agents has led to a significant progress in preparing a large family of thiolated Au clusters with atomically precise structures. Very recently, the rich coordination chemistry of thiolates has been successfully applied to enrich the structures of heterometallic nanocluster as well. Several series of heterometallic nanoclusters having a core–shell distribution pattern of metal atoms have been prepared and structurally identified. The core–shell distribution of metals in thiolated heterometallic nanoclusters is quite different from the stacked ordering in phosphine-stabilized clusters, further demonstrating the critical role of surface-capping ligands in engineering the structures of heterometallic nanoclusters. Together with the combined use of phosphines and thiolates as the capping agents, introducing extra binding motifs such as pyridyl has been also demonstrated highly effective to adjust the available surface binding sites and thus control the number of secondary metal atoms that can be incorporated on the clusters' surfaces. Such a strategy provides an alternative and powerful approach to finely tune the surface structure of heterometallic nanoclusters and the metal exposure as well. To summarize up to present, the critical role of capping ligands on the structural engineering of heterometallic nanocluster has been widely recognized.

With the synthesis and structural identification of more organic-protected heterometallic nanoclusters, the general rules on how ligands influence their surface structure have been established. The surface binding structures of these atomically precise heterometallic nanoclusters are expected to provide ideal molecular structure models for their nanoparticulate analogues. These structure models would be important to help optimizing the functionalities, particularly surface-/interface-dependent properties of heterometallic nanoparticles. However, before such a goal can be achieved, further research efforts are still needed to systematically prepare functional heterometallic nanoclusters, carry out structure–property relation studies, and apply the understanding in the heterometallic nanoparticulate systems.

REFERENCES

1. Ferrando R, Jellinek J, Johnston RL. Nanoalloys: from theory to applications of alloy clusters and nanoparticles. *Chem Rev* 2008;**108**:845–910.
2. Wu BH, Zheng NF. Surface and interface control of noble metal nanocrystals for catalytic and electrocatalytic applications. *Nano Today* 2013;**8**:168–97.
3. Wang D, Li Y. Bimetallic nanocrystals: liquid-phase synthesis and catalytic applications. *Adv Mater* 2011;**23**:1044–60.
4. Tedsree K, Li T, Jones S, Chan CWA, Yu KMK, Bagot PAJ, et al. Hydrogen production from formic acid decomposition at room temperature using a Ag–Pd core–shell nanocatalyst. *Nat Nanotechnol* 2011;**6**:302–7.
5. Alayoglu S, Eichhorn B. Rh–Pt bimetallic catalysts: synthesis, characterization, and catalysis of core–shell, alloy, and monometallic nanoparticles. *J Am Chem Soc* 2008;**130**:17479–86.
6. Alayoglu S, Nilekar AU, Mavrikakis M, Eichhorn B. Ru–Pt core–shell nanoparticles for preferential oxidation of carbon monoxide in hydrogen. *Nat Mater* 2008;**7**:333–8.
7. Alayoglu S, Zavalij P, Eichhorn B, Wang Q, Frenkel AI, Chupas P. Structural and architectural evaluation of bimetallic nanoparticles: a case study of Pt–Ru core–shell and alloy nanoparticles. *ACS Nano* 2009;**3**:3127–37.
8. Nilekar AU, Alayoglu S, Eichhorn B, Mavrikakis M. Preferential CO oxidation in hydrogen: reactivity of core–shell nanoparticles. *J Am Chem Soc* 2010;**132**:7418–28.

9. Yan J, Zhang X, Akita T, Haruta M, Xu Q. One-step seeding growth of magnetically recyclable Au@Co core−shell nanoparticles: highly efficient catalyst for hydrolytic dehydrogenation of ammonia borane. *J Am Chem Soc* 2010;**132**:5326−7.
10. Hong JW, Kim D, Lee YW, Kim M, Kang SW, Han SW. Atomic-distribution-dependent electrocatalytic activity of Au−Pd bimetallic nanocrystals. *Angew Chem Int Ed Engl* 2011;**50**:8876−80.
11. Vargas MD, Nicholls JN. High-nuclearity carbonyl clusters: their synthesis and reactivity. *Adv Inorg Chem Radiochem* 1986;**30**:123−222.
12. Macchi P, Sironi A. Chemical bonding in transition metal carbonyl clusters: complementary analysis of theoretical and experimental electron densities. *Coord Chem Rev* 2003;**238−239**:383−412.
13. Femoni C, Iapalucci MC, Kaswalder F, Longoni G, Zacchini S. The possible role of metal carbonyl clusters in nano-science and nanotechnologies. *Coord Chem Rev* 2006;**250**:1580−604.
14. Teo BK, Zhang H. Polyicosahedricity: icosahedron to icosahedron of icosahedra growth pathway for bimetallic (Au-Ag) and trimetallic (Au-Ag-M; M = Pt, Pd, Ni) supraclusters; synthetic strategies, site preference, and stereochemical principles. *Coord Chem Rev* 1995;**143**:611−36.
15. Pignolet LH, Aubart MA, Craighead KL, Gould RA, Krogstad DA, Wiley JS. Phosphine-stabilized, platinum-gold and palladium-gold cluster compounds and applications in catalysis. *Coord Chem Rev* 1995;**143**:219−63.
16. Brust M, Walker M, Bethell D, Schiffrin DJ, Whyman R. Synthesis of thiol-derivatised gold nanoparticles in a two-phase liquid−liquid system. *J Chem Soc Chem Commun* 1994:801−2.
17. Kurashige W, Yamaguchi M, Nobusada K, Negishi Y. Ligand-induced stability of gold nanoclusters: thiolate versus selenolate. *J Phys Chem Lett* 2012;**3**:2649−52.
18. Teo BK, Zhang H. Design and synthesis of macrocyclic multidentate polyphosphine ligands with nano-sized cavities for the encapsulation of vertex-sharing polyicosahedral mixed-metal nanoclusters. *Inorg Chim Acta* 2001;**317**:1−11.
19. Qian H, Zhu M, Wu Z, Jin R. Quantum sized gold nanoclusters with atomic precision. *Acc Chem Res* 2012;**45**:1470−9.
20. Häkkinen H. The gold-sulfur interface at the nanoscale. *Nat Chem* 2012;**4**:443−55.
21. Teo BK, Zhang H, Shi XB. Molecular architecture of a novel vertex-sharing biicosahedral cluster $(p\text{-}Tol_3P)_{10}Au_{13}Ag_{12}Br_8(PF_6)$ containing a staggered staggered staggered configuration for the 25-atom metal framework. *Inorg Chem* 1990;**29**:2083−91.
22. Teo BK, Shi X, Zhang H. Rotamerism and roulettamerism of vertex-sharing biicosahedral supraclusters: synthesis and structure of $[(Ph_3P)_{10}Au_{13}Ag_{12}C_{18}](SbF_6)$. *J Clust Sci* 1993;**4**:471−6.
23. Teo BK, Shi XB, Zhang H. Cluster of clusters: structure of a novel cluster $(Ph_3P)_{10}Au_{13}Ag_{12}Br_8(SbF_6)$ containing an exact staggered eclipsed staggered metal configuration. Evidence of icosahedral units as building-blocks. *J Am Chem Soc* 1991;**113**:4329−31.
24. Teo BK, Dang H, Campana CF, Zhang H. Synthesis, structure and characterization of $(MePh_2P)_{10}Au_{12}Ag_{13}Br_9$: the first example of a neutral bi-icosahedral Au−Ag cluster with a novel seven-membered satellite ring of bridging ligands. *Polyhedron* 1998;**17**:617−21.
25. Teo BK, Zhang H. Cluster of clusters: structure of a new 25-metal-atom cluster $(p\text{-}Tol_3P)_{10}Au_{13}Ag_{12}Cl_7(SbF_6)_2$ containing a nearly staggered eclipsed staggered metal configuration and 5 doubly bridging ligands. *Inorg Chem* 1991;**30**:3115−6.

26. Teo BK, Keating K. Novel triicosahedral structure of the largest metal alloy cluster: $[(Ph_3P)_{12}Au_{13}Ag_{12}Cl_6]^{m+}$. *J Am Chem Soc* 1984;**106**:2224–6.
27. Teo BK, Zhang H. Molecular machines: molecular-structure of $(p\text{-}Tol_3P)_{10}Au_{13}Ag_{12}Cl_8(PF_6)$– a cluster with a biicosahedral rotor-like metal core and an unusual arrangement of bridging ligands. *Angew Chem Int Ed Engl* 1992;**31**:445–7.
28. Teo BK, Shi XB, Zhang H. Cluster rotamerism of a 25-metal-atom cluster $[(Ph_3P)_{10}Au_{13}Ag_{12}Br_8]^+$ monocation: a molecular rotary unit. *J Chem Soc Chem Comm* 1992:1195–6.
29. Teo BK, Zhang H. Synthesis and structure of a neutral trimetallic biicosahedral cluster, $(Ph_3P)_{10}Au_{11}Ag_{12}Pt_2Cl_7$. A comparative study of molecular and crystal structures of vertex-sharing biicosahedral mixed-metal nanoclusters. *J Clust Sci* 2001;**12**:349–82.
30. Teo BK, Zhang H, Shi X. Design, synthesis, and structure of the largest trimetallic cluster, $[(Ph_3P)_{10}Au_{12}Ag_{12}PtCl_7]Cl$: the first example of a trimetallic biicosahedral supracluster and its implication for the vertex-sharing polyicosahedral growth of the gold/silver/platinum ternary cluster system. *J Am Chem Soc* 1993;**115**:8489–90.
31. Teo BK, Hong MC, Zhang H, Huang DB. Cluster of clusters: structure of the 37-atom cluster $[(p\text{-}Tol_3P)_{12}Au_{18}Ag_{19}Br_{11}]^{2+}$ and a novel series of supraclusters based on vertex-sharing icosahedra. *Angew Chem Int Ed Engl* 1987;**26**:897–900.
32. Teo BK, Hong M, Zhang H, Huang D, Shi X. Cluster of clusters: structure of a novel 38-atom cluster $(p\text{-}Tol_3P)_{12}Au_{18}Ag_{20}Cl_{14}$. *J Chem Soc Chem Comm* 1988:204–6.
33. Teo BK, Shi XB, Zhang H. Clusters of clusters. 25. Synthesis and structure of a new 38-metal-atom cluster $[(Ph_3P)_{14}Au_{18}Ag_{20}Cl_{12}]Cl_2$ and its implications with regard to intra-cavity chemistry on metal cluster surfaces. *Inorg Chem* 1993;**32**:3987–8.
34. Teo BK, Zhang H, Shi XB. Cluster of clusters: a modular approach to large metal-clusters – structural characterization of a 38-atom cluster $(p\text{-}Tol_3P)_{12}Au_{18}Ag_{20}Cl_{14}$ based on vertex-sharing triicosahedra. *J Am Chem Soc* 1990;**112**:8552–62.
35. Teo BK, Zhang H, Shi X. Site preference in vertex-sharing polyicosahedral supraclusters containing groups 10 and 11 metals and their bonding implications: syntheses and structures of the first Au-Ag-M (M = Pt, Ni) biicosahedral clusters $[(Ph_3P)_{10}Au_{12}Ag_{12}PtCl_7]Cl$ and $[(Ph_3P)_{10}Au_{12}Ag_{12}NiCl_7](SbF_6)$. *Inorg Chem* 1994;**33**:4086–97.
36. Teo BK. Cluster of clusters: a new series of high nuclearity Au-Ag clusters. *Polyhedron* 1988;**7**:2317–20.
37. Teo BK, Zhang H. Clusters of clusters: self-organization and self-similarity in the intermediate stages of cluster growth of Au-Ag supraclusters. *Proc Natl Acad Sci* 1991;**88**:5067–71.
38. Teo BK, Zhang H. Molecular and crystal structure of an iodo-gold-silver-platinum cluster, $(Ph_3P)_6Au_6Ag_6Pt(AgI_3)_2(THF)_2$, with a trimetallic icosahedral core capped with two planar AgI_3 units. *J Organomet Chem* 2000;**614**:66–9.
39. Nunokawa K, Ito M, Sunahara T, Onaka S, Ozeki T, Chiba H, et al. A new 19-metal-atom cluster $[(Me_2PhP)_{10}Au_{12}Ag_7(NO_3)_9]$ with a nearly staggered-staggered M_5 ring configuration. *Dalton Trans* 2005:2726–30.
40. Ito LN, Johnson BJ, Mueting AM, Pignolet LH. Heterobimetallic gold-palladium phosphine cluster complexes. X-ray crystal and molecular structure of $[Au_8Pd(PPh_3)_8](NO_3)_2$. *Inorg Chem* 1989;**28**:2026–8.
41. Heaven MW, Dass A, White PS, Holt KM, Murray RW. Crystal structure of the gold nanoparticle $[N(C_8H_{17})_4][Au_{25}(SCH_2CH_2Ph)_{18}]$. *J Am Chem Soc* 2008;**130**:3754–5.
42. Zhu M, Aikens CM, Hollander FJ, Schatz GC, Jin R. Correlating the crystal structure of a thiol-protected Au_{25} cluster and optical properties. *J Am Chem Soc* 2008;**130**:5883–5.

43. Bakr OM, Amendola V, Aikens CM, Wenseleers W, Li R, Dal Negro L, et al. Silver nanoparticles with broad multiband linear optical absorption. *Angew Chem Int Ed Engl* 2009;**48**:5921–6.
44. Fields-Zinna CA, Crowe MC, Dass A, Weaver JE, Murray RW. Mass spectrometry of small bimetal monolayer-protected clusters. *Langmuir* 2009;**25**:7704–10.
45. Negishi Y, Kurashige W, Niihori Y, Iwasa T, Nobusada K. Isolation, structure, and stability of a dodecanethiolate-protected Pd_1Au_{24} cluster. *Phys Chem Chem Phys* 2010;**12**:6219–25.
46. Niihori Y, Matsuzaki M, Pradeep T, Negishi Y. Separation of precise compositions of noble metal clusters protected with mixed ligands. *J Am Chem Soc* 2013;**135**:4946–9.
47. Qian H, Jiang D, Li G, Gayathri C, Das A, Gil RR, et al. Monoplatinum doping of gold nanoclusters and catalytic application. *J Am Chem Soc* 2012;**134**:16159–62.
48. Yang HY, Wang Y, Huang HQ, Gell L, Lehtovaara L, Malola S, et al. All-thiol-stabilized Ag_{44} and $Au_{12}Ag_{32}$ nanoparticles with single-crystal structures. *Nat Comm* 2013;**4**:2422.
49. Yang HY, Wang Y, Yan JZ, Chen X, Zhang X, Häkkinen H, et al. Structural evolution of atomically precise thiolated bimetallic $[Au_{12+n}Cu_{32}SR_{30+n}]^{4-}$ ($n = 0, 2, 4, 6$) nanoclusters. *J Am Chem Soc* 2014;**136**:7197–200.
50. Zhang X, Yang HY, Zhao XJ, Wang Y, Zheng NF. The effects of surface ligands and counter cations on the stability of anionic thiolated $M_{12}Ag_{32}$ (M = Au, Ag) nanoclusters. *Chin Chem Lett* 2014;**25**:839–43.
51. Zheng NF, Fan J, Stucky GD. One-step one-phase synthesis of monodisperse noble-metallic nanoparticles and their colloidal crystals. *J Am Chem Soc* 2006;**128**:6550–1.
52. Ito LN, Sweet JD, Mueting AM, Pignolet LH, Schoondergang MFJ, Steggerda JJ. Heterobimetallic gold-platinum phosphine complexes. X-ray crystal and molecular structures of $[(PPh_3)Pt(AuPPh_3)_6](NO_3)_2$ and $[(PPh_3)(CO)Pt(AuPPh_3)_6](PF_6)_2$. *Inorg Chem* 1989;**28**:3696–701.
53. Kanters RPF, Bour JJ, Schlebos PPJ, Bosman WP, Behm H, Steggerda JJ, et al. Structure and properties of the hydride-containing cluster ion $[Pt(H)(PPh_3)(AuPPh_3)_7]^{2+}$. *Inorg Chem* 1989;**28**:2591–4.
54. Udayabhaskararao T, Sun Y, Goswami N, Pal SK, Balasubramanian K, Pradeep T. Ag_7Au_6: a 13-atom alloy quantum cluster. *Angew Chem Int Ed Engl* 2011;**51**:2155–9.
55. Wang S, Meng X, Das A, Li T, Song Y, Cao T, et al. A 200-fold quantum yield boost in the photoluminescence of silver-doped Ag_xAu_{25-x} nanoclusters: the 13th silver atom matters. *Angew Chem Int Ed Engl* 2014;**53**:2376–80.
56. Kanters RPF, Schlebos PPJ, Bour JJ, Bosman WP, Smits JMM, Beurskens PT, et al. Electrophilic addition reactions of Ag^+ with $Pt(AuPPh_3)_8$ and $[Pt(CO)(AuPPh_3)_8]^{2+}$. Crystal structures of $[Pt(AgNO_3)(AuPPh_3)_8](NO_3)_2$ and $[Pt(CO)(AgNO_3)(AuPPh_3)_8](NO_3)_2$. *Inorg Chem* 1990;**29**:324–8.
57. Krogstad DA, Konze WV, Pignolet LH. Synthesis and characterization of a series of triethylphosphine-ligated Pt–Au cluster compounds. X-ray crystal and molecular structure of $[Pt(AuPEt_3)_9](NO_3)_3$. *Inorg Chem* 1996;**35**:6763–71.
58. Copley RCB, Mingos DMP. Synthesis and structural characterization of $[Au_9Ag_4Cl_4(PMePh_2)_8][C_2B_9H_{12}]\cdot H_2O\cdot 0.5CH_2Cl_2$: the first example of an icosahedral silver-gold cluster. *J Chem Soc Dalton Trans* 1992:1755–6.
59. Krogstad DA, Young Jr VG, Pignolet LH. Synthesis and characterization of a series of diphosphine ligated Pt–Au cluster compounds. *Inorg Chim Acta* 1997:19–32.
60. Spivak GJ, Vittal JJ, Puddephatt RJ. Heteronuclear hexaplatinum cluster complexes: structure as a function of electron count. *Inorg Chem* 1998;**37**:5474–81.

61. Copley RCB, Mingos DMP. Synthesis and characterization of the centred icosahedral cluster series $[Au_9M^{IB}{}_4Cl_4(PMePh_2)_8][C_2B_9H_{12}]$, where M^{IB} = Au, Ag or Cu. *J Chem Soc Dalton Trans* 1996:491–500.
62. Kappen TGMM, Schlebos PPJ, Bour JJ, Bosman WP, Smits JMM, Beurskens PT, et al. Synthesis and characterization of platinum-gold clusters fused to tetrahedral copper units. Crystal structures of $[Pt(PPh_3)(AuPPh_3)_6(Cu_4Cl_3PPh_3)](NO_3)$ and $[Pt(PPh_3)(AuPPh_3)_6(Cu_4I_3)](NO_3)$. *J Am Chem Soc* 1995;**117**:8327–34.
63. Kappen TGMM, Schlebos PPJ, Bour JJ, Bosman WP, Smits JMM, Beurskens PT, et al. New hydride-containing mixed metal-gold phosphine clusters. Crystal structures of $[Pt(H)(CuCl)(AuPPh_3)_8](NO_3)$ and $[Pt(H)(CuCl)_2(AuPPh_3)_8](NO_3)$. *Inorg Chem* 1995;**34**:2133–42.
64. Kappen TGMM, Schlebos PPJ, Bour JJ, Bosman WP, Beurskens G, Smits JMM, et al. Cluster growth: some representative reactions. Crystal structures of $[Pt(H)(AgNO_3)(AuPPh_3)_8](NO_3)$ and $[Pt(H)(AgNO_3)_2(AuPPh_3)_8](NO_3)$. *Inorg Chem* 1995;**34**:2121–32.
65. Gould RAT, Craighead KL, Wiley JS, Pignolet LH. Addition of metallic mercury to platinum- and palladium-gold cluster compounds. X-ray crystal and molecular structures of $[(Hg)_2Pd(AuPPh_3)_8](NO_3)_2$ and $[(Hg)Pt(AuPPh_3)_8](NO_3)_2$. *Inorg Chem* 1995;**34**:2902–13.
66. Laupp M, Strähle J. $[(Ph_3PAu)_6(dppeAu_2)(AuCl)_4Pd]$, an icosahedral Au_{12} cluster with a central Pd atom. *Angew Chem Int Ed Engl* 1994;**33**:207–9.
67. Kappen TGMM, Schlebos PPJ, Bour JJ, Bosman WP, Smits JMM, Beurskens PT, et al. A new 25-metal-atom supracluster: synthesis, characterization, and crystal structure of the ternary cluster compound $[Pt_2(AuPPh_3)_{10}Ag_{13}Cl_7]$. *Inorg Chem* 1994;**33**:754–8.
68. Holzer M, Strähle J, Baum G, Fenske D. Synthese und struktur von $[(Ph_3PAu)_6Co(CO)_2](PF_6)$ und $[(Ph_3PAu)_7Co(CO)_2](PF_6)_2$. *Z Anorg Allg Chem* 1994;**620**:192–8.
69. Gould RAT, Pignolet LH. Heterometallic gold-platinum-mercury and Au-Pt-Hg-cobalt phosphine-stabilized cluster compounds. X-ray crystal and molecular structure of $[(PPh_3)Pt(AuPPh_3)_6(HgNO_3)]NO_3$. *Inorg Chem* 1994;**33**:40–6.
70. Sunick DL, White PS, Schauer CK. Phosphorus-mediated gold aggregation on a metal cluster: syntheses, structures, dynamic behavior, and ligand addition reactions of the clusters $(\mu\text{-}H)_nFe_3(CO)_9P[Au(PR_3)]_{3-n}$ ($n = 0, 1, 2$). *Inorg Chem* 1993;**32**:5665–75.
71. Brown SSD, Salter ID, Adatia T. The heteronuclear cluster chemistry of the group IB metals. Part 16. Metal exchange reactions in group IB metal heteronuclear cluster compounds. Synthesis and structural characterization of the trimetallic hexanuclear cluster compounds $[MM'Ru_4(\mu_3\text{-}H)_2\{\mu\text{-}Ph_2P(CH_2)PPh_2\}(CO)_{12}]$(M = Cu, M′ = Ag or Au; m = Ag, M′ = Au; $n = 1$ or 2) and X-ray crystal structure of $[AuCuRu_4(\mu_3\text{-}H)_2\{\mu\text{-}Ph_2P(CH_2)_2PPh_2\}(CO)_{12}]$. *J Chem Soc Dalton Trans* 1993:559–66.
72. Breuer M, Strähle J. Clusteraufbau durch photolyse von azidokomplexen des platins und golds. Synthesen und kristallstrukturen von $[(Ph_3PAu)_6(AuCl)_3Pt(CO)]$, $[(dppe)PtCo_2(CO)_7]$ und $[(Ph_3PAu)_4Pt(dppe)](PF_6)_2$. *Z Anorg Allg Chem* 1993;**619**:1564–74.
73. Mielcke J, Strähle J. $[(Ph_3PAu)_6Mn(CO)_3]^+$, a cluster with pentagonal-bipyramidal structure. *Angew Chem Int Ed Engl* 1992;**31**:464–6.
74. Schoondergang MFJ, Bour JJ, Van Strijdonck GPF, Schlebos PPJ, Bosman WP, Smits JMM, et al. Reactivity of 16- and 18-electron mixed platinum-gold clusters with halides. Crystal and molecular structures of $[Pt(PPh_3)(AuPPh_3)_6(AuCl_3)\cdot 2C_4H_{10}O$. *Inorg Chem* 1991;**30**:2048–52.
75. Schoondergang MFJ, Bour JJ, Schlebos PPJ, Vermeer AWP, Bosman WP, Smits JMM, et al. Synthesis and characterization of copper- and silver-containing platinum-gold cluster compounds. X-ray crystal structure of $[Pt(CuCl)(AuPPh_3)_8](NO_3)_2\cdot CH_3OH$. *Inorg Chem* 1991;**30**:4704–9.

76. Ito LN, Felicissimo AMP, Pignolet LH. Heterobimetallic gold-palladium cluster complexes. X-ray crystal and molecular structures of $[(CO)Pd(AuPPh_3)_8](NO_3)_2$ and $[(P(OCH_3)_3)Pd(AuPPh_3)_6(AuP(OCH_3)_3)_2](NO_3)_2$. *Inorg Chem* 1991;**30**:988–94.
77. Bour JJ, Van den Berg W, Schlebos PPJ, Kanters RPF, Schoondergang MFJ, Bosman WP, et al. Platinum-gold-mercury $PtAu_8Hg_2$ and $PtAu_7Hg_2$ cluster Compounds: X-ray structure of $[Pt(AuPPh_3)_8(Hg)_2](NO_3)_4 \cdot 3CH_2Cl_2$. *Inorg Chem* 1990;**29**:2971–5.
78. Kanters RPF, Schlebos PPJ, Bour JJ, Bosman WP, Behm HJ, Steggerda JJ. Reaction of CO and an isonitrile with $PtAu_8(PPh_3)_8{}^{3+}$. Crystal structure of $Pt(CO)(AuPPh_3)_8(NO_3)_2 \cdot 2C_4H_{10}O \cdot 2CH_2Cl_2$. *Inorg Chem* 1988;**27**:4034–7.
79. Smith DE, Welch AJ, Treurnicht I. Puddephatt RJA fluxional high-nuclearity gold-platinum cluster: structure of $[Au_6Pt(CC\text{-}t\text{-}Bu)(PPh_3)_7]^+[Au(CC\text{-}t\text{-}Bu)_2]$. *Inorg Chem* 1986;**25**:4616–7.
80. Van der Velden JWA, Bour JJ, Bosman WP, Noordik JH. Reactions of cationic gold clusters with Lewis bases. Preparation and X-ray structure investigation of $[Au_8(PPh_3)_7](NO_3)_2 \cdot 2CH_2Cl_2$ and $Au_6(PPh_3)_4[Co(CO)_4]_2$. *Inorg Chem* 1983;**22**:1913–8.
81. Qian H, Eckenhoff WT, Zhu Y, Pintauer T, Jin R. Total structure determination of thiolate-protected Au_{38} nanoparticles. *J Am Chem Soc* 2010;**132**:8280–1.
82. Fuhr O, Dehnen S, Fenske D. Chalcogenide clusters of copper and silver from silylated chalcogenide sources. *Chem Soc Rev* 2013;**42**:1871–906.
83. Anson CE, Eichhöfer A, Issac I, Fenske D, Fuhr O, Sevillano P, et al. Synthesis and crystal structures of the ligand-stabilized silver chalcogenide clusters $[Ag_{154}Se_{77}(dppxy)_{18}]$, $[Ag_{320}(StBu)_{60}S_{130}(dppp)_{12}]$, $[Ag_{352}S_{128}(S_tC_5H_{11})_{96}]$, and $[Ag_{490}S_{188}(S_tC_5H_{11})_{114}]$. *Angew Chem Int Ed Engl* 2008;**47**:1326–31.
84. Fenske D, Anson CE, Eichhöfer A, Fuhr O, Ingendoh A, Persau C, et al. Syntheses and crystal structures of $[Ag_{123}S_{35}(S_tBu)_{50}]$ and $[Ag_{344}S_{124}(S_tBu)_{96}]$. *Angew Chem Int Ed Engl* 2005;**44**:5242–6.
85. Yang HY, Lei J, Wu BH, Wang Y, Zhou M, Xia AD, et al. Crystal structure of a luminescent thiolated Ag nanocluster with an octahedral $Ag_6{}^{4+}$ core. *Chem Commun (Camb)* 2013;**49**:300–2.
86. Yang HY, Wang Y, Zheng NF. Stabilizing subnanometer Ag(0) nanoclusters by thiolate and diphosphine ligands and their crystal structures. *Nanoscale* 2013;**5**:2674–7.
87. Chanaka K, Amala D. $(AuAg)_{144}(SR)_{60}$ alloy nanomolecules. *Nanoscale* 2011;**3**:3064–7.
88. Jupally VR, Dass A. Synthesis of $Au_{130}(SR)_{50}$ and $Au_{130-x}Ag_x(SR)_{50}$ nanomolecules through core size conversion of larger metal clusters. *Phys Chem Chem Phys* 2014;**16**:10473–9.
89. Harkness KM, Tang Y, Dass A, Pan J, Kothalawala N, Reddy VJ, et al. $Ag_{44}(SR)_{30}{}^{4-}$: a silver–thiolate superatom complex. *Nanoscale* 2012;**4**:4269–74.
90. Negishi Y, Arai R, Niihori Y, Tsukuda T. Isolation and structural characterization of magic silver clusters protected by 4-(*tert*-Butyl) benzyl mercaptan. *Chem Commun (Camb)* 2011;**47**:5693–5.
91. Rao TUB, Pradeep T. Luminescent Ag_7 and Ag_8 clusters by interfacial synthesis. *Angew Chem Int Ed Engl* 2010;**49**:3925–9.
92. Wu Z, Lanni E, Chen W, Bier ME, Ly D, Jin R. High yield, large scale synthesis of thiolate-protected Ag_7 clusters. *J Am Chem Soc* 2009;**131**:16672–4.
93. Desireddy A, Conn BE, Guo J, Yoon B, Barnett RN, Monahan BM, et al. Ultrastable silver nanoparticles. *Nature* 2013;**501**:399–402.
94. Yang HY, Wang Y, Lei J, Shi L, Wu XH, Makinen V, et al. Ligand–stabilized $Au_{13}Cu_x$ ($x = 2, 4, 8$) bimetallic nanoclusters: ligand engineering to control the exposure of metal sites. *J Am Chem Soc* 2013;**135**:9568–71.

95. Negishi Y, Munakata K, Ohgake W, Nobusada K. Effect of copper doping on electronic structure, geometric structure, and stability of thiolate-protected Au_{25} nanoclusters. *J Phys Chem Lett* 2012;**3**:2209–14.
96. Kurashige W, Munakata K, Nobusada K, Negishi Y. Synthesis of stable Cu_nAu_{25-n} nanoclusters ($n = 1-9$) using selenolate ligands. *Chem Commun (Camb)* 2013;**49**:5447–9.
97. Dharmaratne AC, Dass A. $Au_{144-x}Cu_x(SC_6H_{13})_{60}$ nanomolecules: effect of Cu incorporation on composition and plasmon-like peak emergence in optical spectra. *Chem Commun (Camb)* 2014;**50**:1722–4.
98. Kumara C, Aikens CM, Dass A. X-ray crystal structure and theoretical analysis of $Au_{25-x}Ag_x(SCH_2CH_2Ph)_{18}^-$ alloy. *J Phys Chem Lett* 2014;**5**:461–6.

Chapter 5

Structure Determination by Single Crystal X-ray Crystallography

Thomas W. Ni, Marcus A. Tofanelli and Christopher J. Ackerson[1]
Department of Chemistry, Colorado State University, Fort Collins, CO, USA
[1]*Corresponding author: E-mail: ackerson@colostate.edu*

Chapter Outline

5.1 INTRODUCTION

Francis Crick is quoted, "If you want to understand function, study structure."[1] Indeed in the case of thiolate-protected metal clusters, total structure determination revolutionized fundamental and functional understanding of these materials, in large part by elucidating their surprisingly complex surface structure. The revolution prompted by single crystal structures extends to understanding stability trends,[2] optical properties,[3] ligand exchange,[4,5] chirality,[6–8] and catalysis.[9,10]

Many methods are fully capable of providing structural information of nanoscale objects, including biomolecules, metal clusters, and soft materials (Figure 1). Probe sources can be X-rays, electrons, and neutrons on targets that can be powders,[11] solutions, or single crystals.[12] Data are typically recorded as a diffraction pattern, because lenses that can focus X-rays and

Frontiers of Nanoscience, Vol. 9. http://dx.doi.org/10.1016/B978-0-08-100086-1.00005-1

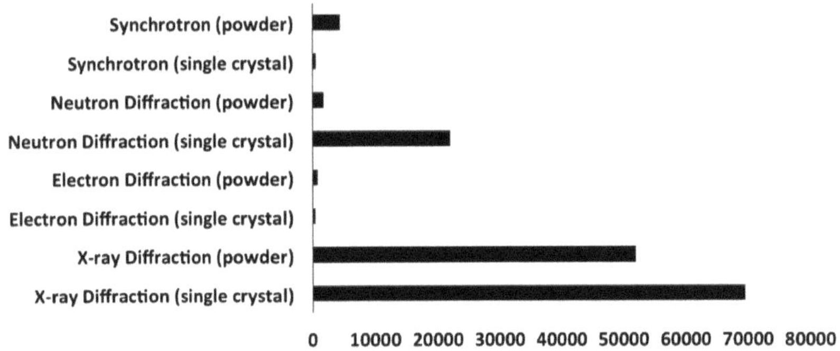

FIGURE 1 Structures solved by different types of diffraction. Data collated from Inorganic Crystal Structure Database.

neutrons are not available, which would otherwise allow direct imaging. Analysis, for instance, by pair distribution function (PDF) is becoming increasingly useful for validating proposed structures.[13,14] In the case of electron microscopy, the possibility of focusing electrons with electromagnetic lenses allows for collection of data in either diffractive or direct imaging modes, and both methods are useful for proposing atomistic models.[15–17]

Single crystal X-ray crystallography, however, is responsible for most widely accepted atomic models of molecular structure. The X-ray diffractometers required for data collection on single crystals are widely available, and the breadth of chemical samples now solved by single crystal X-ray methods is impressive, ranging from small organic molecules and inorganic coordination complexes to some of the largest discrete molecules in biological systems, such as the 70S ribosome macromolecular complex.[18]

The chief drawback to single crystal X-ray methods is the requirement for high-quality single crystals. Indeed, many molecules cannot form crystals that are sufficiently well ordered to diffract to "atomic resolution" (generally defined as 3.3 Å spatial resolution or better for macromolecules, where 1.2 Å or better is typical of small molecules). Failure to form X-ray quality single crystals may arise from impurities in a preparation, from conformational heterogeneity, or from an inability to pack into a well-ordered lattice. When adequately ordered single crystals of a nominally molecular sample cannot be formed, electron microscopy and powder diffraction methods can provide structural information.

Indeed, early structural models of thiolate-protected clusters were based on high-resolution electron microscopy and powder diffraction methods. These methods allowed a preliminary picture of ligated metal clusters as compounds in which all metal atoms were packed into high-symmetry gold cores of

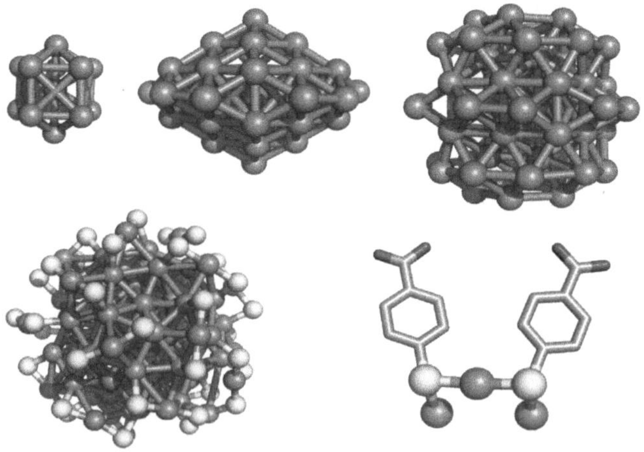

FIGURE 2 Shell-by-shell construction of $Au_{102}(p\text{-MBA})_{44}$. Figure shows a 13-atom truncated decahedron (top left), a 49-atom truncated decahedron (top center), the 79-atom "grand core" of D_{5h} symmetry (top right), the cluster including all Au and S atoms (bottom left, C_2 symmetry), and an image of the "staple" motif including the organic ligand (bottom right).

truncated O_h, truncated D_h, or full I_h geometry[19–21] with organothiolate ligands bonded to vertex, edge, and face sites. There was a surprising set of information missing in these early models regarding, especially, the surface structure of these clusters.

The single crystal structure of $Au_{102}(p\text{-MBA})_{44}$[12] provoked a sea change in the understanding of the structure of ligand-protected metal clusters. This first X-ray crystal structure of a thiolate-protected gold cluster revealed surprisingly that 23 of the 102 atoms in the cluster were not part of a core with fivefold symmetry, as suggested in the previously isolated "14-kDa" compound.[22] Rather, the outermost 23 atoms are held in SR–Au–SR and SR–Au–SR–Au–SR units colloquially referred to as staples or semi-rings (Figure 2). This finding, validated a previous theoretical prediction, the so-called "divide-and-protect" model,[23] in which the Au atoms in ligated clusters are divided into core and protecting shell atoms.

This X-ray crystal structure provoked a sea change in structural study of thiolate-protected gold clusters. Several more structures were solved by single crystal methods (Table 1, vide infra) also show the universality of the divide-and-protect motif, with the outermost gold atoms in these structures being held in SR–Au–SR, SR–Au–SR–Au–SR, and SR–Au–SR–Au–SR–Au–SR units.[24,25] That powder diffraction and high-resolution transmission electron microscopy missed such a key aspect of the structure of protected metal clusters highlights the current necessity of single crystal X-ray methods for generating reliable structural models of nanomaterials.

There is evidence that powder diffraction, electron microscopy, and even theory will improve to compete with or even surpass single crystal X-ray method for generation of reliable atomistic structures.

In the case of electron microscopy, it is increasingly clear that the high electron doses typically used in imaging experiments perturb the samples examined, resulting in inaccurate structural models. Recent work adapting "low-dose" methods from single particle microscopy[26] used by structural biologists shows high promise for preserving the fidelity of the inorganic atoms in a nanoparticle structure.[17] This still imposes substantial monodispersity requirements on the target samples, although certainly avoids the more stringent requirement for single crystallinity.

In the case of powder diffraction, increasingly sophisticated methods based on PDF analysis are good for validating proposed models.[11] Experiments are typically carried out at a synchrotron; Home sources using Mo or Ag radiation result in less accuracy due to the wavelength of the X-ray (Ag = 0.4859 Å, Mo = 0.6199 Å). This wavelength restriction arises from momentum transfer, Q, which is given by $Q = 4\pi \sin(\theta)/\lambda$.[27] Thus, X-rays or neutrons of $\lambda = 0.12–0.27\ \text{Å}^{-1}$ are preferable to give high accuracy, generally necessitating synchrotron sources.

PDF is very good at looking at local structure where atoms may be disordered in a random fashion.[28] Especially in cases where there are atoms of low atomic number that are close to each other where other techniques cannot be utilized. PDF is not as good at determining long-range order where traditional crystallographic techniques are typically used, although the method is good for model validation of long-range order.

Despite the exciting progress shown recently in electron microscopy and powder diffraction methods, single crystal X-ray methods are currently the most widely accepted means of producing atomic structures of protected metal clusters.

5.2 STRUCTURE DETERMINATION BY SINGLE CRYSTAL X-RAY CRYSTALLOGRAPHY

X-ray crystallography differs from more easily understood methods for resolving microscopic objects, such as optical microscopy, in two ways. First, since it is currently impossible to focus X-rays in the same way that light can be focused with an optical lens or electrons can be focused with an electromagnetic lens, X-ray data must be collected as a diffraction pattern. Computer software subsequently plays the role of simulating a lens, allowing the reconstruction in "real space" of the microscopic object. Second, in order to achieve sufficient "signal," it's insufficient to collect data on single copy of a molecule. Instead, the signals of a large ensemble of molecules are averaged by incorporating the molecules into a well-ordered array—i.e., a single crystal.

Here we give an overview of single crystal X-ray methods as applied to protected gold, silver, and alloy clusters. The section is focused on challenges unique to the class of molecules and practical aspects of structures solved so far, as well as anticipation of how practical aspects may evolve if even larger structures are solved. Fundamental aspects of X-ray crystallography as a method may be found elsewhere. *Crystallography Made Crystal Clear*[29] and *Crystal Structure Determination*[30] are excellent texts for understanding fundamental theoretical aspects of the method and are approachable texts for nonexperts. *Crystal Structure Refinement*[31] is a more advanced and practical reference for hands-on crystal structure determination.

Historically, the practice of single crystal X-ray crystallography is split into "small-molecule" and "macromolecular" crystallography traditions. While the underlying fundamental aspects of the work—involving the growing single crystals, acquiring X-ray diffraction patterns (i.e., reflections), and analyzing the indexed reflection data to produce an atomistic model are identical in the big picture, the greater difficulties encountered in solving macromolecule structure result in a set of parallel methods, tailored for more challenging macromolecular crystallography problems. A commonly suggested demarcation between small-molecule and macromolecular crystallography is 500 nonhydrogen atoms. The cluster compounds that are the focus of this chapter range in size from 170 atoms for $Au_{18}(SR)_{14}$ to 344, 542, and 705 for the "giant cluster" compounds $Ag_{44}(p\text{-MBA})_{30}$, $Au_{102}(p\text{-MBA})_{44}$, and $Au_{133}(SR)_{52}$.[12,32,33] Given the sizes of these molecules, it is unsurprising that both small-molecule and macromolecular practices have proven useful in structure solution and evaluation.

Compared to macromolecular crystallography, small molecules have smaller unit cells and typically form larger crystals. The crystals are usually better ordered (i.e., have smaller measured mosaicity values) and as a consequence of these properties, small-molecule crystals typically produce strong diffraction to high spatial resolution (often 1.0 Å or better). The critical step in solving any crystal structure is the determination of phase information that is lost as a result of diffraction[29]; the high quality of the data and comparatively simple contents of the unit cell for small-molecule crystallography usually enable phasing by direct methods. The importance and widespread use of direct phasing methods, which rely largely on mathematical identities relating the intensities of certain sets of reflections to their phase resulted in recognition of Hauptman and Karle with the 1985 Nobel Prize in chemistry. Largely because the structures are easy to solve with direct phasing and high-quality data generally defined as diffraction to 1.2 Å resolution or better, small-molecule crystallography is now "routine" with over 750,000 structures determined so far and additional 40,000 structures deposited annually in the Cambridge Structural Database, where most small-molecule structures are archived.

The ability to solve macromolecular crystal structures, generally biomacromolecules such as proteins and nucleic acids, was pioneered by the 1962 Nobel Prize winning works of Kendrew and Perutz. Compared to small-molecule crystallography, macromolecular crystals typically contain much larger unit cells and tend to be much smaller crystals. In addition, crystalline disorder (i.e., mosaicity) tends to be greater for macromolecular crystals. As a result of this, the spatial resolution of diffraction tends to be lower. Due to the greater complexity of the contents of the unit cell and the lower resolution, the direct phasing methods that have essentially solved the phase problem for small-molecule crystallography are rarely applicable in macromolecular crystallography. Instead, phases are derived historically from Patterson methods, anomalous dispersion methods or molecular replacement, all of which are more applicable to the more complicated, lower resolution data typical of macromolecules.[29] Despite great efforts in making macromolecular crystallography as routine as small-molecule crystallography, the methods are inherently more labor-intensive and this is partially reflected in the smaller number of solved structures compared to small-molecule crystallography. As of this writing, there are just over 95,000 macromolecular structures derived by single crystal X-ray methods deposited in the Protein Data Bank, where most macromolecular structures are archived. In other words, at current rates of structure generation, there are as many small-molecule structures solved in 2.5 years as there are macromolecular structures solved in the past 50 years.

One of the largest practical differences between small-molecule and macromolecular crystallography is the time spent for optimizing crystallization conditions and the sheer number of crystals that are routinely screened in each "tradition" in order to find a crystal that diffracts well. In macromolecular crystallography, it's usual to attempt hundreds to thousands of individual crystallization conditions in a search for the highest quality crystals. It is also common to screen hundreds of crystals before finding a handful of crystals which can produce datasets that are adequate for processing into structure solutions. Indeed, for $Au_{102}(SR)_{44}$ crystals, a macromolecular approach to data collection was required; Only 3−5% of crystals screened for diffraction will diffract well enough for a structure to be solved for $Au_{102}(SR)_{44}$ and its derivatives. Similar efforts are implied for the giant $Pd_{165}(CO)_x(PEt)_{30}$ cluster in the statement in the report of that crystal structure, "The Goddess Fortuna recently revisited our laboratory in that crystals…were again isolated and its crystal structure redetermined."[34]

These sorts of empirical efforts in macromolecular crystallography to optimize and discover crystals that diffract well are unusual for small-molecule crystallography. Effort to make macromolecular structures more routine include crystallization robots that screen the hundreds to thousands of possible crystallization conditions observed as sometimes successful in macromolecular crystallography, as well as beamline robots to automate crystal screening, and the CCP4 software package to streamline structure solution.

The first crystal structure of a thiol-protected gold nanocluster was solved in the laboratory of Kornberg, who earlier won the 2006 Nobel Prize for solving the X-ray crystal structure of RNA polymerase II, which is still the largest protein structure solved by X-ray methods. A hybrid of macromolecular and small-molecule methods was used to solve the crystal structure of the first resolved cluster—Au_{102}(*p*-MBA). Most subsequent structures were solved with methods more typical of small-molecule X-ray diffraction (XRD), including the notable and recent $Au_{133}(SR)_{52}$ structures.[33,35]

Because of the sizes of the molecules studied, the history of both small and macromolecular methods in the field, and the anticipation that macromolecular methods may be very important for solving the structures of even larger thiolate-protected gold clusters, we discuss and contextualize both traditions throughout this chapter.

The process of solving a single crystal X-ray structure generally proceeds as shown in the flow chart of Figure 3. A homogenous sample of the target molecule is prepared, from which crystals can be prepared. Crystals of that material are grown from the preparation, from which X-ray diffraction patterns can be recorded. The diffraction patterns are analyzed to produce a structure as depicted in Figure 3. Here we discuss each of these steps, focusing especially on the steps of generating X-ray quality crystals and processing and evaluating single crystal X-ray data in a practical sense.

5.2.1 Preparing Homogeneous Material

Synthesis of well-defined protected metal clusters is covered in Chapters 2 and 3. Thus the discussion here focuses on aspects of synthesis and purification that are most relevant for structural work.

Etching or size-focusing is the most widespread means of synthesizing protected metal clusters that can be solved by X-ray crystallography. First described by Schaaff and Whetten[36] and later refined independently by Jin[37,38] and Dass[39,40] this approach is one in which excess thiolate ligand etches away less stable materials, generally leaving behind especially stable, "magic" number clusters. The results of etching type synthesis now account for the majority of single crystal X-ray structures (Table 1).

Alternatively, some monolayer-protected clusters (MPCs) can be synthesized directly,[41–43] without an etching step. Purification of these clusters from by-products is often a larger challenge in direct synthesis. Direct synthetic methods enabled the first X-ray crystal structure,[12] but have not proven to be as generalizable as etching methods.

In either case, it's generally advisable that the synthesis and purification result in homogeneous material prior to attempting crystallization. It's not a strict requirement, however. Notably, the original crystallization of $Au_{102}(SR)_{44}$ was from a solution in which it was only a minor component, where the primary

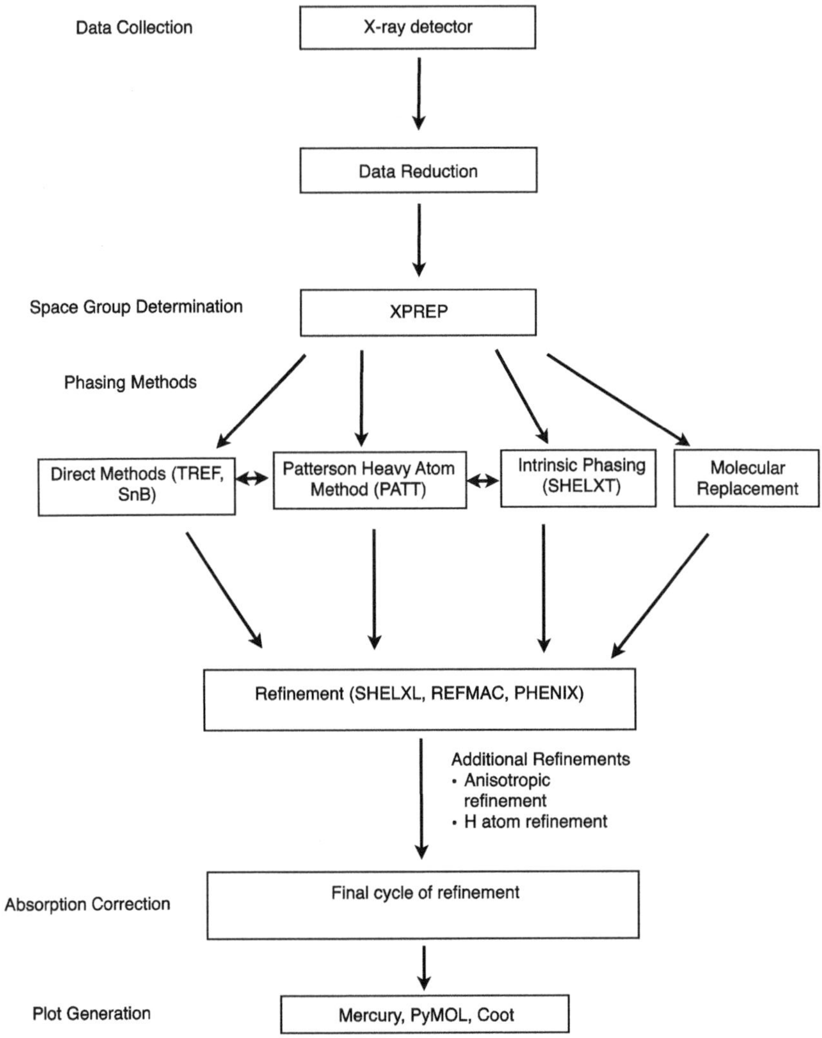

FIGURE 3 Flow chart schematic showing the basic steps of solving a single crystal structure.

component in the solution was $Au_{144}(SR)_{60}$. The serendipitous "purification by crystallization" is not a reliable method for other structures, however.

5.2.2 Growing Crystals

The largest bottleneck in generating single crystal X-ray structures is the production of X-ray quality single crystals. Most success found with phenyl-containing ligands, although recent structures of clusters protected by ethanethiolate,[44] cyclohexanethiolate,[45] and *t*-butyl thiolate[33,35] demonstrates that

TABLE 1 The Known-to-Date Crystal Structures of Thiolate-Protected Gold Clusters

Compound	Space Group	X-ray Source	*R*-factor	Citations	Isotropic/Anisotropic
$Au_{133}(TBBT)_{52}$	P-1	CuKα	15.51%	35	Au only
$Au_{133}(TBBT)_{52}$	C2/c	MoKα	15.91%	33	Au only
$Au_{102}(p\text{-MBA})_{44}$	C2/c	Synchrotron (1.0362 Å, 0.9795 Å)	8.1%	12	Au only
$Au_{102}(p\text{-MBA})_{40}(4BBT)_4$	C2/c	Synchrotron (0.827 Å)	8.4%	5	Au, S, and Br
$Ag_{44}(SPhF_2)_{30}$	P-1	MoKα	5.67%	72	All but H
$Ag_{44}(p\text{-MBA})_{30}$	R-3c	CuKα	5.2%	55	Ag, S, some of the C
$Au_{12}(Ag)_{32}(SPhF_2)_{30}$	P-1	MoKα	9.41%	72	All but H
$Au_{39}(PPh_3)_{14}Cl_6$	C2/c	MoKα	9.2%	73	None
$Au_{38}(SC_2H_4Ph)_{24}$	P-1	MoKα	24.25%	24	Au and S
$Au_{38-x}Ag_x(SCH_2CH_2Ph)_{24}$	P-1	Synchrotron (0.7749 Å)	9.79%	74	Au, Ag, S
$Au_{36}(SPh\text{-tBu})_{24}$	C2/c	CuKα	13.02%	75	Au, S, some C
$Au_{28}(SPh\text{-tBu})_{20}$	P-1	CuKα	17.49%	76	Au, S

Continued

TABLE 1 The Known-to-Date Crystal Structures of Thiolate-Protected Gold Clusters—cont'd

Compound	Space Group	X-ray Source	*R*-factor	Citations	Isotropic/Anisotropic
$Au_{25}(C_2H_4Ph)_{18}^-$	P-1	MoKα	6.87%	57	All but H
$Au_{25}(SC_2H_4Ph)_{18}^-$	P-1	MoKα	5.3%	56	Au, S, some C
$Au_{25}(SC_2H_4Ph)_{18}^0$	Pccn	MoKα	5.16%	71	All but H
$Au_{25}(PET)_{16}(4BBT)_2$	P-1	Synchrotron (0.918 Å)	7.06%	69	All but H
$[Au_{25}(SePh)_{18}]^-$	P-1	MoKα	12.33%	77	All but H
$Au_{25}(SEt)_{18}$	P-1	CuKα	7.47%	44	All but H
$Au_{24}(SCH_2Ph\text{-}tBu)_{20}$	P-1	CuKα	9.16%	78	Au, S, some C
$Au_{24}(SePh)_{20}$	P-1	MoKα	24.65%	79	All but H
$[Au_{23}(SC_6H_{11})_{16}]^-$	Ccca	CuKα	14.62%	45	All but H and N
$Au_{20}(SPh\text{-}tBu)_{16}$	$P222_1$	CuKα	13.03%	25	All but H
$Au_{20}([(C_6H_5)_2PCH_2CH_2]_3P)_4Cl_4$	Pbca	Synchrotron (0.7749 Å)	7.87%	80	All but H
$Au_{18}(SC_6H_{11})_{14}$	P-1	CuKα	14.35%	32	All but H

ligands with a rigid aromatic ring are not requisite for forming the crystal contacts needed to stabilize a crystal of clusters. Crystallization approaches differ when using water-soluble clusters as opposed to organosoluble clusters. In crystallization of aqueous material, the well-developed tools of protein crystallography can be used, whereas for crystallization of organosoluble material, there are more limited tools available. As a general statement, however, the tools developed for protein crystallography have so far had limited success in the growth of water-soluble clusters.

5.2.2.1 Water-Soluble Material

The first crystal structure of a thiolate-protected gold cluster solved was that of $Au_{102}(SR)_{44}$. The structure was solved by a lab with substantial expertise in challenging protein crystal structures.[46,47] The lab was interested in using exactly defined thiolate-protected clusters in biological structure determination by transmission electron microscopy[48] and had become expert in the direct synthesis of thiolate-protected clusters as well as their rigid conjugation to proteins.[42,49–51] As structure of thiolate-protected gold clusters was then a matter of conjecture, the lab began applying the macromolecular crystallographic methods familiar within that lab to the mystery of the structure of thiolate-protected clusters.

A common approach for discovering conditions in which proteins will form high-quality crystals is sparse matrix screening.[52,53] The method is one in which hundreds to thousands of crystallization trial conditions are attempted essentially simultaneously, either in 96-well, 384-well, or other large sample number formats. Each crystallization condition contains a discrete set of solutes, allowing systematic variation of pH, salt concentration, salt valency, and other additives including alcohols, glycols, and other high and low molecular weight molecules that may influence the formation of crystals.[52,53] After crystals are set up, typically in optically appropriate containers for downstream examination of crystals by microscopes, the crystallization attempts are periodically examined and scored. Preliminary conditions that give crystals in a coarse grain or sparse matrix screen can be used as an anchor for fine-grain screening of crystallization conditions. As presently deployed, the method is a quick and wide test of different pH, salts, and precipitants using a relatively small amount of sample. The method is exceptionally useful in protein crystallography, and is also now adapted for robotic crystal screening.

For crystallography of gold clusters, it has so far produced crystallization conditions amenable only for $Au_{102}(SR)_{44}$ and the ligand-exchanged version of this compound. Other thiolate-protected clusters have been given substantial effort in sparse matrix screening. Most notably $Au_{144}(p\text{-MBA})_{60}$ prepared by many different labs is now well known to produce hexagonal plate crystals in many conditions[42,54] but these crystals despite substantial efforts invariably produce diffraction to insufficient resolution to allow crystallographic solution. Diffraction to 5 Å resolution is frequently observed for these crystals which also appear from visual inspection of their diffraction pattern to suffer substantial twinning, making further processing difficult.

The other water-soluble MPC solved crystallographically, M_4Ag_{44} (p-MBA)$_{30}$, was prepared into single crystals by a method typical for crystallizing small molecules—slow evaporation of solvent from a concentrated solution. In this case, evaporation was perhaps especially slow and controlled, as rhombus-shaped crystals appeared only after 1−3 days of this treatment.[55] Despite the comparatively limited success of protein crystallographic methods for discovering crystal growth conditions, we anticipate that future water-soluble structures will be enabled by sparse matrix style screening of conditions that allow the growth of crystals of water-soluble nanoclusters, in part because the surface structure of thiolate-protected clusters is topologically complex in a manner reminiscent of the topological complexity of proteins. This topological complexity may require unexpected cofactors for crystallization into well-ordered crystalline lattices, where the unexpected cofactors can be discovered systematically by methods already pioneered for protein crystallography.

5.2.2.2 *Organosoluble Material*

Most examples of crystallization of organosoluble thiolate-protected clusters proceed by simple growth of crystals from saturated or supersaturated solutions of MPCs. The means by which a saturated or supersaturated solution is generated can vary. In some instances, it's a simple generation of a supersaturated solution followed by waiting, as in the case of $Au_{25}(PET)_{18}$.[56,57] More commonly, an unsaturated solution is made to be saturated by vapor diffusion of a poor solvent into a good solvent. For instance in the crystallization of $Au_{23}(SR)_{15}$, a highly concentrated solution of $Au_{23}(SR)_{15}$ was made—at 4 mg/mL, followed by vapor diffusion of pentane into that solution over 1−2 days, resulting in X-ray quality crystals.[45] In the case of $Au_{25}(Set)_{18}$, vapor diffusion of diisopropyl poor solvent into a solution of toluene containing the clusters resulted in crystals.[44] Several other examples of crystallization by vapor diffusion are in the literature.

5.2.3 Collecting Single Crystal X-ray Diffraction Data

Technical details of data collection such as exposure time and oscillation angle for crystallography of protected metal clusters do not differ substantially from other single crystal experiments, can depend more upon the X-ray source and on the specific nature of the crystal, including the crystalline symmetry (i.e., space group). Details of collection can be readily found elsewhere. This section focuses primarily on selection of X-ray sources for the collection of single crystal data.

Collection of single crystal diffraction data is done with either "home source" X-ray generators based on most commonly Cu or Mo X-ray tubes or for more challenging structures with synchrotron X-ray sources. Home sources are widespread and found at many universities, and for most nonbiological molecules represent the X-ray source for which first attempts at collecting data are made. In most cases for thiolate-protected clusters, data collection on home-source diffractometers, often with assistance from a local full-time crystallographer, results in data sets of sufficient quality to solve the single crystal structure. When crystal quality is insufficient (i.e., when not enough reflections can be recorded or when crystals do not diffract to sufficiently high resolution, or when multiple wavelengths are needed), synchrotron X-ray sources may be employed. The demand from biological macromolecular crystallography for synchrotron sources has provoked a proliferation of specialized single crystal synchrotron beamlines that are relatively straightforward to access, typically through an application process.

Synchrotron sources provide substantially higher flux of X-rays. For small crystals and/or crystals containing large unit cells, synchrotron X-ray sources can provide the flux "needed" to record a complete data set—either by allowing collection of more reflections or higher resolution data in a reasonable amount of time. To illustrate this, the continuous time required to collect an $Au_{102}(p\text{-MBA})_{44}$ data set on a home source can stretch into weeks, while a data set can be recorded in a few minutes on a synchrotron source. Synchrotron sources also have the advantage of wavelength tunability, whereas home sources are limited to the single approximately monochromatic (K-edge-check) of Mo or Cu X-ray tubes. Wavelength tunability may be advantageous in allowing refinement of the positions of "heteroatoms"—such as Br atoms in incoming ligands after ligand exchange or the location of dopant atoms, such as Pt or Pd in $Au_{24}Pd(SR)_{18}$,[58,59] and $Au_{36}Pd_2$. For instance, the wavelength can be tuned to the Pd K-edge, and the subsequent absorption of X-rays by Pd allows refinement of the location of these atoms in a difference map. Here the difference map is considered as the maps generated by two different wavelengths. While the use of synchrotron sources is so far useful in just a handful of structures, we anticipate that solution of clusters larger than $Au_{102}(SR)_{44}$, such as $Au_{144}(SR)_{60}$ or $Au_{329}(SR)_{84}$,[60] and their alloys will require similarly large flux to Au_{102}, as those structures are also likely to produce crystals with large unit cells and small overall crystal dimensions.

Data collection at synchrotron sources is now typically done through remote control of beamline robots.[61] In other words, crystals are grown, mounted, and frozen in a wet lab, and then shipped in a dry dewar to a synchrotron beamline. At the beamline, a beamline scientist mounts the crystals in a robot, and the end user can then collect data on each crystal, controlling the robot and beamline remotely from the user's home institution.

If diffraction is of sufficient quality—i.e., a complete data set can be taken at ~3.0 Å resolution or better, without obvious twinning of spots, then the set of reflections can be indexed, a preliminary unit cell determined, and the task of calculating a structure from the recorded diffraction pattern can begin. The task of indexing reflections is generally handled by manufacturer provided software at home source X-ray facilities; for synchrotron sources, other software can be used such as XRD.[62]

5.2.4 Solving Single Crystal Structures of Protected Metal Clusters

All but a handful of single crystal structures of thiolate-protected clusters are solved through the commercially available SHELXTL software, which is based on SHELX[63] but with a convenient graphical user interface provided by X-ray diffractometer vendor Bruker. This approach is common for most small molecules that are solved crystallographically.

"Solving" a crystal structure refers to the process of interpreting a set of reflections into an atomic model. In essence, the reflections are converted into an electron density map by a Fourier transform. However, because phase information is lost, some additional steps are made to recover enough phase information to build an initial model; the main hurdle in solving a crystal structure is obtaining phase information such that an initial model can be built. Once an initial solution is found, successive rounds of refinement and model building allow the production of a refined model. Structures taking a "small-molecule approach" almost universally use the phasing options built into SHELXTL software, which are now so well implemented as to be almost invisible. The two implemented phasing methods are direct methods and Patterson methods; thus an initial electron density map (a Fourier synthesis of the initially phased reflection intensities) is easy to generate and the majority of hands-on work is in model building and refinement.

Unlike other crystal structures detailed in this chapter, $Au_{102}(SR)_{44}$ was phased not by direct methods in SHELXTL, but rather by the stand-alone Shake and Bake (SnB) software.[64] This software determines initial phases by checking the success of multiple initial phase trials and iterating the phase solution from the most successful initial trial. Such an algorithm is also implemented in SHELXD.

All of the phasing software used so far successfully for solution of the clusters outlined in Table 1 require comparatively high resolution (1.2 Å or better) data. We expect that as larger clusters are solved, crystal quality will limit diffraction to poorer resolution that is still solvable (3.3 to 1.5 Å resolution). Such structures may need to use phasing methods such as Patterson methods or, if the molecules are not centrosymmetric (i.e., lacking an inversion center) anomalous dispersion methods.[65] One should note, however, that if macromolecular software packages are used for structure solution, the space groups that are not encountered in biological crystallography, such as centrosymmetric space groups are often not well implemented. Only 65 space groups are encountered in biological crystallography; the other 165 space groups may be poorly supported by software written to solve biological molecules.

Into this electron density map, a model of the cluster (or some part of the cluster) can be built. For most small molecules built in SHELXTL, this involves assigning the Q-peaks in XSHELL. For macromolecules (and indeed the initial $Au_{102}(SR)_{44}$ structure) model building is popularly done in Coot.[66] One advantage of Coot over XSHELL is the ability to view a contour map of the electron density. After as much of the model as can be fit into the electron density map is placed, the model is refined against the electron density, usually as a least squares refinement (i.e., SHELXL). Subsequent cycles of model building and refinement continue until no further improvements can be made to the model, typically as judged by a metric such as the R and/or wR_2 value.

R-factor measures the discrepancy between a crystallographic model and the experimental diffraction data. $R = \frac{\sum |F_{obs} - F_{calc}|}{\sum |F_{obs}|}$, where F is the amplitude of the structure factor calculated from a model and those from the original X-ray diffraction data. Depending on program used, will usually give you two different R-factors. R_{work} and R_{free}, R_{work} is the R value for all of the observations while R_{free} is only a random 5% of the data that are not used in the refinement, but create a metric for overrefinement; briefly, it is possible to refine a model in such a way that improves the R-value, and divergence between R and R_{free} values indicates that the model is overrefined.[67]

Because this is not an automatically calculated or reported value for small-molecule crystallography in SHELX, the reporting of this value is not yet common in the crystallography of metal clusters. However, for lower-resolution data sets, it is an important metric to report.

5.2.4.1 Refining Disorder

Refining crystallographic disorder is an important topic for both developing models of mixed metal alloy clusters and also for examining ligand-exchange structures. Classically, crystallographic disorder is categorized in three ways: substitutional disorder, positional disorder, and solvent disorder. All three forms of disorder are encountered in crystal structures of ligand-protected clusters, with substitutional disorder being the most important kind of disorder for understanding fundamental scientific questions in ligated metal clusters.

Positional disorder occurs when one part of a structure is observed in two or more positions. In ligated metal clusters, it is most commonly encountered in ligand shells, where one ligand may be disordered over two or more positions. Indeed, some p-MBA ligands may be nearly freely rotating between the carboxylic acid and sulfur atom. Because the light atoms in the ligand shell are harder to locate in difference maps (because the light atoms of the ligand shell account for a miniscule amount of electron density and can be partially obscured by "Fourier ripples" emanating from heavy metals in the map) it is rare to attempt refinement of positional disorder of ligands. Indeed, in many structures, the ligand shell is only partially resolved. Positional disorder is present, but often not modeled because of the difficulties inherent in building the model for the ligand shell.

Solvent disorder is observed in ligated metal cluster structures because larger molecules very often accommodate large solvent channels. Solvent disorder is almost never refined in macromolecular crystallography. While solvent disorder (large solvent channels) often results in high-level CheckCIF alerts, these are easily explainable because large solvent channels with accompanying disorder are expected. In small-molecule crystallography, solvent disorder can be refined with the SQUEEZE command in PLATON.[68] While this can better account for all the electron density in a model, it is a controversial practice because it is in essence refining the data against the model.

Substitutional disorder occurs when two or more different atom types (or ligand types) occupy the same position in a crystal structure. In refinement, relative occupancies of each atom type or ligand type are produced. In the ligated cluster literature, substitutional disorder refinement is used to refine the occupancy of ligands in exchange structures.[5,69] In ligand-exchange structures, brominated incoming ligands were used because the electron density of Br is much easier to pick out in an electron density map compared to lighter elements. Now there are two examples also in which the occupancy of Au and Ag in bimetallic clusters was refined.[70] Indeed, questions about the locations and relative occupancy of metal atom sites by alloyed atoms are at the forefront of research now, as well as questions regarding the fluxionality of the ligand shell. Thus it is expected that structures that refine and quantify substitutional disorder will become increasingly better represented in the literature of crystal structures of ligated metal clusters.

5.2.4.2 Evaluating Solutions

Standard values reported for X-ray crystal structures include R/R_1, wR_2, "Goodness of fit" (GooF), and x. R is defined above and is a useful metric for evaluating model quality. For small molecules, an R value of 5% or smaller is considered good for a typical structure; for macromolecules, an R value of 20% or smaller is considered good. In the crystallography of ligated metal clusters, reported R values range from 5.2% to 25.4%, with an average R value for all structures published so far of 11.25%. In this metric, ligated cluster structures clearly fall between "small-molecule" and "macromolecular" ideals. Another reported metric, GooF, is based on the squared intensities where $\text{GooF} = S = \sqrt{\sum \frac{w(F_{\text{obs}}^2 - F_{\text{calc}}^2)}{n-p}}$ where w is the weight term, n is the number of reflections, and p is the number of parameters. GooF and the restrained GooF should come close to unity, and if it does not is generally an indication that there are inadequacies in the model (this is normally seen in macromolecular models).

Per standard practice, all single crystal structures are validated through CheckCIF/PLATON prior to publication.[68] This automated vetting of structures can highlight problems that are not immediately apparent in the statistical checks on structure outlined above. Checks on structural factors, molecular geometry, space group symmetry, and anisotropic displacement parameters are made. It is a useful tool in determining errors or areas of potential concern.

5.3 EXAMPLES

Although the single crystal X-ray structures allowed profound new structural understanding, they were initially slow to be solved. For the 5 years spanning 2007–2011, only three single crystal structures were solved, beginning with $Au_{102}(SR)_{44}$ in 2007, $Au_{25}(SR)_{18}$ in two oxidation states in 2008 and $Au_{38}(SR)_{24}$ in 2010.[12,24,56,57,71] These initial structures reveal the basic anatomy of thiolate-protected cluster compounds now observed in almost all thiolate-protected gold clusters.

The calendar year 2012 marked a turning point in which crystal structure determination became more routine; 2014 emerged as watershed year with a total of 11 structures published in that year alone.

Table 1 summarizes all the known-to-date crystal structures of thiolate-protected gold clusters, ranging in size from $Au_{18}(SR)_{14}$ to $Au_{133}(SR)_{52}$. Also shown here are alloy and ligand-exchange clusters. Early on, all clusters solved were protected by aromatic ligands, but this changed in 2013 with the observation of cyclohexanethiolate-protected $Au_{23}(SR)_{16}$. $Au_{25}(SR)_{18}$ structures are the most represented basic formulation, with solutions of this structure in multiple oxidation states, as an Au-Ag alloy core and with both ethanethiolate ligands and as a ligand-exchanged structure. Figure 4 shows example structures for five of the compounds.

5.4 SUMMARY AND PROSPECTS

There are 12 nuclearities so far resolved, ranging in size from $Au_{18}(SR)_{14}$ to $Au_{133}(SR)_{52}$. Alloy clusters are resolved with Ag for Au_{25}, Au_{38}, and Au_{44} clusters. Ligand exchange is resolved for Au_{25} and Au_{102} clusters. Multiple oxidation states are resolved for Au_{25}.

The frontier of the field is in two areas—first, resolving even larger clusters and second determining the structure of alloy clusters. Several candidate clusters larger than $Au_{133}(SR)_{52}$ are reported, including $Au_{144}(SR)_{60}$, $Au_{329}(SR)_{84}$, and $Au_{\sim 500}$. There is some suggestion that larger clusters may be impossible to make as exact species, meaning that it may be difficult to

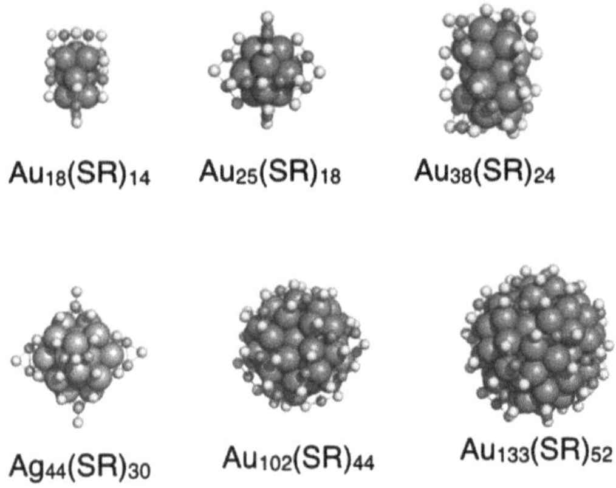

FIGURE 4 Examples of single crystal structures. Ligand atoms (other than sulfur) are omitted for clarity.

prepare an "atomically precise" solution from which a single crystal can be grown. The difficulty presented by the $Au_{144}(SR)_{60}$ structure, which remains unsolved despite efforts of many groups, also dims the prospects for solving larger clusters. Solving clusters larger than Au_{133}, however, is not without precedent. The large Pd/Pt clusters solved by Dahl, as well as the giant polyoxometalate clusters solved by others, are examples of crystallography on even larger cluster materials.[34,81,82] In solving the structures of larger nanoclusters, it is likely that macromolecular methods will find more use, as the data are likely to be of lower quality, potentially making determining initial phases difficult.

Advances in X-ray methods may also impact the structure determination of giant metal clusters. In a broad sense, "brighter and brighter" X-ray sources enable successful data collection on smaller and smaller crystals. The "brightest" proposed X-ray sources, such as the Linac Coherent Light Source (LCLS) can theoretically collect atomic resolution diffraction from a series of single molecule data collection events,[83] although the greatest successes with the LCLS and similar light sources are for "nanocrystals."[84] Structural information of larger materials, however, may ultimately come from recently developed electron microscopy methods or other methods which do not require single crystals.[17]

The other frontiers in the area involve resolving ligand exchange and resolving the positions of mixed metals in alloys. These will both presumably rely on refinement of disorder in metal clusters, in either the organic or inorganic parts of the structure. The refinement of positions of Pd and Cu atoms should be as accomplishable as refinement of positions of Ag atoms, which is already done. The refinement of positions of Pt atoms may be more difficult because of the similar electron densities of Pt and Au; use of tunable radiation at synchrotrons, with radiation tuned to the absorbance edges of either Pt or Au may allow localization of Pt or other atoms similar in electron density.

Overall, X-ray crystallography has proven indispensible to understanding the structure of thiolate-protected metal clusters. A mixture of both small-molecule and macromolecular methods are used to solve structures in this subfield. While there are challenges in the practice compared to small-molecule structures, future more challenging structures are likely to be methodologically informed by macromolecular methods.

REFERENCES

1. Crick F. *What mad pursuit: a personal view of scientific discovery.* New York: Basic Books; 2008.
2. Walter M, Akola J, Lopez-Acevedo O, Jadzinsky PD, Calero G, Ackerson CJ, et al. A unified view of ligand-protected gold clusters as superatom complexes. *Proc Natl Acad Sci USA* 2008;**105**:9157–62.
3. Wyrwas RB, Alvarez MM, Khoury JT, Price RC, Schaaf TG, Whetten RL. The colours of nanometric gold – optical response functions of selected gold-cluster thiolates. *Eur Phys J D* 2007;**43**:91–5.
4. Hostetler M, Templeton A, Murray RW. Dynamics of place-exchange reactions on monolayer-protected gold cluster molecules. *Langmuir* 1999;**15**:3782–9.
5. Heinecke CL, Ni TW, Malola S, Mäkinen V, Wong OA, Häkkinen H, et al. Structural and theoretical basis for ligand exchange on thiolate monolayer protected gold nanoclusters. *J Am Chem Soc* 2012;**134**:13316–22.
6. Oila MJ, Koskinen AMP. Chirally modified gold nanoparticles: nanostructured chiral ligands for catalysis. *Arkivoc* 2006;**xv**:76–83.
7. Knoppe S, Dolamic I, Bürgi T. Racemization of a chiral nanoparticle evidences the flexibility of the gold-thiolate interface. *J Am Chem Soc* 2012;**134**:13114–20.
8. Noguez C, Garzon IL. Optically active metal nanoparticles. *Chem Soc Rev* 2009;**38**:757–71.
9. Sardar R, Funston A, Mulvaney P, Murray RW. Gold nanoparticles: past, present, and future. *Langmuir* 2009;**25**:13840–51.
10. Li G, Jin R. Atomically precise gold nanoclusters as new model catalysts. *Acc Chem Res* 2013;**46**:1749–58.
11. Billinge SJL, Levin I. The problem with determining atomic structure at the nanoscale. *Science* 2007;**316**:561–5.
12. Jadzinsky PD, Calero G, Ackerson CJ, Bushnell DA, Kornberg RD. Structure of a thiol monolayer-protected gold nanoparticle at 1.1 Å resolution. *Science* 2007;**318**:430–3.
13. Juhás P, Cherba DM, Duxbury PM, Punch WF, Billinge SJL. Ab initio determination of solid-state nanostructure. *Nature* 2006;**440**:655–8.
14. Simpson CA, Farrow CL, Tian P, Billinge SJL, Huffman BJ, Harkness KM, et al. Tiopronin gold nanoparticle precursor forms aurophilic ring tetramer. *Inorg Chem* 2010;**49**:10858–66.
15. Schaaff T, Shafigullin MN, Khoury JT, Vezmar I, Whetten RL, Cullen WG, et al. Isolation of smaller nanocrystal-Au molecules: robust quantum effects in optical spectra. *J Phys Chem B* 1997;**101**:7885–91.

16. Bahena D, Bhattarai N, Santiago U, Tlahuice A, Ponce A, Bach SBH, et al. STEM electron diffraction and high-resolution images used in the determination of the crystal structure of the $Au_{144}(SR)_{60}$ cluster. *J Phys Chem Lett* 2013;**4**:975–81.
17. Azubel M, Koivisto J, Malola S, Bushnell D, Hura GL, Koh AL, et al. Nanoparticle imaging. Electron microscopy of gold nanoparticles at atomic resolution. *Science* 2014;**345**:909–12.
18. Schuwirth BS, Borovinskaya MA, Hau CW, Zhang W, Vila-Sanjuro A, Holton JM, et al. Structures of the bacterial ribosome at 3.5 Å resolution. *Science* 2005;**310**:827–34.
19. Whetten R, Khoury JT, Alvarez MM, Murthy S, Vezmar I, Wang ZL, et al. Nanocrystal gold molecules. *Adv Mater* 1996;**8**:428–33.
20. Alvarez MM, Khoury JT, Schaaff TG, Shafigullin M, Vezmar I, Whetten RL. Critical sizes in the growth of Au clusters. *Chem Phys Lett* 1997;**266**:91–8.
21. Cleveland C, Landman U, Shafigullin M, Stephens P, Whetten RL. Structural evolution of larger gold clusters. *Z Phys D At Mol Clust* 1997;**40**:503–8.
22. Cleveland C, Landman U, Schaaff TG, Shafigullin MN, Stephens PW, Whetten RL. Structural evolution of smaller gold nanocrystals: the truncated decahedral motif. *Phys Rev Lett* 1997;**79**:1873–6.
23. Häkkinen H, Walter M, Grönbeck H. Divide and protect: capping gold nanoclusters with molecular gold-thiolate rings. *J Phys Chem B* 2006;**110**:9927–31.
24. Qian H, Eckenhoff WT, Zhu Y, Pintauer T, Jin R. Total structure determination of thiolate-protected Au_{38} nanoparticles. *J Am Chem Soc* 2010;**132**:8280–1.
25. Zeng C, Liu C, Chen Y, Rosi NL, Jin R. Gold-thiolate ring as a protecting motif in the $Au_{20}(SR)_{16}$ nanocluster and implications. *J Am Chem Soc* 2014;**136**:11922–5.
26. Taylor KA, Glaeser RM. Retrospective on the early development of cryoelectron microscopy of macromolecules and a prospective on opportunities for the future. *J Struct Biol* 2008;**163**: 214–23.
27. Billinge SJL, Kanatzidis MG. Beyond crystallography: the study of disorder, nanocrystallinity and crystallographically challenged materials with pair distribution functions. *Chem Commun* 2004:749–60.
28. Petkov V, Billinge SJL, Shastri SD, Himmel B. Polyhedral units and network connectivity in calcium aluminosilicate glasses from high-energy X-ray diffraction. *Phys Rev Lett* 2000;**85**: 3436–9.
29. Rhodes G. *Crystallography made crystal clear: a guide for users of macromolecular models*. 3rd ed. Massachusetts: Academic Press; 2006.
30. Massa W. *Crystal structure determination*. Germany: Springer; 2004.
31. Müller P, Herbst-Irmer R, Spek AL, Schneider TR, Sawaya M. *Crystal structure refinement*. England: Oxford University Publishing; 2006.
32. Das A, Liu C, Byun HY, Nobusada K, Zhao S, Rosi N, et al. Structure determination of $[Au_{18}(SR)_{14}]$. *Angew Chem Int Ed* 2015;**54**:3140–4.
33. Dass A, Theivendran S, Nimmala PR, Kumara C, Jupally VR, Fortunelli A, et al. $Au_{133}(SPh\text{-}t\mathrm{Bu})_{52}$ nanomolecules: X-ray crystallography, optical, electrochemical, and theoretical analysis. *J Am Chem Soc* 2015;**137**:4610–3.
34. Mednikov EG, Jewell MC, Dahl LF. Nanosized $(\mu_{12}\text{-Pt})Pd_{164-x}Pt_x(CO)_{72}(PPh_3)_{20}$ ($x \approx 7$) containing Pt-centered four-shell 165-Atom Pd-Pt core with unprecedented intershell bridging carbonyl ligands: comparative analysis of icosahedral shell-growth patterns with geometrically related $Pd_{145}(CO)_x(PEt_3)_{30}$ ($x \approx 60$) containing capped three-shell Pd_{145} core. *J Am Chem Soc* 2007;**129**:11619–30.
35. Zeng C, Chen Y, Kirschbaum K, Appavoo K, Sfeir MY, Jin R. Structural patterns at all scales in a nonmetallic chiral $Au_{133}(SR)_{52}$ nanoparticle. *Sci Adv* 2015;**1**:e1500045–51.

36. Schaaff TG, Whetten RL. Controlled etching of Au:SR cluster compounds. *J Phys Chem B* 1999;**103**:9394–6.
37. Jin R, Qian H, Wu Z, Zhu Y, Zhu M, Mohanty A, et al. Size focusing: a methodology for synthesizing atomically precise gold nanoclusters. *J Phys Chem Lett* 2010;**1**:2903–10.
38. Qian H, Zhu Y, Jin R. Size-focusing synthesis, optical and electrochemical properties of monodisperse $Au_{38}(SC_2H_4Ph)_{24}$ nanoclusters. *ACS Nano* 2009;**3**:3795–803.
39. Nimmala PR, Yoon B, Whetten RL, Landman U, Dass A. $Au_{67}(SR)_{35}$ nanomolecules: characteristic size-specific optical, electrochemical, structural properties and first-principles theoretical analysis. *J Phys Chem A* 2013;**117**:504–17.
40. Dass A. Faradaurate nanomolecules: a superstable plasmonic 76.3 kDa cluster. *J Am Chem Soc* 2011;**133**:19259–61.
41. Levi-Kalisman Y, Jadzinsky PD, Kalisman N, Tsunoyama H, Tsukuda T, Bushnell DA, et al. Synthesis and characterization of $Au_{102}(p\text{-}MBA)_{44}$ nanoparticles. *J Am Chem Soc* 2011;**133**:2976–82.
42. Ackerson CJ, Jadzinsky PD, Sexton JZ, Bushnell DA, Kornberg RD. Synthesis and bioconjugation of 2 and 3 nm-diameter gold nanoparticles. *Bioconjug Chem* 2010;**21**:214–8.
43. Wong OA, Compel WS, Ackerson CJ. Combinatorial discovery of cosolvent systems for production of narrow dispersion thiolate-protected gold nanoparticles. *ACS Comb Sci* 2015;**17**:11–8.
44. Dainese T, Antonello S, Gascón JA, Pan F, Perera NV, Ruzzi M, et al. $Au_{25}(SEt)_{18}$, a nearly naked thiolate-protected Au_{25} cluster: structural analysis by single crystal X-ray crystallography and electron nuclear double resonance. *ACS Nano* 2014;**8**:3904–12.
45. Das A, Li T, Nobusada K, Zeng C, Rosi NL, Jin R. Nonsuperatomic $[Au_{23}(SC_6H_{11})_{16}]^-$ nanocluster featuring bipyramidal Au_{15} kernel and trimeric $Au_3(SR)_4$ motif. *J Am Chem Soc* 2013;**135**:18264–7.
46. Cramer P, Bushnell DA, Fu J, Gnatt AL, Maier-Davis B, Thompson NE, et al. Architecture of RNA polymerase II and implications for the transcription mechanism. *Science* 2000;**288**: 640–9.
47. Gnatt AL, Cramer P, Fu J, Bushnell DA, Kornberg RD. Structural basis of transcription: an RNA polymerase II elongation complex at 3.3 Å resolution. *Science* 2001;**292**:1876–82.
48. Jensen GJ, Kornberg RD. Single-particle selection and alignment with heavy atom cluster-antibody conjugates. *Proc Natl Acad Sci USA* 1998;**95**:9262–7.
49. Ackerson CJ, Sykes MT, Kornberg RD. Defined DNA/nanoparticle conjugates. *Proc Natl Acad Sci USA* 2005;**102**:13383–5.
50. Ackerson CJ, Jadzinsky PD, Jensen GJ, Kornberg RD. Rigid, specific, and discrete gold nanoparticle/antibody conjugates. *J Am Chem Soc* 2006;**128**:2635–40.
51. Ackerson CJ, Jadzinsky PD. *Monolayer-protected gold clusters: improved synthesis and bioconjugation*. 2007. US 8304257 B2.
52. Wooh JW, Kidd RD, Martin JL, Kobe B. Comparison of three commercial sparse-matrix crystallization screens. *Acta Crystallogr D Biol Crystallogr* 2003;**59**:769–72.
53. Thakur AS, Robin G, Guncar G, Saunders NFW, Newman J, Martin JL, et al. Improved success of sparse matrix protein crystallization screening with heterogeneous nucleating agents. *PLoS One* 2007;**2**:e1091.
54. Koivisto J, Salorinne K, Mustalahti S, Lahtinen T, Malola S, Häkkinen H, et al. Vibrational perturbations and ligand–layer coupling in a single crystal of $Au_{144}(SC_2H_4Ph)_{60}$ nanocluster. *J Phys Chem Lett* 2014;**5**:387–92.
55. Desireddy A, Conn BE, Guo J, Yoon B, Barnett RN, Monahan BM, et al. Ultrastable silver nanoparticles. *Nature* 2013;**501**:399–402.

56. Heaven MW, Dass A, White PS, Holt KM, Murray RW. Crystal structure of the gold nanoparticle $[N(C_8H_{17})_4][Au_{25}(SCH_2CH_2Ph)_{18}]$. *J Am Chem Soc* 2008;**130**:3754–5.
57. Zhu M, Aikens CM, Hollander FJ, Schatz GC, Jin R. Correlating the crystal structure of a thiol-protected Au_{25} cluster and optical properties. *J Am Chem Soc* 2008;**130**:5883–5.
58. Christensen SL, MacDonald MA, Chatt A. Dopant location, local structure, and electronic properties of $Au_{24}Pt(SR)_{18}$ nanoclusters. *J Phys Chem C* 2012;**116**:26932–7.
59. Parker JF, Fields-Zinna CA, Murray RW. The story of a monodisperse gold nanoparticle: $Au_{25}L_{18}$. *Acc Chem Res* 2010;**43**:1289–96.
60. Kumara C, Dass A. $Au_{329}(SR)_{84}$ nanomolecules: compositional assignment of the 76.3 kDa plasmonic faradaurates. *Anal Chem* 2014;**86**:4227–32.
61. González A, Moorhead P, McPhillips SE, Song J, Sharp K, Taylor JR, et al. Web-Ice: integrated data collection and analysis for macromolecular crystallography. *J Appl Crystallogr* 2008;**41**:176–84.
62. Kabsch W. XDS. *Acta Crystallogr D Biol Crystallogr* 2010;**66**:125–32.
63. Sheldrick GM. A short history of SHELX. *Acta Crystallogr A Found Crystallogr* 2008;**64**: 112–22.
64. Xu H, Weeks CM. Rapid and automated substructure solution by Shake-and-Bake. *Acta Crystallogr D Biol Crystallogr* 2008;**64**:172–7.
65. McCoy AJ, Grosse-Kuntsleve RW, Adams PD, Winn MD, Storoni LC, Read RJ. Phaser crystallographic software. *J Appl Crystallogr* 2007;**40**:658–74.
66. Emsley P, Lohkamp B, Scott WG, Cowtan K. Features and development of coot. *Acta Crystallogr D Biol Crystallogr* 2010;**66**:486–501.
67. Brünger AT. Free *R* value: a novel statistical quantity for assessing the accuracy of crystal structures. *Nature* 1992;**355**:472–5.
68. Spek AL. Single-crystal structure validation with the program PLATON. *J Appl Crystallogr* 2003;**36**:7–13.
69. Ni TW, Tofanelli MA, Phillips BD, Ackerson CJ. Structural basis for ligand exchange on $Au_{25}(SR)_{18}$. *Inorg Chem* 2014;**53**:6500–2.
70. Kumara C, Aikens CM, Dass A. X-ray crystal structure and theoretical analysis of $Au_{25-x}Ag_x(SCH_2CH_2Ph)_{18}^-$ alloy. *J Phys Chem Lett* 2014;**5**:461–6.
71. Zhu M, Eckenhoff WT, Pintauer T, Jin R. Conversion of anionic $[Au_{25}(SCH_2CH_2Ph)_{18}]^-$ cluster to charge neutral cluster via air oxidation. *J Phys Chem C* 2008;**112**:14221–4.
72. Yang H, Wang Y, Huang H, Gell L, Lehtovaara L, Malola S, et al. All-thiol-stabilized Ag_{44} and $Au_{12}Ag_{32}$ nanoparticles with single-crystal structures. *Nat Commun* 2013;**4**:2422.
73. Teo B, Shi X, Zhang H. Pure gold cluster of 1:9:9:1:9:9:1 layered structure – a novel 39-metal-atom cluster $[(Ph_3P)_{14}Au_{39}Cl_6]Cl_2$ with an interstitial gold atom in a hexagonal antiprismatic cage. *J Am Chem Soc* 1992;**114**:2743–5.
74. Kumara C, Gagnon KJ, Dass A. X-ray crystal structure of $Au_{38-x}Ag_x(SCH_2CH_2Ph)_{24}$ alloy nanomolecules. *J Phys Chem Lett* 2015;**6**:1223–8.
75. Zeng C, Qian H, Li T, Li G, Rosi NL, Yoon B, et al. Total structure and electronic properties of the gold nanocrystal $Au_{36}(SR)_{24}$. *Angew Chem Int Ed* 2012;**51**:13114–8.
76. Zeng C, Li T, Das A, Rosi NL, Jin R. Chiral structure of thiolate-protected 28-gold-atom nanocluster determined by X-ray crystallography. *J Am Chem Soc* 2013;**135**:10011–3.
77. Song Y, Zhong J, Yang S, Wang S, Cao T, Zhang J, et al. Crystal structure of $Au_{25}(SePh)_{18}$ nanoclusters and insights into their electronic, optical and catalytic properties. *Nanoscale* 2014;**6**:13977–85.
78. Das A, Li T, Li G, Nobusada K, Zeng C, Rosi NL, et al. Crystal structure and electronic properties of a thiolate-protected Au_{24} nanocluster. *Nanoscale* 2014;**6**:6458–62.

79. Song Y, Wang S, Zhang J, Kang X, Chen S, Li P, et al. Crystal structure of selenolate-protected $Au_{24}(SeR)_{20}$ nanocluster. *J Am Chem Soc* 2014;**136**:2963–5.
80. Wan XK, Lin ZW, Wang QM. Au_{20} nanocluster protected by hemilabile phosphines. *J Am Chem Soc* 2012;**134**:14750–2.
81. Tran N, Powell D, Dahl L. Nanosized $Pd_{145}(CO)_x(PEt3)_{30}$ containing a capped three-shell 145-atom metal-core geometry of pseudo icosahedral symmetry. *Angew Chem Int Ed* 2000;**39**:4121–5.
82. Miras HN, Cooper GJT, Long DL, Bögge H, Müller A, Streb C, et al. Unveiling the transient template in the self-assembly of a molecular oxide nanowheel. *Science* 2010;**327**:72–4.
83. Bogan MJ, Benner WH, Boutet S, Rohner U, Frank M, Barty A, et al. Single particle X-ray diffractive imaging. *Nano Lett* 2008;**8**:310–6.
84. Kern J, Alonso-Mori R, Tran R, Hattne J, Gildea RJ, Echols N, et al. Simultaneous femtosecond X-ray spectroscopy and diffraction of photosystem II at room temperature. *Science* 2013;**340**:491–5.

Chapter 6

Atomic-Scale Structure Analysis by Advanced Transmission Electron Microscopy

Zhi Wei Wang* and Richard E. Palmer[§,1]
*Beijing Institute of Nanoenergy and Nanosystems, Chinese Academy of Sciences, Beijing, China;
[§]Nanoscale Physics Research Laboratory, School of Physics and Astronomy, University of Birmingham, Birmingham, UK
[1]Corresponding author: E-mail: R.E.Palmer@bham.ac.uk

Chapter Outline

6.1 INTRODUCTION

Metal nanoclusters/nanoparticles have generated considerable scientific and technological interest due to their important role in bridging different fields of physics and chemistry as well as their increasing applications, and potential applications, in catalysis, photovoltaics, biological labeling and sensing, etc.[1–7] With recent technical developments, precise size-controlled synthesis of nanoclusters has been successfully achieved with a resolution even down to

Frontiers of Nanoscience, Vol. 9. http://dx.doi.org/10.1016/B978-0-08-100086-1.00006-3

single atoms for both physical and chemical routes.[8,9] The accurate structural determination of nanoparticles is critical to understanding the dependence of their properties on cluster size, shape, and atomic arrangements. Using first principles and semiempirical potential treatments, theoretical work has been able to predict the atomic structures with lowest energy for a range of small nanoparticles.[10,11] However, direct experimental measurements of nanoparticle structures have lagged far behind the theoretical calculations. Because of their ultrasmall size, nanoparticles can exhibit continuous structural instabilities when exposed to heat or irradiation, which makes experimental accurate structural determination extremely difficult.[12,13]

To address these challenges, great endeavour has been made in recent years, leading to some notable progress in the structural determination of nanoparticles, especially gold clusters. X-ray crystallography is an experimental technique which has now been successfully applied to the structural characterization of protected nanoclusters. Combined with theoretical calculations, atomic structures of a range of small nanoparticles have been determined this way, such as thiolate-protected Au_{102}, Au_{25}, and Au_{38} clusters.[14–18] However, the extensive application of X-ray crystallography as a characterization approach is limited, because it remains a general challenge to produce sufficiently high-quality cluster crystals from a solvent. As a powerful characterization tool providing both chemical and structural information, transmission electron microscope (TEM) has been widely applied in a variety of research fields spanning physics, chemistry, materials science, nanoscience, biology, etc.[19–22] Via TEM characterization, accurate structural determination can be achieved not only for bulk metals featuring regular atomic arrangements, but also for amorphous-like materials such as glass as well as impurity segregation at grain boundaries or interfaces.[23,24] TEM is also frequently employed in nanocluster research, and in the last decade three-dimensional structural determination has been demonstrated successfully for a range of metal nanoparticles with different size and composition.[21,25]

In this chapter, we will first give a very brief introduction to the TEM technique, and then focus on structural characterizations of small gold nanoclusters. Note here that our definition of "small" and "ultrasmall" nanoclusters is that they contain less than 1000 and 100 atoms, corresponding to a maximum diameter of ~3.2 and ~1.5 nm in the case of gold nanoclusters, respectively. We will discuss the challenges faced when using TEM to study small nanoclusters, principally the effect of electron beam irradiation, the methods developed to address this issue, and some important accomplishments in determining the structures of nanoclusters, including both size-selected (bare) clusters and ligand-protected nanoclusters. The aim of the chapter is not to provide a comprehensive review about this rapidly growing research field, but to give a basic introduction to TEM and show by examples how it can be adapted to study small nanoclusters that exhibit strong dynamic behavior under electron beam irradiation.

6.2 TRANSMISSION ELECTRON MICROSCOPY

Modern transmission electron microscopes differ significantly from the original version invented by Knoll and Ruska[26] in the early 1930s in terms of instrumental performance, functionality, etc. The first microscope,[26] built up in an extension of earlier work to optimize cathode-ray oscilloscopes, offered limited spatial resolution that was actually not better than optical microscopes. After the unremitting efforts of the past 80 years, the current-generation TEM incorporating aberration correctors has reached an imaging resolution at the sub-angstrom level.[23] In addition, it is no longer a simple magnifying instrument, but can offer other important information about elemental composition, chemical bonding, and electronic structures through combination with energy-dispersive X-ray (EDS) spectroscopy, electron energy loss spectroscopy (EELS), etc.[19,27] Furthermore, with the advancement of in situ TEM techniques, we are now able to investigate samples in liquid or gas environment, at various temperatures (via heating/cryogenic holders) or under the influence of applied fields or forces (e.g., mechanical stressing).[28,29]

Now, we will briefly describe the TEM technique, including equipment components, image formation, and some recent technical and methodological developments that are highly relevant to the topic of nanocluster characterization. Readers who would like to gain a detailed understanding of TEM can consult the textbooks and review articles given at the end of this chapter.[19–23,30,31] A simplified schematic diagram of a TEM imaging system is shown in Figure 1. Two different kinds of electron gun are generally used in TEM for generating a beam of electrons, thermionic and field emission sources. The latter features higher brightness and better electron coherence and is thus more often used in high-resolution imaging. The electrons are accelerated usually to 100–300 kV, which corresponds to electron wavelengths of 3.7–2.0 pm (with relativistic effects considered). Two kinds of imaging modes are then available in most modern TEMs—conventional transmission electron microscopy (CTEM) and scanning transmission electron microscopy (STEM).

In CTEM mode, a (nearly) parallel, broad beam is formed before the specimen via an illumination system generally consisting of a group of condenser lenses, an objective lens pre-field, and a mini lens, as shown schematically in Figure 1(a). As the electrons pass through the specimen, they are scattered once or repeatedly involving kinematical or dynamical events. The objective lens post-field forms the first image in its imaging plane (the location of selected-area diffraction aperture in Figure 1(a)), which is subsequently magnified further by intermediate and projector lenses below the objective lens and recorded by electron detectors such as a charge-coupled device. In CTEM mode, the phase-contrast imaging, also called high-resolution TEM (HRTEM) imaging, is widely used in studying the lattice structure of thin specimens when the microscope conditions are properly

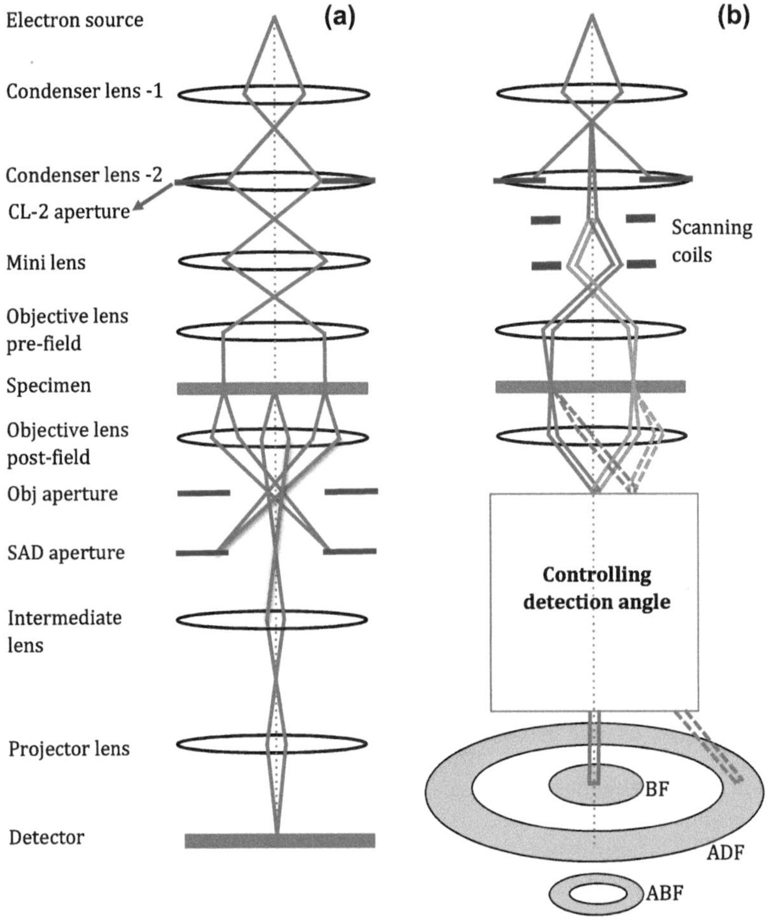

FIGURE 1 Schematic layout of transmission electron microscope operated in (a) conventional transmission electron microscopy and (b) scanning transmission electron microscopy modes. CL-2: condenser lens-2; Obj: objective; SAD: selected-area diffraction; ADF: annular dark field; BF: bright field; ABF: annular bright field.

adjusted. The image wave function $\Psi(x, y)$ can be written as the convolution of a specimen function $s(x, y)$ and point-spread function $p(x, y)$[19,30]:

$$\begin{aligned}\Psi(x,y) &= s(x,y)\otimes p(x,y)\\ &= \exp[-i\sigma\phi(x,y)-\mu(x,y)]\otimes FT\{A(u,v)\exp[i\chi(u,v)]\}\end{aligned} \tag{1}$$

where σ denotes a scale factor $(\pi/\lambda E)$, $\phi(x, y)$ the projection potential of the specimen, $\mu(x, y)$ the absorption function, $\chi(u, v)$ the aberration function, and $A(u, v)$ the aperture function. λ: wavelength; E: acceleration voltage; FT denotes fourier transformation.

Under the weak phase object approximation and ignoring the $\mu(x, y)$ and $A(u, v)$ functions, the image intensity $I(x, y)$ is given by

$$I(x,y) = |s(x,y) \otimes p(x,y)|^2 = 1 + 2\sigma\phi(x,y) \otimes FT[\sin\chi(u,v)] \quad (2)$$

When the imaging defocus (Δf) and spherical aberration (C_s) are considered only,

$$\chi(u,v) = \pi\Delta f\lambda\left(u^2+v^2\right) + \frac{1}{2}\pi C_s\lambda^3\left(u^2+v^2\right)^2 \quad (3)$$

Equation (2) shows clearly that the intensity transfer function ($2\sin\chi(u, v)$) oscillates with the spatial frequency of the samples, which thus significantly complicates the process of phase-contrast imaging. The image contrast can be dramatically influenced by both microscope conditions (spherical aberration, wavelength, defocus, etc.) and sample conditions (thickness, defects, etc.). Therefore, to interpret the HRTEM data in terms of projected atomic potentials precisely, image simulations are often required to make a comparative analysis, even under the kinematic conditions. Figure 1(a) also shows that a diffraction pattern (DP) is formed in the back focal plane of the objective lens (the location of obj aperture in Figure 1(a)). Measurement of a DP is widely employed in the determination of crystal structures, specimen orientation, the measurement of lattice constants, etc.

In STEM mode, the electron beam is converged into a small probe at the specimen, as shown in Figure 1(b). STEM images are obtained by scanning the electron probe across the specimen point by point, in a fashion similar to scanning electron microscopy. The first STEM was designed and constructed by Manfred von Ardenne in the late 1930s,[32] shortly after the invention of CTEM, and the performance of STEM was greatly improved by Albert Crewe and his colleagues in the 1960s–1970s.[33,34] STEM has been applied extensively in many different research areas. The current popularity of STEM is partly attributed to the introduction of high-angle annular dark field (HAADF) detectors, which record those electrons scattered incoherently to high angles by the specimen.[35–41] The intensity in incoherent imaging[23,30] has a form of

$$I'(x,y) = |s(x,y)|^2 \otimes |p(x,y)|^2 \quad (4)$$

Here, $p(x, y)$ denotes a probe function in STEM imaging. It is seen clearly from Eqn (4) that HAADF-STEM data are much simpler to interpret due to the incoherent nature of imaging electrons. Both theoretical and experimental studies have confirmed that HAADF intensity (or scattering cross section) depends monotonically on sample thickness (t), and at least for sufficiently thin samples, follows an approximately linear relationship.[23,42,43] In addition, the electron-scattering cross section depends on the atomic number (Z) in the form Z^α, where the exponent α would be 2 for unscreened Rutherford scattering. This indicates that the HAADF signal itself contains chemical information about the sample, and HAADF imaging is also called Z-contrast imaging.

In general, bright features in HAADF images mean either higher (mean) atomic numbers or thick sample regions. It is worth noting that, to obtain true *Z*-contrast images, sufficiently large collection angles have to be used to minimize the contributions from diffraction contrast and strain contrast, especially for thick samples in which multiple electron scattering occurs.[42] The HAADF technique also has an attractive compatibility with EELS spectroscopy, generally detected at small scattering angles, in that one can record both imaging and spectral signals simultaneously from the sample region of interest. Recently, STEM-based annular bright-field (ABF) imaging[44,45] has drawn some attention, owing to its ability to visualize, in addition to heavy elements, light (low *Z*) elements that are very hard to detect in HAADF-STEM imaging due to very weak scattering. According to the reciprocity theorem, ABF imaging corresponds to hollow-cone illumination imaging in CTEM. Experimentally, ABF image signal can be obtained via a bright-field detector by masking its central detection region.

6.2.1 Aberration Correction

The spatial resolution of TEM is determined by a combination of electron diffraction and lens aberrations. If the diffraction limit alone is considered, the achievable resolution should be on the picometer scale for TEM images obtained with an electron energy of hundreds of kilovolts. In practice, no TEM has approached such values due to the malign influence of lens aberrations. The aberration function can be expressed using the following expansion,[46]

$$\chi(\theta,\varphi) = A + \frac{\theta^{N+1}}{N+1}\left[C_{NSa}\cos(S\varphi) + C_{NSb}\sin(S\varphi)\right] \quad (5)$$

where A is an initial constant term, θ an axial angle, and φ azimuthal angle. $S = 0,2,4,6\ldots N+1$ or $1,3,5,7\ldots N+1$ when N is odd or even, respectively. The typical coefficients include C_{10} (defocus), C_{12} (twofold astigmatism), C_{21} (coma), and C_{30} (also C_s, third-order spherical aberration). The dominant aberration in uncorrected TEM is the spherical aberration (C_s). Spherical aberration causes high-angle rays to converge more strongly than those closer to the optical axis, and as a consequence the image of a point object is not a point but a disk in the Gaussian image plane. Scherzer proved in 1936 that spherical aberration is intrinsically positive for rotationally symmetric electron lenses and also proposed in 1947 several methods for aberration correction.[47,48] However, a significant breakthrough was not obtained until the 1990s, when great progress was made in improving computational performance, mechanical fabrication precision, and the stability of the electronics.[46] Nowadays, there are mainly two practical C_s correction systems developed both through the use of multipole lenses, double-hexapole and quadrupole-octupole systems.[49–51] Figure 2 shows schematic drawing of a hexapole STEM corrector system.[52] The two hexapole lenses (HP1 and HP2) are coupled through a double transfer lenses (TL21 and TL22)

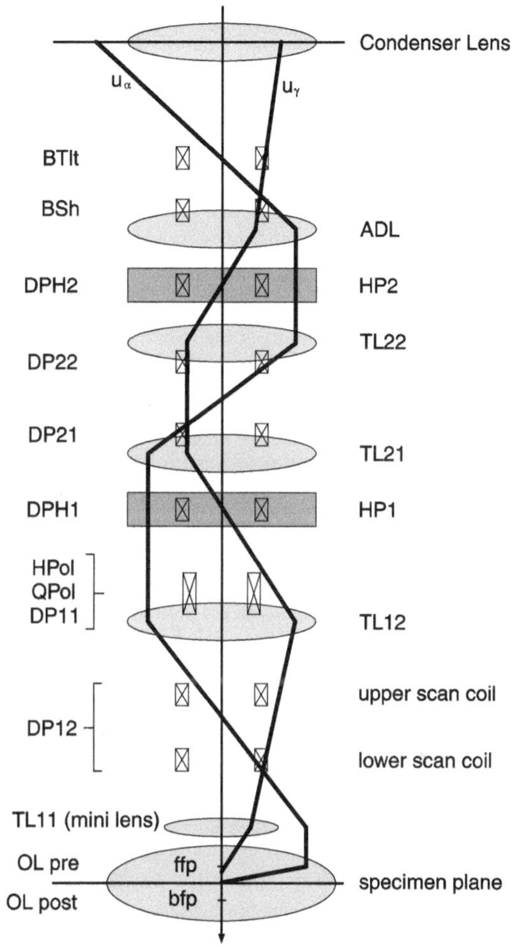

FIGURE 2 Schematic layout of a hexapole C_S-corrector system for scanning transmission electron microscopy, consisting of transfer lenses (TL), an adapter lens (ADL), hexapole lenses (HP), deflectors (DP), beam tilt (BTlt) and beam shift (BSh) coils, and stigmators (QPol, HPol). u_α and u_γ indicate the trajectories of an axial ray and a selected field ray through the corrector system, respectively. *Reproduced from Ref. 52 with permission.*

such that (1) the hexapole lenses impart negative spherical aberrations to the wavefront which compensate for the positive C_S of the objective lens, and (2) the primary aberrations (threefold astigmatism) of the hexapole elements can cancel each other. In addition, the transfer lenses TL11 (or mini lens) and TL12 are added to make the midplane of the hexapole fields and the coma-free plane of objective lens optically conjugated, and some other (weakly excited) multipole elements for the purpose of the residual aberration corrections and beam alignments within the corrector. So far, most commercial C_S-corrected TEMs have adopted the double-hexapole system, often fitted to e.g., FEI and JEOL microscopes.

Aberration-corrected TEM is a revolutionary technology which offers several distinct advantages. The most direct benefit is of course the improvement in spatial resolution. For example, Erni et al. demonstrated that sub-0.5 Å resolution could be obtained in STEM with a fifth-order C_S corrector.[53] This improved spatial resolution can significantly increase the number of available crystal projection axes, which is especially useful for studying low-symmetry structures, interface structures, etc. Secondly, uncorrected TEM suffers serious contrast delocalization due to spherical aberrations. The delocalization may be reduced using appropriate objective defocus values, but the effect is very limited. C_S-corrected TEM can dramatically remove the contrast delocalization,[50] thus allowing for the detailed analysis of atomic structures at discontinuities such as interfaces and surfaces. Thirdly, C_S correction increases the probe current by at least one order of magnitude at a given resolution,[54] and thus significantly improves the detection sensitivity of EELS or EDS in the STEM. This is especially important for column-by-column compositional analysis and for identifying trace elements segregating at grain boundaries with atomic resolution. Fourthly, with C_S-correction, it is possible to employ a large objective pole-piece gap or a lower acceleration voltage in C_S-corrected TEM while maintaining atomic resolution, which is highly advantageous for, respectively, developing in situ viewing techniques or imaging light (low Z) materials at low voltages to reduce knock-on effects.

6.2.2 3D Structural Determination with Atomic Resolution

A TEM image is actually formed as a two-dimensional (2D) projection of the three-dimensional (3D) structure of the specimen. Obtaining full 3D structures with atomic resolution has been a long-standing dream, and has remained extremely challenging. Tomographic reconstruction is the method used most often to extract 3D information. It combines multiple projection images to generate a 3D image of the sample through a reconstruction algorithm. The reconstruction precision is determined predominantly by the tilt range and tilt increment. In practice, the tilt range is usually limited to $\sim \pm 80°$ for single-tilt TEM holders due to shadowing effects, which produce a "missing wedge" effect and decrease the image quality of 3D reconstructions.[22] Solutions proposed to reduce or remove this effect include using double-tilt holders or fabricating "needle-shaped" specimens without tilting limits. Fiducial markers (typically gold nanoparticles) are frequently added to the specimen to perform more accurate alignment of the images recorded in the tilt series.[55] Previously, the tomographic reconstruction method was mainly used for generating the 3D morphology of biological specimens, but now, with the improvement of instrumental resolution and computational methods, it appears possible to

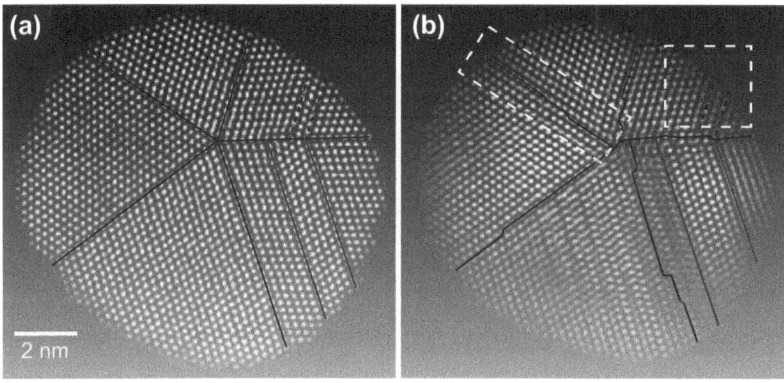

FIGURE 3 Scanning transmission electron microscopy imaging and tomographic reconstructions of a multiply twinned Pt nanoparticle. (a) Experimental projection in x–y plane. (b) A 2.6-Å thick internal slice of the reconstructed nanoparticle, showing the existence of atomic steps at twin boundaries (red lines) that are not visible to the eyes in the experimental projection (a). The subgrain boundaries (blue lines) are two lattice spacings wider than those in (a). (see color plate). *Reproduced from Ref. 56 with permission.*

achieve atomic-scale 3D reconstructions for some objects. Chen et al.[56] recently showed, by applying 3D Fourier filtering together with an equal-slope tomography method, that three-dimensional, atomic-resolution STEM images can be successfully obtained for a multiply twinned Pt nanoparticle supported on silicon nitride membrane, as shown in Figure 3. Nearly all the atoms in the particle can be identified in the reconstructed image, allowing direct visualization of atomic steps at 3D twin boundaries, and 3D core structures of edge and screw dislocations. In addition, Van Aert et al. reported an atomic-resolution 3D reconstruction of an embedded Ag nanoparticle, based on discrete tomography, statistical parameter estimation, and some prior knowledge about the nanoparticle's crystallographic characteristics via HAADF-STEM imaging.[57]

For the proper use of the tomographic reconstruction method, the sample has to remain stable during the acquisition of the necessary series of tilt images. However, this condition is very difficult to satisfy for radiation-sensitive samples, such as supported small nanoparticles, that often exhibit continuous movement and/or structural fluctuations under electron beam irradiation.[12,13] Thus, one would ideally hope to determine 3D atomic structures from just one single image. Based on the monotonic mass–intensity relationship in Z-contrast imaging, atom-counting approaches have been developed to extract depth information from the sample.[58–61] Combined with 2D atomic projection images, this should in principle allow for 3D structural determination with atomic resolution from a single HAADF image. Indeed, by this method 3D atomic structures have been successfully identified for some small nanoparticles in combination with image simulations.[62,63] This single-image-based reconstruction method should also be valuable for studying in situ the behaviors of nanoparticles exposed to gas, thermal, or radiation environments under which structural transformations may occur.[28,64] However, since it is really very challenging to determine uniquely the 3D location of *every* atom from just one single projection image, obtaining as much a priori knowledge as possible about the atomic configuration of the sample is important in achieving the best structural determination.

6.3 ATOMIC STRUCTURE AND DYNAMICS OF SMALL NANOCLUSTERS

Structural characterization of nanoparticles/nanoclusters via transmission electron microscopy has a lengthy history. For large and stable nanoparticles (thousands of atoms or more) that exhibit a marked crystalline nature, one may carry out the characterization using the same TEM techniques developed for foil samples. In this study, we will focus in contrast on small nanoclusters consisting of less than 1000 atoms. These clusters are inherently unstable under electron beam irradiation and other external excitations, so it is a challenge to determine precisely their 3D structures. Nevertheless, some progress has been achieved via the advanced characterization approaches developed recently. In these studies, gold is the most commonly studied material, so Au nanoparticles will form the centerpiece of our discussion. Of course the methods developed on gold clusters should be applicable to other nanoparticles too.

In this section, we will first present structural studies of mass-selected (bare) gold clusters and demonstrate a number of phenomena, including size-dependent structures, kinetic and thermodynamic effects, and structural fluctuations. Because of the precise mass-selection and ligand-free surfaces, size-selected clusters can provide useful archetypes in developing our understanding of the atomic structures of nanosystems. We will then move on to the structural characterization of thiolate-protected gold nanoclusters, giving several specific examples to show the achievements recently made in determining the 3D structures of protected nanoclusters via optimized imaging approaches. The effect of thiolate ligands on cluster structures will also be discussed. It is worth mentioning that although the spatial resolution in C_S-corrected TEM has routinely reached the sub-angstrom level, the total structure determination of ligand-protected metal nanoclusters remains very challenging. With the advanced techniques developed recently, the determination of the locations of each atom in the metal cores seems to be achievable, but not for the ligand shells. The major reason is that the ligands, usually consisting only of light elements (e.g., S, C, and H) give rise to very weak image contrast due to their small electron scattering cross sections. Thus, it is hard to distinguish the ligands above the signal due to the metal atoms. The idea of enhancing the ligand signal levels by increasing of the electron dose does not work well since the ligands are very sensitive to electron irradiation.

Therefore, with current TEM technology, the total structural determination of thiolate-protected clusters is mainly performed by probing the location of each metal atom, followed by a manual assignment of ligand locations made on the basis of the priori knowledge on Au–S bonding motifs and/or theoretical models and calculations.

6.3.1 Size-Dependent Structure and Dynamics

Recent advances in cluster beam source technology have enabled the production and deposition of precisely mass-selected metal clusters.[65–68] For example, size-selected gold clusters can be prepared with a radio-frequency magnetron plasma sputtering, gas condensation cluster beam source combined with a lateral time-flight mass selector, then deposited onto copper TEM grids covered with amorphous carbon under "soft-landing" conditions.[62,63,69] Systematic studies show that the geometric structures (and thus atomic arrangements) of the nanoclusters differ dramatically from their bulk counterparts. Translational-symmetry-forbidden structural configurations, such as decahedral (Dh) and icosahedral (Ih) motifs, as well as low-symmetry, amorphous-like structures, are often found in such nanoscale systems. They also display dynamic behavior under electron beam irradiation or other excitation conditions. Transformations/fluctuations of the atomic structures are frequently observed, generally at an increasing rate with decreasing cluster size.

6.3.1.1 Polymorphism and Equilibration (100–1000 Atoms)

For Au nanoclusters within the size range 100–1000 atoms, Ih, Dh, and face-centered cubic (fcc) polyhedral structures are the three commonly observed atomic configurations, all of which may coexist in clusters of the same size due either to small energy differences or to kinetic trapping effects.[10] Experimentally, the single-shot HAADF-STEM intensity map approach, as introduced in Section 6.2.2, can be used to determine the 3D atomic structures of nanoclusters in combination with STEM simulations.

Figure 4(a–d) shows typical HAADF images characteristic of the three main ordered structures for size-selected Au_{923} clusters.[69] Since the nominal mass resolution employed was $M/\Delta M = 20$, the "Au_{923}" clusters actually contains 923 ± 23 atoms (923 is a nominal "magic number" for the icosahedron, Ino-decahedron, and cuboctahedron). The geometric structures of the nanoclusters are determined via atomic resolution, single-shot HAADF imaging, in combination with HAADF simulations (Figure 4(e–h)). The simulations were performed using the multislice electron scattering method[70–72] and based on the standard atomic models of Ino-Dh, cuboctahedron, and Ih. Note that the Marks truncated decahedron, formed by introducing reentrant facets at the "twin interfaces", has lower energy than the Ino-Dh in some calculations.[10,13] The Marks-type reconstruction is indeed observed in Au_{923}

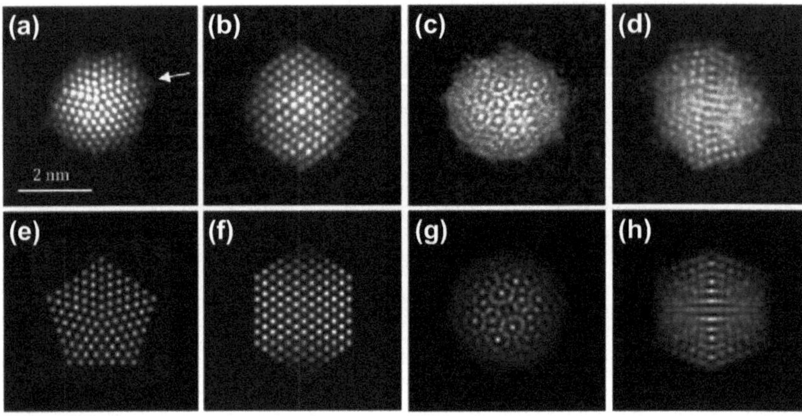

FIGURE 4 Representative high-angle annular dark field scanning transmission electron microscopy (HAADF-STEM) images and the corresponding simulations of size-selected Au_{923} clusters. (a–d) The experimental HAADF images for (a) a decahedron along a fivefold axis; (b) fcc polyhedron along the <110> axis; (c and d) icosahedron along a fivefold axis and a twofold axis, respectively. (e–h) The simulated images performed using the multislice method for the standard atomic models of the Ino-decahedron, cuboctahedron, and icosahedron, respectively. The arrow in (a) marks the edge where no reentrant structure is present. *Reproduced from Ref. 69 with permission.*

clusters as shown in Figure 4(a), but there are still some edge(s) which remain unreconstructed, possibly because 923 is not a magic number for the Marks-Dh (the nearest magic number is 887), but kinetic trapping is an important factor too.

The three typical structures, Ih, Dh, and fcc, have also been observed in other gold nanoclusters, such as size-selected Au_{309},[62] as shown in Figure 5. The coexistence of a range of structural isomers of the same size clusters is a general phenomenon for metal nanoclusters prepared by either physical or

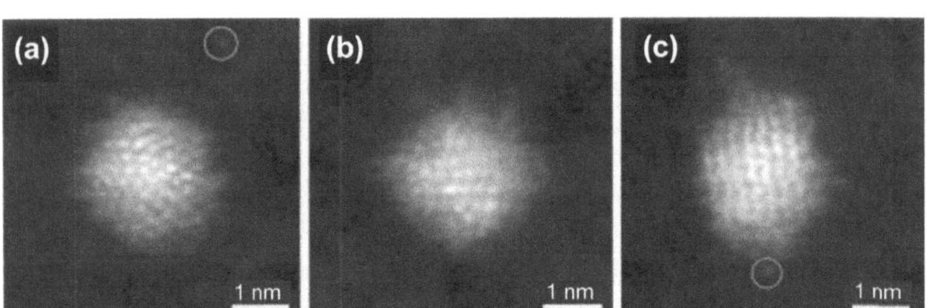

FIGURE 5 Atomic-resolution high-angle annular dark field scanning transmission electron microscopy images of size-selected Au_{309} clusters. Various cluster projection shapes are clearly visible, (a) pentagon, (b) square, and (c) hexagon. The circles in (a) and (c) mark single atoms around the nanoclusters. *Reproduced from Ref. 62 with permission.*

chemical routes.[10,13] The growth environments in which nanoparticles are produced are hardly ideal in terms of thermodynamic equilibrium, so the formation of atomic structures is very likely to be determined by a combination of both kinetic and thermodynamic factors. By varying the growth conditions (and thus changing kinetic effect), it is possible to tune the structural populations of nanoclusters. For example, it is found that the proportion of the Ih isomer significantly depends on the magnetron power and condensation length which are used in generating size-selected Au_{932} clusters, and at a specific condition no Ih clusters are observed from a statistical investigation.[73]

It is important to determine experimentally the most stable atomic structures for comparison with the lowest energy (ground state) structures that are a standard output of density functional theory (DFT) calculations. To obtain these most stable structures, nanoparticles are usually annealed using in situ TEM sample holders[29] or through ex situ heating followed by TEM imaging.[74] Figure 6 shows the statistical structural populations of gold nanoparticles on amorphous carbon over a wide size range (3−18 nm) obtained by the latter method.[74] The nanoparticles are prepared by cooling gold vapor using helium gas but without mass selection. It is found that a proportion of Ih nanoparticles convert to Dh over a wide size range for cluster samples annealed to below the bulk melting point, indicating the Dh is a more stable ("lower energy") isomer than Ih in this case. For Au nanoparticles with diameters of 3−5 nm, no structural transformation from either Ih or Dh to fcc was observed.

In fact, the most stable structures of small nanoparticles can be also achieved simply by electron beam irradiation. Indeed, small nanoclusters often change their structures during continuous TEM viewing.[12,13] By harnessing the effect of electron irradiation, the structural stability of size-selected Au_{923} clusters was studied systematically in C_s-corrected STEM via a collection of image sequences for a large set of the clusters.[69] Each image series (movie) was recorded with a field of view of 10.5 × 10.5 nm, an electron dose of $2.4 \times 10^4\ e^-/Å^2$/frame, and an acquisition time of 0.8 s/frame. The data are given in Figure 7, which show that almost all the Ih clusters converted to Dh (~2/3) or fcc (~1/3) clusters, while Dh and fcc clusters basically remain unchanged during the irradiation period of up to 400 s. This result clearly manifests that Ih is not a stable structural isomer, similar to the finding from the thermal annealing method shown in Figure 6. It was also found that, once the structural transformations from Ih to Dh or fcc occurred, the clusters appeared rather stable; i.e., further transformations were not observed.

However, the effects of electron beam irradiation on the structural transformations of the Au clusters seem somewhat different from the effects of thermal annealing, Figure 6, discussed above. The observed difference mainly lies in the transformation from Ih to fcc, which was observed only in the electron irradiation method. The mechanisms of electron beam interaction with the samples have been widely studied, but a quantitative understanding is

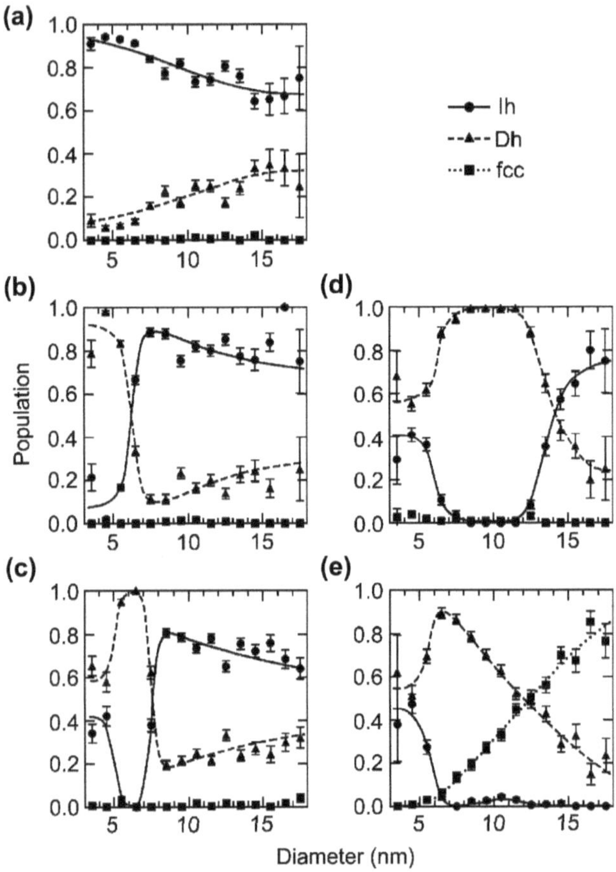

FIGURE 6 Systematic study of the structural stability of gold nanoparticles via the annealing method. Populations of different structures of gold clusters as a function of size in the 3–18 nm range. (a) Before annealing and (b–e) after annealing at (b) 1173 K, (c) 1223 K, (d) 1273 K, and (e) 1373 K (bulk melting point 1337 K). *Reproduced from Ref. 74 with permission.*

still lacking. In general, both direct (momentum transfer) and indirect (local heating) effects are expected.[13,75] It is challenging to measure accurately the temperature of the sample under electron beam irradiation. The studies show that the temperature elevation is estimated to be several tens of degree for small gold clusters (<3 nm) and up to 800° for large nanoparticles (>10 nm) on carbon supports.[12,76] Considering the rather small cluster size (~3.1 nm for Au_{923}) and carbon substrate used in the electron beam irradiation method, the temperature elevation due to beam heating should be the modest. Thus, the momentum transfer in electron irradiation may play a major role in driving the Ih to fcc structural transformation of gold nanoclusters, a proposition which requires further investigation including, ideally, theoretical calculations.

Initial state	After irradiation	Examples	
Ih: 42	Dh: 29 FCC: 12 Ih: 1	Ih->Dh	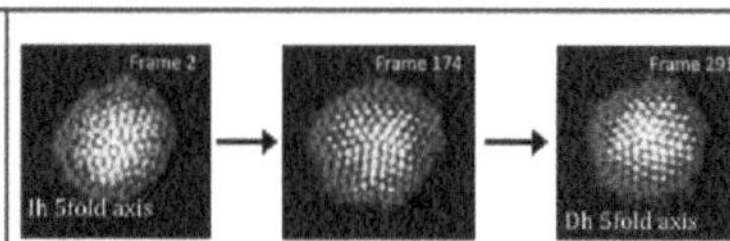
		Ih->FCC	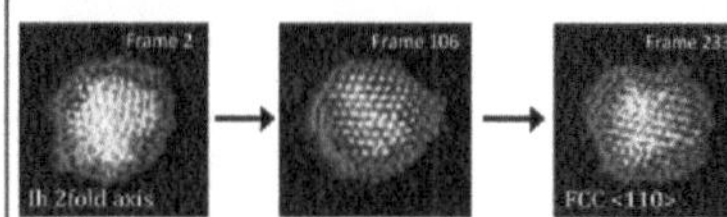
Dh: 18	Dh: 18	Dh unchanged	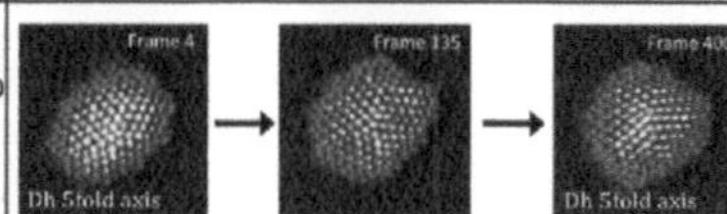
FCC: 19	FCC: 18 Dh: 1	FCC unchanged	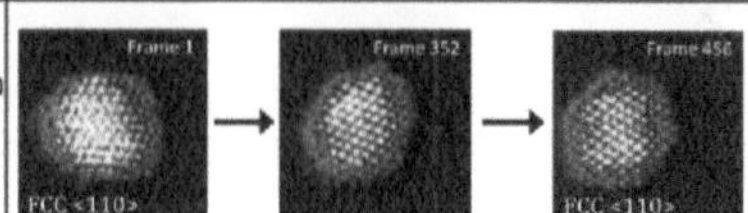

FIGURE 7 Statistical investigation of structural transformations and stability of 79 size-selected Au_{923} clusters. Left columns show the structural populations of Au_{923} clusters initially and after electron beam irradiation. Right panel shows examples of structural transformation or structural persistence. *Reproduced from Ref. 69 with permission.*

6.3.1.2 Low-Symmetry, Fluctuating Structures (<100 Atoms)

For ultrasmall clusters (<100 atoms), theoretical calculations generally predict that there exist a plentiful range of morphologies and structural categories such as flakes and cages.[10,11] It is of great interest to determine the structures of such clusters via direct TEM imaging. However, as mentioned in Section 6.3.1.1, there is an unavoidable electron beam–sample interaction problem in TEM imaging. This issue will become more serious as the clusters get smaller. Figure 8 shows, by way of example, several individual frames from a time-lapse series of HAADF images of an $Au_{55\pm1}$ cluster deposited on the

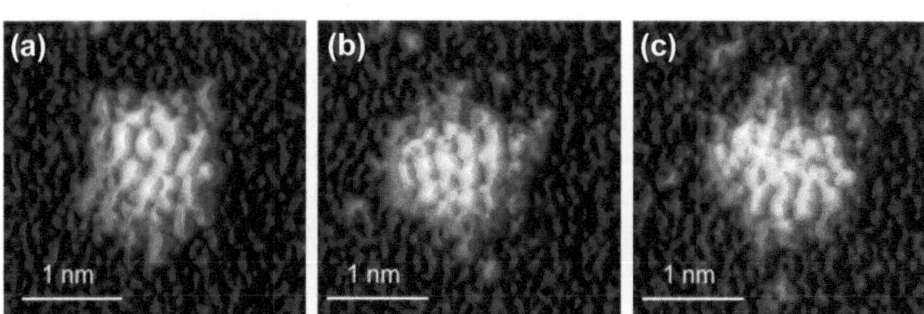

FIGURE 8 Individual frames from a sequence of high-angle annular dark field scanning transmission electron microscopy (HAADF-STEM) images of a size-selected Au_{55} cluster. (a) Frame 4 (b) Frame 8, (c) Frame 11. Significant variations in both the intensities and positions of the atomic columns are clearly visible. (see color plate). *Reproduced from Ref. 63 with permission.*

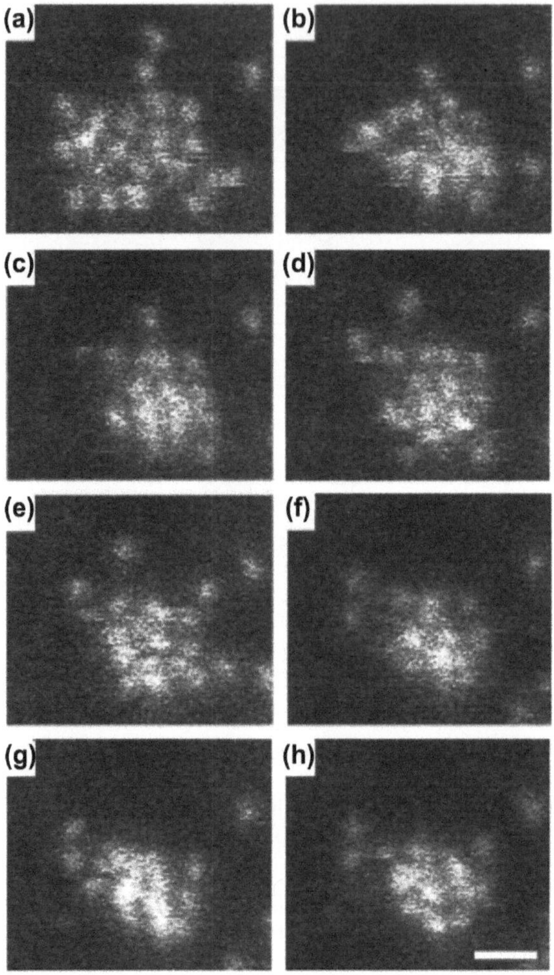

FIGURE 9 A series of high-angle annular dark field scanning transmission electron microscopy (HAADF-STEM) images of a Ge cluster consisting of ~25 atoms. (a–h) Eight consecutive frames were acquired with a frame time of 0.1 s. The images show a largely 2D configuration transforming into a 3D configuration. The scale bar corresponds to 0.5 nm. *Reproduced from Ref. 77 with permission.*

amorphous carbon support.[63] It can be seen clearly that there are significant variations in the intensities and positions of the atomic columns from one image to another, which arise from structural fluctuations and/or the rotational/rolling movement of the cluster during the acquisition of the image series. Structural fluctuation is a ubiquitous phenomenon for ultrasmall nanoclusters. Figure 9 shows a series of HAADF images recorded in a Ge cluster consisting of ~25 atoms, produced in a cluster source and deposited without size selection on TEM grids covered with ultrathin amorphous carbon layer.[77]

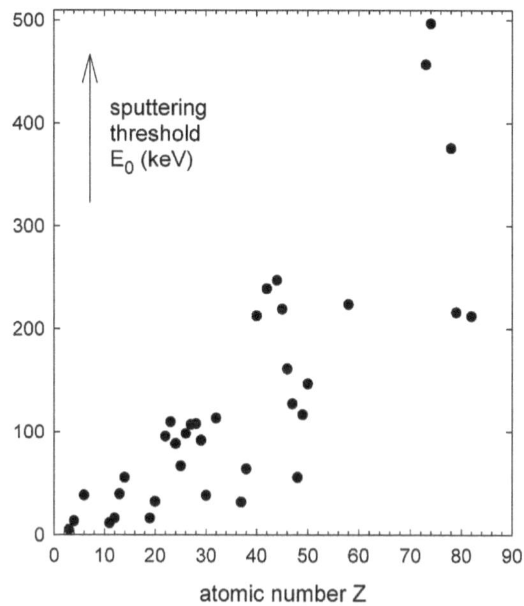

FIGURE 10 Calculated threshold energy for onset of electron sputtering in solid elements. *Reproduced from Ref. 75 with permission.*

The eight consecutive frames again display significant variations in the shape and atomic arrangements, similar to the Au_{55} clusters.

If the cluster structure can be modified by the electron beam, it is natural to ask whether the surface atoms of small nanoclusters are likely to be detached by electron beam irradiation? Egerton et al. calculated the threshold incident energy for the onset of electron sputtering, as summarized in Figure 10, which shows that the threshold energy is generally larger than 200 kV for heavy elements (with atomic numbers > 60), indicating that Au nanoclusters should survive 200 kV TEM imaging.[75] This issue has also been investigated experimentally, through serial imaging of Au_{20} clusters on amorphous carbon support at 200 kV in STEM.[78] Figure 11 shows an integrated intensity analysis of a series of HAADF images of a size-selected $Au_{20\pm1}$ cluster acquired with a relatively low electron dose ($\sim 8.8 \times 10^3$ $e^-/Å^2$ per frame). No obvious intensity decrease can be seen, which suggests that no significant sputtering of surface atoms occurs in such small Au clusters under these experimental conditions. However, under intense irradiation conditions, e.g., by placing the probe on the clusters without scanning, the clusters could be destroyed in a short period of time. Thus, a gentle beam during TEM imaging is necessary to retain the integrity of the clusters.

It is evident from our discussions that significantly more effort (in terms of imaging and interpretation) is required to determine the structures of ultrasmall nanoclusters than large nanoparticles and bulk materials. The conventional

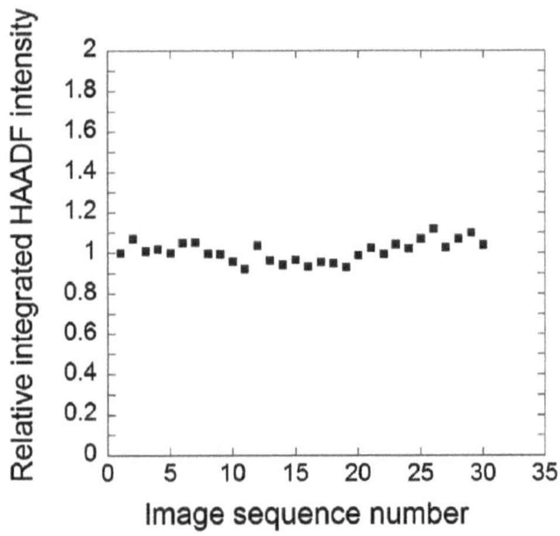

FIGURE 11 The integrated high-angle annular dark field scanning transmission electron microscopy (HAADF-STEM) intensity of an Au_{20} cluster on carbon support as a function of time in a series of HAADF-STEM images, recorded with relatively low electron dose ($\sim 8.8 \times 10^3$ $e^-/Å^2/$frame). *Reproduced from Ref. 78 with permission.*

characterization approach routinely used for the latter, i.e., titling the samples to low-index zone axes to facilitate structural assignments, obviously does not apply to ultrasmall clusters which exhibit dynamic behavior under the electron beam. In addition, many ultrasmall nanoclusters feature low-symmetry or amorphous-like structures,[79,80] which can be very difficult to distinguish from the high-symmetry isomers viewed "off-axis."

To address this issue, we need to extract structural data from single-shot HAADF-STEM images of low-symmetry structures and of high-symmetry structures with arbitrary orientations. The key step in reaching this goal is to perform image simulations based on a variety of candidate atomic structure models to generate a "simulation atlas" that covers all the possible orientations of these isomers. The key question is: can we distinguish low-symmetry structures from high-symmetry isomers orientated "off-axis" via their single-shot images? We consider Au_{55} clusters as an example.[63] Figure 12 shows some example images of (free) Au_{55} clusters from the simulation atlas obtained based on three candidates, highly ordered structures—Ih, Dh, and cuboctahedral atomic models—as well as a hybrid model containing both Ih character and a large (111) facet, with predicted lower energy.[80]

A careful analysis of the HAADF-STEM simulation atlas for Au_{55} clusters reveals the criteria to be used to make structural assignments. The cuboctahedral isomers exhibit the simplest projection features, which are dominated by either two-dimensional or unidirectional fringe patterns. For the "multiple-twin" isomers Dh and Ih, the simulations performed along or near their fivefold axes tend to show the most easily identifiable projection patterns; one sees five

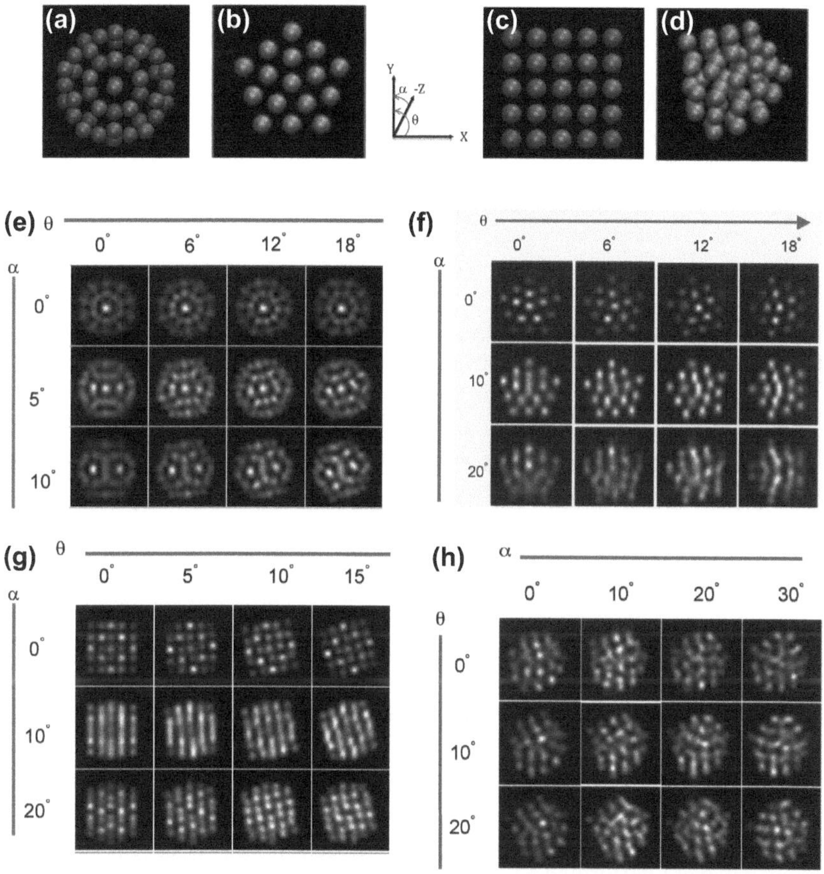

FIGURE 12 Systematic high-angle annular dark field scanning transmission electron microscopy (HAADF-STEM) simulation of size-selected Au_{55} clusters. (a—d) show atomic models of icosahedral, Ino-decahedral, cuboctahedral, and hybrid Au_{55} clusters, respectively. (e—h), some example images from HAADF simulation atlas obtained for the models (a—d), respectively. *Reproduced from Ref. 63 with permission.*

cyclically symmetric twins for Dh, each of which contains crystalline {111} and {200} fringes, and concentric or "double-circle" features for Ih. Overall, the projections of these three high-symmetry models exhibit rather simple and regular projection patterns. By contrast, the more amorphous-like hybrid isomer presents much more complex structural characteristics. In general, the projection patterns of the hybrid isomer look significantly messier owing to its more random atomic arrangements and complex potential configurations (in such case direct channels lack for electron beam to pass through). In addition, the hybrid isomer shows noncircular projection outlines in some cases. Thus, different isomers present different structural projection features, which allow one to identify and distinguish them without too much ambiguity. In other words, with the help of a simulation atlas, it is possible to make rather reliable structural assignments for single snapshot images of ultrasmall nanoclusters.

The electron beam irradiation, or even ambient temperatures in the absence of irradiation, may cause an ultrasmall cluster to explore different configurations in the multidimensional potential energy surface as a function of time. Thus it is clearly best if one can simulate the HAADF-STEM images from a large number of different atomic models, preferably on a frame-by-frame basis. In practice, this approach may run up against the limits of available computational power. However, one can at least identify whether certain energetically favorable structures predicted by theory are consistent with the experiments or not, although this is also not an easy task for the reason given as follows. There may exist some imperfection in the details of the match between the experimental images and the corresponding simulation images obtained from the structural models of free clusters without considering the effect of substrates and, perhaps more importantly, the effect of electron beam irradiation. For this reason, it is challenging to make a quantitative comparison analysis between the experimental and simulated images, and currently the structural characterization is mainly performed based on a comparison of the relative intensities and positions of atomic columns in the clusters. However, with the future development of theoretical calculation approaches by including all the practical experimental factors, it will be possible to make a quantitative analysis and structural assignments by computer programs automatically. We will take Au_{55} and Au_{20} as examples to demonstrate this application of the simulation atlas approach.

The atomic structures of size-selected Au_{55} clusters on carbon film support have been systematically investigated via C_s-corrected STEM imaging in combination with multislice image simulations. By carefully analyzing the single snapshot images with reference to the simulation atlas, structural assignments were performed successfully for the Au_{55} clusters.[63] The structural analysis does not show the existence of any of the three high-symmetry isomers, Ih, Dh, and fcc. A certain proportion of the cluster image sequences observed exhibit features consistent with the simulated images for the hybrid isomer in an appropriate orientation. Figure 13(a–c) shows several such images by way of example. Note that some imperfection in the match between the experimental image and corresponding simulation is expected due to the reasons we have discussed. Moreover, a statistical investigation shows that the majority of the images cannot be assigned neither to any of the three high-symmetry models nor to hybrid model. Some examples of these "uncertain" isomers are given in Figure 13(d–f). Thus, although the hybrid structure is the most commonly observed model structure, it does not dominate the data. Other cluster images often display low-symmetry structural characteristics, perhaps closer in character to the hybrid model than the three high-symmetry structural models considered.

In the case of Au_{20} clusters, it was previously proposed that the tetrahedral pyramid structure (fcc segment) is highly favored on energetic grounds.[81,82]

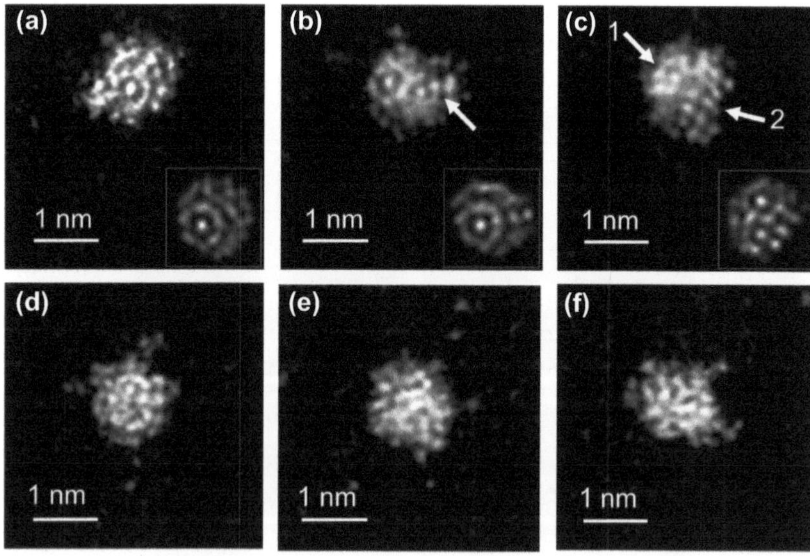

FIGURE 13 High-angle annular dark field scanning transmission electron microscopy (HAADF-STEM) images extracted from the sequences recorded for six different size-selected Au_{55} clusters. (a–c) Hybrid-type structures; the insets are the simulations obtained using the Au_{55} hybrid model for the appropriate orientations. (d–f) Amorphous-like structures that are not consistent with any of the models investigated (i.e., Ih, Dh, fcc, and the hybrid). *Reproduced from Ref. 63 with permission.*

To investigate this proposal via direct imaging, a systematic HAADF-STEM investigation was carried out with the reference to the simulation atlas obtained based on the tetrahedral Au_{20} model.[78] Some of the frames from a sequence of HAADF-STEM images of one Au_{20} cluster on amorphous carbon support are shown in Figure 14(a–d), which clearly exhibits the structural fluctuations of the Au_{20} cluster. However, the cluster in Figure 14(c) does display a triangular shape, consistent with the STEM simulation of the proposed tetrahedral cluster near the <110> axis (see insets). The simulation atlas also shows that nontriangular shapes should also arise in the projections of the tetrahedral Au_{20} model. These are indeed sometimes observed in HAADF imaging of the Au_{20} clusters; Figure 14(f) is an example. Therefore, the existence of the proposed tetrahedral Au_{20} isomer was confirmed through systematic HAADF imaging and simulations. However, the statistical measurements of the Au_{20} clusters show that tetrahedral projections do not dominate the observed images (they represent ~5%), and most of frames present low-symmetry or disordered structures.

Taking together the statistical investigations of the Au_{55} and Au_{20} clusters we have discussed, we may draw some preliminary conclusions about the structural characterization of ultrasmall metal clusters via TEM imaging with

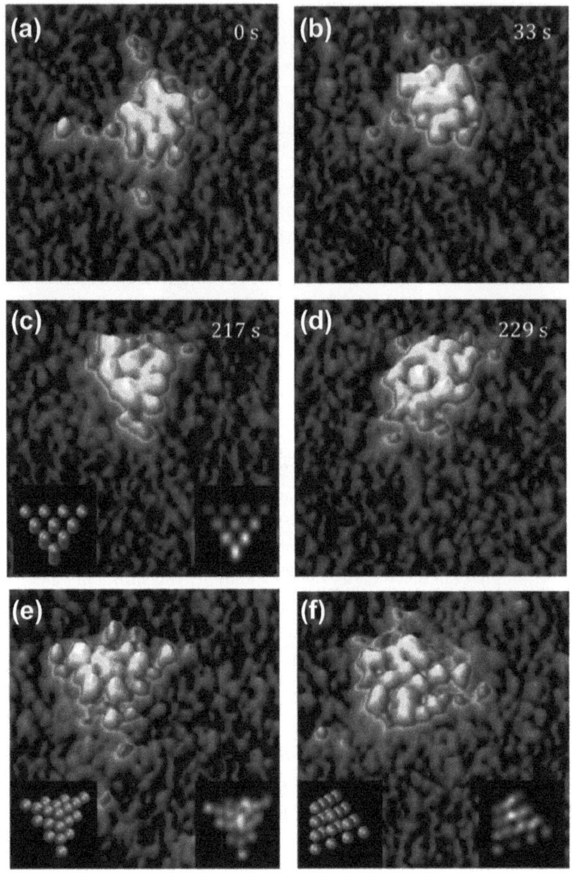

FIGURE 14 3D intensity plots of high-angle annular dark field scanning transmission electron microscopy (HAADF-STEM) images of size-selected Au_{20} clusters. (a–d) Individual frames from a time-lapse series of HAADF images of one Au_{20} cluster; (e and f) images of two other Au_{20} clusters. The image size is 2.8 × 2.8 nm in each case. The insets on the right of (c), (e), and (f) are simulations based an Au_{20} fcc tetrahedron model with orientations shown in the corresponding insets on the left. (see color plate). *Reproduced from Ref. 78 with permission.*

gentle electron beam as follows. (1) Ultrasmall nanoclusters exhibit significant structural fluctuations in time-lapse TEM imaging; (2) high-symmetry structural forms sometimes arise in imaging; but (3) a majority of frames present low-symmetry or disordered configurations.

6.3.2 Thiolate-Protected Gold Clusters

Ligand-protected metal nanoparticles present distinctive physical and chemical properties, for which the protection layers are partly responsible.[1,83,84] It was believed that the protection ligands are only weakly bound to metal core and thus do not significantly influence the structural configurations of the clusters. This is supported by a recent C_S-corrected TEM study of $Au_{55}(PPh_3)_{12}Cl_6$ clusters on carbon film supports.[85] It is found that the phosphine-protected Au_{55} clusters mainly feature the hybrid and amorphous structural motifs, which is thus consistent with the structures of size-selected bare Au_{55} clusters as discussed in Section 6.3.1.2. However, this conclusion does not apply to thiolate-protected gold clusters, which are a focus of attention because of their stability and unique properties.[11,84] It is now accepted that the preferred structures of these clusters are strongly affected by a combination of electron shell closing in the Au core and a complex Au—thiolate shell (also called "divide-and-protect" bonding motif), which significantly deviates from the traditional idea of a sharp boundary between a raft of alkanethiol molecules and nanoparticle facets.[86–88]

As discussed in Section 6.3.1.2, the high dynamic behavior of ultrasmall gold nanoclusters under the electron beam creates a challenge for precise structural determination in the TEM. This problem might in principle be even more serious for ligand-protected gold nanoclusters, due to the radiation-sensitive light ligand materials around the metal cores. Recently, there has been some progress in developing the optimized imaging conditions to obtain high-quality TEM images for structural determination. Next, we will describe the technical advancements in this area and their application to several particular ligand-protected gold clusters.

6.3.2.1 *Imaging at Cryogenic Temperatures: Monolayer-Protected Au_{38}*

Electron beam-induced local heating cannot be avoided during TEM imaging. The beam-heating issue is a particular concern for ligand-protected clusters, since the molecular ligands do not usually have good thermal conductivity. One method of addressing the issue is of course to counteract it by cooling the samples during TEM imaging.

Using a cryogenic TEM sample holder, a systematic study has been performed for hexanethiolate monolayer-protected Au_{38} (MP-Au_{38}) clusters via C_S-corrected STEM imaging under optimized imaging conditions.[89] The samples were first coated with a thin film of carbon to ensure a good sample-substrate thermal contact and to prevent migration of clusters. A liquid nitrogen holder was used to cool the samples down to −164 °C to minimize instability due to beam-induced thermal heating. A gentle electron beam was then used for fast recording a series of images (generally several frames) with an acquisition speed of 0.2 s/frame for each cluster. Figure 15 shows two series of HAADF images of two MP-Au_{38} clusters recorded with all these techniques employed. It is clear that changes of the atomic arrangements between consecutive images still occur, indicating that structural fluctuations cannot be entirely removed even under the "optimized" conditions including cryogenic cooling. It is also noticed that the images recorded with liquid nitrogen cryogenic holders often appear blurred perhaps due to the thermal drift and vibrations caused by liquid holder dewars,

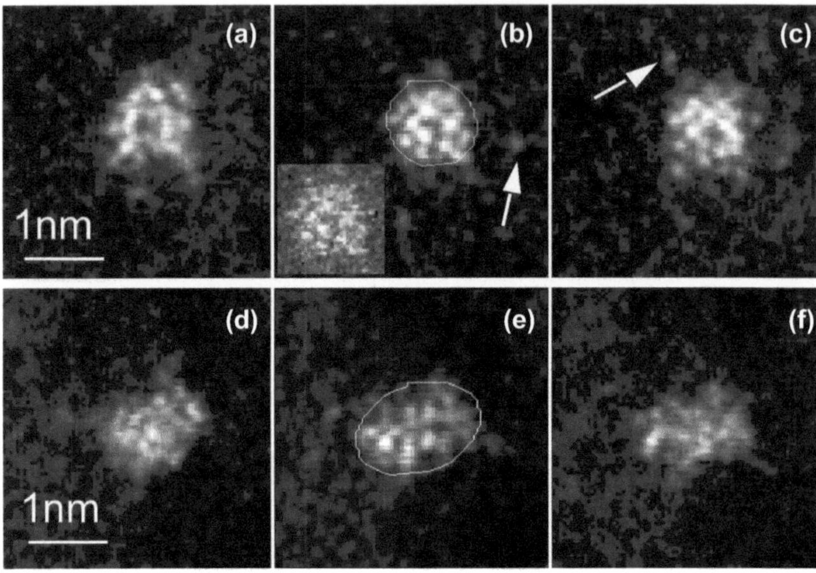

FIGURE 15 High-angle annular dark field scanning transmission electron microscopy (HAADF-STEM) images recorded for hexanethiolate-protected Au_{38} clusters with a liquid nitrogen cryogenic holder and at relatively low electron dose ($\sim 6.9 \times 10^3$ $e^-/Å^2$). Panels (a—c) are a series of images of cluster A, with an approximately circular projected shape; panels (d—f) show cluster B, with a prolate projected shape. The insert in (b) displays the corresponding raw image. The best-fit ellipses used to obtain the aspect ratios are shown in (b) and (e). *Reproduced from Ref. 89 with permission.*

which thus makes the precise structural determination of the nanoclusters very challenging. However, Figure 15 does show that the projected shapes of the MP-Au_{38} clusters remain basically unchanged between frames. A systematic investigation reveals that this shape stability holds true in most cluster image sequences recorded with the optimized imaging conditions in this study. Note that the effect of the total incident electron dose on the cluster stability cannot be ignored even under the cooling conditions, and a continuous image acquisition for long time (e.g., tens of seconds) may modify the morphology of the clusters significantly. However, since a total of several frames (0.2 s/frame) are sufficient to reveal initial shapes of the clusters, the electron beam irradiation issue is not a big concern for the purpose of the shape study.

It is clear that the shapes of clusters A and B in Figure 15 differ significantly from each other, and a measurement shows that they have aspect ratios of 1.09 and 1.49, respectively. Statistical analysis of the shape population for monolayer-protected Au_{38} clusters shows that a large proportion of clusters exhibit a nonspherical, prolate projected shape. The data can be explained well with reference to the DFT calculations. Two low-energy structures were reported based on the "divide-and-protect" motif; model (a) containing a central Au_{14} core[90] and model (b) containing a bi-icosahedral Au_{23} core, with

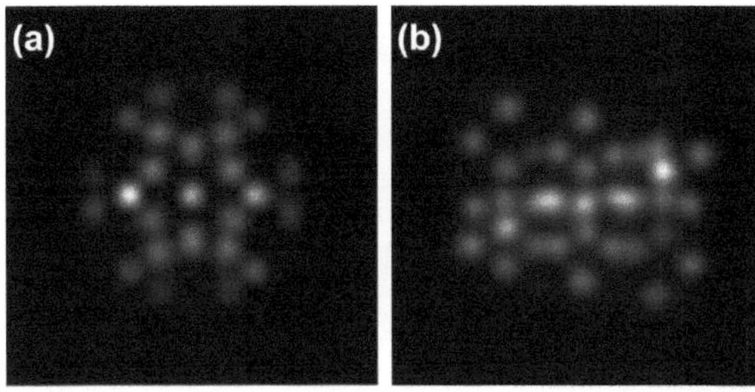

FIGURE 16 Simulated high-angle annular dark field scanning transmission electron microscopy (HAADF-STEM) images of MP-Au_{38} clusters, based on models of (a) a central Au_{14} core,[90] with a quasi-spherical shape, and (b) a bi-icosahedral Au_{23} core,[17] with a nanorod shape. The aspect ratio is 1.02 and 1.58 for (a) and (b), respectively. *Reproduced from Ref. 89 with permission.*

slightly lower energy.[17] Due to steric considerations, different arrangements of the gold–thiolate ligand shell can exist in model (b), but a chiral arrangement of the protecting $Au_x(SR)y$ units was found to have the lowest potential energy. The model (a) presents a quasi-spherical shape, while the model (b) exhibits a prolate shape with aspect ratio as large as ~1.6. This is seen clearly in the simulations of the HAADF-STEM images in Figure 16. The MP-Au_{38} clusters observed experimentally to present prolate shapes and large aspect ratios are then consistent with model (b) and its bi-icosahedral core. The smaller aspect ratios correspond either to the compact projections of bi-icosahedral model (b) along the appropriate orientation, or to model (a) with its quasi-spherical Au core. The consistency between the experimental images and theoretical calculations confirms the existence of prolate MP-Au_{38} clusters. Such prolate shapes are also observed in thiolate-stabilized $Au_{40}(SR)_{24}$ clusters consisting of an Au_{26} core.[91]

6.3.2.2 Low-Kilovolt STEM Nanobeam Diffraction: $Au_{144}(SR)_{60}$

Nanobeam electron diffraction (NBD) is an experimental technique originally designed for working in CTEM mode, but recent technical developments have made it also compatible with STEM.[92–94] Alloyeau et al.[94] showed that, by fine-tuning the microscope illumination lenses in STEM, a nearly parallel nanoprobe ~1 nm in diameter can be formed, suitable for generating nanobeam (i.e., local) DPs. Although the STEM spatial resolution is significantly decreased due to the broadened probe used, there is no difficulty in distinguishing and identifying the positions of nanoparticles if their distribution density is not too high. Applying this technique to a structural study of CoPt nanoparticles led to an observation of size effects on the order–disorder phase transition temperature of the particles.[94]

Bahena et al. recently demonstrated that the STEM-NBD technique could be employed to explore the structure of thiolate-protected gold nanoclusters.[95] With an aberration-corrected STEM operating at 80 kV, they

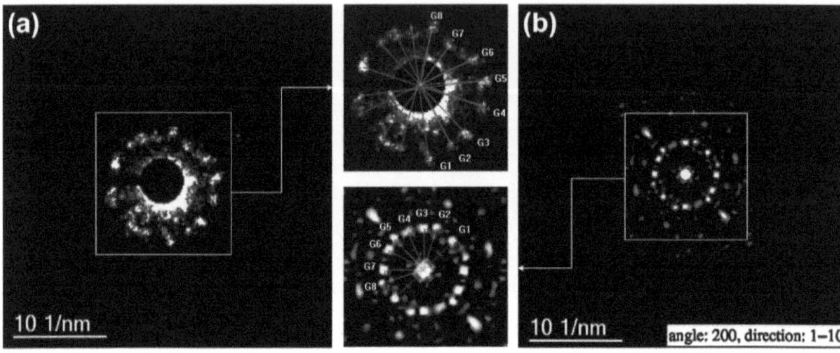

FIGURE 17 The experimental and theoretical STEM nanobeam electron diffraction (NBD) patterns of $Au_{144}(SR)_{60}$ nanoclusters. Sixteen reflections are present in each of the patterns. (a) Experimental NBD pattern and (b) simulated electron diffraction pattern. *Reproduced from Ref. 95 with permission.*

recorded the local DPs of individual $Au_{144}(SR)_{60}$ nanoclusters on holey carbon film supports, which were found to remain basically stable for a period of several seconds. The DPs were interpreted with the help of DFT calculations and diffraction simulations. A structural model was constructed based on a priori knowledge obtained from nuclear magnetic resonance experiments, which consists of a Mackay-icosahedral core of 54 Au atoms surrounded by a 60-Au atom shell bonded to 30 SR–Au–SR "staple" units. The model was theoretically optimized and then used to simulate the DPs. A comparison between the simulated and experimental DPs revealed a good match in terms of the number of diffracted beam and their angles, as shown in Figure 17.

6.3.2.3 Low-Dose TEM Imaging: $Au_{68}(SR)_{32}$

In biological cryo-TEM imaging, a minimal electron dose (e.g., 1–10 $e^-/Å^2$) is commonly used to minimize the effect of electron beam damage.[96] The TEM images recorded with such low dose are very noisy and provide very little interpretable contrast information. However, this issue is addressed by averaging the images of many molecules (believed to be identical) to improve the signal-to-noise ratio dramatically. For example, by using the low-dose TEM technique, a 3.3-Å resolution structure of an aquareovirus was reported by Zhang et al., which provides new insights into the mechanism of cell entry.[97] Again, Evans et al. demonstrated a 0.16-nm spatial resolution for an organic sample (a two-dimensional paraffin crystal) using low-dose aberration-corrected cryo-electron microscopy.[98] Further applications of this low-dose, averaging technique to electron irradiation-sensitive samples are of course expected.

Azubel et al. recently introduced the low-dose TEM technique into the study of ligand-protected gold nanoclusters, enabling the accurate structural

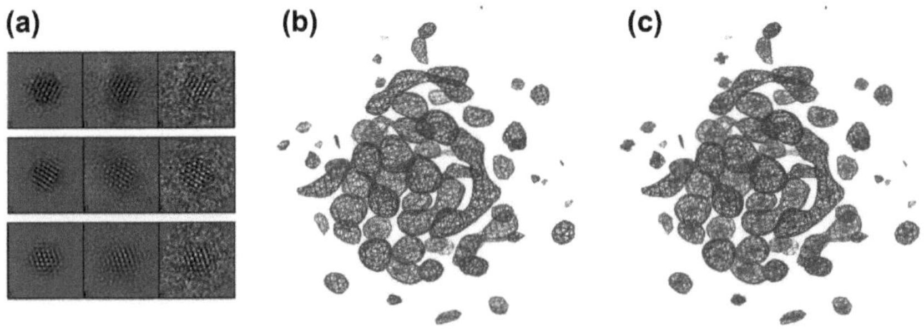

FIGURE 18 Three-dimensional reconstruction of 3-mercaptobenzoic acid-protected Au_{68} nanoparticle structure from low-dose TEM images. (a) Representative components of the reconstruction. (Left) Back projection from the reconstruction; (middle) corresponding class average of the EM images; (right) TEM images. (b) and (c), electron density map, blue mesh, with the locations of atomic coordinates for gold atoms marked by pink stars in (c). (see color plate). *Reproduced from Ref. 99 with permission.*

determination of 3-mercaptobenzoic acid (3-MBA)-protected Au_{68} nanoparticles with atomic resolution.[99] They recorded the images of nearly 1000 nanoparticles with an aberration-corrected TEM at a minimal acquisition time required to record a signal above the noise. Figure 18(b) shows the electron density map built up by averaging over the Au_{68} nanoparticles (supposed identical). It contains 68 peaks that can be assigned to the 68 Au atoms, as marked with the pink stars in Figure 18(c). The analysis shows that the atomic arrangement of the gold core of the nanoparticles can be described as such: a 13-atom cuboctahedron lies in the interior, which is surrounded by 24 Au atoms located in the fcc-like high-symmetry sites and another 31 Au atoms that do not obey the fcc packing rule. Since the location of S atoms cannot be determined by the TEM imaging, they are manually added to the Au core based on a priori knowledge about Au–S motifs, which formed a structural model. It was then relaxed with DFT to a local energy minimum (assuming a molecular formula of $Au_{68}(SH)_{32}$). A comparative analysis between the models before and after relaxation displays a close similarity in both the locations of the Au atoms and the Au–S bonding motifs. Thus, the structure model built based on TEM technique is supported well by DFT calculations. Interestingly, the proposed structure of the 3-MBA-protected Au_{68} nanoparticles does not show a filled electronic shell as found in some other thiolated-protected ultrasmall gold nanoclusters.[11]

The low-dose TEM image reconstruction approach appears to be a powerful technique for structural characterization of nanoclusters, which hold promise for a wide application in the future. To have this approach work effectively, one needs to make sure that all the nanoparticles used for the image averaging should be identical (the same isomer).

6.4 SUMMARY AND PROSPECTS

Small nanoclusters (<1000 atoms) exhibit a rich array of fascinating and important physical and chemical properties that strongly depend on their atomic structures. Transmission electron microscopy has now shown itself to be a powerful tool for structural characterization of such clusters. With the recent technical advances in TEM, including aberration correction and 3D imaging methods, structural determination has been successfully performed for a range of size-selected and ligand-protected gold nanoclusters. The approach applies not only to high-symmetry structures but also to low-symmetry or rather disordered atomic configurations.

However, the challenge to obtain a complete structure determination for small nanoclusters remains. The major issue is the effect of electron beam irradiation on the sample stability. One solution may lie in the ongoing technical developments aimed at correcting higher-order spherical aberrations and chromatic aberrations, which may allow small nanoclusters to be imaged at very low voltage TEM (e.g., 10–20 kV) to reduce the knock-on effect, while atomic resolution is retained. Another solution is to employ the single-shot dynamical TEM[100] or ultrafast TEM[101,102] imaging technique. In this technique, an (ultra)short laser pulse is used to generate a packet of electrons, which are employed to image the specimen. The signal obtained this way is likely to emerge more quickly than irradiation damage occurs, making it possible to determine the initial structures of nanoclusters. At present, limited spatial resolution (~10 nm in 15 nanosecond exposures[100]) due to Coulomb repulsion between electrons is a major limitation for applications of this single-shot dynamic/ultrafast TEM, but it is promising to improve the spatial resolution significantly with future technical developments. Due to the low electron scattering cross sections of light atoms, the direct atomic-resolution imaging of ligands is still extremely challenging. One possible solution appears to be the ABF imaging technique in the STEM as it provides a good signal for both heavy and light atoms simultaneously. Assuming that the electron beam damage issue can be removed by one new TEM imaging method or another, the in situ structural dynamics of small nanoclusters induced by chemical reactions at liquid or gas environments will become accessible, as well as an understanding of the dynamic phenomena occurring during growth and catalytic processes.

REFERENCES

1. Daniel MC, Astruc D. Gold nanoparticles: assembly, supramolecular chemistry, quantum-size-related properties, and applications toward biology, catalysis, and nanotechnology. *Chem Rev* 2004;**104**:293–346.
2. Whetten RL, Khoury JT, Alvarez MM, Murthy S, Vezmar I, Wang ZL, et al. Nanocrystal gold molecules. *Adv Mater* 1996;**8**:428–33.
3. Schreiber F. Structure and growth of self-assembling monolayers. *Prog Surf Sci* 2000; **65**:151–256.

4. Love JC, Estroff LA, Kriebel JK, Nuzzo RG, Whitesides GM. Self-assembled monolayers of thiolates on metals as a form of nanotechnology. *Chem Rev* 2005;**105**:1103–70.
5. Murray RW. Nanoelectrochemistry: metal nanoparticles, nanoelectrodes, and nanopores. *Chem Rev* 2008;**108**:2688–720.
6. Vajda S, Pellin MJ, Greeley JP, Marshall CL, Curtiss LA, Ballentine GA, et al. Subnanometre platinum clusters as highly active and selective catalysts for the oxidative dehydrogenation of propane. *Nat Mater* 2009;**8**:213–6.
7. Habibpour V, Song MY, Wang ZW, Cookson J, Brown CM, Bishop PT, et al. Novel powder-supported size-selected clusters for heterogeneous catalysis under realistic reaction conditions. *J Phys Chem C* 2012;**116**:26295–9.
8. Palmer RE, Pratontep S, Boyen HG. Nanostructured surfaces from size-selected clusters. *Nat Mater* 2003;**2**:443–8.
9. Li G, Jin RC. Atomically precise gold nanoclusters as new model catalysts. *Acc Chem Res* 2013;**46**:1749–58.
10. Baletto F, Ferrando R. Structural properties of nanoclusters: energetic, thermodynamic, and kinetic effects. *Rev Mod Phys* 2005;**77**:371–422.
11. Häkkinen H. Atomic and electronic structure of gold clusters: understanding flakes, cages and superatoms from simple concepts. *Chem Soc Rev* 2008;**37**:1847–59.
12. Iijima S, Ichihashi T. Structural instability of ultrafine particles of metals. *Phys Rev Lett* 1986;**56**:616–9.
13. Marks LD. Experimental studies of small particle structures. *Rep Prog Phys* 1994; **57**:603–49.
14. Jadzinsky PD, Calero G, Ackerson CJ, Bushnell DA, Kornberg RD. Structure of a thiol monolayer-protected gold nanoparticle at 1.1 Å resolution. *Science* 2007;**318**:430–3.
15. Heaven MW, Dass A, White PS, Holt KM, Murray RW. Crystal structure of the gold nanoparticle $[N(C_8H_{17})_4][Au_{25}(SCH_2CH_2Ph)_{18}]$. *J Am Chem Soc* 2008;**130**:3754–5.
16. Zhu M, Aikens CM, Hollander FJ, Schatz GC, Jin RC. Correlating the crystal structure of a thiol-protected Au_{25} cluster and optical properties. *J Am Chem Soc* 2008;**130**:5883–5.
17. Lopez-Acevedo H, Tsunoyama T, Tsukuda H, Häkkinen H, Aikens CM. Chirality and electronic structure of the thiolate-protected Au_{38} nanocluster. *J Am Chem Soc* 2010; **132**:8210–8.
18. Qian HF, Eckenhoff WT, Zhu Y, Pintauer T, Jin RC. Total structure determination of thiolate-protected Au_{38} nanoparticles. *J Am Chem Soc* 2010;**132**:8280–1.
19. William DB, Cater CB. *Transmission electron microscopy*. 2nd ed. New York: Springer; 2009.
20. Yao N, Wang ZL. *Handbook of microscopy for nanotechnology*. Kluwer Academic Publishers; 2005.
21. Yang JC, Small MW, Grieshaber RV, Nuzzo RG. Recent developments and applications of electron microscopy to heterogeneous catalysis. *Chem Soc Rev* 2012;**41**:8179–94.
22. Van Tendeloo G, Bals S, Van Aert S, Verbeeck J, Van Dyck D. Advanced electron microscopy for advanced materials. *Adv Mater* 2012;**24**:5655–75.
23. Pennycook SJ, Nellist PD. *Scanning transmission electron microscopy*. New York: Springer; 2011.
24. Huang PY, Kurasch S, Alden JS, Shekhawat A, Alemi AA, McEuen PL, et al. Imaging atomic rearrangements in two-dimensional silica glass: watching silica's dance. *Science* 2003;**342**:224–7.
25. Li ZY, Johnston RL, Wilcoxon JP, editors. *Metal nanoparticles and nanoalloys*. Amsterdam: Elsevier; 2012. p. 213–47.

26. Knoll M, Ruska E. Das elektronenmikroskop. *Z fur Phys* 1932;**78**:318–39.
27. Egerton RF. *Electron energy loss spectroscopy in transmission electron microscopy*. 2nd ed. New York: Plenum; 1996.
28. Dehm G, Howe JM. *In-situ electron microscopy: applications in physics, chemistry and materials science*. Weinheim: Wiley-VCH; 2012.
29. Barnard AS, Young NP, Kirkland AI, van Huis MA, Xu H. Nanogold: a quantitative phase map. *ACS Nano* 2009;**3**:1431–6.
30. Cowley J. *Diffraction physics*. 3rd ed. Amsterdam: Elsevier; 1995.
31. Reimer L, Kohl H. *Transmission electron microscopy physics of image formation*. 5th ed. Berlin: Springer; 2008.
32. von Ardenne M. Das Elektronen-Rastermikroskop. *Z fur Phys* 1938;**109**:553–72.
33. Crewe AV, Wall J, Welter LM. A high-resolution scanning transmission electron microscope. *J Appl Phys* 1968;**39**:5861–8.
34. Crewe AV, Langmore JP, Isaacson MS, Siegel B, Beaman D, editors. *Physical aspects of electron microscopy and microbeam analysis*. New York: John Wiley; 1975. p. 47–62.
35. Zeitler E, Thomson MGR. Scanning transmission electron microscopy. *Optik* 1970; **31**:359–66.
36. Crewe AV, Wall J, Langmore J. Visibility of single atoms. *Science* 1970;**168**:1338–40.
37. Treacy MMJ, Howie A, Wilson CJ. Z-contrast of platinum and palladium catalysts. *Philos Mag A* 1978;**38**:569–85.
38. Howie A. Image contrast and localized signal selection techniques. *J Microsc* 1979; **117**:11–23.
39. Liu J, Cowley JM. High-resolution scanning transmission electron microscopy. *Ultramicroscopy* 1993;**52**:335–46.
40. Nellist PD, Pennycook SJ. Incoherent imaging using dynamically scattered coherent electrons. *Ultramicroscopy* 1999;**78**:111–24.
41. Muller DA. Structure and bonding at the atomic scale by scanning transmission electron microscopy. *Nat Mater* 2009;**8**:263–70.
42. Hartel P, Rose H, Dinges C. Conditions and reasons for incoherent imaging in STEM. *Ultramicroscopy* 1996;**63**:93–114.
43. Wang ZW, Li ZY, Park S, Abdela A, Tang D, Palmer RE. Quantitative Z-contrast imaging in the scanning transmission electron microscope with size-selected clusters. *Phys Rev B* 2011;**84**:073408.
44. Findlay SD, Shibata N, Sawada H, Okunishi E, Kondo Y, Yamamoto T, et al. Robust atomic resolution imaging of light elements using scanning transmission electron microscopy. *Appl Phys Lett* 2009;**95**:191913.
45. Ishikawa R, Okunishi E, Sawada H, Kondo Y, Hosokawa F, Abe E. Direct imaging of hydrogen-atom columns in a crystal by annular bright-field electron microscopy. *Nat Mater* 2011;**10**:278–81.
46. Bleloch A, Ramasse Q, Brydson R, editors. *Aberration-corrected analytical electron microscopy*. Wiley; 2011. p. 55–87.
47. Scherzer O. Uber einige Fehler von Elektronenlinsen. *Z Phys* 1936;**101**:114–32.
48. Scherzer O. Sparische und chromatische Korrektur von Electronen-Linsen. *Optik* 1947; **2**:114–32.
49. Haider M, Rose H, Uhlemann S, Schwan E, Kablus B, Urban K. A spherical-aberration-corrected 200 kV transmission electron microscope. *Ultramicroscopy* 1998;**75**:53–60.
50. Haider M, Uhlemann S, Schwan E, Rose H, Kabius B, Urban K. Electron microscopy image enhanced. *Nature* 1998;**392**:768–9.

51. Batson PE, Dellby N, Krivanek OL. Sub-Ångström resolution using aberration corrected electron optics. *Nature* 2002;**418**:617–20.
52. Müller H, Uhlemann S, Hartel P, Haider M. Advancing the hexapole C_s-corrector for the scanning transmission electron microscope. *Microsc Microanal* 2006;**12**:442–55.
53. Erni R, Rossell M, Kisielowski C, Dahmen U. Atomic-resolution imaging with a sub-50-pm electron probe. *Phys Rev Lett* 2009;**102**:096101.
54. Dellby M, Krivanek OL, Nellist PD, Batson PE, Lupini AR. Progress in aberration-corrected scanning transmission electron microscopy. *J Electron Microsc* 2001;**50**:177–85.
55. Arkill1 KP, Mantell JM, Plant SR, Verkade P, Palmer RE. Using size-selected gold clusters on graphene oxide films to aid cryo-transmission electron tomography alignment. *Sci Rep* 2015;**5**:9234.
56. Chen CC, Zhu C, White ER, Chiu CY, Scott MC, Regan BC, et al. Three-dimensional imaging of dislocations in a nanoparticle at atomic resolution. *Nature* 2013;**496**:74–7.
57. Van Aert S, Batenburg KJ, Rossell MD, Erni R, Van Tendeloo G. *Nature* 2011;**470**:374–7.
58. Singhal A, Yang JC, Gibson JM. STEM-based mass spectroscopy of supported Re clusters. *Ultramicroscopy* 1997;**67**:191–206.
59. LeBeau JM, Findlay SD, Allen LJ, Stemmer S. Standardless atom counting in scanning transmission electron microscopy. *Nano Lett* 2010;**10**:4405–8.
60. Young N, Li ZY, Chen Y, Palomba S, Di Vece M, Palmer RE. Weighing supported nanoparticles: size-selected clusters as mass standards in nanometrology. *Phys Rev Lett* 2008;**101**:246103.
61. Wang ZW, Toikkanen O, Yin F, Li ZY, Quinn BM, Palmer RE. Counting the atoms in supported, monolayer-protected gold clusters. *J Am Chem Soc* 2010;**132**:2854–5.
62. Li ZY, Young NP, Di Vece M, Palomba S, Palmer RE, Bleloch AL, et al. Three-dimensional atomic-scale structure of size-selected gold nanoclusters. *Nature* 2008;**451**:46–8.
63. Wang ZW, Palmer RE. Experimental evidence for fluctuating, chiral-type Au_{55} clusters by direct atomic imaging. *Nano Lett* 2012;**12**:5510–4.
64. Hu KJ, Plant SR, Ellis PR, Brown CM, Bishop PT, Palmer RE. The effects of 1-pentyne hydrogenation on the atomic structures of size-selected Au_N and Pd_N ($N = 923$ and 2057) nanoclusters. *Phys Chem Chem Phys* 2014;**16**:26631–7.
65. von Issendorff B, Palmer RE. A new high transmission infinite range mass selector for cluster and nanoparticle beams. *Rev Sci Instrum* 1999;**70**:4497–501.
66. Pratontep S, Preece P, Xirouchaki C, Palmer RE, Sanz-Navarro CF, Kenny SD, et al. Scaling relations for implantation of size-selected Au, Ag and Si clusters into graphite. *Phys Rev Lett* 2003;**90**:055503.
67. Pratontep S, Carroll SJ, Xirouchaki C, Streun M, Palmer RE. Size-selected cluster beam source based on radio frequency magnetron plasma sputtering and gas condensation. *Rev Sci Instrum* 2005;**76**:045103.
68. Yin F, Wang ZW, Palmer RE. Controlled formation of mass-selected Cu–Au core-shell cluster beams. *J Am Chem Soc* 2011;**133**:10325–7.
69. Wang ZW, Palmer RE. Determination of the ground-state atomic structures of size-selected Au nanoclusters by electron-beam-induced transformation. *Phys Rev Lett* 2012;**108**:245502.
70. Cowley JM, Moodie AF. The scattering of electrons by atoms and crystals. I. A new theoretical approach. *Acta Crystallogr* 1957;**10**:609–19.
71. Kirkland EJ, Loane RF, Silcox J. Simulation of annular dark field STEM images using a modified multislice method. *Ultramicroscopy* 1987;**23**:77–96.
72. Wang ZL, Cowley JM. Simulating high-angle annular dark-field STEM images including inelastic thermal diffuse-scattering. *Ultramicroscopy* 1989;**31**:437–54.

73. Plant SR, Cao L, Palmer RE. Atomic structure control of size-selected gold nanoclusters during formation. *J Am Chem Soc* 2014;**136**:7559–62.
74. Koga K, Ikeshoji T, Sugawara K. Size- and temperature-dependent structural transitions in gold nanoparticles. *Phys Rev Lett* 2004;**92**:115507.
75. Egerton RF, Li P, Malac M. Radiation damage in the TEM and SEM. *Micron* 2004; **35**:399–409.
76. Ascencio JA, Perez-Alvarez M, Tehuacanero S, Jose-Yacaman M. Experimental and theoretical studies of instabilities of metal nanoparticles: a new kind of quasimelting. *Appl Phys A* 2001;**73**:295–300.
77. Bals S, Van Aert S, Romero CP, Lauwaet K, Van Bael MJ, Schoeters B, et al. Atomic scale dynamics of ultrasmall germanium clusters. *Nat Commun* 2012;**3**:897.
78. Wang ZW, Palmer RE. Direct atomic imaging and dynamical fluctuations of the tetrahedral Au_{20} cluster. *Nanoscale* 2012;**4**:4947–9.
79. Huang W, Ji M, Dong CD, Gu X, Wang LM, Gong XG, et al. Relativistic effects and the unique low-symmetry structures of gold nanoclusters. *ACS Nano* 2008;**2**:897–904.
80. Lopez-Lozano X, Perez LA, Garzon IL. Enantiospecific adsorption of chiral molecules on chiral gold clusters. *Phys Rev Lett* 2006;**97**:233401.
81. Li J, Li X, Zhai HJ, Wang LS. Au_{20}: a tetrahedral cluster. *Science* 2003;**299**:864–7.
82. Gruene P, Rayner DM, Redlich B, van der Meer AFG, Lyon JT, Meijer G, et al. Structures of neutral Au_7, Au_{19}, and Au_{20} clusters in the gas phase. *Science* 2008;**321**:674–6.
83. Merrifield RC, Wang ZW, Palmer RE, Lead JR. Synthesis and characterization of polyvinylpyrrolidone coated cerium oxide nanoparticles. *Environ Sci Technol* 2013;**47**: 12426–33.
84. Maity P, Xie SH, Yamauchi M, Tsukuda T. Stabilized gold clusters: from isolation toward controlled synthesis. *Nanoscale* 2012;**4**:4027–37.
85. Jian N, Stapelfeldt C, Hu KJ, Fröba M, Palmer RE. Hybrid atomic structure of the schmid cluster $Au_{55}(PPh_3)_{12}Cl_6$ resolved by aberration-corrected STEM. *Nanoscale* 2015;**7**:885–8.
86. Walter M, Akola J, Lopez-Acevedo O, Jadzinsky PD, Calero G, Ackerson CJ, et al. *Proc Natl Acad Sci USA* 2008;**105**:9157–62.
87. Pei Y, Gao Y, Zeng XC. Structural prediction of thiolate-protected Au_{38}: a face-fused bi-icosahedral Au core. *J Am Chem Soc* 2008;**130**:7830–2.
88. Akola J, Walter M, Whetten RL, Häkkinen H, Grönbeck H. On the structure of thiolate-protected Au_{25}. *J Am Chem Soc* 2008;**130**:3756–7.
89. Wang ZW, Toikkanen O, Quinn BM, Palmer RE. Real-space observation of prolate monolayer-protected Au_{38} clusters using aberration-corrected scanning transmission electron microscopy. *Small* 2011;**7**:1542–5.
90. Häkkinen H, Walter M, Grönbeck H. Divide and protect: capping gold nanoclusters with molecular gold-thiolate rings. *J Phys Chem B* 2006;**110**:9927–31.
91. Malola S, Lehtovaara L, Knoppe S, Hu KJ, Palmer RE, Bürgi T, et al. $Au_{40}(SR)_{24}$ cluster as a chiral dimer of 8-electron superatoms: structure and optical properties. *J Am Chem Soc* 2012;**134**:19560–3.
92. He H, Nelson C. A method of combining STEM image with parallel beam diffraction and electron-optical conditions for diffractive imaging. *Ultramicroscopy* 2007;**107**:340–4.
93. Kolb U, Gorelik T, Kubel C, Otten MT, Hubert D. Towards automated diffraction tomography: Part I-data acquisition. *Ultramicroscopy* 2007;**107**:507–13.
94. Alloyeau D, Ricolleau C, Oikawa T, Langlois C, Le Bouar Y, Loiseau A. STEM nanodiffraction technique for structural analysis of CoPt nanoparticles. *Ultramicroscopy* 2008; **108**:656–62.

95. Bahena D, Bhattarai N, Santiago U, Tlahuice A, Ponce A, Bach SBH, et al. STEM electron diffraction and high-resolution images used in the determination of the crystal structure of the $Au_{144}(SR)_{60}$ cluster. *J Phys Chem Lett* 2013;**4**:975–81.
96. Hoenger A. High-resolution cryo-electron microscopy on macromolecular complexes and cell organelles. *Protoplasma* 2014;**251**:417–27.
97. Zhang X, Jin L, Fang Q, Hui WH, Zhou ZH. 3.3 Å cryo-EM structure of a nonenveloped virus Reveals a priming mechanism for cell entry. *Cell* 2010;**141**:472–82.
98. Evans JE, Hetherington C, Kirkland A, Chang LY, Stahlberg H, Browning N. Low-dose aberration corrected cryo-electron microscopy of organic specimens. *Ultramicroscopy* 2008;**108**:1636–44.
99. Azubel M, Koivisto J, Malola S, Bushnell D, Hura GL, Koh AL, et al. Electron microscopy of gold nanoparticles at atomic resolution. *Science* 2014;**345**:909–12.
100. Browning ND, Bonds MA, Campbell GH, Evans JE, LaGrange T, Jungjohann KL, et al. Recent developments in dynamic transmission electron microscopy. *Curr Opin Solid State Mater Sci* 2012;**16**:23–30.
101. Zewail AH. Four-dimensional electron microscopy. *Science* 2010;**328**:187–93.
102. Flannigan DJ, Zewail AH. 4D electron microscopy: principles and applications. *Acc Chem Res* 2012;**45**:1828–39.

Chapter 7

Structure Prediction by Density Functional Theory Calculations

Qing Tang and De-en Jiang[1]
Department of Chemistry, University of California, Riverside, CA, USA
[1]*Corresponding author: E-mail: de-en.jiang@ucr.edu*

Chapter Outline

7.1 INTRODUCTION

Ligand-protected metal nanoclusters/nanoparticles, especially the gold nanoclusters, constitute an attractive field of research in both chemistry and nanomaterials owing to their extraordinary stability, interesting structural and physicochemical properties, as well as potential applications in many fields such as catalysis, biosensing, drug delivery, luminescence, and molecular electronics.[1–9] Thiolates (−SR) are the most commonly used protective ligands for gold nanoclusters, and an increasingly large number of atomically precise thiolate−gold nanoclusters (denoted as $Au_n(SR)_m$) with size less than 2 nm have been isolated and characterized.[10]

The chemical compositions of $Au_n(SR)_m$ nanoclusters are easy to get from mass spectrometry, but their precise atomic structures are hard to determine because of the experimental difficulty in growing high-quality single crystals. So far only nine $Au_n(SR)_m$ compositions have been crystallized and structurally resolved by X-ray crystallography: $Au_{102}(SR)_{44}$,[11] $Au_{25}(SR)_{18}^-$,[12,13]

Frontiers of Nanoscience, Vol. 9. http://dx.doi.org/10.1016/B978-0-08-100086-1.00007-5

$Au_{38}(SR)_{24}$,[14] $Au_{36}(SR)_{24}$,[15] $Au_{28}(SR)_{20}$,[16] $Au_{23}(SR)_{16}^-$,[17] $Au_{30}(SR)_{18}$,[18] $Au_{24}(SR)_{20}$,[19] $Au_{24}(SeR)_{20}$,[20] and $Au_{20}(SR)_{16}$[21] (the black triangle shown in Figure 1). For most $Au_n(SR)_m$ nanoclusters, their atomic structures remain unknown and no structure models have been proposed (the red triangle shown in Figure 1).[1,22–36] For some of the $Au_n(SR)_m$ compositions, including $Au_{144}(SR)_{60}$,[37] $Au_{44}(SR)_{28}$,[38,39] $Au_{40}(SR)_{24}$,[40] $Au_{19}(SR)_{13}$,[41] $Au_{18}(SR)_{14}$,[42] and $Au_{15}(SR)_{13}$,[43,44] structural models have been proposed based on density functional theory (DFT) (the blue rectangle in Figure 1).

Considering the relatively slow progress in experimental crystallization and X-ray total structure determination of $Au_n(SR)_m$ species, how to predict the structures for known compositions becomes a major motivation but also a grand challenge for the computational chemists. As a matter of fact, in parallel with the experimental progress, the theoretical efforts based on DFT have contributed significantly to the understanding of the electronic structures and cluster stabilities by superatomic complex concept.[45–52] Particularly, the successful prediction on the atomic structures of $Au_{25}(SR)_{18}^-$[53] and $Au_{38}(SR)_{24}$[54,55] (both were confirmed later by experiments[12–14]) makes DFT an important research tool to understand and predict structures of $Au_n(SR)_m$ clusters.

For the structure prediction of thiolate–gold nanoclusters based on DFT computation, a major hypothesis or guiding principle, based on the known structures (e.g., $Au_{102}(SR)_{44}$, $Au_{25}(SR)_{18}^-$, $Au_{38}(SR)_{24}$), is that the structure of $Au_n(SR)_m$ clusters is composed of a high-symmetry gold core protected by $Au_x(SR)_{x+1}$ staple motifs or oligomers.[56,57] During the past few years, this hypothesis has been applied to structural predictions of a variety of $Au_n(SR)_m$ clusters. Generally, for computational efficiency, the experimental R-group is usually simplified as CH_3 to first quickly search the energetically low-lying isomers from many candidates. Note that the simplified $Au_n(SCH_3)_m$ cluster model sometimes may not yield the true structure of $Au_n(SR)_m$ cluster for the real R-group, especially for the sterically hindered bulky ligands. Recent

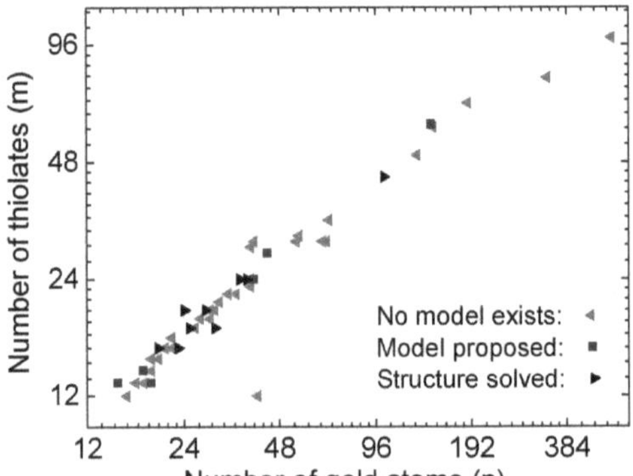

FIGURE 1 $Au_n(SR)_m$ clusters of known compositions.

experiments by Jin,[58] Tracy,[59] and Tsukuda[60] groups revealed that using large, bulky thiols in the synthesis of gold nanoparticles would result in the change in core sizes and interfacial structures, as well as in the electronic structures and optical properties. Therefore, it is necessary to consider the effect of real experimental ligands on the atomic structure, core geometry, and isomer stability of specific $Au_n(SR)_m$ clusters.[61–63]

Going beyond Au and thiolate, other metals such as Ag[64,65] and Ni[66] or other ligands[67–72] have been explored to make atomically precise metal clusters. Particularly, the alkynyl group (R−C≡C−), where R can be an aliphatic or aromatic group, has attracted great attention recently as a new way to protect gold surfaces by forming strong Au−C bonds, leading to the so-called "organogold" structures.[73,74] For instance, the aliphatic alkynes ($HC{\equiv}C(CH_2)_nCH_3$, $n = 5, 7, 9, 11$) can form highly ordered self-assembly monolayers on Au(111) surface.[75] Tsukuda and coworkers used aromatic alkynes such as phenylacetylene (PA-H) or 9-ethynyl-phenanthrene (EPT-H) as the stabilizers and synthesized a series of organogold clusters, e.g., $Au_{34}(PA)_{16}$, $Au_{43}(PA)_{22}$, $Au_{54}(PA)_{26}$, $Au_{30}(EPT)_{13}$, and $Au_{35}(EPT)_{18}$.[76,77] Among them, $Au_{54}(PA)_{26}$ was found to be a magic organogold cluster, having higher stability than the other compositions.[78] These new ligand-protected metal nanoclusters open up an exciting area for the researchers to reveal their interfacial bonding and structures.

Ligand-protected metal clusters are a unique class of chemical species. They are made in wet chemistry and have well-defined structures. Hence to predict their structures is both challenging and tantalizing for computational chemists. In this chapter, we will recapitulate the recent progress on structure prediction of thiolate-protected gold clusters. We will also discuss the interfacial structures of organogold clusters, as a first step toward their structure prediction.

7.2 STRUCTURAL SEARCH

7.2.1 Staple Hypothesis

The X-ray total structure determination of the $Au_{102}(SR)_{44}$ cluster in 2007 has greatly advanced our understanding about the thiolated gold clusters.[11] The solved structure showed that $Au_{102}(SR)_{44}$ comprises a 79-atom gold core protected by 19 [RS−Au−SR] monomer and 2 [RS−Au−SR−Au−SR] dimer units (Figure 2). In 2008, six [RS−Au−SR−Au−SR] dimer units were found to protect an icosahedral Au_{13} core in the solved $Au_{25}(SR)_{18}$ structure.[12,13] These [RS−Au−SR] monomer and [RS−Au−SR−Au−SR] dimer units can be generalized to the $[RS(AuSR)_x]$ ($x = 1, 2, 3,\ldots$) oligomers, or staple motifs, that bind to two surface atoms of the gold core through their two terminal RS− groups. The staple motifs are found to be the key interfacial structures of thiolated gold nanoclusters, and serve as a prevailing structural hypothesis for

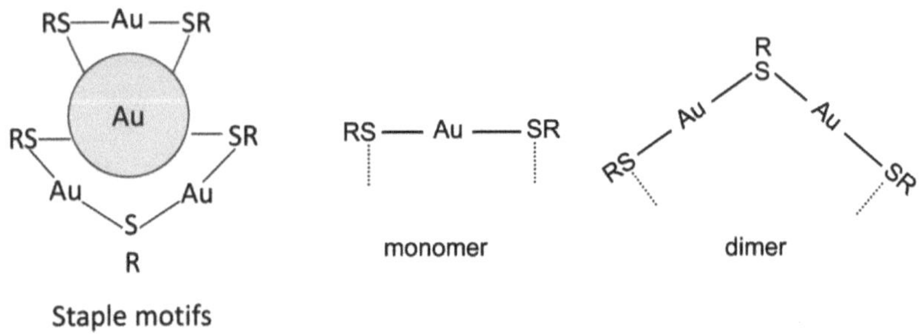

FIGURE 2 Schematic for the core and staple motifs construction of $Au_n(SR)_m$ (left), and the monomer and dimer staple motifs (right). *Ref. 56, reprinted with permission. Copyright © 2013 Royal Society of Chemistry.*

predicting the atomic structures of $Au_n(SR)_m$ clusters.[79,80] The main point of the staple hypothesis, in the spirit of the "divide-and-protect" concept,[81] has been that the structure of any $Au_n(SR)_m$ composition comprises a gold core stabilized by staple motifs in the protective layer. Clusters with lower Au/SR ratios typically need longer $RS(AuSR)_x$ staples ($x \geq 3$). The Au core usually has high symmetry, such as the centered icosahedral Au_{13} in $Au_{25}(SR)_{18}$ and the face-sharing bi-icosahedral Au_{23} in $Au_{38}(SR)_{24}$.

The protection of the gold core by the staple motifs has three contributions to its structural stability: (1) formation of the linear RS−Au−SR bond in $RS(AuSR)x$ motifs, (2) the bonding between terminal thiolates and the surface Au atoms of the core, (3) the Au−Au interaction between the Au core and Au in the $RS(AuSR)_x$ staple motifs. Figure 3 shows a simple computational test to demonstrate the energetic stability of forming a staple motif. We started with a bare Au_{38} cluster in the O_h symmetry, and next we added two CH_3S- groups

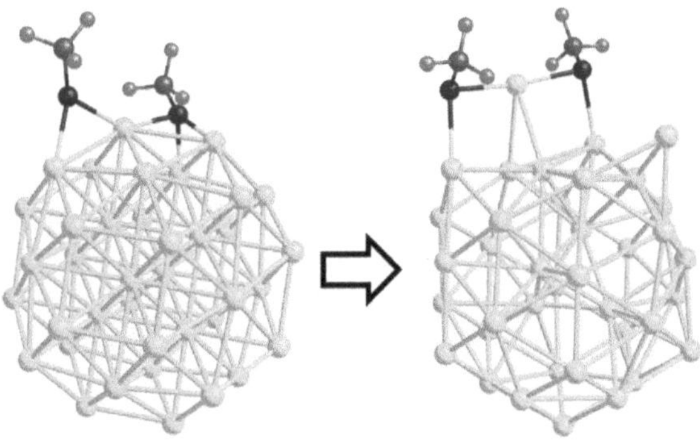

FIGURE 3 Placing two CH_3S- groups next to a gold atom on Au_{38} (initial structure, left) leads to spontaneous formation of the $CH_3S-Au-SCH_3$ motif (final structure, right) after DFT geometry optimization. *Ref. 79, reprinted with permission. Copyright © 2008 American Chemical Society.*

next to one Au atom on the surface and allowed the system to relax. After structural optimization, the Au atom in-between the two CH_3S- groups is pulled out of the outer shell, and the linear $CH_3S-Au-SCH_3$ staple motif forms spontaneously, accompanied by a drastic change to the structure of the underlying cluster. All three contributions to the stability from the staple motif (linear $S-Au_{staple}-S$ bond, $S-Au_{core}$ bond, and $Au_{staple}-Au_{core}$ interaction) can be clearly seen in the final relaxed structure.

In a similar spirit, Tsukuda et al.[36] and Pei et al.[54] proposed the structural principles for building up $Au_n(SR)_m$ clusters from monomer and dimer motifs. Pei et al. predicted the structure of the $Au_{38}(SR)_{24}$ cluster as a face-fused bi-icosahedral Au_{23} core protected by three monomeric staples (RS−Au−SR, $Au(SR)_2$) and six dimeric staples (RS−Au−RS−Au−RS, $Au_2(SR)_3$), which was quiet close to the experimental X-ray structure of $Au_{38}(SR)_{24}$, except that the arrangement of the dimeric staples was not correctly predicted.[54]

The staple motif is a key feature of the gold−thiolate interface.[82] Is there a similar motif at the gold−alkynyl interface for organogold clusters and surfaces? We examined the organogold interfacial structures of PhC≡C− ligands on Au(111) surface and on Au_{20} cluster.[83] Our studies showed that at the PhCC/Au interface, the PhCC− ligands form a similar PhCC−Au−CCPh staple motif as the thiolate groups. On the Au(111) surface, formation of the $PhCC-Au_{adatom}-CCPh$ motif is energetically preferred when the Au adatom is present, resembling the RS−Au−SR motif on Au(111) (Figure 4, inset). Different from the RS−Au−SR staple motifs formed via the Au−S σ bonds, the $PhCC-Au_{adatom}-CCPh$ staple motif consists of both the σ bonding of the terminal C to the staple Au atom and the π bonding of the unsaturated C≡C bond to a surface Au atom. We called this new type of staple motif with the involvement of π-type bonding as the "π staple motif." Figure 4 shows that if one starts with one Au adatom and two isolated PhCC− groups on Au(111), formation of the $PhCC-Au_{adatom}-CCPh$ π motif lowers the energy of the

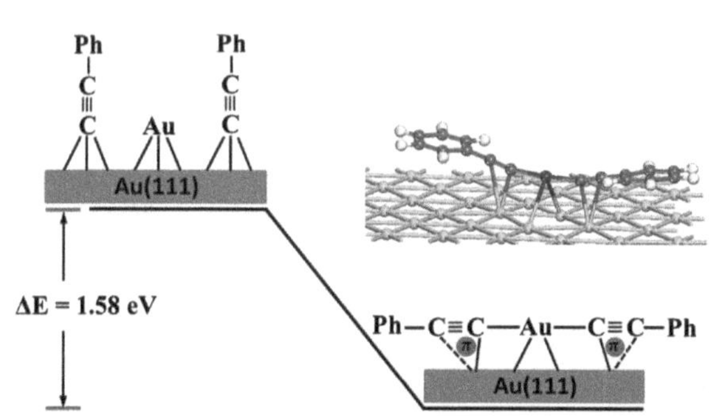

FIGURE 4 Schematic showing the energetic of formation of the $PhCC-Au_{adatom}-CCPh$ π motif relative to the isolated Au adatom and PhCC− groups on Au(111). The inset is a side view of the $PhCC-Au_{adatom}-CCPh$ π staple motif on Au(111). *Ref. 83, reprinted with permission. Copyright © 2014 American Chemical Society.*

system significantly by about 1.58 eV. This simple energetic plot indicates the thermodynamic driving force for the formation of π staple motifs at the PhCC/Au interface. Actually, the π staple motif is stabilized due both to the C≡C/Au π interaction and to the Ph/surface van der Waals (vdW) interaction, in addition to the Au−C σ interaction.

As found at the PhCC/Au(111) interface, geometry optimization and simulated annealing show that this new type of PhCC−Au−CCPh π staple motif is also the preferred bonding mode of the PhC≡C− groups on the Au_{20} cluster (Figure 5(a)), and the C≡C/Au π interaction drives the formation of the π staple motif, even without the help of the vdW interaction between the phenyl rings and gold facets. The unique π-type bonding at the PhCC/Au(111) or PhCC/Au_{20} interface is attributed to the back bonding of the d electrons of the underlying substrate Au atom donated to the π^* orbitals of C≡C bonds. As we discussed above for thiolated gold nanoclusters, longer staple motifs such as the dimer motif [RS−Au−SR−Au−SR] are common. So how about the PhCC−Au system? Interestingly, we found that the dimeric [PhCC−Au−CCPh−Au−CCPh] π staple motif (Figure 5(b)) could also exist as an interfacial motif of Au_{20} cluster. Such a dimeric π staple motif is energetically more stable than the monomeric π staple motif plus one isolated PhCC− group on Au_{20}. As a matter of fact, the dimeric [PhCC−Au−CCPh−Au−CCPh] π staple motif has been observed in the recent single crystal X-ray total structure determination of a mixed-ligand-protected Au_{19} cluster, $[Au_{19}(PhC{\equiv}C)_9(Hdppa)_3]^{2+}$ (Hdppa = N,N-Bis(diphenylphosphino)amine), wherein three V-shaped PhCC−Au−CCPh−Au−CCPh motifs are present in the protective shell of this cluster.[84] Since the staple hypothesis has been very successfully applied in the structure prediction for thiolated gold nanoclusters, the idea of alkynyl-gold π staple motif would give us the insight to predict the possible atomic structures of the series of organogold nanoclusters as identified by Tsukuda and coworkers[76–78] This is a potential gold mine to prospect.

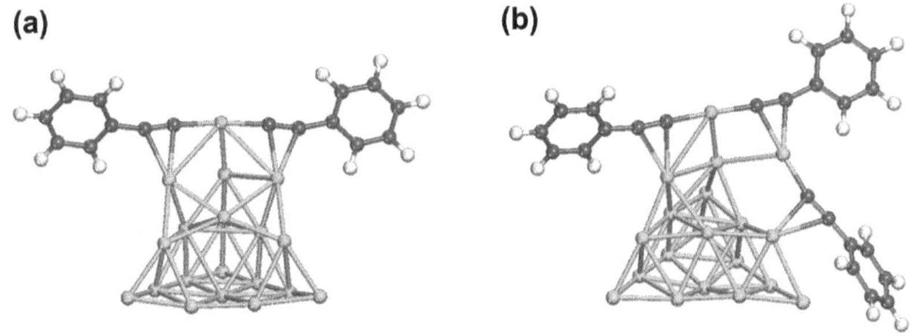

FIGURE 5 Geometry structures of monomeric [PhCC−Au−CCPh] (a) and dimeric [PhCC−Au−CCPh−Au−CCPh] (b) π staple motifs on Au_{20} cluster. *Ref. 83, reprinted with permission. Copyright © 2014 American Chemical Society.*

7.2.2 Staple Fitness

The "staple motif" idea suggests that the $Au_n(SR)_m$ clusters are most likely to have a structure with a high-symmetry core capped by staple motifs such as monomer, dimer, or even longer motifs. One big challenge in structure prediction for $Au_n(SR)_m$ clusters is to come up with a correct Au core structure. In most cases, one often relies on the Cambridge Cluster Database to search the candidates for the core structure, usually the structure with high symmetry and high stability is picked up as the Au core.[85] Besides the challenge in proposing a correct model for the core, another challenge in understanding the thiolate–gold clusters lies in that, if a core geometry is already given or guessed correctly, how do we then predict the way the staple motifs arrange themselves on the core surface? Hence, the idea of "staple fitness" was proposed to address this issue.[41]

The basic idea for staple fitness is that, if we have a high-symmetry core and only the monomer and/or dimer staple motifs are considered as the protecting units, it is straightforward to partition the gold atoms into the core and the staple motifs satisfying a given composition.[41] Since each staple motif is connected to the core surface via two terminal RS groups, the possible arrangements of staple motifs can be directly mapped to a purely combinatorial problem of pairing up dots on a surface. Considering the huge amount of possible combinations, how can we select the right ones out of all the possibilities? Actually the distance between two Au atoms connected to the staple motif (usually in the range of 3.5–5.5 Å) should be greater than the nearest neighbor surface Au–Au distance (2.8 Å). Based on this geometrical constraint, we can preclude lots of combinations, and this initial screening process can be called as "pruning." For the remaining structures survived from "pruning," we then ranked them by using the total pair distance, which is defined as the sum of all the interpair distances for a given combinations of paring up N dots. We hypothesize that the combination with the best staple fitness corresponds to the greatest total pair distance where the steric repulsion among the staple motifs would be minimal.

With the "staple fitness" concept, we can correctly arrive at the experimental structures of $Au_{25}(SR)_{18}$ and $Au_{38}(SR)_{24}$ based on the icosahedral Au_{13} core and the bi-icosahedral Au_{23} core, respectively. In Figure 6(a) and (c), we showed the model combinations with the best fitness (the greatest total pair distance) for the $Au_{25}(SR)_{18}$ and $Au_{38}(SR)_{24}$ clusters, and their corresponding experimental structures are shown in Figure 6(b) and (d).

The advantage of using the "staple fitness" approach is that it can significantly accelerate the explorations of the candidate structures of thiolate–gold clusters for a given core, since only the simple arithmetic operation is applied. By combining the concept of staple fitness with DFT computations, Jiang et al. have then applied this idea to predict structures for two known compositions, $Au_{19}(SR)_{13}$ and $Au_{40}(SR)_{24}$.[41]

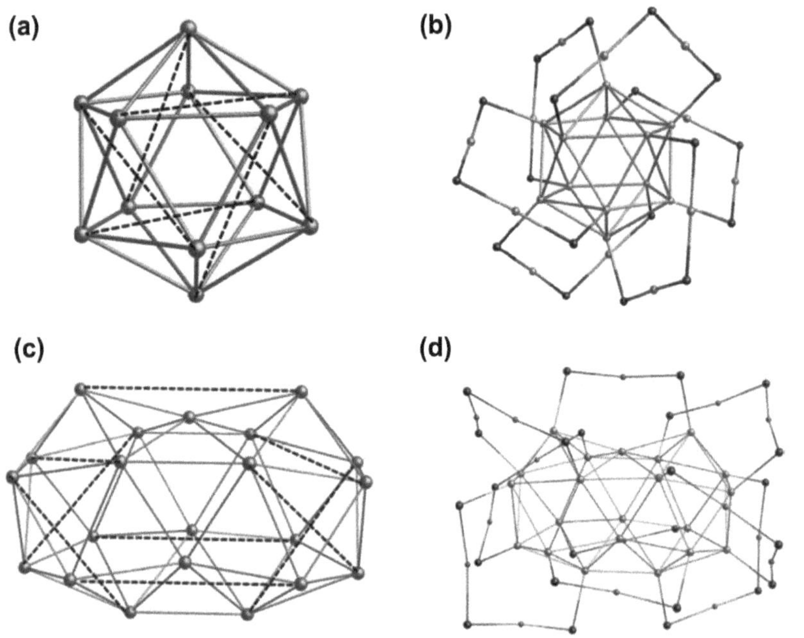

FIGURE 6 (a) The combination of dividing the 12 surface Au atoms into 6 pairs (each pair is represented by a dashed line) on an icosahedron that has the greatest total pair distance; (b) the experimental structure of $Au_{25}(SR)_{18}$. The Au atom at the icosahedral center is not shown. (c) The fittest combination of pairing up 18 atoms on the surface of a bi-icosahedral core; (d) the experimental structure of $Au_{38}(SR)_{24}$. The two Au atoms at the two icosahedral centers are not shown for clarity. *Ref. 41, reprinted with permission. Copyright © 2011 WILEYVCH Verlag GmbH & Co. KGaA, Weinheim.*

Based on the staple motif hypothesis, one can partition $Au_{19}(SR)_{13}$ into two possible scenarios: (1) either a Au_{11} core protected by two monomers and three dimers or (2) a Au_{12} core protected by five monomers and one dimer. Scenario 1 should be more viable than scenario 2 since the gold cluster with a small size usually prefers more V-shaped dimer motifs than the monomer. For scenario 1, there are 10 terminal SR thiolates, thus one Au atom in the Au_{11} core is most likely to form the center of the core. Next, we considered 11 high-symmetry structures for the Au_{11} core and for each core, we employed the staple fitness method to find the fittest combination. For the 11 fittest candidates of $Au_{19}(SCH_3)_{13}$ clusters, we optimized their structures with DFT and then compared their energetics. The most stable configuration corresponds to a centered, defective icosahedron with C2 symmetry (Figure 7(a)), where the two monomers cap the two opposite concave regions of the core, while the three dimers protect the convex regions (Figure 7(b)). The Au–S framework of the obtained $Au_{19}(SCH_3)_{13}$ cluster has a quasi-twofold symmetry (Figure 7(c)). This structure model gives an excellent agreement with the

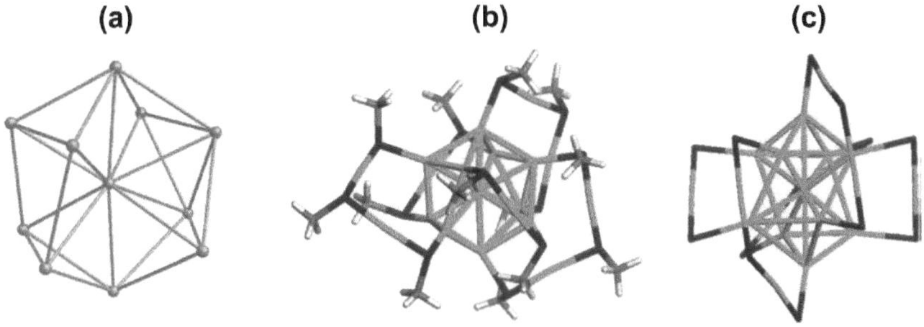

FIGURE 7 (a) The centered, defective icosahedron with C2 symmetry as the Au_{11} core. (b) The optimized structure of $Au_{19}(SCH_3)_{13}$ corresponding to the fittest combination. (c) The Au–S framework shows a quasi-twofold symmetry. Au, green (gray in print versions); S, blue (black in print versions). *Ref. 41, reprinted with permission. Copyright © 2011 WILEYVCH Verlag GmbH & Co. KGaA, Weinheim.*

experiment in terms of the powder X-ray diffraction[86] and is also further supported by the analysis of the extended X-ray absorption fine structure.[87]

Another example associated with the application of staple fitness method is the structure prediction of $Au_{40}(SR)_{24}$ cluster. $Au_{40}(SR)_{24}$ was first experimentally isolated in 2010,[23] and this cluster draws great interest since it has only two more Au atoms but the same number of RS– groups in comparison with $Au_{38}(SR)_{24}$. In 2012, Häkkinen et al. proposed several structural models for $Au_{40}(SR)_{24}$, their lowest energy structure is composed of a nonchiral Au_{26} bi-icosahedral cluster core protected by 6 RS–Au–SR monomers and 4 RS–Au–SR–Au–SR dimer units.[40] Based on the staple fitness concept, Jiang et al. also studied the structure of $Au_{40}(SR)_{24}$, and a promising structure model for $Au_{40}(SR)_{24}$ was discovered.[56] This model has an Au_{25} core protected by three monomers and six dimers (Figure 8). Actually, the Au_{25} core is built upon the face-sharing bi-icosahedral Au_{23} core of $Au_{38}(SR)_{24}$ with two extra Au atoms located in the waist region of the prolate core (the two extra Au atoms are highlighted in gray in Figure 8). The arrangement of three dimers on one side of the Au_{25} core is similar to that of $Au_{38}(SR)_{24}$, while on the other side, it is quite different. This structure model is more stable than the model proposed by Häkkinen et al. (about −0.53 eV more stable at the DFT-TPSS (Tao, Perdew, Staroverov, and Scuseria) level[88]), indicating that $Au_{40}(SR)_{24}$ could possibly have a Au_{25} core plus three monomers and six dimers instead of a Au_{26} core plus six monomers and four dimers.

7.2.3 Interlocked Feature

The interlocked staple motifs are another interesting structure feature of thiolate–gold nanoclusters, which usually occur for $Au_n(SR)_m$ nanoclusters with relatively low Au/SR ratios. Actually the interlock-like motifs have been previously found in the [Au(I)–SR]x oligomers such as $Au_{10}(SR)_{10}$ and

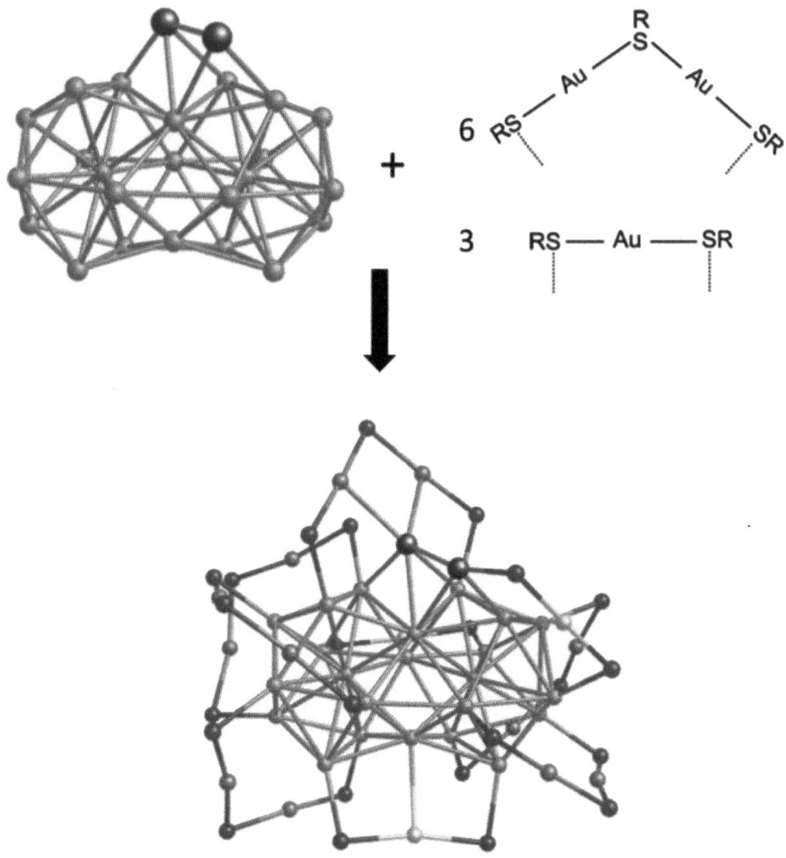

FIGURE 8 Structure model for $Au_{40}(SR)_{24}$ based on a Au_{25} core protected by six dimer and three monomer motifs. The Au_{25} core is built upon the face-sharing bi-icosahedral Au_{23} core of $Au_{38}(SR)_{24}$, and the extra two Au atoms are highlighted in gray. (see color plate). *Ref. 56. Reprinted with permission. Copyright © 2013 Royal Society of Chemistry.*

$Au_{12}(SR)_{12}$,[89–91] and the interlocking arrangements between two motifs can greatly increase the stability of the cluster system. Besides the [Au(I)−SR]*x* oligomers, this type of binding motifs have been recently proposed in the structure models for $Au_{24}(SR)_{20}$ cluster predicted by Pei et al.[92] and for $Au_{22}(SR)_{18}$ predicted by Jiang et al.[93] In Pei's most promising model (Figure 9), $Au_{24}(SR)_{20}$ is found to consist of a prolate Au_8 core covered by two sets of symmetrically interlocked $[Au_3(SR)_4]$ trimer and $[Au_5(SR)_6]$ pentamer motifs. The prolate Au_8 core with roughly D_{2d} symmetry can be viewed as two tetrahedral Au_4 units cross-joined together without sharing any Au atoms. Each pair of trimer and pentamer motifs is interlocked together like a chain, forming a catenane-like configuration. In fact, the theoretical model proposed

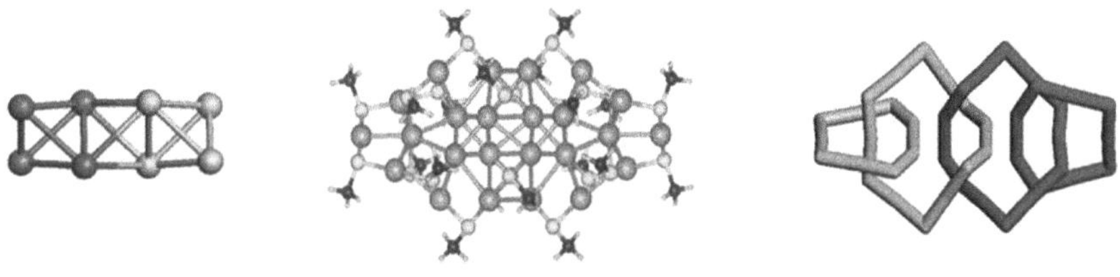

FIGURE 9 The prolate Au_8 core of $Au_{24}(SCH_3)_{20}$ with two tetrahedral Au_4 units cross-joined together (left panel). Optimized structure of the lowest lying isomer for $Au_{24}(SCH_3)_{20}$ (middle panel). The methyl groups are removed for clarity in the topological structural model (right panel). The Au, S, C, and H atoms are in khaki, yellow, gray, and white colors, respectively (middle panel). (see color plate). *Ref. 92, reprinted with permission. Copyright © 2012 American Chemical Society.*

for $Au_{24}(SR)_{20}$ herein has been confirmed to be the recently resolved structure of a selenolate-capped $Au_{24}(SeR)_{20}$ nanocluster (R = C_6H_5).[20] Interestingly, in a more recent work, Jin et al. reported the X-ray structure determination of another $Au_{24}(SR)_{20}$ cluster (R = $-CH_2Ph-{}^tBu$).[19] Distinct from the predicted structure for $Au_{24}(SCH_3)_{20}$ or the resolved crystal structure for $Au_{24}(SeC_6H_5)_{20}$, $Au_{24}(SCH_2Ph-{}^tBu)_{20}$ is shown to be composed of an anti-prismatical Au_8 core (with roughly C_{2h} symmetry) protected by four tetrameric $[Au_4S_5]$ staple motifs, wherein the four $[Au_4S_5]$ motifs form two sets of interlocked catenane-like $[Au_4S_5][Au_4S_5]$ staple motifs. This indicates that $Au_n(SR)_m$ clusters with the same chemical compositions but different protecting ligands may have different atomic structures, and the ligand effect will be discussed in our later section. The observation of interlocked structure in the experimentally resolved $Au_{24}(SeC_6H_5)_{20}$ and $Au_{24}(SCH_2Ph-{}^tBu)_{20}$ further suggests that this type of structural motif might be prevalent in small-sized thiolate–gold clusters with low Au–to–thiolate ratios.

Furthermore, another recently synthesized and composition-determined $Au_{22}(SR)_{18}$ cluster by Xie et al. has also aroused great research interest.[93] From a compositional perspective, $Au_{22}(SR)_{18}$ has the same number of thiolates as the well-known $Au_{25}(SR)_{18}$ cluster but with a different number of Au atoms. Moreover, $Au_{22}(SR)_{18}$ has the same number of Au atoms as the previously characterized $Au_{22}(SR)_{16}$[1] and $Au_{22}(SR)_{17}$[1] clusters but with a different number of thiolate ligands. Despite their similarity in compositions and molecular sizes, their optical absorption characteristics differ significantly. The $Au_{22}(SR)_{18}$ species exhibits strong red emission under UV illumination (at ~665 nm), whereas the other three neighboring clusters show only very weak luminescence. The highly luminescent feature of $Au_{22}(SR)_{18}$ implies that this cluster might have a distinctly different structure from those weakly luminescent clusters.

Based on the staple motif hypothesis, the $Au_{22}(SR)_{18}$ cluster with $4e^-$ electron shell configuration is most likely to have a Au_8 core capped by two trimeric $[Au_3(SR)_4]$ and two tetrameric $[Au_4(SR)_5]$ staple motifs. Under this

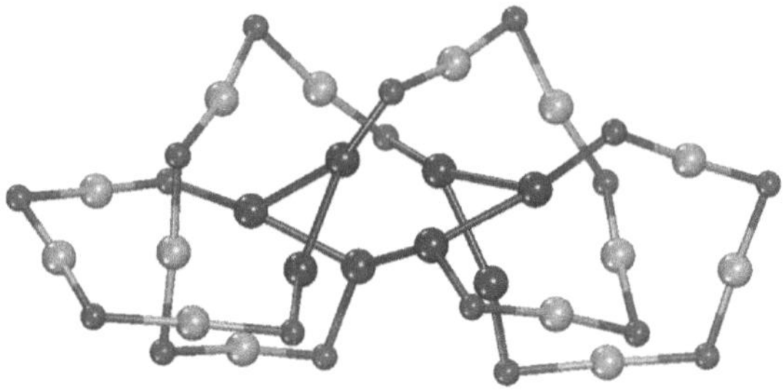

FIGURE 10 The structure model of $Au_{22}(SCH_3)_{18}$ predicted by DFT. The purple balls represent the eight Au atoms in the core, and the yellow balls represent the Au atoms in the staple motifs, while the red balls represent S atoms. All other atoms (carbon, hydrogen) have been omitted for clarity. (see color plate). *Ref. 93, reprinted with permission. Copyright © 2014 American Chemical Society.*

scenario, DFT was then used to explore the possible isomers of $Au_{22}(SR)_{18}$, and the most stable isomer is shown in Figure 10.[93] This proposed model corresponds to a prolate Au_8 core protected by two pairs of interlocked $[Au_3(SR)_4]$ and $[Au_4(SR)_5]$ motifs. In light of the relative stability, as measured by the interconversion mediated by $Au_{10}(SR)_{10}$ cluster, this model is significantly more stable than the optimized theoretical models of $Au_{18}(SR)_{14}$[42] and $Au_{20}(SR)_{16}$,[94] though slightly less stable than that of $Au_{24}(SR)_{20}$.[92] The computed optical absorption spectrum for this model shows reasonable agreement with the experimental one. What is more, the predicted isomer is also verified experimentally via X-ray absorption fine structure analysis, and the local bonding environment of the theoretical model is in good agreement with the measured values.

7.2.4 Longer Motifs and Ring Hypothesis

As mentioned above, as a main guiding principle and structural rule, the "staple hypothesis" has been widely applied for the geometrical construction of thiolate–gold nanoclusters with precise molecular compositions but has not yielded to total structure determination. Although the linear monomer and V-shaped dimer motifs without interlocking bonding dominate in large- or medium-size $Au_n(SR)_m$ nanoclusters, longer protecting staples (or extended staple motifs) such as $Au_3(SR)_4$ trimer, $Au_4(SR)_5$ tetramer, and $Au_5(SR)_6$ pentamer as well as the interlocked staple motifs are often needed when the thiolated gold clusters are small (less than 25 Au atoms), owing to the increasing surface curvature of the gold core with decreasing the nanocluster size. In addition, the ring-like $(Au-SR)x$ motif could exist as a new type of protecting unit for the ultrasmall nanoclusters, as demonstrated in the theoretically proposed structure for $Au_{15}(SR)_{13}$[43] and the experimentally solved single crystal structure for $Au_{20}(SR)_{16}$.[21]

For $Au_{15}(SR)_{13}$, what interests us here is that it is the smallest thiolated gold cluster ($Au_n(SR)_m$, where $n > m$) experimentally identified so far, in a relatively large yield and high stability. Below this size, only [Au(I)−SR]*x* oligomers can be obtained in large yields, such as $Au_{10}(SR)_{10}$, $Au_{11}(SR)_{11}$, and $Au_{12}(SR)_{12}$. This implies that the structure of $Au_{15}(SR)_{13}$, which remains still elusive until now, may hold the key to understand the origin of the nucleation and growth process of an Au core starting from the core-free gold−thiolate complexes to the core−shell gold nanoclusters.

To predict the structure of $Au_{15}(SR)_{13}$ cluster, Jiang et al. performed an extensive DFT study of the possible isomers for $Au_{15}(SCH_3)_{13}$, and about 55 guessed isomers are explored for geometry optimizations.[43] These initial structures are constructed mainly based on three strategies: (1) deriving from the previous models of other clusters (e.g., $Au_{24}(SR)_{20}$ and $Au_{12}(SR)_9$) by adding or pruning off Au atoms and −SR ligands, (2) based on the core-staple hypothesis, and (3) based on a new hypothesis that takes into account the [Au−SR]*x* cyclic motifs. After an extensive structural exploration, a novel structure with high stability evolves out: a Au_4 nucleus or core protected by a cyclic [Au(I)−SR] pentamer, [$Au_5(SR)_5$ ring motif] together with two regular [$Au_3(SR)_4$] trimer staple motifs (Figure 11, left panel). In this structure, the $Au_5(SR)_5$ ring is interlocked with one of the two trimer staples. Note that the interlocked catenane-like structure is a unique feature of homoleptic [Au(I)−SR]*x* oligomers (e.g., $Au_{10}(SR)_{10}$, $Au_{11}(SR)_{11}$, $Au_{12}(SR)_{12}$) and $Au_n(SR)_m$ clusters with low Au/SR ratios. This model is very interesting since for the first time, a cyclic [Au(I)−SR]*x* complex or ring-like structure is found to participate in the bonding of thiolate−gold protective layer in addition to the

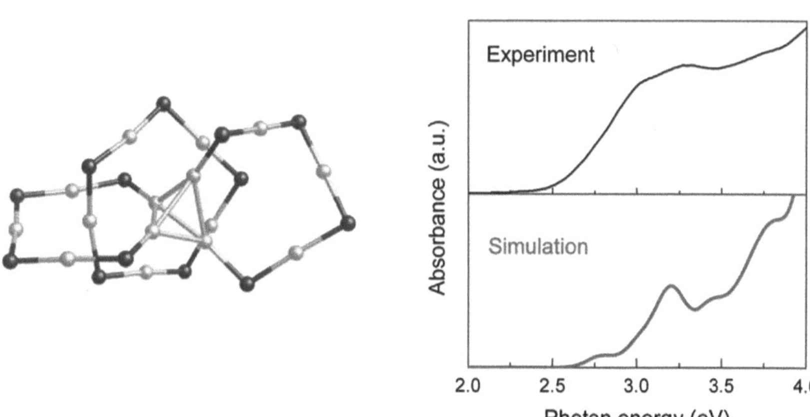

FIGURE 11 The Au−S framework of the most stable structure found for $Au_{15}(SR)_{13}$, the $-CH_3$ groups are omitted here (left). Simulated optical absorption spectrum of the $Au_{15}(SCH_3)_{13}$ cluster in comparison with the experimental spectrum of $Au_{15}(SG)_{13}$. *Ref. 43, reprinted with permission. Copyright © 2013 American Chemical Society.*

common staple motifs. Compared with the "staple-only" motifs, such a "ring and staple" hybrid motif is energetically more stable by 0.24 eV (relative to the energy of the best model based on the staple-only hypothesis). Note that the interaction of the cyclic $Au_5(SR)_5$ ring with the Au_4 core mainly derives from the $Au_{ring}-Au_{core}$ binding, and no thiolate ligand in the ring is bonded with the Au atoms of the Au_4 core. This is in stark contrast with the common staple motifs where the two terminal thiolates of a staple motif would form bonding with the surface Au atoms of gold core. In terms of the structural stability, the presence of the ring structure can significantly stabilize the cluster, and the computed interaction energy between the cyclic pentamer and the rest of the cluster (about 2.1 eV) is quite strong. Thus for the case of low Au/SR ratio, $Au_{15}(SR)_{13}$ cluster, instead of using much longer staple motifs in the ligand shell, it prefers the ring-like structure. In addition, this model for $Au_{15}(SR)_{13}$ indicates that it forms a hybrid composition between a homoleptic cyclic $[Au(I)-SR]x$ complex and the regular $Au_n(SR)_m$ clusters. This suggests that $Au_{15}(SR)_{13}$ might be the first transitional structure from the $[Au(I)-SR]x$ complex or oligomers to the regular $Au_n(SR)_m$ $(n > m)$ nanoclusters that only have the staple motifs, which is consistent with the fact that $Au_{15}(SR)_{13}$ is the smallest magic cluster identified from the experiment. The simulated optical absorption spectrum for $Au_{15}(SR)_{13}$ exhibits an onset of absorption around 2.6 eV and a major peak at 3.2 eV, in good agreement with the measured onset at 2.55 eV and main peak at 3.28 eV (Figure 11, right panel). This indicates that the proposed model should be a viable structure for $Au_{15}(SR)_{13}$. Further, the predicted structure for $Au_{15}(SR)_{13}$ also suggests that the nucleation process starts from the Au_4 core, which might be achieved by the controlled reduction of a smaller cyclic [Au(I)$-$SR] complex such as a tetramer or pentamer.

The "ring and staple motifs" model as proposed for $Au_{15}(SR)_{13}$ has extended the general guiding principle based on "staple hypothesis" for the structure prediction of small $Au_n(SR)_m$ clusters. In fact, the ring-type motif has been observed experimentally in the recently solved total structure of $Au_{20}(SR)_{16}$ nanocluster (where R = Ph$-t-$Bu), which comprises a vertex-sharing bi-tetrahedral Au_7 core protected by an octameric ring motif ($Au_8(SR)_8$) plus one trimeric $Au_3(SR)_4$ and two monomeric $Au(SR)_2$ staple motifs (Figure 12(a)).[21] Structurally, the vertex-sharing Au_7 core has two tetrahedra rotating by $\sim 60°$ with respect to each other to form a staggered configuration. The octameric ring motif adopts a chair-like conformation and circles the whole Au_7 core. This is the first time that a ring-like bonding motif is discovered in small gold nanoclusters by experiment, which illustrates that $Au_{20}(SR)_{16}$, as in the case of $Au_{15}(SR)_{13}$, can be regarded as another transitional composition between Au(I)$-$SR complexes and $Au_n(SR)_m$ nanoclusters. The $Au_8(SR)_8$ ring interacts with the Au_7 core mainly through $Au_{ring}-Au_{core}$ bonding ($\sim$3.2 Å), which is only about 5% longer than the Au$-$Au distance in the bulk phase. This indicates a strong binding interaction between $Au_8(SR)_8$ ring and Au kernel. Note that in

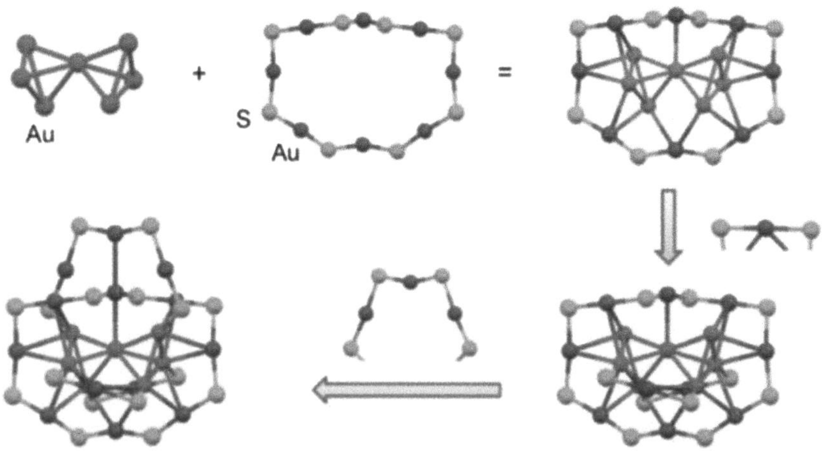

FIGURE 12 Anatomy of the structure of $Au_{20}(SPh{-}t{-}Bu)_{16}$: Au_7 kernel protected by the octameric ring motif and then by the trimeric and monomeric staple motifs. Color codes: magenta (gray in print versions), Au in the kernel; blue (dark gray in print versions), Au on the surface; yellow (light gray in print versions), S. *Ref. 21, reprinted with permission. Copyright © 2014 American Chemical Society.*

earlier reports, both Pei et al.[94] and Jiang et al.[95] independently presented theoretical predictions on the structure of $Au_{20}(SR)_{16}$. Their predictions are based on the staple-only hypothesis, and their most viable models under this hypothesis are both found to contain four trimeric $[Au_3(SR)_4]$ staple motifs protecting a prolate Au_8 core. The experimental X-ray determination of $Au_{20}(SR)_{16}$ might imply that the ring motif, as a new and "smart" type of surface-protecting unit, may prevail in ultrasmall nanoclusters.

7.2.5 vdW Interaction or Steric Effect on Structural Prediction

For the structure prediction of thiolate−gold nanoclusters, another important point we need to take into account is the ligand effect, which plays a vital role in dictating the geometrical construction and isomer stabilities of the gold nanoclusters. Currently, the majority of theoretical studies reported in literature simply use CH_3 to represent the experimental R-groups for computational efficiency. This simplification has an underlying assumption that the isomer stability of a specific $Au_n(SR)_m$ cluster does not depend on the R-group. This assumption, however, cannot be always true in every case. To what extent this assumption is valid remains to be explored. As a general trend, for ligands of small or moderate bulkiness, one can obtain the same composition or even the same bonding structure for different ligands. However, in the case of sterically large, bulky ligands, the interfacial motif and the composition or even the geometric structures will be different, as demonstrated by the recent experiments.[58−60] An important factor associated with the ligand effect is the vdW interaction within the ligands, which can be otherwise viewed as the steric effect of thiolate ligands. When $-CH_3$ groups are used to replace the real experimental −R groups for structure prediction, the vdW interaction among $-CH_3$ groups can be completely ignored. However, when the real SR ligands are taken into account for comparing the isomer stability, the vdW effect or steric effect should not be neglected. Herein an interesting question arises: will the cluster structure change according to the ligand used, with the molecular composition being the same?

For the above question, we hypothesize here that the relative stability and optimal structure of $Au_n(SR)_m$ isomers may depend on the protecting ligands used.[96] As a testing example, the recent progress on $Au_{24}(SR)_{20}$ clusters provides a perfect case to test our hypothesis. Early in 2010, Zhu et al. reported the composition characterization of the $Au_{24}(SCH_2CH_2Ph)_{20}$ cluster, but its precise structure is not known at that time.[97] In 2012, Pei et al. reported the first major structural search of $Au_{24}(SR)_{20}$ based on the simplified $Au_{24}(SCH_3)_{20}$ model (see Section 7.2.3),[92] and a viable structure (we call it the "P" isomer) is proposed: a prolate Au_8 core (near-D_{2d} symmetry) capped by two sets of interlocked trimeric $[Au_3(SCH_3)_4]$ and pentameric $[Au_5(SCH_3)_6]$ staples (as shown in Figure 13(a)). This model is found to be in good agreement with the

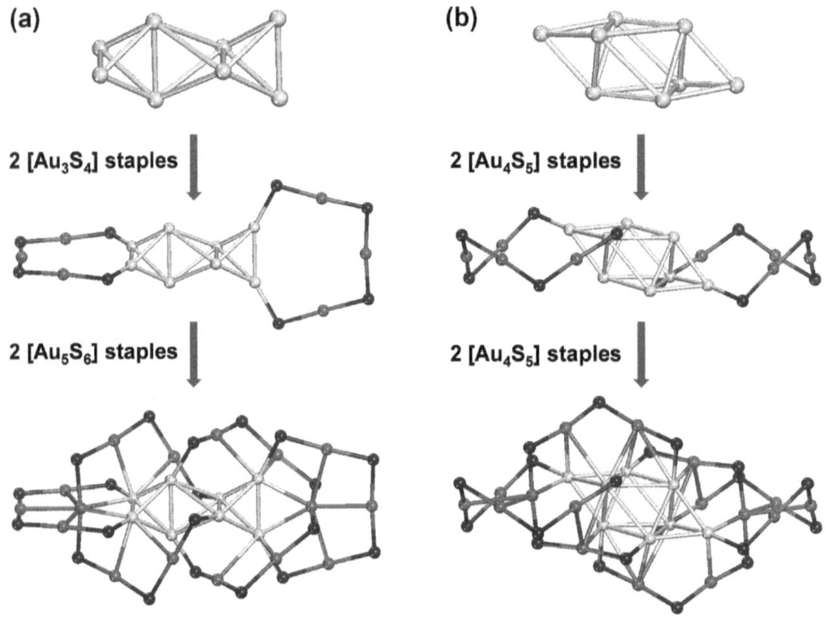

FIGURE 13 The Au−S frameworks of considered two $Au_{24}(SR)_{20}$ isomers: (a) isomer P, written as $Au_8[Au_3(SR)_4]_2[Au_5(SR)_6]_2$; (b) isomer J, written as $Au_8[Au_4(SR)_5]_4$. For color modes, pink (light gray in print versions), kernel Au; yellow (gray in print versions), staple Au; green (black in print versions), S; the C and H atoms are not shown here. *Ref. 96, reprinted with permission. Copyright © 2014 Royal Society of Chemistry.*

experimentally solved structure of $Au_{24}(SeR)_{20}$ nanocluster ($R = -C_6H_5$).[20] More recently, Jin et al. solved the single crystal structure of $Au_{24}(SR)_{20}$, where $R = -CH_2Ph-{}^tBu$.[19] Different from the $Au_{24}(SCH_3)_{20}$ P model, $Au_{24}(SCH_2Ph-{}^tBu)_{20}$ comprises an antiprismatical Au_8 core (near-C_{2h} symmetry) protected by four tetrameric Au_4S_5 staples or two sets of symmetrically interlocked tetramer motifs (shown in Figure 13(b); we call it the "J" model). We can clearly see that $Au_{24}(SCH_3)_{20}$ and $Au_{24}(SCH_2Ph-{}^tBu)_{20}$ have different core structure and protecting motifs. Would $Au_{24}(SCH_2CH_2Ph)_{20}$ prefer the P structure or the J structure?

After configuration sampling, the relative energies of the two isomers with the highest stability for $R = -CH_3$, $-CH_2CH_2Ph$, and $-CH_2Ph-{}^tBu$ are summarized in Table 1 (the energy is computed at the *meta*-generalized gradient approximation (GGA) TPSS functional plus def2-SV(P) basis sets). For the small $-CH_3$ ligand, the ΔE_{total} between isomer P and isomer J is small, implying the comparable stabilities of both isomers. For $R = CH_2CH_2Ph$, isomer P is energetically more favorable, while for $R = CH_2Ph-{}^tBu$, the J isomer is more stable. This reversal of stability confirms our hypothesis that the isomer stability depends on the ligand used. It also indicates that the synthesized $Au_{24}(SCH_2CH_2Ph)_{20}$ cluster is likely to have a structure like that of the P isomer.

Further, by partitioning the total energy of $Au_{24}(SCH_2CH_2Ph)_{20}$ and $Au_{24}(SCH_2Ph-{}^tBu)_{20}$ isomers into DFT and vdW contributions, one can see that the preference of the P isomer for $Au_{24}(SCH_2CH_2Ph)_{20}$ is due to its much stronger vdW interaction, which overwhelms the penalty in DFT energy. Differently, in the case of $Au_{24}(SCH_2Ph-{}^tBu)_{20}$, the two isomers have almost the same vdW contribution, but the J isomer has a much lower DFT energy, resulting in its higher stability. From $R = CH_2CH_2Ph$ to $R = CH_2Ph-{}^tBu$, the average vdW energy per carbon atom ($\overline{E}_{vdw}$, eV) for the same P or J isomer becomes weakened (Table 1), indicating the larger steric effect of the tail t–Bu group.

By further analyzing the core structure, we found that the average core Au–Au bond length of the P isomer is contracted by about 0.83% when changing the R group from CH_3 to $CH_2Ph-{}^tBu$. For the J isomer, the average core Au–Au bond is elongated by about 0.17% when changing R from CH_3 to $R = CH_2Ph-{}^tBu$. This indicates that the bulky $-CH_2Ph-{}^tBu$ group leads to greater perturbation to the core of the P isomer than to that of the J isomer. In other words, the J isomer with less steric effect can better accommodate the bulky $CH_2Ph-{}^tBu$ group, which might be the reason why the J isomer is preferred for $R = CH_2Ph-{}^tBu$.

7.2.6 Fcc Core

The reported crystal structures of $Au_n(SR)_m$ nanoclusters often have highly symmetric but non-face-centered cubic (fcc) cores, e.g., the Au_{79} decahedron

TABLE 1 The Energy Difference between Two $Au_{24}(SR)_{20}$ Isomers (P and J) for Different R Groups. For R = CH_2CH_2Ph and $CH_2Ph{-}^{t}Bu$, the Energy Difference Is Further Broken Down to DFT and vdW Contributions, as Well as the Computed Average vdW Energy per Carbon Atom ($\overline{E}_{vdw}$, eV). The HOMO-LUMO Gaps (eV) Are Also Shown

$Au_{24}(SR)_{20}$	ΔE_{total} (eV)	ΔE_{dft} (eV)	ΔE_{vdw} (eV)	$\overline{E}_{vdw}$ (eV)		Gap (eV)	
	$E_{total(J)} - E_{total(P)}$	$E_{dft(J)} - E_{dft(P)}$	$E_{vdw(J)} - E_{vdw(P)}$	Isomer P	Isomer J	Isomer P	Isomer J
R = CH_3	0.13	0.13	–	–	–	1.70	1.96
R = CH_2CH_2Ph	1.60	−0.72	2.32	−0.195	−0.181	1.63	1.90
R = $CH_2Ph{-}^{t}Bu$	−1.02	−0.92	−0.10	−0.178	−0.178	1.75	1.93

Ref. 96. Reprinted with permission. Copyright © 2014 Royal Society of Chemistry.

in $Au_{102}(SR)_{44}$,[11] the Au_{13} icosahedron in $Au_{25}(SR)_{18}$,[12,13] the Au_{23} face-sharing bi-icosahedron in $Au_{38}(SR)_{24}$,[14] the Au_8 prolate[20] or antiprismatic geometry[19] in $Au_{24}(SR)_{20}$, and the Au_7 vertex-sharing bi-tetrahedron in $Au_{20}(SR)_{16}$.[21] Interestingly, the crystallized $Au_{36}(SR)_{24}$ cluster (R refers to 4-tert-butylbenzenethiolate, denoted as SPh−^{t}Bu) discovered by Jin et al. has a new type of nanocrystalline fcc Au_{28} kernel, with a truncated tetrahedron appearance exposing (111) and (100) facets (Figure 14).[15,98] This is the first observation of a fcc structure in small $Au_n(SR)_m$ nanoclusters ($n < 100$). Surprisingly, a new type of thiolate binding motif, the bridging motif, was also discovered for the first time, where 12 out of the 24 thiolates are bonded directly to the underlying Au atoms in a simple bridging mode (as opposed to the staple motifs with the participation of Au atoms), while the remaining 12 thiolates are configured in the form of 4 dimeric $[Au_2(SR)_3]$ staple motifs. A more detailed inspection of the protecting units revealed that the four dimeric staples reside the four (111) facets along the bisecting line, while each of the six (100) facets is connected by two bridging thiolates. Thus, the structure formula of $Au_{36}(SR)_{24}$ could be denoted as $Au_{28}[Au_2(SR)_3]_4(SR)_{12}$, and the whole Au−S framework of the $Au_{36}(SR)_{24}$ cluster forms a D_{2d} atomic symmetry.

The measured optical HOMO-LUMO gap of $Au_{36}(SPh{-}^tBu)_{24}$ cluster is about 1.7 eV, larger than that measured for $Au_{25}(SR)_{18}$ (1.3 eV) and $Au_{38}(SR)_{34}$ (0.9 eV). A large HOMO-LUMO gap endows a high stability for this cluster. Following the successful crystallization and structure elucidation of $Au_{36}(SR)_{24}$, Jin et al. further solved the single crystal structure of another fcc core cluster, $Au_{28}(SR)_{20}$ (R refers to −Ph−^{t}Bu),[16] which is a second example of fcc gold nanoclusters, featuring a fcc Au_{20} kernal (two inter-penetrating cuboctahedra) protected by four dimeric $[Au_2(SR)_3]$ staples and eight bridging thiolates, with the formula written as $Au_{20}[Au_2(SR)_3]_4(SR)_8$.

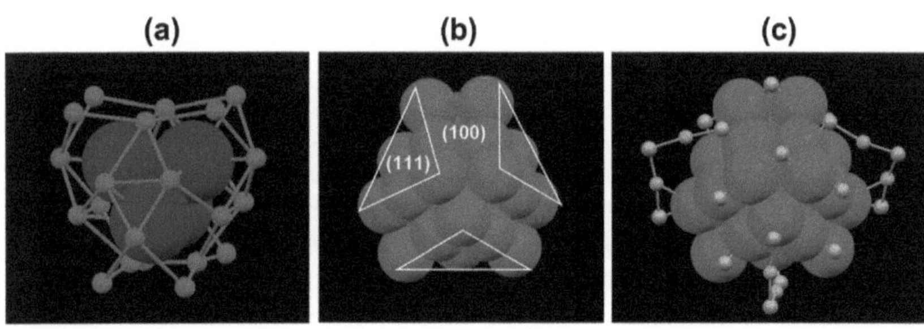

FIGURE 14 Tetrahedral structure of the Au_{28} kernel and surface protection by thiolate ligands. (a) The inner Au_4 tetrahedron of Au_{28} kernel is shown in space-filled fashion, (b) the two-shell Au_{28} kernel exhibits bulk-like (111) and (100) faces, (c) the surface protection of the Au_{28} tetrahedron by four dimeric staples (containing 12 thiolates) on (111) and 12 bridging thiolates on (100). The Ph−^{t}Bu tail groups are omitted. Color labels: yellow (light gray in print versions) for sulfur, other colors are for gold atoms. *Ref. 15, reprinted with permission. Copyright © 2012 WILEYVCH Verlag GmbH & Co. KGaA, Weinheim.*

The structural insight gained from Au_{36} and Au_{28} clusters leads to prediction of another elusive gold cluster, $Au_{44}(SR)_{28}$. The earlier reported $Au_{44}(SR)_{28}$ nanocluster is a magic number gold cluster protected by all-aromatic thiophenolate ligands (−SPh).[99] Jin et al. recently reported the synthesis of $Au_{44}(SR)_{28}$ protected by SPh−^{t}Bu groups.[100] $Au_{44}(SR)_{28}$ can be compositionally viewed as the evolution of the Au_{28} or Au_{36} cluster via addition of $[Au_8(SR)_4]$ units, $[Au_{28}(SR)_{20} + [Au_8(SR)_4] \rightarrow Au_{36}(SR)_{24} + [Au_8(SR)_4] \rightarrow Au_{44}(SR)_{28}]$. The interesting $Au_8(SR)_4$ increment might indicate the inherent structural connection among these three clusters. Based on the structural construction rules learned from $Au_{28}(SR)_{20}$ and $Au_{36}(SR)_{24}$ (their common features: an unique two-shell fcc kernel composed of interpenetrating cuboctahedra and a ligand shell composed of bridging thiolates and dimeric motifs), Jin et al. proposed a possible structural model for $Au_{44}(SR)_{28}$: an Au_{36} fcc kernel protected by 4 dimer staples (residing on the four (111) faces of the kernel) and 16 bridging thiolates (residing on 16 squares of the (100) faces of the kernel).[100]

Interestingly, immediately after Jin et al.'s report, Pei et al. reported the structure prediction of $Au_{44}(SR)_{28}$ (R is simplified as CH_3) by DFT,[38] and the framework of Pei's model is actually the same as the structural construction predicted by Jin et al. In Pei's work, they proposed a possible Au_{26} core for $Au_{44}(SR)_{28}$, which can be viewed as two Au_{14} cores fused together by sharing a gold dimer, with four symmetrically distributed gold adatoms on four hexagonal Au_7 facets. Based on the proposed Au_{26} core, $Au_{44}(SR)_{28}$ can then be partitioned into $Au_{26}[Au_2(SR)_3]_8[Au(SR)_2]_2$ according to the staple motif hypothesis, where only the monomeric and dimeric motifs are considered. After a systematic structural search of the possible isomers, they found a low-lying structure containing a D_2-symmetric Au−S framework (Figure 15). The eight $[Au_2(SR)_3]$ dimers are added upon four hexagonal Au_7 facets and four rhombus Au_4 units, and the two $[Au(SR)_2]$ monomers are added on the two sides of the Au_{26} core. Like the two-shell fcc-type Au_{20} kernel in Au_{28} cluster and the Au_{28} kernel in Au_{36} cluster, Au_{44} cluster contains a two-shell fcc-type Au_{36} kernel as well (the first-shell Au_{26} core plus the second-shell 10 Au atoms contributed from the staple motifs), which can be regarded as the fusion of two Au_{20} kernels sharing square (100) facets. In this case, similar to the structural construction of $Au_{38}(SR)_{24}$, the formula of $Au_{44}(SR)_{28}$ cluster can be alternatively viewed as the combination of a Au_{36} fcc kernel plus 4 dimeric staples residing at the four (111) facets and 16 bridging ligands at the (100) facets $[Au_{36}[Au_2(SR)_3]_4(SR)_{16}]$. This predicted model in its neutral charging state can well reproduce the experimental optical absorption gap (~1.5 eV) and the feature absorption peaks at about 2.2 and 3.2 eV. Further, the simulated circular dichroism spectra indicate the $Au_{44}(SR)_{28}$ cluster exhibits strong chiral responses in the range of excitation energy of 1.5−3.5 eV. The predicted chiral feature of Au_{44} cluster needs to be further confirmed by experiment.

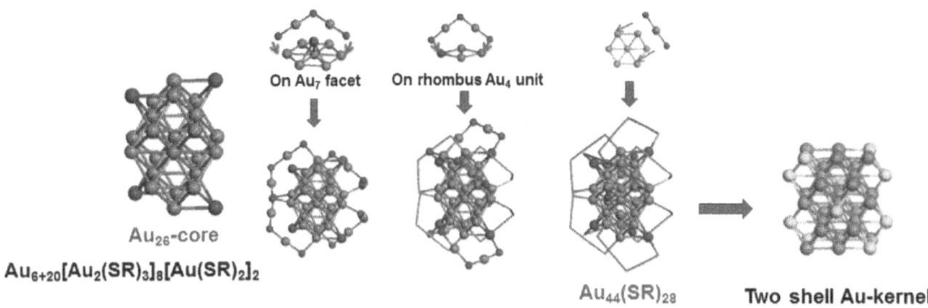

FIGURE 15 Proposed Au_{26} core and assembly process of $Au_{44}(SR)_{28}$. The right panel shows the two-shell fcc-type Au_{36} kernel, which is composed of the first-shell Au_{26} core (yellow balls) and the second-shell 10 Au atoms (pink balls) contributed from the staple motifs. (see color plate). *Ref. 38, reprinted with permission. Copyright © 2013 American Chemical Society.*

As shown above, the series of solved $Au_{36}(SR)_{24}$ and $Au_{28}(SR)_{20}$ as well as the predicted $Au_{44}(SR)_{28}$ constitutes a special family of thiolated gold clusters with unique fcc structure as the internal kernels. The fcc core might be an important hypothesis for predicting even larger $Au_n(SR)_m$ nanoclusters.

7.3 SUMMARY AND PROSPECTS

In this chapter, we discussed the recent progress in the structural prediction and understanding of the ligand-protected gold nanoclusters. We focus our attention on the widely studied thiolate–gold clusters and the recently emerged organogold nanoclusters. A leading principle to predict the atomic structure of $Au_n(SR)_m$ nanoclusters is the staple hypothesis, including an internal Au core and protecting staple motifs. A key challenge for staple hypothesis is to guess a correct core structure. Once given the core structure, we can then apply the staple fitness concept to predict how the staple motifs are optimally arranged around the core surface. The staple motifs are also the key interfacial structures of organogold surfaces (π staple motif), but the interfacial bonding is slightly different from the thiolated gold surfaces (σ staple motif), where the unsaturated C≡C bond is also involved in the bonding to surface Au atom. The π staple motif offers us an important insight for structure prediction of organogold nanoclusters. More future efforts are needed to reveal the structural and fundamental properties of this new type of gold nanoclusters.

For the structures of thiolate–gold nanoclusters, usually the protecting layer is exclusively composed of the monomeric and/or dimeric motifs when the cluster size is large. As the cluster size decreases (typically less than 25 Au atom, with low Au/SR ratios), longer staple motifs (such as trimer, tetramer, pentamer) and interlocked motifs are needed to stabilize the cluster (such as $Au_{24}(SR)_{20}$, $Au_{22}(SR)_{18}$). Moreover, the discovery of cyclic ring motifs in ultrasmall clusters such as $Au_{15}(SR)_{13}$ and $Au_{20}(SR)_{16}$ indicated that these small clusters could be the transitional structures from core-free Au(I)−SR complexes to the common core−shell $Au_n(SR)_m$ nanoclusters. As an extended guiding principle of the "staple" hypothesis, the "ring and staple" hypothesis could be applied to structure prediction of ultrasmall nanoclusters. Furthermore, the discovery of unusual fcc-type core resembling the bulk-like Au surface in $Au_{36}(SR)_{24}$, $Au_{28}(SR)_{20}$, and $Au_{44}(SR)_{28}$ offers important implications for larger-sized gold nanoclusters.

Another important aspect worthy of special attention is the steric effect of ligands. Computationally, one always simplifies the experimental SR groups to CH_3. This assumption, however, might not be valid for the bulky ligands, since the vdW interaction or steric effect of large ligands plays an important role in dictating the relative stability of cluster isomers.

There are lots of work and challenges facing the theorists in the field of ligand-protected gold nanoclusters. Currently most structure predictions for gold nanoclusters using DFT are based on the hypothesis learned from the experimental structure, which is feasible for predicting small nanoclusters but is still time-consuming. In future we envision a force field-based approach as a more efficient method for structure prediction of $Au_n(SR)_m$ nanoclusters.

ACKNOWLEDGMENTS

This work was supported by the University of California, Riverside. This research used resources of the National Energy Research Scientific Computing Center, a DOE Office of Science User Facility supported by the Office of Science of the U.S. Department of Energy under Contract No. DE-AC02-05CH11231.

REFERENCES

1. Negishi Y, Nobusada K, Tsukuda T. Glutathione-protected gold clusters revisited: bridging the gap between gold(I)-thiolate complexes and thiolate-protected gold nanocrystals. *J Am Chem Soc* 2005;**127**:5261−70.
2. Love JC, Estroff LA, Kriebel JK, Nuzzo RG, Whitesides GM. Self-assembled monolayers of thiolates on metals as a form of nanotechnology. *Chem Rev* 2005;**105**:1103−69.
3. Maity P, Xie S, Yamauchi M, Tsukuda T. Stabilized gold clusters: from isolation toward controlled synthesis. *Nanoscale* 2012;**4**:4027−37.
4. Yau SH, Varnavski O, Goodson T. An ultrafast look at Au nanoclusters. *Acc Chem Res* 2013;**46**:1506−16.
5. Yamazoe S, Koyasu K, Tsukuda T. Nonscalable oxidation catalysis of gold clusters. *Acc Chem Res* 2013;**47**:816−24.
6. Jin R. Atomically precise metal nanoclusters: stable sizes and optical properties. *Nanoscale* 2015;**7**:1549−65.
7. Jin RC. Quantum sized, thiolate-protected gold nanoclusters. *Nanoscale* 2010;**2**:343−62.
8. Häkkinen H. The gold-sulfur interface at the nanoscale. *Nat Chem* 2012;**4**:443−55.
9. Tsukuda T. Toward an atomic-level understanding of size-specific properties of protected and stabilized gold clusters. *Bull Chem Soc Jpn* 2012;**85**:151−68.

10. Jin R, Qian H, Wu Z, Zhu Y, Zhu M, Mohanty A, et al. Size focusing: a methodology for synthesizing atomically precise gold nanoclusters. *J Phys Chem Lett* 2010;**1**:2903–10.
11. Jadzinsky PD, Calero G, Ackerson CJ, Bushnell DA, Kornberg RD. Structure of a thiol monolayer-protected gold nanoparticle at 1.1 Å resolution. *Science* 2007;**318**:430–3.
12. Heaven MW, Dass A, White PS, Holt KM, Murray RW. Crystal structure of the gold nanoparticle $[N(C_8H_{17})_4][Au_{25}(SCH_2CH_2Ph)_{18}]$. *J Am Chem Soc* 2008;**130**:3754–5.
13. Zhu M, Aikens CM, Hollander FJ, Schatz GC, Jin R. Correlating the crystal structure of a thiol-protected Au_{25} cluster and optical properties. *J Am Chem Soc* 2008;**130**: 5883–5.
14. Qian H, Eckenhoff WT, Zhu Y, Pintauer T, Jin R. Total structure determination of thiolate-protected Au_{38} nanoparticles. *J Am Chem Soc* 2010;**132**:8280–1.
15. Zeng C, Qian H, Li T, Li G, Rosi NL, Yoon B, et al. Total structure and electronic properties of the gold nanocrystal $Au_{36}(SR)_{24}$. *Angew Chem Int Ed* 2012;**51**:13114–8.
16. Zeng C, Li T, Das A, Rosi NL, Jin R. Chiral structure of thiolate-protected 28-gold-atom nanocluster determined by X-ray crystallography. *J Am Chem Soc* 2013;**135**:10011–3.
17. Das A, Li T, Nobusada K, Zeng C, Rosi NL, Jin R. Nonsuperatomic $[Au_{23}(SC_6H_{11})_{16}]^-$ nanocluster featuring bipyramidal Au_{15} kernel and trimeric $Au_3(SR)_4$ motif. *J Am Chem Soc* 2013;**135**:18264–7.
18. Crasto D, Malola S, Brosofsky G, Dass A, Häkkinen H. Single crystal XRD structure and theoretical analysis of the chiral $Au_{30}S(S\text{-}t\text{-}Bu)_{18}$ cluster. *J Am Chem Soc* 2014;**136**: 5000–5.
19. Das A, Li T, Li G, Nobusada K, Zeng C, Rosi NL, et al. Crystal structure and electronic properties of a thiolate-protected Au_{24} nanocluster. *Nanoscale* 2014;**6**:6458–62.
20. Song Y, Wang S, Zhang J, Kang X, Chen S, Li P, et al. Crystal structure of selenolate-protected $Au_{24}(Ser)_{20}$ nanocluster. *J Am Chem Soc* 2014;**136**:2963–5.
21. Zeng C, Liu C, Chen Y, Rosi NL, Jin R. Gold–thiolate ring as a protecting motif in the $Au_{20}(SR)_{16}$ nanocluster and implications. *J Am Chem Soc* 2014;**136**:11922–5.
22. Dass A. Mass spectrometric identification of $Au_{68}(SR)_{34}$ molecular gold nanoclusters with 34-electron shell closing. *J Am Chem Soc* 2009;**131**:11666–7.
23. Qian H, Zhu Y, Jin R. Isolation of ubiquitous $Au_{40}(SR)_{24}$ clusters from the 8 kDa gold clusters. *J Am Chem Soc* 2010;**132**:4583–5.
24. Yu Y, Chen X, Yao Q, Yu Y, Yan N, Xie J. Scalable and precise synthesis of thiolated Au_{10-12}, Au_{15}, Au_{18}, and Au_{25} nanoclusters via PH controlled CO reduction. *Chem Mater* 2013;**25**:946–52.
25. Qian H, Jin R. Controlling nanoparticles with atomic precision: the case of $Au_{144}(SCH_2CH_2Ph)_{60}$. *Nano Lett* 2009;**9**:4083–7.
26. Nimmala PR, Dass A. $Au_{36}(Sph)_{23}$ nanomolecules. *J Am Chem Soc* 2011;**133**:9175–7.
27. Negishi Y, Sakamoto C, Ohyama T, Tsukuda T. Synthesis and the origin of the stability of thiolate-protected Au_{130} and Au_{187} clusters. *J Phys Chem Lett* 2012;**3**:1624–8.
28. Qian H, Zhu Y, Jin R. Atomically precise gold nanocrystal molecules with surface plasmon resonance. *Proc Natl Acad Sci U. S. A.* 2012;**109**:696–700.
29. Qian H, Jin R. Synthesis and electrospray mass spectrometry determination of thiolate-protected $Au_{55}(SR)_{31}$ nanoclusters. *Chem Commun* 2011;**47**:11462–4.
30. Nimmala PR, Yoon B, Whetten RL, Landman U, Dass A. $Au_{67}(SR)_{35}$ nanomolecules: characteristic size-specific optical, electrochemical, structural properties and first-principles theoretical analysis. *J Phys Chem A* 2013;**117**:504–17.
31. Lavenn C, Albrieux F, Bergeret G, Chiriac R, Delichere P, Tuel A, et al. Functionalized gold magic clusters: $Au_{25}(Sphnh_2)_{17}$. *Nanoscale* 2012;**4**:7334–7.

32. Zeng C, Chen Y, Li G, Jin R. Magic size $Au_{64}(S\text{-}C\text{-}C_6H_{11})_{32}$ nanocluster protected by cyclohexanethiolate. *Chem Mater* 2014;**26**:2635–41.
33. Kumara C, Dass A. $Au_{329}(SR)_{84}$ nanomolecules: compositional assignment of the 76.3 kDa plasmonic faradaurates. *Anal Chem* 2014;**86**:4227–32.
34. Li G, Zeng C, Jin R. Thermally robust $Au_{99}(Sph)_{42}$ nanoclusters for chemoselective hydrogenation of nitrobenzaldehyde derivatives in water. *J Am Chem Soc* 2014;**136**:3673–9.
35. Tsunoyama R, Tsunoyama H, Pannopard P, Limtrakul J, Tsukuda T. MALDI mass analysis of 11 kDa gold clusters protected by octadecanethiolate ligands. *J Phys Chem C* 2010;**114**:16004–9.
36. Chaki NK, Negishi Y, Tsunoyama H, Shichibu Y, Tsukuda T. Ubiquitous 8 and 29 kDa gold:alkanethiolate cluster compounds: mass-spectrometric determination of molecular formulas and structural implications. *J Am Chem Soc* 2008;**130**:8608–10.
37. Lopez-Acevedo O, Akola J, Whetten RL, Grönbeck H, Häkkinen H. Structure and bonding in the ubiquitous icosahedral metallic gold cluster $Au_{144}(SR)_{60}$. *J Phys Chem C* 2009;**113**:5035–8.
38. Pei Y, Lin S, Su J, Liu C. Structure prediction of $Au_{44}(SR)_{28}$: a chiral superatom cluster. *J Am Chem Soc* 2013;**135**:19060–3.
39. Jiang DE, Walter M, Akola J. On the structure of a thiolated gold cluster: $Au_{44}(SR)_{28}^{2-}$. *J Phys Chem C* 2010;**114**:15883–9.
40. Malola S, Lehtovaara L, Knoppe S, Hu KJ, Palmer RE, Bürgi T, et al. $Au_{40}(SR)_{24}$ cluster as a chiral dimer of 8-electron superatoms: structure and optical properties. *J Am Chem Soc* 2012;**134**:19560–3.
41. Jiang DE. Staple fitness: a concept to understand and predict the structures of thiolated gold nanoclusters. *Chem Eur J* 2011;**17**:12289–93.
42. Tlahuice A, Garzon IL. On the structure of the $Au_{18}(SR)_{14}$ cluster. *Phys Chem Chem Phys* 2012;**14**:3737–40.
43. Jiang DE, Overbury SH, Dai S. Structure of $Au_{15}(SR)_{13}$ and its implication for the origin of the nucleus in thiolated gold nanoclusters. *J Am Chem Soc* 2013;**135**:8786–9.
44. Tlahuice-Flores A, Jose-Yacaman M, Whetten RL. On the structure of the thiolated Au_{15} cluster. *Phys Chem Chem Phys* 2013;**15**:19557–60.
45. Walter M, Akola J, Lopez-Acevedo O, Jadzinsky PD, Calero G, Ackerson CJ, et al. A unified view of ligand-protected gold clusters as superatom complexes. *Proc Natl Acad Sci U. S. A.* 2008;**105**:9157–62.
46. Gao Y, Shao N, Zeng XC. Ab initio study of thiolate-protected Au_{102} nanocluster. *ACS Nano* 2008;**2**:1497–503.
47. Jiang DE, Whetten RL, Luo W, Dai S. The smallest thiolated gold superatom complexes. *J Phys Chem C* 2009;**113**:17291–5.
48. Li Y, Galli G, Gygi F. Electronic structure of thiolate-covered gold nanoparticles: $Au_{102}(MBA)_{44}$. *ACS Nano* 2008;**2**:1896–902.
49. Tofanelli MA, Ackerson CJ. Superatom electron configuration predicts thermal stability of $Au_{25}(SR)_{18}$ nanoclusters. *J Am Chem Soc* 2012;**134**:16937–40.
50. Cheng L, Ren C, Zhang X, Yang J. New insight into the electronic shell of $Au_{38}(SR)_{24}$: a superatomic molecule. *Nanoscale* 2013;**5**:1475–8.
51. Reimers JR, Wang Y, Cankurtaran BO, Ford MJ. Chemical analysis of the superatom model for sulfur-stabilized gold nanoparticles. *J Am Chem Soc* 2010;**132**:8378–84.
52. Akola J, Kacprzak KA, Lopez-Acevedo O, Walter M, Grönbeck H, Häkkinen H. Thiolate-protected Au_{25} superatoms as building blocks: dimers and crystals. *J Phys Chem C* 2010;**114**:15986–94.

53. Akola J, Walter M, Whetten RL, Häkkinen H, Grönbeck H. On the structure of thiolate-protected Au_{25}. *J Am Chem Soc* 2008;**130**:3756–7.
54. Pei Y, Gao Y, Zeng XC. Structural prediction of thiolate-protected Au_{38}: a face-fused bi-icosahedral Au core. *J Am Chem Soc* 2008;**130**:7830–2.
55. Lopez-Acevedo O, Tsunoyama H, Tsukuda T, Häkkinen H, Aikens CM. Chirality and electronic structure of the thiolate-protected Au_{38} nanocluster. *J Am Chem Soc* 2010;**132**: 8210–8.
56. Jiang DE. The expanding universe of thiolated gold nanoclusters and beyond. *Nanoscale* 2013;**5**:7149–60.
57. Pei Y, Zeng XC. Investigating the structural evolution of thiolate protected gold clusters from first-principles. *Nanoscale* 2012;**4**:4054–72.
58. Krommenhoek PJ, Wang J, Hentz N, Johnston-Peck AC, Kozek KA, Kalyuzhny G, et al. Bulky adamantanethiolate and cyclohexanethiolate ligands favor smaller gold nanoparticles with altered discrete sizes. *ACS Nano* 2012;**6**:4903–11.
59. Zeng C, Liu C, Pei Y, Jin R. Thiol ligand-induced transformation of $Au_{38}(SC_2H_4Ph)_{24}$ to Au_{36}(Sph-t-Bu)$_{24}$. *ACS Nano* 2013;**7**:6138–45.
60. Nishigaki JI, Tsunoyama R, Tsunoyama H, Ichikuni N, Yamazoe S, Negishi Y, et al. A new binding motif of sterically demanding thiolates on a gold cluster. *J Am Chem Soc* 2012;**134**:14295–7.
61. Jung J, Kang S, Han YK. Ligand effects on the stability of thiol-stabilized gold nanoclusters: $Au_{25}(SR)^{-}_{18}$, $Au_{38}(SR)_{24}$, and $Au_{102}(SR)_{44}$. *Nanoscale* 2012;**4**:4206–10.
62. Goel S, Velizhanin KA, Piryatinski A, Ivanov SA, Tretiak S. Ligand effects on optical properties of small gold clusters: a TDDFT study. *J Phys Chem C* 2012;**116**:3242–9.
63. Tlahuice-Flores A, Whetten RL, Jose-Yacaman M. Ligand effects on the structure and the electronic optical properties of anionic $Au_{25}(SR)_{18}$ clusters. *J Phys Chem C* 2013;**117**: 20867–75.
64. Zheng K, Yuan X, Goswami N, Zhang Q, Xie J. Recent advances in the synthesis, characterization, and biomedical applications of ultrasmall thiolated silver nanoclusters. *RSC Adv* 2014;**4**:60581–96.
65. Yang H, Wang Y, Huang H, Gell L, Lehtovaara L, Malola S, et al. All-thiol-stabilized Ag_{44} and $Au_{12}Ag_{32}$ nanoparticles with single-crystal structures. *Nat Commun* 2013;**4**:2422.
66. Ji J, Wang G, Wang T, You X, Xu X. Thiolate-protected Ni_{39} and Ni_{41} nanoclusters: synthesis, self-assembly and magnetic properties. *Nanoscale* 2014;**6**:9185–91.
67. Kurashige W, Yamazoe S, Kanehira K, Tsukuda T, Negishi Y. Selenolate-protected Au_{38} nanoclusters: isolation and structural characterization. *J Phys Chem Lett* 2013;**4**:3181–5.
68. Wan XK, Lin ZW, Wang QM. Au_{20} nanocluster protected by hemilabile phosphines. *J Am Chem Soc* 2012;**134**:14750–2.
69. Chen J, Zhang QF, Bonaccorso TA, Williard PG, Wang LS. Controlling gold nanoclusters by diphospine ligands. *J Am Chem Soc* 2013;**136**:92–5.
70. Gutrath BS, Oppel IM, Presly O, Beljakov I, Meded V, Wenzel W, et al. $Au_{14}(Pph_3)_8(NO_3)_4$: an example of a new class of $Au(NO_3)$-ligated superatom complexes. *Angew Chem Int Ed* 2013;**52**:3529–32.
71. Wan XK, Yuan SF, Lin ZW, Wang QM. A chiral gold nanocluster Au_{20} protected by tetradentate phosphine ligands. *Angew Chem Int Ed* 2014;**53**:2923–6.
72. Kurashige W, Yamaguchi M, Nobusada K, Negishi Y. Ligand-induced stability of gold nanoclusters: thiolate versus selenolate. *J Phys Chem Lett* 2012;**3**:2649–52.
73. Zhang S, Chandra KL, Gorman CB. Self-assembled monolayers of terminal alkynes on gold. *J Am Chem Soc* 2007;**129**:4876–7.

74. Ford MJ, Hoft RC, Mcdonagh A. Theoretical study of ethynylbenzene adsorption on Au(111) and implications for a new class of self-assembled monolayer. *J Phys Chem B* 2005;**109**:20387–92.
75. Zaba T, Noworolska A, Bowers CM, Breiten B, Whitesides GM, Cyganik P. Formation of highly ordered self-assembled monolayers of alkynes on Au(111) substrate. *J Am Chem Soc* 2014;**136**:11918–21.
76. Maity P, Tsunoyama H, Yamauchi M, Xie SH, Tsukuda T. Organogold clusters protected by phenylacetylene. *J Am Chem Soc* 2011;**133**:20123–5.
77. Maity P, Takano S, Yamazoe S, Wakabayashi T, Tsukuda T. Binding motif of terminal alkynes on gold clusters. *J Am Chem Soc* 2013;**135**:9450–7.
78. Maity P, Wakabayashi T, Ichikuni N, Tsunoyama H, Xie SH, Yamauchi M, et al. Selective synthesis of organogold magic clusters $Au_{54}(C{\equiv}Cph)_{26}$. *Chem Commun* 2012;**48**:6085–7.
79. Jiang DE, Tiago ML, Luo W, Dai S. The "staple" motif: a key to stability of thiolate-protected gold nanoclusters. *J Am Chem Soc* 2008;**130**:2777–9.
80. Jiang DE, Luo W, Tiago ML, Dai S. In search of a structural model for a thiolate-protected Au_{38} cluster. *J Phys Chem C* 2008;**112**:13905–10.
81. Häkkinen H, Walter M, Grönbeck H. Divide and protect: capping gold nanoclusters with molecular gold–thiolate rings. *J Phys Chem B* 2006;**110**:9927–31.
82. Jiang DE, Dai S. Diffusion of the linear $CH_3S{-}Au{-}SCH_3$ complex on Au(111) from first principles. *J Phys Chem C* 2009;**113**:3763–6.
83. Tang Q, Jiang DE. Insights into the Phc≡C/Au interface. *J Phys Chem C* 2015;**110**:10804–10.
84. Wan XK, Tang Q, Yuan SF, Jiang DE, Wang QM. Au_{19} nanocluster featuring a V-shaped alkynyl-gold motif. *J Am Chem Soc* 2015;**137**:652–5.
85. The Cambridge Cluster Database. http://www-wales.ch.cam.ac.uk/ccd.html.
86. Wu Z, Macdonald MA, Chen J, Zhang P, Jin R. Kinetic control and thermodynamic selection in the synthesis of atomically precise gold nanoclusters. *J Am Chem Soc* 2011;**133**:9670–3.
87. Chevrier DM, Macdonald MA, Chatt A, Zhang P, Wu Z, Jin R. Sensitivity of structural and electronic properties of gold–thiolate nanoclusters to the atomic composition: a comparative X-ray study of $Au_{19}(SR)_{13}$ and $Au_{25}(SR)_{18}$. *J Phys Chem C* 2012;**116**: 25137–42.
88. Tao J, Perdew JP, Staroverov VN, Scuseria GE. Climbing the density functional ladder: nonempirical meta-generalized gradient approximation designed for molecules and solids. *Phys Rev Lett* 2003;**91**:146401.
89. Wiseman MR, Marsh PA, Bishop PT, Brisdon BJ, Mahon MF. Homoleptic gold thiolate catenanes. *J Am Chem Soc* 2000;**122**:12598–9.
90. Grönbeck H, Walter M, Häkkinen H. Theoretical characterization of cyclic thiolated gold clusters. *J Am Chem Soc* 2006;**128**:10268–75.
91. Shao N, Pei Y, Gao Y, Zeng XC. Onset of double helical structure in small-sized homoleptic gold thiolate clusters. *J Phys Chem A* 2009;**113**:629–32.
92. Pei Y, Pal R, Liu C, Gao Y, Zhang Z, Zeng XC. Interlocked catenane-like structure predicted in $Au_{24}(SR)_{20}$: implication to structural evolution of thiolated gold clusters from homoleptic gold(I) thiolates to core-stacked nanoparticles. *J Am Chem Soc* 2012;**134**:3015–24.
93. Yu Y, Luo Z, Chevrier DM, Leong DT, Zhang P, Jiang DE, et al. Identification of a highly luminescent $Au_{22}(SG)_{18}$ nanocluster. *J Am Chem Soc* 2014;**136**:1246–9.
94. Pei Y, Gao Y, Shao N, Zeng XC. Thiolate-protected $Au_{20}(SR)_{16}$ cluster: prolate Au_8 core with new $[Au_3(SR)_4]$ staple motif. *J Am Chem Soc* 2009;**131**:13619–21.
95. Jiang DE, Chen W, Whetten RL, Chen Z. What protects the core when the thiolated Au cluster is extremely small? *J Phys Chem C* 2009;**113**:16983–7.

96. Tang Q, Ouyang R, Tian Z, Jiang DE. Ligand effect on isomer stability of $Au_{24}(SR)_{20}$ clusters. *Nanoscale* 2015;**7**:2225–9.
97. Zhu M, Qian H, Jin R. Thiolate-protected $Au_{24}(SC_2H_4Ph)_{20}$ nanoclusters: superatoms or not? *J Phys Chem Lett* 2010;**1**:1003–7.
98. Das A, Liu C, Zeng C, Li G, Li T, Rosi NL, et al. Cyclopentanethiolato-protected $Au_{36}(SC_5H_9)_{24}$ nanocluster: crystal structure and implications for the steric and electronic effects of ligand. *J Phys Chem A* 2014;**118**:8264–9.
99. Price RC, Whetten RL. All-aromatic, nanometer-scale, gold-cluster thiolate complexes. *J Am Chem Soc* 2005;**127**:13750–1.
100. Zeng C, Chen Y, Li G, Jin R. Synthesis of a $Au_{44}(SR)_{28}$ nanocluster: structure prediction and evolution from $Au_{28}(SR)_{20}$, $Au_{36}(SR)_{24}$ to $Au_{44}(SR)_{28}$. *Chem Commun* 2014;**50**:55–7.

Chapter 8

Electronic Structure: Shell Structure and the Superatom Concept

Hannu Häkkinen
Department of Physics, Nanoscience Center, University of Jyväskylä, Jyväskylä, Finland; Department of Chemistry, Nanoscience Center, University of Jyväskylä, Jyväskylä, Finland
E-mail: Hannu.j.hakkinen@jyu.fi

Chapter Outline

8.1 INTRODUCTION

Fermionic shell structure is a fundamental physical manifestation of quantum confinement and symmetry. It is a widely appearing phenomenon in fermion systems from the atomic nucleus to an ordinary atom to nanometer-size metal clusters. For decades, it has been a central concept in physics and chemistry of gas-phase metal clusters. It has helped to organize a vast amount of data on stability and chemical reactivity of clusters as a function of cluster size to understandable trends. In recent years, ever-increasing numbers of chemically

Frontiers of Nanoscience, Vol. 9. http://dx.doi.org/10.1016/B978-0-08-100086-1.00008-7

stabilized ("passivated") metal clusters have appeared, and successful structural characterizations have opened a fascinating view on how matter organizes in the nanometer scale. The large database of solved structures has given a treasure trove for theoretical analysis. It has been confirmed numerous times that electron shell structure does pay a role and is a useful concept for understanding electronic and optical properties of chemically stabilized metal clusters (the so-called monolayer-protected clusters or MPCs).

This chapter is not intended to be a comprehensive review on MPCs. Several recent, more comprehensive reviews are available, concentrating on both experimental[1–4] and theoretical[5–10] aspects. Rather, it discusses a concept that has become quite central in the field during the recent years, namely the concept of "superatom." It is shown how this concept is a natural extension of the theory of electron shell structure, which has been a central theme in the physics and chemistry of gas-phase metal clusters for decades. As such this chapter is complementary to Chapters 7 and 9 that discuss structural aspects and structural prediction (7) and details of electronic structure, optical properties and chirality (9).

8.2 ELECTRON SHELLS

8.2.1 Noninteracting Electrons in a Potential Well

Shell structure, i.e., bunching of eigenenergy levels into groups of (nearly) degenerate energy, is a fundamental property of fermions confined in small systems. It is observed for nucleons in atomic nuclei, electrons in ordinary atoms, in metal nanoclusters, and in semiconductor quantum dots, and arises from the symmetry properties of the Schrödinger equation governing the solutions to their quantum states. Historically, the concept of fermionic shell structure and symmetry principle in quantum mechanics was tied together in the Nobel Prize of Physics 1963. Half of the Prize was given to Eugene Paul Wigner "for his contribution to the theory of the atomic nucleus and the elementary particles, particularly through the discovery and application of fundamental symmetry principles" and the other half was shared by Maria Goeppert Mayer and J. Hans D. Jensen "for their discoveries concerning nuclear shell structure."[11] In fact, theories first worked out for nuclear shell structure in 1950s greatly fertilized the field of metal cluster physics in 1980s and 1990s.

The basics of the shell structure can be grasped by considering quantum mechanics of noninteracting fermions in simple forms of radial potentials, solved routinely as an exercise in undergraduate physics courses. Figure 1 shows the degeneracies ($2(2l+1)$) of solutions to the radial Schrödinger equation for the 3D infinite spherical well with a "square box shape," organized to the increasing order in energy according to the angular momentum (l) subshells.[12] Note that it is customary to label each set of subshells by

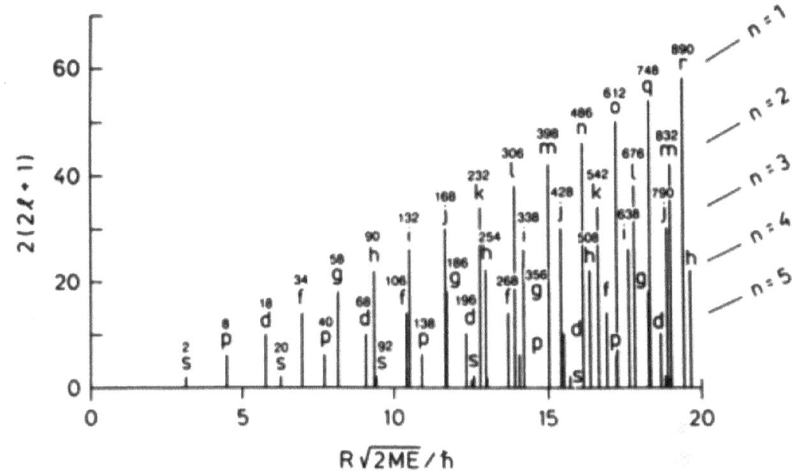

FIGURE 1 Shell structure of noninteracting fermions in an infinite spherical well. Solutions to the radial Schrödinger equation are grouped into subshells according to the radial quantum number (l) with degeneracy of $2(2l+1)$. The cumulative number of fermions needed to fill all levels up to each subshell is shown. The shells are ordered according to an increasing momentum (x-axis). *Reproduced with permission from Ref. 12. Copyright American Chemical Society 1991.*

a "principal" quantum number starting always from $n = 1$ (1s, 1p, 1d,..., 2s, 2p, 2d,...). This notation differs from electron shells in hydrogen-like atoms where additional degeneracies arise due to the $1/r$ potential between electrons and the point-like nucleus. Figure 1 clearly shows that angular momentum subshells at higher energies start to group together; this is seen already in the set of 2d-1h-3s subshells that accommodate electrons between the major shell-closing ("magic") numbers of 58 and 92. Furthermore, details of the radial potential affect sensitively this grouping. For start, the infinite harmonic well ($V \propto r^2$) has highly degenerate levels at strong magic numbers of 40, 70, 112, 168..., but when going to anharmonic shapes toward the square well, many of the subshells split and group together in a new way. This gives rise to the often-quoted series of magic numbers 2, 8, 18, 20, 34, 40, 58,... (Figure 2).

Deformation from 3D radial symmetry modifies the shell structure in a different way.[13,14] Reducing the spherical symmetry to spheroidal, two of the principal axes of the cluster remain the same while the third one is different. This gives rise to prolate (cigar-like) and oblate (disk-like) shapes. In the field of nuclear physics, Nilsson[15] worked out the effect of spheroidal deformations to nucleon shell structure in 1950s and similar ideas were applied for alkali metal clusters by Clemenger in 1980s.[16] The "Clemenger–Nilsson" model was successful in explaining the observed relative abundance (based on cluster energetics and stability) of sodium clusters in the early molecular beam experiments in 1980s.[17] Shape deformations are discussed more below in the context of models for interacting electron gas.

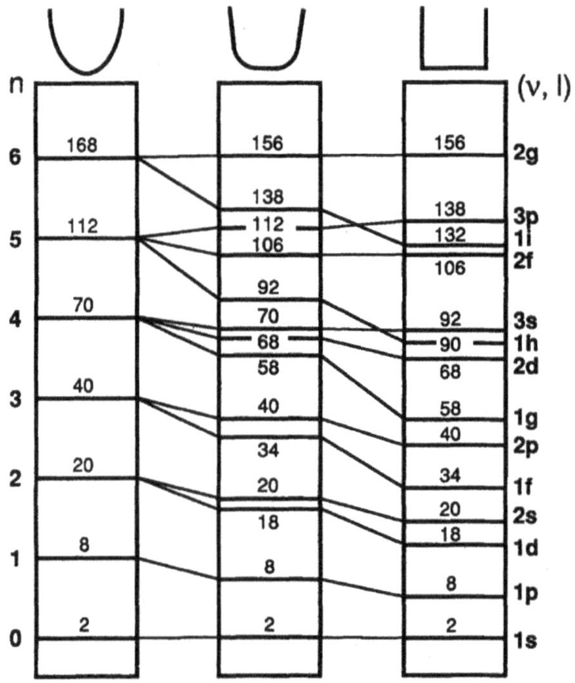

FIGURE 2 Energy levels of noninteracting fermions in various 3D infinite potential wells (from left: harmonic, intermediate anharmonic, square-well). The numbers above levels denote cumulative shell-filling electron counts. Quantum numbers (principal and angular momentum) specifying each shell are given on the right. *Reproduced with permission from Ref. 13. Copyright American Physical Society 1993.*

8.2.2 Interacting Electron Gas—The Jellium Model

The Hamiltonian problem for condensed matter systems

$$H\Psi = E\Psi \tag{1}$$

is easy to write up but impossible to solve accurately. The reason lies in Ψ which is the hugely complicated many-body wave function of the system, depending in principle on the positions of all the atomic nuclei $\{R_{i,\ i=1,\ldots,\mathrm{Na}}\}$ and all the electrons $\{r_{j,\ j=1,\ldots,\mathrm{Ne}}\}$:

$$\Psi = \Psi(R_1,\ R_2,\ldots,R_{\mathrm{Na}};\quad r_1,\ r_2,\ldots,\ r_{\mathrm{Ne}}) \tag{2}$$

Even after applying the Born–Oppenheimer approximation (decouple nuclear and electronic motion and just solve the electronic system for each fixed configuration of nuclei), the problem is unsolvable due to the many-body nature of the electron–electron interaction. Since 1960s, the density functional theory (DFT) has facilitated a convenient treatment of this many-body problem, by reducing the number of important degrees of freedom from $3N_e$ to just

3—defining the density of the electrons $\rho(r)$ in the ground state as the key variable instead of Ψ. An excellent review of how various theoretical treatments have been applied in the study of metal clusters is provided by Brack,[14] here the discussion focuses only on the key points to treat the electron–electron interactions in the electron gas model and what additional insight the electron gas models are able to yield beyond the "toy models" discussed above.

In DFT, the total energy of the electron–ion system in the electronic ground state is a functional of the (valence) electron density $\rho(r)$ of the ground state:

$$E[\rho] = T_s[\rho] + E_{xc}[\rho] + \int \left\{ V_I(r)\rho(r) + \frac{1}{2}\rho(r)\left[e^2 \int \frac{\rho(r')}{|r-r'|} d^3r' \right] \right\} d^3r + E_I \tag{3}$$

where T_s is the kinetic energy of a single-particle (Kohn–Sham, KS) reference system with the same electron density as the true system, E_{xc} is the quantum-mechanical exchange-correlation energy, V_I is the Coulomb energy between ion cores and valence electrons, the fourth term is the Coulomb self-energy of the valence electrons and E_I is the energy of the ion system. The electron density is expressed as a sum over occupied states whose wave functions are represented by the KS virtual orbitals $\varphi(r)$

$$\rho(r) = \sum_{i=1}^{wN} |\varphi_i(r)|^2 \tag{4}$$

where w is the number of valence electrons per atom in the system of N atoms. In the limit of weak ion core–valence electron interactions (good approximation for instance for alkali metals), the density of the (positive) ion cores is approximated as constant in space over the volume of the cluster, with the normalization that it produces the total positive background charge that neutralizes the total negative charge of the valence electrons

$$\int \rho(r) d^3r = wN \tag{5}$$

This approximation is commonly termed as the *jellium* model—perhaps due to the fact that the electron gas is living in a "jelly-like" medium made out of averaged ion density instead of discrete ions. This model, with several sophistications, was broadly applied in many of the pioneering theoretical studies in the field of metal cluster physics in 1980s and early 1990s.[14]

Apart from simple nonspherical deformations such as spheroidal shapes, it is also of interest to study electron shell structure in jellium systems where the background density has various polyhedral shapes. This is relevant particularly for ligand-stabilized metal clusters where the metal cores often assume polyhedral shapes due to constraints from the ligand shell. Due to numerically efficient DFT codes, it is now possible to solve the KS equations virtually for any shape of the background jellium density. This was recently done in our group by using the real-space DFT code GPAW (grid-based projector-augmented wave method).[18–20]

Figure 3 shows symmetry analyses of KS states solved self-consistently for five common polyhedra and their comparison to a sphere. The systems were solved for 92 electron states inside the polyhedral, out of which 58 states were occupied providing an electron density corresponding to the density of bulk gold, and the electron–electron interaction was treated by the local density approximation (LDA). The symmetry of the KS wave functions $\varphi_i(r)$ was analyzed by decomposing them into angular momenta by calculating weights $c_{i,l}(R_0)$

$$c_{i,l}(R_0) = \sum_m \int_0^{R_0} r^2 \mathrm{d}r \left|\phi_{i,lm}(r)\right|^2 \tag{6}$$

where

$$\phi_{i,lm}(r) = \int \mathrm{d}\hat{r} Y^*_{lm}\left(\hat{r}\right)\varphi_i\left(\vec{r}\right) \tag{7}$$

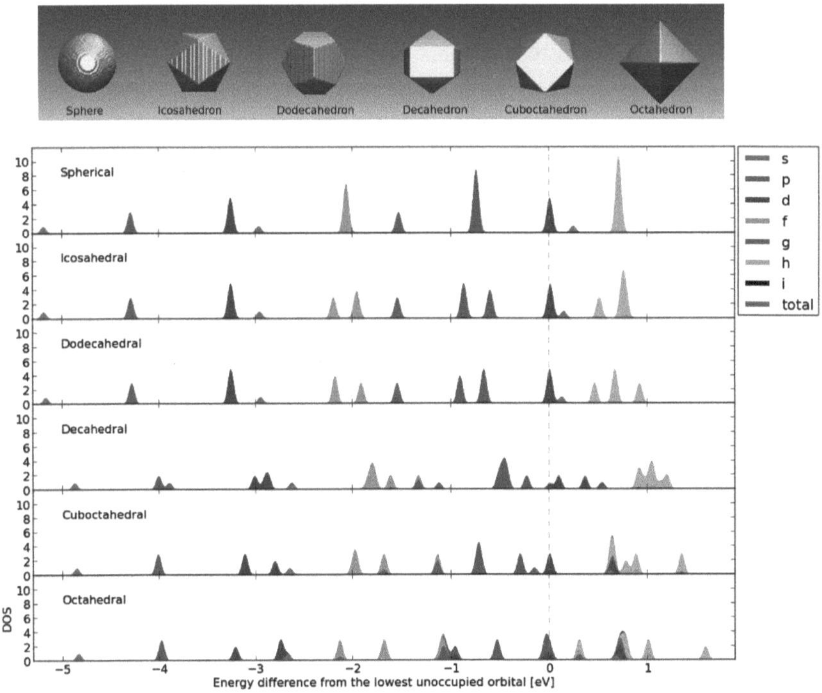

FIGURE 3 Electron shells obtained in a fixed-background density functional theory–local density approximation (DFT-LDA) jellium calculation for various shapes (shown on top) of the background density. The systems contain 58 electrons corresponding to the 6s-electron density of bulk gold. The LUMO state ("59th electron") is marked by a dashed vertical line (Ref. 19). (see color plate).

is the projection of a given wave function onto spherical harmonics Y_{lm}. Here R_0 is the radius of a sphere where the symmetry analysis was done.

58 and 92 are magic shell-closing numbers for the infinite spherical well as shown in Figure 2. Figure 3 shows that all the considered polyhedra (icosahedron, dodecahedron, decahedron, cuboctahedron, and octahedron) preserve the same shell ordering as seen in the spherical system: 1s, 1p, 1d, 2s, 1f, 2p, 1g for the occupied states and 2d, 3s, 1h for the unoccupied states. However, the polyhedral shapes split many of the angular momentum shells. 1p and 2p shells are split in the decahedron, 1d and 2d are split in decahedron, cuboctahedron, and octahedron, and 1f, 1g, 1h shells are split in all polyhedra. Cuboctahedron and octahedron mix some of the states with different angular momenta together.

DFT—jellium calculations and state symmetry analysis can be done also for quasi-2D systems. Figure 4 shows circular, hexagonal, and triangular disks with a thickness corresponding to two monolayers of gold containing 60 electrons at the density of bulk gold, with several unoccupied states solved as well. State symmetries can be identified by using a similar decomposition into angular momenta as for 3D systems. Figure 5 shows an example for the circular disk and Figure 6 illustrates several occupied and unoccupied states for the three considered shapes. It is easy to visually identify most of the low angular momentum states in Figure 6. It is important to realize that the flat systems obey drastically different shell-closing sequence from 3D systems. The overall degeneracy of the states is either 1 (s-states) or 2 (all the higher states), giving a sequence of "magic numbers," for instance, for a circular disk as 2, 6, 10, 12, 16, 20, 24, and so on.

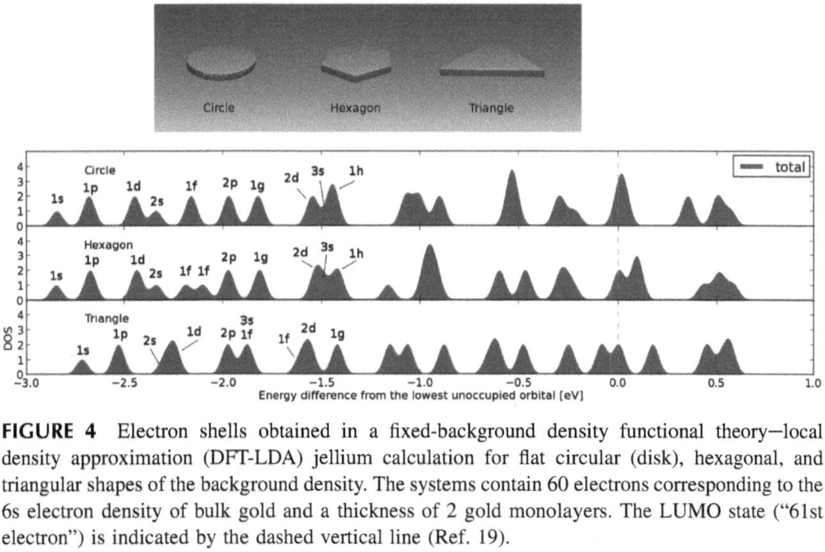

FIGURE 4 Electron shells obtained in a fixed-background density functional theory—local density approximation (DFT-LDA) jellium calculation for flat circular (disk), hexagonal, and triangular shapes of the background density. The systems contain 60 electrons corresponding to the 6s electron density of bulk gold and a thickness of 2 gold monolayers. The LUMO state ("61st electron") is indicated by the dashed vertical line (Ref. 19).

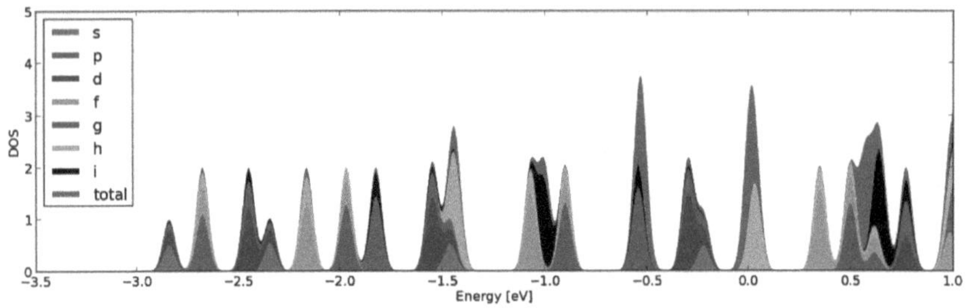

FIGURE 5 Electron shells of the circular flat electron cluster of 60 electrons shown in Figure 4, projected to angular momenta. The LUMO state is at zero energy as in Figure 4 (Ref. 19). (see color plate).

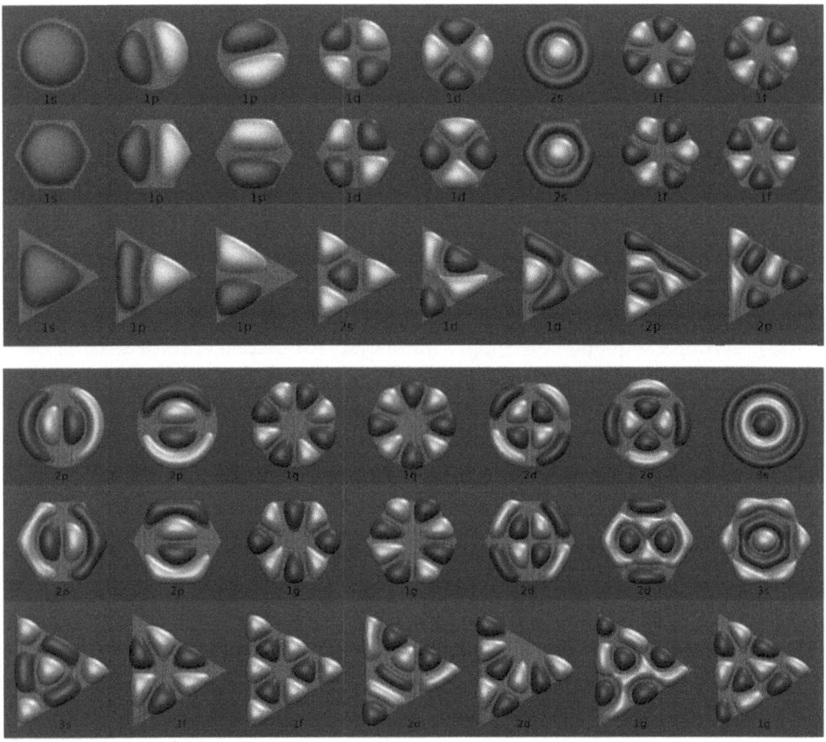

FIGURE 6 The occupied states of the flat clusters shown in Figure 4 (Ref. 19).

As the examples above show, considering electron shells in different predefined shapes of the positive background density gives useful information on the effect of the shape on the splitting of angular momentum subshells. However, one can ask an important question: given the number of electrons, what is the optimal shape of the electron cluster, corresponding to the minimal energy? A conventional approach to this question involves studies of a large number of different fixed shapes and making conclusions from comparison of their total energies. This approach is completely analogous to a study into optimal structure (lowest-energy isomer) of clusters with real atoms. However, an elegant method development in mid-1990s introduced the "ultimate jellium" model where the positive background density is allowed to freely deform during the iterative solution of the electronic problem so that in the converged solution the system is charge-neutral in each point in space.[21] This brings about a special situation, where all the Coulomb energies in the system (electron–electron, electron–ion, and ion–ion) cancel exactly and the energetics is defined only by the quantum mechanical kinetic energy of the KS electrons and the exchange-correlation energy:

$$E[\rho] = T_s[\rho] + E_{xc}[\rho]. \tag{8}$$

Another special feature of this model is that there are no free parameters, and the bulk limit of the model gives a unique "jellium material" with the electron density parameter of $r_s = 4.18$ Bohr units (r_s is defined as the radius of a sphere that holds one electron). This density (most closely found in the real alkali metals, sodium and potassium) minimizes the total energy per electron in an infinite bulk electron gas. A slight drawback of the model is that only neutral systems can be considered due to the requirement of charge cancellation by electrons and the background at each point in space.

Figure 7 shows the optimal shapes of electron clusters from 2 to 22 electrons, calculated from the ultimate jellium model with LDA and taking spin polarization into account. It is important to realize that the obtained shapes are free from any symmetry constraints, i.e., all possible multipolar deformations can occur. The evolution of the shape for the smallest clusters from 2 to 8 electrons is easily understood by considering shapes of the single-particle s, p_x, p_y, and p_z orbitals. 3- and 4-electron clusters have 1 and 2 electrons in one of the p-orbitals—say p_x, which makes the electron density axially symmetric and prolate. The initial symmetry degeneration of p orbitals is here broken by the Jahn–Teller deformation which favors shape deformation rather than a symmetric shape with paramagnetic spin configuration (such as occupation of spin triplet $1s^2\ 1p_x^{2/3}p_y^{2/3}p_z^{2/3}$ for a spherical 4-electron cluster). The 5th and 6th electrons go to the p_y orbital, making the 6-electron cluster axially symmetric oblate. Note that Figure 4 indicates 6 as the magic electron number for circular, hexagonal, and triangular disks as well—there the orbital filling is identical to this case. Clusters with 7 and 8 electrons fill the remaining p_z orbital which makes the 8-electron cluster a magic, spherical $1s^2\ 1p^6$ octet.

Clusters with 9–20 electrons can be understood in terms of filling 1d and 2s orbitals. Generally, elongated shapes are observed in the beginning of the d-shell filling up to the mid-shell 13 and 14 electron clusters, after which the shapes turn oblate. Some of the clusters however are missing both axial and inversion symmetry. Likewise, 21- and 22-electron clusters can be understood in terms of the 1f shell filling.

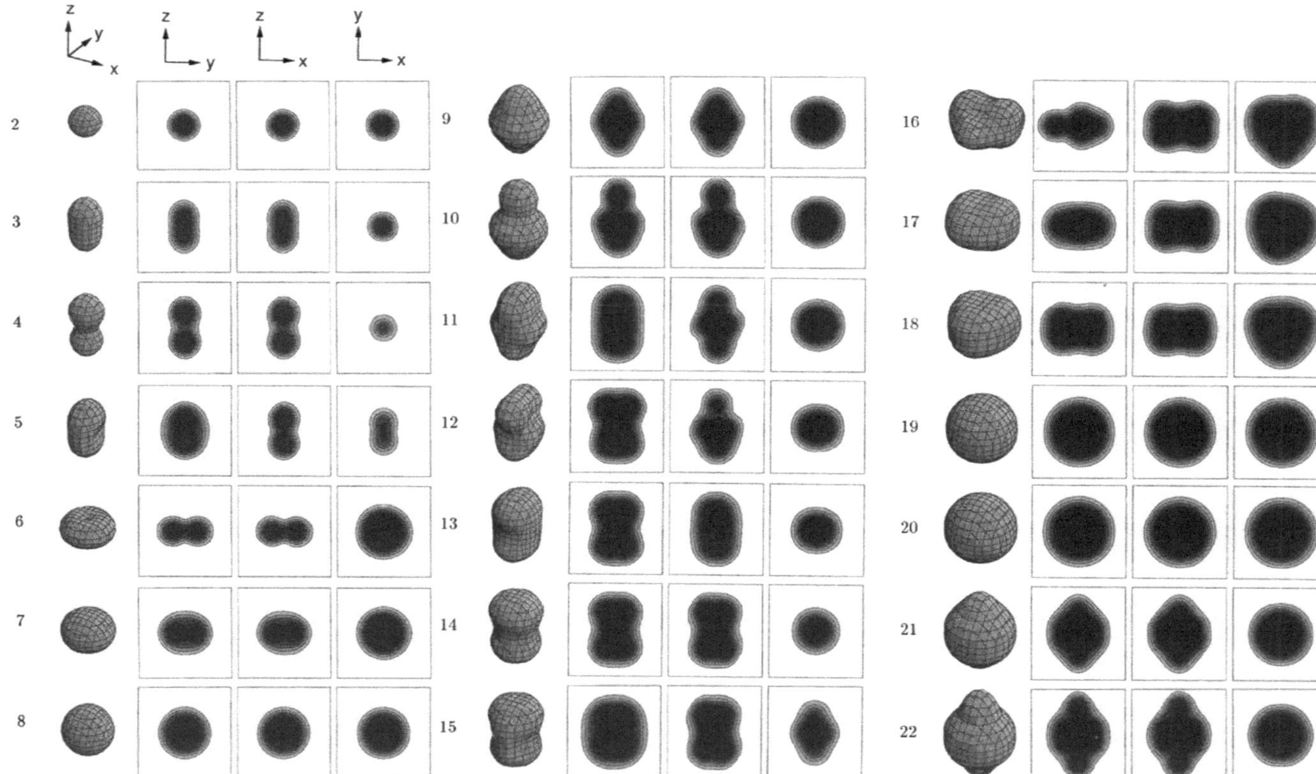

FIGURE 7 Ground state densities of electron cluster from 2 to 22 electrons, obtained from the ultimate jellium model. The 3D constant density surface is plotted at the value of 38% of the bulk density (Ref. 21).

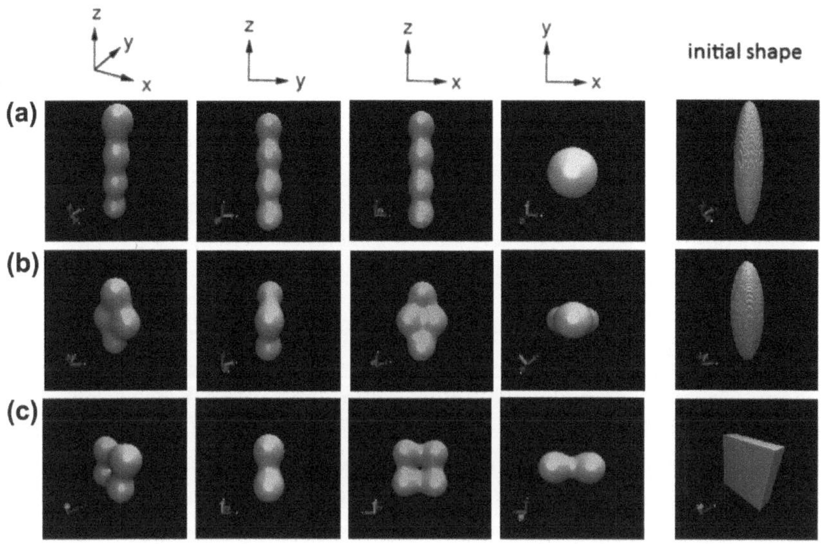

FIGURE 8 Three shape isomers for the 8-electron cluster in the ultimate jellium model. The converged shapes were started from electron densities shown on the right. The energies per electron are (a): 1.831 eV; (b): 1.855 eV; (c): 1.841 eV; as compared to −1.915 eV for the ground state sphere shown in Figure 6 (Ref. 20).

Shape isomers exist and were largely discussed in the original paper.[21] For instance, two shapes were obtained with an almost degenerate energy for the 18-electron cluster: a nonspherical shape as shown in Figure 7 and a spherical shape with a reduced density at the center ("hollow-like"). Figure 8 demonstrates that even the 8-electron cluster, assumed usually as the spherical octet configuration discussed above, can have several isomeric shapes that are energetically quite close to each other. The shapes shown in Figure 8 suggest that these isomeric clusters could be understood as "clusters of clusters," i.e., consisting of linked, spherical 2-electron systems. For the discussion in Section 8.3.4, it is important to realize already here that a cluster with a "magic" number of electrons may not always be spherical.

8.3 CONCEPT OF A SUPERATOM

8.3.1 Little History

Although no formal definition exists for a "superatom," it is widely understood as being a (meta)stable group of atoms, either "made by nature" or "engineered by man," that has at least some collective properties that are reminiscent of an ordinary atom. It is very interesting to realize that scientists contributing to great advances of chemistry in late 1800s, such as Loschmidt, Kekule, and Couper, regarded benzene as a kind of "superatom": it was known to be a molecule of six carbons but of unknown structure, be very stable and still chemically modifiable via associating side groups.[22] This period of time coincides with the very development of the periodic table itself by Mendeleev! Much later, in 1980s, Watanabe and Inoshita suggested the concept of a "superatom" in the field of optoelectronics as a conceptually new semiconductor heterostructure, where a spherical volume is doped by donor atoms. The released electrons from the donors are bound in the vicinity of the donors by the screened Coulomb field and occupy atom-like orbitals.[23]

The community of researchers studying metal clusters in the gas phase has used the superatom concept in conjunction with electronic shell theory successfully since 1980s.[13,17,24] It is interesting to note that the electronic shell theory has been adjusted from the nuclear shell theory where the 1963 Nobel Prize in physics was given to its developers M. Goeppert-Mayer and J.H.D. Jensen. The seminal work by the Knight group[17] observed enhanced stability of neutral sodium clusters for sizes that correspond to 8, 20, 40, 58,... free electrons (originating from the Na(3s) atomic orbital). The observations were elegantly explained by considering free electrons in a 3D spherical model potential, where angular momentum shells are filled with an Aufbau principle $1s^2$ $1p^6$ $1d^{10}$ $2s^2$ $1f^{14}$ $2p^6$ $1g^{18}$... (note that it is customary to label the first appearing angular momentum shell with a principal number $n = 1$ contrary to the electronic shells in atoms where the notation follows the symmetries of the solution to the central point-like Coulomb potential $1/r$). The approximation of "free" electrons in metal clusters works best for so-called "simple" metals such as alkali, alkali-earth, and noble metals where the electrons in the outermost atomic s-orbital are thought to form the "electron gas" delocalized over the volume of the cluster. The numerical treatment of the Schrödinger equation becomes efficient if the positive ion cores are approximated by a smooth background of positive charge, giving rise to the "jellium" model. It is important to realize, however, that a similar shell structure builds up naturally even in an infinitely deep 3D spherical potential simply due to the symmetry of the potential, Pauli principle, and fermionic nature of the electrons.[12] It is the symmetry of the problem, not the interaction between electrons and the cluster (or electron–electron interactions), that leads to the existence of the shell structure.

The initial observation of the shell structure for alkali metal clusters and its obvious correlation to the cluster stability (clusters with a filled electron shell or a "magic" electron number were found to be significantly more abundant than clusters with a neighboring size) created a lot of further experimental and theoretical work to understand chemical reactivity and cluster–cluster interactions by considering the "super-valence" (for a recent review see Ref. 25). For instance, clusters with a magic 8-electron configurations were thought to be chemically more inert that 7- or 9-electron systems. A very interesting experimental observation in 1989 reinforced this thinking: an electronically magic (40 free electrons) Al_{13}^- anion was found to be resistive to etching by oxygen in gas-phase experiments.[26] Ideas were presented for assembling crystals with novel properties that are made from stable clusters as building blocks.[27] However it was soon realized that the stability of metal clusters even with a magic electron number can depend strongly on the chemical environment. As an example, an early DFT study showed that the magic Na_8 cluster remains "magic" when it is placed on an insulating NaCl(001) surface, but reacts immediately with the metal Na surface or with another Na_8 cluster.[28] This and other early studies manifested a simple truth: all nanoscale matter is metastable with respect to its bulk counterparts (materials have an ultimate tendency to coalesce). A major question in nanoscale materials science is then how one can learn to control the (meta)stability and to understand the principle by which nanoscale matter organizes into metastable assemblies. Monolayer-protected metal clusters (MPCs) offer intriguing examples of varying degrees of metastability; consequently, understanding factors contributing to the stability is crucial for engineering optimal assemblies with desired size and functionality for their stable use as well-behaving components in applications. In recent years, the concept of the superatom (or "superatom complex" model as sometimes called) has played an important role in this regard.

In the remainder of this Chapter, I will briefly review the superatom concept for MPC's, discuss one example of an interesting thiol-stabilized gold cluster ($Au_{38}(SR)_{24}$), and present remarks relating the composition, shape, and features of electronic structure of 70 MPC compounds, most of which are structurally known.

8.3.2 The Superatom Concept for Monolayer-Protected Metal Clusters (MPCs)

In October 2007, the first-ever total-structure-determination of an all-thiol-protected gold nanoparticle was published by the group of R.D. Kornberg,[29] based on X-ray diffraction at 1.1 Å resolution from single crystals containing a distinct compound with 21 kDa Au core mass, protected by *p*-MBA ligands (*p*-MBA = *para*-mercaptobenzoic acid). Specifically, the composition was determined as $Au_{102}(p\text{-MBA})_{44}$ and the crystal unit cell was observed to contain an enantiomeric pair of these clusters. The structure of $Au_{102}(p\text{-MBA})_{44}$ was shown to exhibit a high degree of fivefold symmetry: a central 49-atom Marks decahedron, capped by two 20-atom caps of C_5 symmetry, extending the number of Au atoms with fivefold rotational symmetry up to 89. Only the 13 Au atoms forming an "equatorial band" were found to be in irregular positions, and their bonding combination with the *p*-MBA ligands is responsible for the overall chirality of the particle.

A subsequent thorough analysis of the atomic structure and full density functional treatment of the electronic structure of the $Au_{102}(p\text{-MBA})_{44}$ cluster (with all of its 762 atoms and 3366 valence electrons) resulted in a clear picture of the identity of the protecting gold–thiolate ligands and the gold core, and the underlying reasons for the thermodynamic stability of this compound.[30] It was found that the atomic structure of the $Au_{102}(p\text{-MBA})_{44}$ compound (Figure 9) consists of an approximately D_{5h}-symmetric Au_{79} metallic core with a

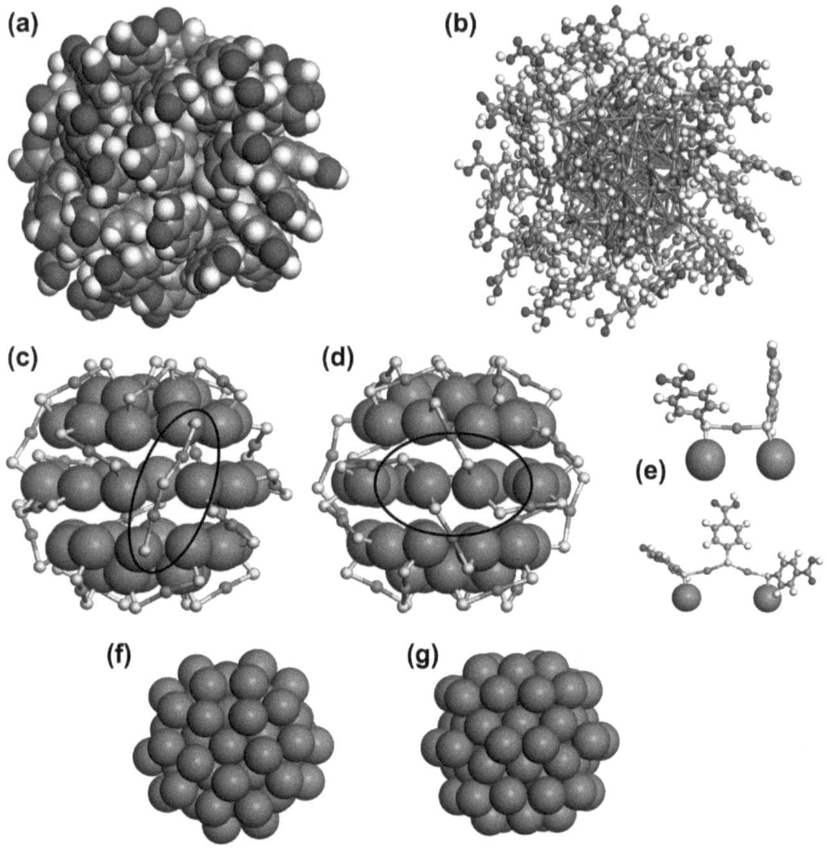

FIGURE 9 Core-shell structure of the $Au_{102}(p\text{-MBA})_{44}$ cluster. (a and b) Space-filling (a) and ball-and-stick (b) representations of the $Au_{102}(p\text{-MBA})_{44}$. Au, orange; S, yellow; C, gray; O, red; H, white. (c and d) Two views of the 40-atom surface of the Au_{79} core, together with the passivating $Au_{23}(p\text{-MBA})_{44}$ mantle. The cationic Au atoms in the mantle are depicted by the smaller orange spheres. The "structure defects" at the core–mantle interface [two Au atoms with two Au–S bonds, and a long $RS{-}(AuSR)_2$ unit] are highlighted by the ellipse. (e) Close-up of the protecting $RS{-}(AuSR)_x$ unit with $x = 1$ or $x = 2$. (f and g) Two views of the Au_{79} core, which has a symmetry of D_{5h} (within 0.4-Å tolerance). (see color plate). *Reproduced with permission from Ref. 30. Copyright American Association for the Advancement of Science 2008.*

protective gold–thiolate layer of composition $Au_{23}(p\text{-MBA})_{44}$. Hence $Au_{102}(p\text{-MBA})_{44}$ is more accurately described in the formulation $Au_{79}[Au_{23}(p\text{-MBA})_{44}]$. The gold atoms in the cluster are in two distinct chemical states: the 79 core Au atoms (Au_{core}) are in a metallic (charge-neutral) state whereas the 23 Au atoms (Au_{ligand}) that belong to protecting $RS(AuSR)_x$ units are oxidized. Consequently, the composition evokes the predicted "divide-and-protect" structure motif.[31] The $Au_{23}(p\text{-MBA})_{44}$ layer can further be decomposed into 19 units with

$x = 1$ and 2 units with $x = 2$, which are anchored to the core via sulfur in atop positions. The $Au_{core}-S-Au_{ligand}$ angle is close to 90° and the Au_{ligand} atoms are linearly coordinated by two sulfurs. We note here that the concept of a "staple" was introduced in Ref. 29 to describe the protecting gold–thiolate unit. A visualization of such "staples" found in $Au_{102}(p\text{-MBA})_{44}$ is shown in Figure 9. However, in the spirit of the "divide-and-protect" scheme, the actual protecting units are rather the $RS(AuSR)_x$ where Au is oxidized. The total composition of the metal core and the ligand layer can thus be written as $Au_{79}[Au(p\text{-MBA})_2]_{19}[Au_2(p\text{-MBA})_3]_2$, signifying the usefulness of the "divide-and-protect" notation to easily distinguish between the two kinds of gold atoms in the system. The total number of the protecting gold–thiolate units (21) is intimately related to the electronic stability of the particle as we will see below.

A confirmation of the metallic character of the Au_{79} core came through analysis of radial difference in the cumulative induced charge when the $Au_{102}(p\text{-MBA})_{44}$ compound was made either cationic or anionic (that is, remove or add one electron, recalculate the electron density, and analyze the radial density difference). In both cases, the major portion (90%) of the induced charge was found in the $Au_{23}(p\text{-MBA})_{44}$ shell (Figure 10). Virtually no change was observed inside a radius of 5 Å and only 10% of the induced charge resides at the interface between the Au_{79} core and the $Au_{23}(p\text{-MBA})_{44}$ protective layer (5 Å $< R <$ 7 Å). Since a metallic cluster accepts charge only at its surface, it could be directly concluded that the electronic structure of the Au_{79} should feature delocalized-electron shell structure, just as the bare free-electron metal clusters in the gas phase.

The calculated HOMO-LUMO gap of the $Au_{102}(p\text{-MBA})_{44}$ cluster was found to be appreciable, about 0.5 eV, indicating a major electron shell closing (Figure 10). The analysis of angular momentum character of the electron states of the bare Au_{79} core and the full compound revealed the exact mechanism of how this shell closing is obtained in the protected cluster.

The bare Au_{79} has an odd number of valence electrons. Its SOMO (singly occupied (highest) molecular orbital) lies in the middle of a band that consists of 34 single-electron states, out of which 2 states show the S-symmetry, 10 states D-symmetry, and 22 states H-symmetry (Figure 10). This set of states is separated from other states by rather large energy gaps: 0.5 eV gap to deeper-lying (occupied) states with a dominant G-character, and 0.33 eV gap to higher-lying (empty) states with I-character. The symmetries and the energy ordering of the states correspond well to the delocalized conduction-electron model where angular symmetries of 3S, 2D, and 1H appear between major electron shell closings at 58 delocalized electrons and at 92 delocalized electrons. In fact, if one does a model calculation for the bare Au_{79} core just considering Au to be 6s-monovalent metal, the situation may be well compared to the classic case of gas-phase sodium clusters: the core supports all the major gaps expected for a (spherical) 79-electron jellium system: those at 2, 8, 18, 20, 34, and 58 electrons (Figure 11).

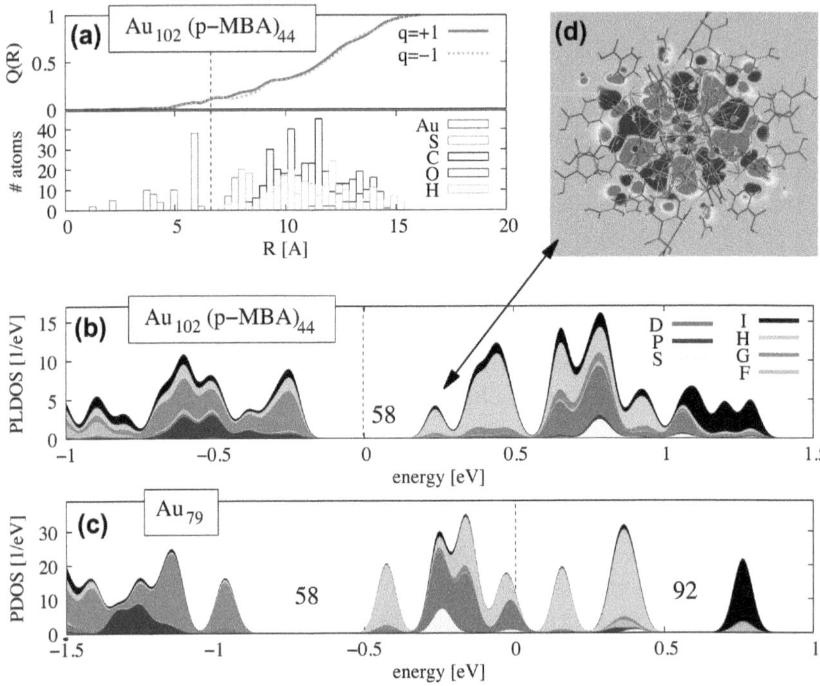

FIGURE 10 Electronic structure analysis of the $Au_{102}(p\text{-MBA})_{44}$ cluster. (a) The radial dependence of the integrated induced charge Q(R) upon removing (red curve) and adding (green curve) one electron to the neutral $Au_{102}(p\text{-MBA})_{44}$ cluster (Upper), and the radial distribution of atoms (Lower). The dashed line indicates a midpoint between the surface of Au_{79} core and the Au−thiolate layer. (b) The angular-momentum-projected local electron density of states (PLDOS) (projection up to the I symmetry) for the Au_{79} core in $Au_{102}(p\text{-MBA})_{44}$. (c) The angular-momentum-projected electron density of states (PDOS) for the bare Au_{79} without the Au−thiolate layer. (d) A cut-plane visualization of the LUMO state of the $Au_{102}(p\text{-MBA})_{44}$ cluster. Note the H symmetry (10 angular nodes) at the interface between the Au_{79} core and the Au−thiolate layer. In (b), the zero energy corresponds to the middle of the HOMO-LUMO gap, whereas in (c), the zero energy is at the HOMO level (dashed lines). For plotting PLDOS/PDOS curves, each individual electron state is displayed by a Gaussian smoothing of 0.03 eV. Shell-closing electron numbers are indicated in (b) and (c). (see color plate). *Reproduced with permission from Ref. 30. Copyright American Association for the Advancement of Science 2008.*

What happens now when the 21 $RS(AuSR)_x$ ($x = 1, 2$) units interact with the Au_{79}? *Formal* charge counting of such units indicates that they would like to attract one electron each out of the gold core (since the $RS(AuSR)_x$ anion is a closed-shell system), making the core to be *formally* $Au_{79}^{(21+)}$, justifying the first conjectures of the (79 − 21 = 58)-electron shell closure in $Au_{102}(p\text{-MBA})_{44}$.[29] In reality, the surface-covalent bonds that form between the surface Au atoms of the core and the S atoms of the units were found to be close to charge-neutral. The net result with respect to the modifications to the conduction-electron shell structure of the initially bare Au_{79} core is still

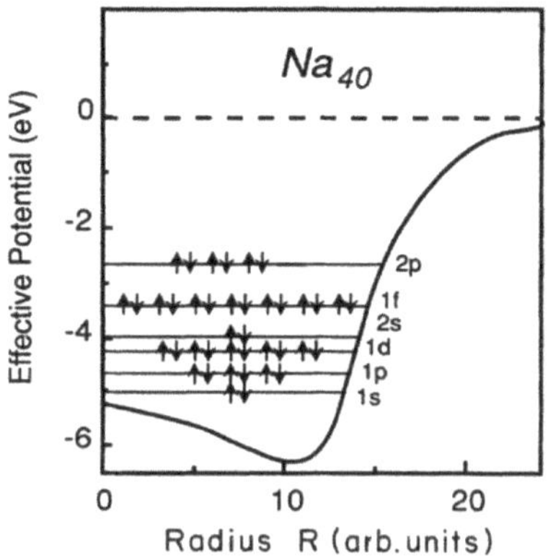

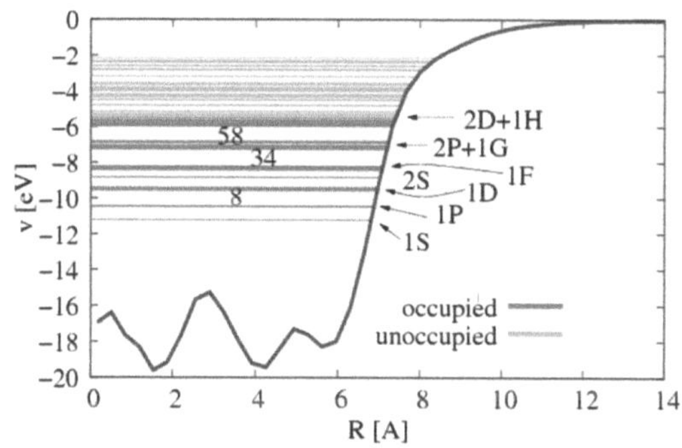

FIGURE 11 Top: self-consistent effective potential of jellium sphere corresponding to Na_{40} with the occupation of electron energy levels. Bottom: radially averaged Kohn−Sham potential of the Au_{79} core in $Au_{102}(p\text{-MBA})_{44}$ (6s-only calculation) with the occupied (red (gray in print versions)) and unoccupied (green (light gray in print versions)) energy levels and their symmetries. *Reproduced with permission from Ref. 6. Copyright Royal Society of Chemistry 2008.*

the same as in the formal charge-counting model: 21 conduction electrons are localized to the covalent bonds with sulfurs, therefore they "disappear" from the occupied electronic shells and rehybridize with sulfur states. Therefore one can well characterize the protected $Au_{102}(p\text{-MBA})_{44}$ electronically as a 58 free-electron system. It is also interesting to note that this picture of modification of electron shells is very rigid: the 0.5 eV gap at 58 electrons (the HOMO-LUMO gap of the full particle) is the same as found already in the bare Au_{79} core. The HOMO-LUMO gap of $Au_{102}(p\text{-MBA})_{44}$ corresponds also to the measured optical gap as has been confirmed spectroscopically.[32,33]

All this analysis can be summarized as follows: the stability of the $Au_{102}(p\text{-MBA})_{44}$ particle is due to several coexisting factors: (1) formation of a compact, symmetric-enough metal core that can support clear electron-shell structure, (2) complete chemical protection (passivation) of the core surface by $RS(AuSR)_x$ units, the number of which has to be "just right" so that (3) a major gap is exposed in the electron shell structure at 58 free electron shell closing in the metal core.

Several other gold-containing MPCs with thiolates and/or phosphines in the ligand shell were analyzed in 2008, and a simple formula was proposed to account for the free-electron count:

$$n_e = N\nu_A - M - q \tag{9}$$

where N is the number of gold atoms in the MPC, ν_A is taken as $\nu_A = 1$ (effective valence), M is the number of thiolates (or other ligands withdrawing/localizing one electron from the gold, such as halides), and q is the MPC total charge (sign included). This formula is easily generalized to other simple metals considering the proper valence, to other ligand types that withdraw more than one electron from the metal (such as iron carbonyls, $Fe(CO)_4$ or $Fe(CO)_3$) or to donor ligands.

8.3.3 The Story of $Au_{38}(SR)_{24}$—from an Interesting but Obscure Object to One of the Best Understood Clusters

The case of the $Au_{38}(SR)_{24}$ is interesting enough to be briefly summarized here, since it provides a nice example on how the field has advances since late 1990s and how theory and experimental work have progressed hand-in-hand, finally resulting in the full characterization of this nanocluster.

Using matrix-assisted laser desorption ionization coupled with mass spectrometry (MALDI-MS), Whetten and coworkers proposed in late 1990s that a thiol-stabilized gold nanoparticle with a metal core mass of ~8 kDa should have approximately 38 Au atoms and 24 thiolates.[34,35] In the following years, confusion arose in the literature when several studies examined a very stable compound with an unknown composition with distinct molecular properties, labeling it as Au_{38}, but in fact now known to be $Au_{25}(SR)_{18}^{-}$.[36] However, in 2008, precise nondestructive electrospray ionization mass spectrometry (ESI-MS) experiments confirmed the assignment $Au_{38}(SR)_{24}$.[37] Fingerprint characteristics were determined from electrochemistry and optical experiments. From electrochemistry experiments, the HOMO-LUMO gap of this species was determined to be 0.9 eV and the estimated charge of the nanoparticle is zero.[1,38–40] Optical absorption spectra showed that the cluster has molecule-like electronic states in accordance with the results of electrochemical experiments. The optical spectrum showed strong low-energy peaks at 1.2, 1.7, and 2.0 eV,[1,37,41,42] distinctly different from the well-characterized, slightly smaller $Au_{25}(SR)_{18}^{-}$.

The first mass spectrometric characterization of the 8 kDa cluster in 1997 inspired early theoretical work to examine possible atomic models and structural and electronic properties of this nanoparticle.[43–47] The first studies with DFT on a model cluster $Au_{38}(SR)_{24}$ examined a truncated-octahedral Au_{38} core passivated by 24 SCH_3 ligands bridging edges of the 6 fcc(100) facets,[43] characterizing the surface-covalent Au–S bond to involve clear charge transfer from Au to S. Later studies considered structural motifs where the Au_{38} core is amorphous and the adsorption sites of the thiolates are irregular, finding this motif to be energetically preferable compared to the truncated-octahedral form.[45–47] In hindsight, the early DFT work suffered from the unbalance of treating the relative strengths of Au–Au and Au–S interactions reliably. In 2006, structural relaxations of $Au_{38}(SR)_{24}$ with improved exchange-correlation functionals lead to a drastically different structure, where a small gold core (Au_{14}) was protected from six sides by oligomeric ring-like $(AuSCH_3)_4$ units, consequently, the composition could be written in a "divide-and-protect" scheme as $Au_{14}[(AuSR)_4]_6$.[31] The gold atoms in the cluster were found to be in two distinct chemical states, the ones in the Au_{14} core were essentially neutral ("metallic") and the ones in the ligand shell positively charged ("oxidized"). Energy comparisons to earlier models showed that, in general, the structure of thiolate-protected clusters involves competition between maximizing the metal–metal Au–Au cohesion and the optimal arrangement of the (strong) surface-covalent Au–S bonds.

The "divide-and-protect" structural scheme (gold in two distinct chemical states, the protecting units involve gold and not only thiolates) introduced by Häkkinen and collaborators first as a model for $Au_{38}(SR)_{24}$ in 2006 received much credibility in 2007 when the first crystal structure of any Au:SR cluster was solved by Kornberg and collaborators.[29] The solved structure of $Au_{102}(p\text{-}MBA)_{44}$ showed a metal core of Au_{79} and a ligand layer that consisted of 19 short $Au(p\text{-}MBA)_2$ and 2 long $Au_2(p\text{-}MBA)_3$ gold–thiolate units, i.e., the composition could be written as $Au_{79}[Au(p\text{-}MBA)_2]_{19}[Au_2(p\text{-}MBA)_3]_2$. Soon thereafter, the crystal structure of $Au_{25}(SR)_{18}^{-}$ ($R = (CH_2)_2Ph$) was solved showing a composition of $Au_{13}[Au_2(SR)_3]_6$.[48,49] These discoveries led to a significant increase in theoretical activity and reexamination of $Au_{38}(SCH_3)_{24}$ in the spirit of the "divide-and-protect" scheme, considering either disordered or symmetric cores surrounded by varying number of the short and long Au:SR units.[37,50] Structural principles based on numbers of short and long units, core size, and number of sulfur anchors could now be made, and it was proposed that $Au_{38}(SCH_3)_{24}$ may consist of an fcc, bi-icosahedral, or bicapped Au_{25} core with nine short and two long motifs. An interesting structural model considered a face-shared bi-icosahedral Au_{23} core surrounded by three short and six long motifs (written as $Au_{23}@[Au(SR)_2]_3[Au_2(SR)_3]_6$) which was found to be substantially lower in energy than all the other structures in the literature at the time. This model also yielded an electrochemical gap that agreed with the experimental one. However, the significant stability of the $Au_{38}(SR)_{24}$ nanocluster, its relation to the electronic structure of the gold core, and its strong chiroptical response remained to be explained.

The final solution to the structure of $Au_{38}(SR)_{24}$ came in 2010 from simultaneous but independent DFT[51] and X-ray crystallography[52] studies. The DFT study considered the earlier proposed $Au_{23}@[Au(SR)_2]_3[Au_2(SR)_3]_6$ structure with a face-sharing bi-icosahedral Au_{23} core.[53] A new low-energy structure was found in that composition where the protecting Au:SR units were in a *chiral arrangement around the core* (Figure 12). This yielded an excellent match between the computed (for $R = C_6H_{13}$) and measured (for $R = C_{12}H_{25}$) powder X-ray

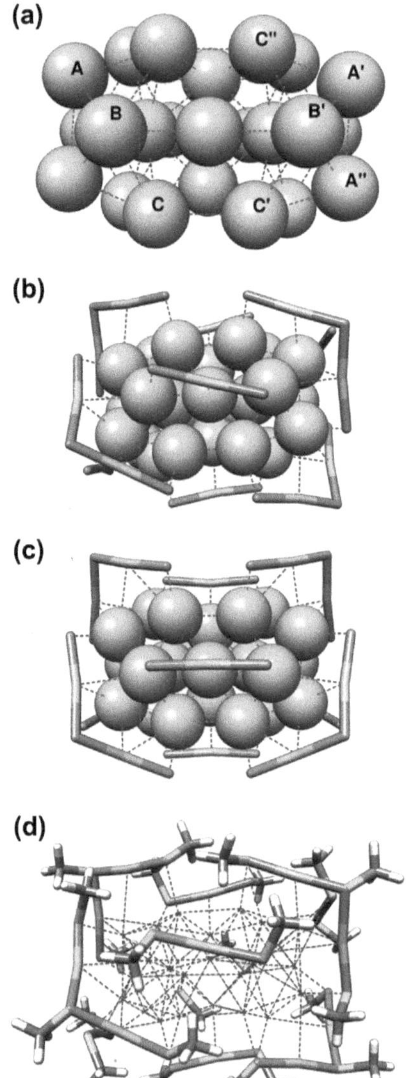

FIGURE 12 (a) Au_{23} core of $Au_{38}(SR)_{24}$. (b) Chiral D_3 arrangement of the Au–S framework units. (c) Nonchiral C_{3h} form of the cluster. (d) Optimized structure or the chiral clusters. *Reproduced with permission from Ref. 51. Copyright American Chemical Society 2010.*

diffraction function. The simultaneous single-crystal X-ray crystallography from the Jin group gave an identical total structure for R = $(CH_2)_2Ph$.[52]

The 8 kDa (Au_{38}) cluster, stabilized by glutathione (SG), was reported in 2000 to yield a very strong optical response in the visible range.[35] This pioneering observation led to several years of literature debate on the possible mechanisms of the optical activity. Glutathione is a chiral molecule by itself but it absorbs in the UV region, so its impact on the optical response of the cluster in the visible region remained a mystery. Two main mechanisms were proposed: (1) the chiral thiol induces a chiral atomic structure of the gold core or (2) the chiral thiol imposes a "chiral footprint" in the electronic structure of the gold, visible also in the low-energy excitations.[54–56] The predicted and observed structure of $Au_{38}(SR)_{24}$[51,52] solved this mystery elegantly. The Au_{23} core of the $Au_{38}(SR)_{24}$ is *achiral* having a D_{3h} idealized symmetry, however, the *chiral* arrangement of the six $Au_2(SR)_3$ and three $Au(SR)_2$ units in the ligand layer lowers the symmetry of the Au–S framework in the whole cluster to D_3. Since the gold–thiolate units contribute to the optical transitions in the visible region, this arrangement also creates the CD signals.

It is also worth noting that the Au(core)–RS–Au(unit)–SR–Au(core) structure of the gold–thiolate units makes each sulfur atom a chiral center, since the bonding environment for S has four nonequivalent directions: (metal) Au in the core, sulfur lone pair, R unit, and the Au(I) atom inside the unit. Thus it can be expected that a "chiral footprint" will be left on all transitions where the sulfur orbitals contribute.

The DFT calculation in 2010[51] predicted that an achiral ligand (such as an alkanethiolate) can create a *chiral* response of the nanocluster due to its binding configuration with gold, even when the metal core is achiral. This prediction was verified when the first enantiomeric separation of $Au_{38}(SR)_{24}$ using the achiral $S(CH_2)_2Ph$ ligand was achieved in 2012.[57] Figure 13 shows that the calculated CD spectrum of $Au_{38}(SCH_3)_{24}$ in the experimental structure matches almost perfectly with the measured one for $Au_{38}(S(CH_2)_2Ph)_{24}$.

Theoretical studies of the electronic structure of $Au_{38}(SR)_{24}$ gave important insight into its role in the stability.[51] The count of free (Au(6s)-derived) electron states is 14 by applying the simple "superatom electron counting rule" from Eqn (9) and Ref. 30. Obviously 14 electrons do not make a shell closing for a spherical electron system since that electron number corresponds to mid-shell 1D filling. However, early studies using simple models already showed that strong nonspherical Jahn–Teller deformations (oblate or prolate) can stabilize a significant HOMO-LUMO gap for 14 electron systems.[13] This stabilization was directly demonstrated in 1995 in atomistic DFT molecular dynamics simulations of a bare Na_{14} cluster where the Na atoms were described with pseudopotentials. Both oblate and prolate shapes, and their dynamical interconversion, were observed in a timescale of 10 ps at elevated temperatures.[58] The electronic structure of $Au_{38}(SR)_{24}$ was interpreted in Ref. 51 by using two complementary models for independent electrons in a

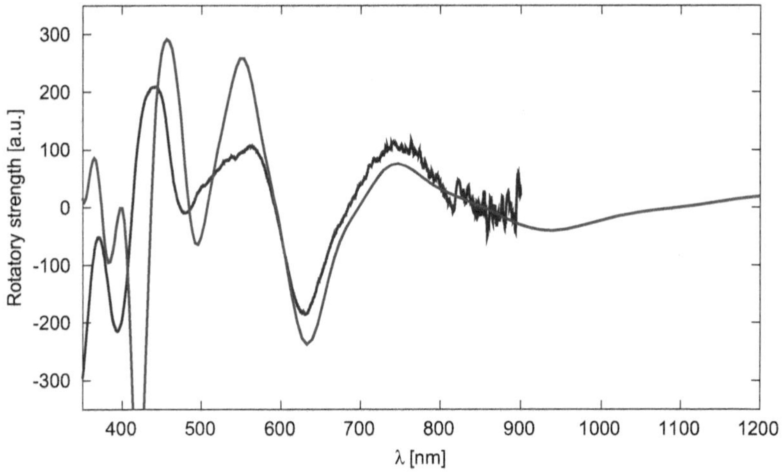

FIGURE 13 The calculated CD spectrum of $Au_{38}(SCH_3)_{24}$ (green (gray in print versions)) from linear response time-dependent density functional theory by using grid-based projector-augmented wave method as compared to the measured CD signal (blue (black in print versions)) from an enantiopure solution of $Au_{38}(PET)_{24}$ (Ref. 57). *Reproduced with permission from Ref. 83. Copyright American Chemical Society 2012.*

model potential: the "particle-in-a-cylinder" model and an anisotropic harmonic oscillator with an elongated shape. These models were motivated by a rod-like shape of the Au_{23} core. Both models reproduced the observed symmetries and energy ordering of the few highest orbitals close to the HOMO level in the atomistic DFT calculation. Opening of the HOMO-LUMO gap was evident in these models by using distortion parameters that corresponded to the atomistic dimensions of the Au_{23} core. A refinement of the anisotropic oscillator model would consider the long axis of the cluster to have two potential minima. This would correspond to an electronic "super-dimer." In that model the valence 14 is thought to be a dimer of 2×7 electrons, i.e., the cluster is analogous to a halogen dimer such as Cl_2.

8.3.4 A Brief Overview of Known MPC Structures

Table 1 lists 70 known MPCs, most of which have been structurally characterized, so that the size and the shape of the metal core, and the composition of the ligand layer are known exactly. The table contains not only MPCs made out of gold metal and thiolate ligands, but also silver, copper, aluminum, and gallium metals, and phosphine, iron carbonyl, halide, phenyl acetylene, trimethyl silane, and pentamethylcyclopentadienyl as ligands. The compounds are shown in the ascending order of the free-electron count (Eqn (9)), that is, showing increasing tendency to "metallicity." The original references to the compounds are given in Table 1 and not repeated here.

TABLE 1 A Representative Sample of Compositionally Known metal MPCs. In Most Cases, Also the Total Structure Is Known as Referenced. n_e is the Free-Electron Count from Eqn (9). For "Method," XRD = Single Crystal X-ray Diffraction, MS = Mass Spectrometry, EM = Electron Microscopy and TM = Theoretical Model

#	n_e		Method	References/ note
1	0	$[Ag_{123}S_{35}(StBu)_{50}]^{3+}$	XRD	59
2	0	$Ag_{154}Se_{77}(dppxy)_{18}$	XRD	60
3	0	$Ag_{188}S_{94}(PR_3)_{30}$	XRD	61
4	0	$Ag_{320}(StBu)_{60}S_{130}(dppp)_{12}$	XRD	60
5	0	$Ag_{344}S_{124}(StBu)_{96}$	XRD	59
6	0	$Ag_{352}S_{128}(StC_5H_{11})_{96}$	XRD	60
7	0	$Ag_{490}S_{188}(StC_5H_{11})_{114}$	XRD	60
8	0	$\{Au_{28}[Fe(CO)_3]_4\ [Fe(CO)_4]_{10}\}^{8-}$	XRD	62
9	2	$Ag_{14}(SR)_{12}(PR)_8$	XRD	63
10	2	$Ag_{16}(SR)_{14}(dppe)_4$	XRD	64
11	2	$\{Au_{34}[Fe(CO)_3]_6\ [Fe(CO)_4]_8\}^{8-}$	XRD	62
12	4	$[Au_7(dppp)_4]^{3+}$	XRD	65
13	4	$[Au_8(dppp)_4Cl_2]^{2+}$	XRD	66
14	4	$Au_{18}(SR)_{14}$	XRD	67,68
15	4	$\{Au_{22}[Fe(CO)_4]_{12}\}^{6-}$	XRD	62
16	4	$Ag_{70}S_{16}(SPh)_{34}(PhCO_2)_4(triphos)_4$	XRD	61
17	6	$\{Au_{21}[Fe(CO)_4]_{10}\}^{5-}$	XRD	62
18	6	$[Au_8(dppp)_4]^{2+}$	XRD	66
19	7	$Au_{25}(PET)_{18}$	XRD	69
20	8	$Au_{11}X_3(PR_3)_7$	XRD	70[a]
21	8	$[Au_{13}Cl_2(PR_3)_{10}]^{3+}$	XRD	71
22	8	$[Au_{25}(PET)_{18}]^-$	XRD	48,49
23	8	$Au_{28}(SR)_{20}$	XRD	72
24	8	$Al_4(C_5Me_5)_4$	XRD	73
25	8	$Al_4(SiC(CH_3)_3)_4$	XRD	74
26	8	$[Au_{13}Cu_2(PR)_6(SPy)_6]^+$	XRD	75

Continued

TABLE 1 A Representative Sample of Compositionally Known metal MPCs. In Most Cases, Also the Total Structure Is Known as Referenced. n_e is the Free-Electron Count from Eqn (9). For "Method," XRD = Single Crystal X-ray Diffraction, MS = Mass Spectrometry, EM = Electron Microscopy and TM = Theoretical Model—cont'd

#	n_e		Method	References/ note
27	8	$[Au_{13}Cu_4(PR_2Py)_4(SR)_8]^+$	XRD	75
28	8	$[Au_{13}Cu_8(SPy)_{12}]^+$	XRD	75
29	8	$[Au_{11}(dpdp)_6]^{3+}$	XRD	76
30	8	$[Ag_{21}(S_2P(O/Pr)_2)_{12}]^+$	XRD	77
31	10	$[Ag_{32}(SR)_{24}(dppe)_5]^{2-}$	XRD	64
32	12	$Au_{30}S(SR)_{18}$	XRD	78,79
33	12	$Au_{36}(SR)_{24}$	XRD	80
34	12	$Au_{30}(SR)_{18}$	MS, TM	78
35	12	$[Au_{23}(C_2Ph)_9(Ph_3P)_6]^{2+}$	XRD	81
36	14	$Au_{38}(SR)_{24}$	XRD	52[b]
37	14	$Au_{38}(SR)_{24}$	XRD	82[b]
38	16	$Au_{40}(SR)_{24}$	MS, TM	83
39	16	$[Au_{24}(PR_3)_{10}(SR)_5Cl_2]^+$	XRD	84
40	16	$[Au_{25}(PR_3)_{10}(SR)_5Cl_2]^{2+}$	XRD	85
41	16	$[Au_{13}Ag_{12}(PR_3)_{10}Br_8]^+$	XRD	86
42	16	$Au_{20}(PP_3)_4Cl_4$	XRD	87
43	16	$Au_{20}(PP_3)_4Cl_4$	XRD	88
44	18	$[Ag_{44}(SR)_{30}]^{4-}$	XRD	89
45	18	$[Ag_{44}(SR)_{30}]^{4-}$	XRD	90
46	18	$[Au_{12}Ag_{32}(SR)_{30}]^{4-}$	XRD	89
47	18	$[Au_{12+n}Cu_{32}(SR)_{30+n}]^{4-}$ ($n = 0,2,4,6$)	XRD	91
48	18	$Au_{24}Ag_{20}(SR)_4(C_2Ph)_{20}Cl_2$	XRD	92
49	22	$Au_{22}(L^8)_6$	XRD	93
50	24	$Au_{18}Ag_{20}(PR_3)_{12}Cl_{14}$	XRD	94
51	34	$[Au_{39}(PR_3)_{14}Cl_6]^-$	XRD	95
52	34	$Au_{68}(SR)_{34}$	MS	96

TABLE 1 A Representative Sample of Compositionally Known metal MPCs. In Most Cases, Also the Total Structure Is Known as Referenced. n_e is the Free-Electron Count from Eqn (9). For "Method," XRD = Single Crystal X-ray Diffraction, MS = Mass Spectrometry, EM = Electron Microscopy and TM = Theoretical Model—cont'd

#	n_e		Method	References/ note
53	34	$[Au_{67}(SR)_{35}]^{2-}$	MS, TM	97
54	36	$Au_{68}(SR)_{32}$	EM	98
55	40	$\{Ge_9[Si(SiMe_3)_3]_3\}^-$	XRD	99
56	40	$SiAl_{14}(C_5Me_5)_6$	XRD	100
57	58	$Au_{102}(p\text{-}MBA)_{44}$	XRD	29
58	58	$Ga_{23}(N(SiMe_3)_2)_{11}$	XRD	101
59	58/59	$Au_{104}(SR)_{45}$	MS, TM	102,103
60	70	$[Ag_{136}(SR)_{65}Br_2]^-$	XRD	104
61	80	$Au_{130}(SC_{12}H_{25})_{50}$	MS, TM	105
62	80/81	$Au_{133}(SPhtBu)_{52}$	XRD	106,107
63	84	$Au_{144}(SR)_{60}$	MS, TM	108,109
64	84	$Au_{144}(SR)_{60}$	EM	110
65	84	$Ag_xAu_{144-x}(SR)_{60}$	MS, TM	111,112
66	119	$Au_{187}(SC_{12}H_{25})_{68}$	MS	105
67	138	$Al_{50}(C_5Me_5)_{12}$	XRD	113
68	~220	$Au_{314}(SR)_{96}$, ~Au_{330}	MS, TM	114–116
69	245	$Au_{329}(SR)_{84}$	MS	103
70	257	$Ag_{374}(SR)_{115}Br_2$	XRD	104

[a]*X = halide, SCN, SR.*
[b]*These two clusters have the same composition but a totally different core structure.*

First, we note that a large number of metal atoms in an MPC compound do not necessarily make the system "metallic." Compounds **1** to **7** have *zero* free electrons even though the largest of them has 490 silver atoms! In fact, they are nanoscale clusters of bulk *silver-sulfide*, stabilized by the surface ligands (thiolates and/or phosphines). Compound **8** belongs to the series of iron-carbonyl-protected gold clusters that all have a very low "metallicity," due to the strongly electron-localizing nature of the ligand ($Fe(CO)_4$ and $Fe(CO)_3$

localize two and four electrons, respectively). The other MPCs in this series are compounds **11**, **15**, and **17**.

For two-electron systems, clusters **9** and **10** were the first ones where a crystallographic characterization of thiolate–phosphine-stabilized silver clusters succeeded recently (2013).

Clusters **12** to **18** represent a diverse group of gold and silver clusters with four or six free electrons. The metal core of these clusters is strongly nonspherical.

There is a wide variety of different 8-electron clusters (**20–30**), where the metal core is generally rather spherical in shape. The metal core is made of gold, intermetallic gold–copper, silver, or aluminum. Of historical note, compound **20** is the famous "undecagold" cluster that was discovered as early as 1969. 10- and 12-electron clusters **31** to **35** have (again) nonspherical metal cores.

Clusters **36** and **37** form an important pair of clusters with an exactly same number of metal (gold) and ligands (thiolate) but drastically different shapes and structure of the metal core. Actually, this is the *first realization of different isomers of* $Au_{38}(SR)_{24}$—although the thiolate is not the same. This is also the first demonstration of the fact that the chemical nature of the thiolate dictates the morphology of the metal core. Both are still 14-electron systems where the electronic structure is stabilized by a strong nonspherical deformation of the metal core.

Clusters **38–43** represent 16-electron systems. Sixteen is an interesting electron numbers since $8 + 8 = 16$, i.e., it is possible to form a dimer of two 8-electron superatoms. It is very interesting to note that clusters **39–41** are exactly like that as confirmed directly from the X-ray total structure and theoretical analysis. Cluster **38** is proposed to be like that, although the crystal structure is still missing. Clusters **42** and **43** are the same compound from two independent studies, and feature an interesting chiral gold core Au_{20}. A recent DFT analysis[117] showed that the $1S^2$ $1P^6$ $1D^4$ $2S^2$ $1D^2$ and the 1D shell is split by the deformation of the core.

Eighteen is a special electron number since it can stabilize a large HOMO-LUMO gap for a *hollow metal cluster* with a superatom configuration $1S^2$ $1P^6$ $1D^{10}$. Clusters **44–48** have just that!

Cluster **50**, as a 24-electron system, is an 8-electron superatom *trimer*.

Clusters **51–53** represent 34-electron systems. Cluster **51** was the largest structurally solved gold cluster until 2007. The total structure of **52** and **53** remain unsolved.

Cluster **54** is a special case so far. The 3D atomic structure of its gold core was recently solved by a low-dose electron microscopy technique.[98] It has a low symmetry core, and its total composition was determined by a combination of mass spectrometry, infrared spectroscopy, and DFT calculations.

Clusters **55** and **56** are interesting 40-electron systems offering examples of MPCs made out of elements in main groups 3 and 4. According to a recent DFT analysis, the ligand layer in **55** *donates* electrons to the metal core to achieve the 40-electron shell closing.[118]

Cluster **57** is the famous $Au_{102}(p\text{-MBA})_{44}$ compound discussed in detail in Section 8.3.2. Another 58-electron system, compound **58**, is known as well, formed by a totally different chemistry. Cluster **59** is believed to be the "organo-soluble counterpart" of the water-soluble **58**, with a very similar total atomic structure.

Clusters **60** and **70** are currently the largest all-silver MPCs with a resolved total structure. Cluster **62** is currently the largest all-gold MPC with a known total structure.

Cluster **67** is currently the largest metal MPC that shows a very clear superatom shell structure and a shell closing at a high electron number (138 electrons), as analyzed from DFT.[119]

Clusters **63–65** belong to the "metal-144 family" where the total structure determination has been elusive despite years of effort from several groups. The cluster is believed to be stabilized by *atomic shell effects*, namely a symmetric total structure having an icosahedral-based motif of the metal core.[109]

8.4 SUMMARY AND PROSPECTS

The field of monolayer-protected metal clusters is currently in a very rapid discovery phase. New experimental and theoretical groups are coming into the field, and new efforts from groups from increasingly diverse backgrounds have fertilized the research greatly. My personal opinion is that just like in the field of gas-phase bare metal clusters, the number of possible metastable metal MPC compounds is virtually infinite when one takes into account the possibility to mix various metals and various ligands. One just has to explore more diverse chemistries than what has been the tradition in the field. Table 1 lists 70 known compounds, but new clusters and new structures are currently discovered with such a pace that most likely several new ones have appeared during the time between finishing this chapter (late spring 2015) and when this book is out of print. Still, the currently known compounds are just a tip of the iceberg. A huge "sea" of possible compounds and structures is waiting to be discovered.

All nanoparticles are metastable, i.e., the thermodynamic ground state consists of the constituents in their bulk phases (metals in their bulk crystal lattice, ligand molecules in their stable, nonactivated forms). Since long-term degradation and coalescence is a real problem, it is an obvious challenge for applications to find the suitable combinations of metals and ligands that produce the optimal MPC stability. It is known for a long time that gold–thiolate chemistry has often this "magic touch"; some of the most stable compounds (such as the $Au_{102}(p\text{-MBA})_{44}$ material in dry form in our laboratory) stay stable in ambient conditions for many years. It is very conceivable that other combinations of metals and ligands can be found that will produce similar superior stability. In this regard, the recent developments for achieving new compounds consisting of mixed noble metal core and other ligands than thiolates seem promising. Very exciting developments are currently taking place to discover the interface between carbon and gold, i.e., without the thiol (sulfur) linkage.

Metal nanoparticles are probably most used in the huge field of catalysis. The potential impact of atomically precise nanometer-size metal MPCs is still very unclear in this regard. In principle, the ligands offer a great variety to control the electronic structure and reactivity of the metal core, but the ligand layer has to also allow for binding of the reactant molecules at the active sites, and the "traffic" of the reactant and product molecules in and out from the system. An alternative route is to use the precise MPCs as starting materials for making very monodisperse nanoparticle model catalysts, but then the ligand layer is usually removed in the process of the catalyst activation.

Perhaps still the most unexplored, but potentially the most promising area of MPC applications relates to the use of atomically precise MPCs in biological environments. Colloidal noble metal nanoparticles have been used for decades, e.g., as markers in electron microscopy[120] and more recently in photothermal therapy applications.[121] Atomically precise 1−2 nm MPCs could bring much more selectivity and control for the binding, as demonstrated recently with functionalized $Au_{102}(p\text{-MBA})_{44}$ clusters that bind selectively on enterovirus capsids, helping to image virus structures.[122] One can envisage using such functionalized MPCs as biocompatible "intelligent" sensors and/or therapy agents.

As the brief discussion in this chapter shows, the role of theoretical work and large-scale computations has been and will be important in the MPC field. Much can be learned from theoretical concepts, particularly from the theory of electronic shell structure and the related superatom concept. Many analogs can be found to similar concepts in bare gas-phase metal clusters, once the effect of the ligand shell is taken into account. Large-scale DFT calculations have been extremely useful for studies of interplay between the atomic structure, electronic structure, and stability of MPCs. In a few cases, DFT calculations have even successfully predicted essentially the correct total structure (the Au−S framework of $Au_{25}(SR)_{18}$ and the full structure of the chiral $Au_{38}(SR)_{24}$). As the computational power increases rapidly, the predictive power of DFT calculations will certainly be applied to "designing" MPCs with desired properties, e.g., via magnetic doping. In summary, the future of the MPC field appears exciting for many years to come.

ACKNOWLEDGMENTS

I am indebted to my excellent students and postdocs in Jyväskylä since 2003 and to my many local and foreign collaborators. Financial support for my research has been provided by the Academy of Finland and the University of Jyväskylä. Supercomputer resources have been provided by CSC—the Finnish IT Center for Science and the EU Partnership for Advanced Computing in Europe (PRACE) and Distributed European Infrastructure for Supercomputing Applications (DEISA) infrastructures.

REFERENCES

1. Sardart R, Funston AM, Mulvaney P, Murray RW. Gold nanoparticles: past, present, and future. *Langmuir* 2009;**25**:13840–51.
2. Jin R. Quantum sized, thiolate-protected gold nanoclusters. *Nanoscale* 2010;**2**:343–62.
3. Tsukuda T. Toward an atomic-level understanding of size-specific properties of protected and stabilized gold clusters. *Bull Chem Soc Jpn* 2012;**85**:151–68.
4. Negishi Y. Toward the creation of functionalized metal nanoclusters and highly active photocatalytic materials using thiolate-protected magic gold clusters. *Bull Chem Soc Jpn* 2014;**87**:375–89.
5. Aikens CM. Electronic structure of ligand-passivated gold and silver nanoclusters. *J Phys Chem Lett* 2011;**2**:99–104.
6. Häkkinen H. Atomic and electronic structure of gold clusters: understanding flakes, cages and superatoms from simple concepts. *Chem Soc Rev* 2008;**37**:1847–59.
7. Häkkinen H. The gold–sulfur interface at the nanoscale. *Nat Chem* 2012;**4**:443–55.
8. Häkkinen H. Theoretical studies of gold nanoclusters in various chemical environments: when the size matters. Chapter 9. In: Louis C, Pluchery O, editors. *Gold nanoparticles for physics, biology and chemistry*. Imperial College Press; 2012.
9. Häkkinen H. Ligand-protected gold nanoclusters as superatoms – insights from theory and computations. Chapter 3. In: Johnston RL, Wilcoxon J, editors. *Metal nanoparticles and nanoalloys. Frontiers of nanoscience*. Elsevier; 2012.
10. Fernando A, Dimuthu KL, Weerawardene W, Karimova NV, Aikens CM. Quantum mechanical studies of large metal, metal oxide, and metal chalcogenide nanoparticles and clusters. *Chem Rev* 2015. http://dx.doi.org/10.1021/cr500506r.
11. http://www.nobelprize.org/nobel_prizes/physics/laureates/1963/.
12. Martin TP, Bergmann T, Görlich H, Lange T. Shell structure of clusters. *J Phys Chem* 1991;**95**:6421–9.
13. deHeer WA. The physics of simple metal clusters: experimental aspects and simple models. *Rev Mod Phys* 1993;**65**:611–76.
14. Brack M. The physics of simple metal clusters: self-consistent jellium model and semiclassical approaches. *Rev Mod Phys* 1993;**65**:677–732.
15. Nilsson SG. Binding states of individual nucleons in strongly deformed nuclei. *Dan Mat Fys Medd* 1955;**29**:16.
16. Clemenger K. Ellipsoidal shell structure in free-electron metal clusters. *Phys Rev B* 1985;**32**:1359–62.
17. Knight WD, Clemenger K, deHeer WA, Saunders WA, Chou MY, Cohen ML. Electronic shell structure and abundances of sodium clusters. *Phys Rev Lett* 1984;**52**:2141–3.
18. Mortensen JJ, Hansen LB, Jacobsen KW. Real-space grid implementation of the projector augmented wave method. *Phys Rev B* 2005;**71**:035109. See also: https://wiki.fysik.dtu.dk/gpaw.
19. Kaappa S. (Häkkinen group, project report in 2014, unpublished).
20. Selenius E. (Häkkinen group, project report in 2014, unpublished).
21. Koskinen M, Lipas PO, Manninen M. Electron-gas clusters: ultimate jellium model. *Z Phys D* 1995;**35**:285.
22. Fleischhacker W, Schönfeld T, editors. *Pioneering ideas for the physical and chemical sciences*. Springer; 1997.
23. Watanabe H, Inoshita T. Superatom: a novel concept in materials science. *Optoelectronics* 1986;**1**:33–40.

24. Koutecky J, Fantucci P. General principles governing structures of small clusters. *Z Phys D* 1986;**3**:147.
25. Jena P. Beyond the periodic table of elements: the role of superatoms. *J Phys Chem Lett* 2013;**4**:1432–42.
26. Leuchtner RE, Harms AC, Castleman Jr AW. Thermal metal cluster anion reactions: behavior of aluminum clusters with oxygen. *J Chem Phys* 1989;**91**:2753–4.
27. Khanna SN, Jena P. Assembling crystals from clusters. *Phys Rev Lett* 1992;**69**:1664–7.
28. Häkkinen H, Manninen M. How "magic" is a magic metal cluster? *Phys Rev Lett* 1996;**76**:1599–602.
29. Jadzinsky PD, Calermo G, Ackerson CJ, Bushnell DA, Kornberg RD. Structure of a thiol monolayer-protected gold nanoparticle at 1.1 Å resolution. *Science* 2007;**318**:430–3.
30. Walter M, Akola J, Lopez-Acevedo O, Jadzinsky PD, Calero G, Ackerson CJ, et al. A unified view of ligand-protected gold clusters as superatom complexes. *Proc Natl Acad Sci USA* 2008;**105**:9157–62.
31. Häkkinen H, Walter M, Grönbeck H. Divide and protect: capping gold nanoclusters with molecular gold–thiolate rings. *J Phys Chem B* 2006;**110**:9927–31.
32. Hulkko E, Lopez-Acevedo O, Koivisto J, Levi-Kalisman Y, Kornberg RD, Pettersson M, et al. Electronic and vibrational signatures of the $Au_{102}(p\text{-}MBA)_{44}$ cluster. *J Am Chem Soc* 2011;**133**:3752–5.
33. Mustalahti S, Myllyperkiö P, Malola S, Lahtinen T, Salorinne K, Koivisto J, et al. Molecule-like photodynamics of $Au_{102}(p\text{-}MBA)_{44}$ nanocluster. *ACS Nano* 2015;**9**:2328–35.
34. Schaaff TG, Shafigullin MN, Khoury JT, Vezmar I, Whetten RL, Cullen WG, et al. Isolation of smaller nanocrystal Au molecules: robust quantum effects in optical spectra. *J Phys Chem B* 1997;**101**:7885–91.
35. Schaaff TG, Whetten RL. Giant gold–glutathione cluster compounds: intense optical activity in metal-based transitions. *J Phys Chem B* 2000;**104**:2630–41.
36. Parker JF, Fields-Zinna CA, Murray RW. The story of a monodisperse gold nanoparticle: $Au_{25}L_{18}$. *Acc Chem Res* 2010;**43**:1289–96.
37. Chaki NK, Negishi Y, Tsunoyama H, Shichibu Y, Tsukuda T. Ubiquitous 8 and 29 kDa gold:alkanethiolate cluster compounds: mass-spectrometric determination of molecular formulas and structural implications. *J Am Chem Soc* 2008;**130**:8608–10.
38. Quinn BM, Liljeroth P, Ruiz V, Laaksonen T, Kontturi K. Electrochemical resolution of 15 oxidation states for monolayer protected gold nanoparticles. *J Am Chem Soc* 2003;**125**:6644–5.
39. Toikkanen O, Ruiz V, Rönnholm G, Kalkkinen N, Liljeroth P, Quinn BM. Synthesis and stability of monolayer-protected Au_{38} clusters. *J Am Chem Soc* 2008;**130**:11049–55.
40. Toikkanen O, Carlsson S, Dass A, Rönnholm G, Kalkkinen N, Quinn BM. Solvent-dependent stability of monolayer-protected Au_{38} clusters. *J Phys Chem Lett* 2010;**1**:32–7.
41. Qian H, Zhu Y, Jin R. Size-focusing synthesis, optical and electrochemical properties of monodisperse $Au_{38}(SC_2H_4Ph)_{24}$ nanoclusters. *ACS Nano* 2009;**3**:3795–803.
42. Wyrwas RB, Alvarez MM, Khoury JT, Price RC, Schaaff TG, Whetten RL. The colours of nanometric gold: optical response functions of selected gold-cluster thiolates. *Eur Phys J D* 2007;**43**:91–5.
43. Häkkinen H, Barnett RN, Landman U. Electronic structure of passivated $Au_{38}(SCH_3)_{24}$ nanocrystal. *Phys Rev Lett* 1999;**82**:3264–7.
44. Andreoni W, Curioni A, Grönbeck H. Density functional theory approach to thiols and disulfides on gold: Au(111) surface and clusters. *Int J Quantum Chem* 2000;**80**:598–608.

45. Garzón IL, Beltran MR, Posada-Amarillas A, Ordejon P, Artacho E, Sanchez-Portal D, et al. Lowest energy structures of gold nanoclusters. *Phys Rev Lett* 1998;**81**:1600–3.
46. Garzón IL, Michaelian K, Beltran MR, Posada-Amarillas A, Ordejon P, Artacho E, et al. Structure and stability of gold nanoclusters: the Au_{38} case. *Eur Phys J D* 1999;**9**:211–5.
47. Garzón IL, Roviera C, Michaelian K, Beltran MR, Ordejon P, Junquera J, et al. Do thiols merely passivate gold nanoclusters? *Phys Rev Lett* 2000;**85**:5250–1.
48. Heaven MW, Dass A, White PS, Holt KM, Murray RW. Crystal structure of the gold nanoparticle $[N(C_8H_{17})_4][Au_{25}(SCH_2CH_2Ph)_{18}]$. *J Am Chem Soc* 2008;**130**:3754–5.
49. Zhu M, Aikens CM, Hollander FJ, Schatz GC, Jin R. Correlating the crystal structure of a thiol-protected Au_{25} cluster and optical properties. *J Am Chem Soc* 2008;**130**:5883–5.
50. a. Jiang D, Tiago M, Luo W, Dai S. The "staple" motif: a key to stability of thiolate-protected gold nanoclusters. *J Am Chem Soc* 2008;**130**:2777–9;
b. Jiang D, Luo W, Tiago ML, Dai S. In search of a structural model for a thiolate-protected Au_{38} cluster. *J Phys Chem C* 2008;**112**:13905–10.
51. Lopez-Acevedo O, Tsunoyama H, Tsukuda T, Häkkinen H, Aikens CM. Chirality and electronic structure of the thiolate-protected Au_{38} nanocluster. *J Am Chem Soc* 2010;**132**:8210–8.
52. Qian H, Eckenhoff WT, Zhu Y, Pintauer T, Jin R. Total structure determination of thiolate-protected Au_{38} nanoparticles. *J Am Chem Soc* 2010;**132**:8280–1.
53. Pei Y, Gao Y, Zeng XC. Structural prediction of thiolate-protected Au_{38}: a face-fused bi-icosahedral Au core. *J Am Chem Soc* 2008;**130**:7830–2.
54. Gautier C, Bürgi T. Chiral gold nanoparticles. *ChemPhysChem* 2009;**10**:483–92.
55. Noguez C, Garzón IL. Optically active metal nanoparticles. *Chem Soc Rev* 2009;**38**:757–71.
56. Knoppe S, Bürgi T. Chirality in thiolate-protected gold clusters. *Acc Chem Res* 2014;**4**:1318–26.
57. Dolamic I, Knoppe S, Dass A, Buergi T. First enantioseparation and circular dichroism spectra of Au_{38} clusters protected with achiral ligands. *Nat Commun* 2012;**3**:798.
58. Häkkinen H, Manninen M. Electronic-structure-induced deformations of liquid metal clusters. *Phys Rev B* 1995;**52**:1540–3.
59. Fenske D, Anson CE, Eichhöfer A, Fuhr O, Ingendoh A, Perseu C, et al. Syntheses and crystal structures of $[Ag_{123}S_{35}(StBu)_{50}]$ and $[Ag_{344}S_{124}(StBu)_{96}]$. *Angew Chem Int Ed* 2005;**44**:5242–6.
60. Anson CE, Eichhöfer A, Issac I, Fenske D, Fuhr O, Sevillano P, et al. Synthesis and crystal structures of the ligand-stabilized silver chalcogenide clusters $[Ag_{154}Se_{77}(dppxy)_{18}]$, $[Ag_{320}(StBu)_{60}S_{130}(dppp)_{12}]$, $[Ag_{352}S_{128}(StC_5H_{11})_{96}]$, and $[Ag_{490}S_{188}(StC_5H_{11})_{114}]$. *Angew Chem Int Ed* 2008;**47**:1326–31.
61. Wang XJ, Langetepe T, Persay C, Kang BS, Sheldrick GM, Fenske D. Syntheses and crystal structures of the new Ag–S clusters $[Ag_{70}S_{16}(SPh)_{34}(PhCO_2)_4(triphos)_4]$ and $[Ag_{188}S_{94}(PR_3)_{30}]$. *Angew Chem Int Ed* 2002;**41**:3818–22.
62. Femoni C, Iapauzzi M, Longoni G, Tiozzo C, Zacchini S. An organometallic approach to gold nanoparticles: synthesis and X-ray structure of CO-protected $Au_{21}Fe_{10}$, $Au_{22}Fe_{12}$, $Au_{28}Fe_{14}$, and $Au_{34}Fe_{14}$ clusters. *Angew Chem Int Ed* 2008;**47**:6666–9.
63. Yang H, Lei J, Wu B, Wang Y, Zhou M, Xia A, et al. Crystal structure of a luminescent thiolated Ag nanocluster with an octahedral Ag_6^{4+} core. *Chem Commun* 2013;**49**:300–2.
64. Yang H, Wang Y, Zheng N. Stabilizing subnanometer Ag(0) nanoclusters by thiolate and diphosphine ligands and their crystal structures. *Nanoscale* 2013;**5**:2674–7.
65. Shichibu Y, Zhang MZ, Kamei Y, Konishi K. $[Au_7]^{3+}$: a missing link in the four-electron gold cluster family. *J Am Chem Soc* 2014;**136**:12892–5.

66. Kamei Y, Schichibu Y, Konsihi K. Generation of small gold clusters with unique geometries through cluster-to-cluster transformations: octanuclear clusters with edge-sharing gold tetrahedron motifs. *Angew Chem Int Ed* 2011;**50**:7442–5.
67. Das A, Liu C, Byun HY, Nobusada K, Zhao S, Rosi N, et al. Structure determination of $[Au_{18}(SR)_{14}]$. *Angew Chem Int Ed* 2015;**54**:3140–4.
68. Chen S, Wang SX, Zhong J, Song YB, Zhang J, Sheng HT, et al. The structure and optical properties of the $[Au_{18}(SR)_{14}]$ nanocluster. *Angew Chem Int Ed* 2015;**54**:3145–9.
69. Zhu MZ, Eckenhoff WT, Pintauer T, Rin JC. Conversion of anionic $[Au_{25}(SCH_2CH_2Ph)_{18}]^-$ cluster to charge neutral cluster via air oxidation. *J Phys Chem C* 2008;**112**:14221–4.
70. McPartlin M, Mason R, Malatesta L. Novel cluster complexes of gold(0)–gold(I). *J Chem Soc D* 1969:334.
71. Briant CE, Theobald BRC, White JW, Bell LK, Mingos DMP, Welch AJ. Synthesis and X-ray structural characterization of the centred icosahedral gold cluster compound $[Au_{13}(PMe_2Ph)_{10}Cl_2](PF_6)_3$; the realization of a theoretical prediction. *J Chem Soc Chem Commun* 1981:201–2.
72. Zeng CJ, Li T, Das A, Rosi NL, Jin RC. Chiral structure of thiolate-protected 28-gold-atom nanocluster determined by X-ray crystallography. *J Am Chem Soc* 2013;**135**:10011–3.
73. Dohmeier C, Robl C, Tacke M, Schnöckel H. The tetrameric aluminum(I) compound $[\{Al(\eta^5\text{-}C_5Me_5)\}_4]$. *Angew Chem Int Ed* 1991;**30**:564–5.
74. Purath A, Dohmeier C, Ecker A, Schnöckel H, Amelunxen K, Passler T, et al. Synthesis and crystal structure of the tetraaluminatetrahedrane $Al_4[Si(t\text{-}Bu)_3]_4$, the second Al_4R_4 compound. *Organometallics* 1998;**17**:1894–6.
75. Yang H, Wang Y, Lei J, Shi L, Wu X, Mäkinen V, et al. Ligand-stabilized $Au_{13}Cu_x$ ($x = 2, 4, 8$) bimetallic nanoclusters: ligand engineering to control the exposure of metal sites. *J Am Chem Soc* 2013;**135**:9568–71.
76. Shichibu Y, Kamei Y, Konishi K. Unique [core+two] structure and optical property of a dodeca-ligated undecagold cluster: critical contribution of the *exo* gold atoms to the electronic structure. *Chem Commun* 2012;**48**:7559–61.
77. Dhayal RS, Liao JH, Liu YC, Chiang MH, Kahlal S, Saillard JY, et al. $[Ag_{21}\{S_2P(OiPr)_2\}_{12}]^+$: an eight-electron superatom. *Angew Chem Int Ed* 2015;**54**:3702–6.
78. Crasto D, Malola S, Brosofsky G, Dass A, Häkkinen H. Single crystal XRD structure and theoretical analysis of the chiral $Au_{30}S(S\text{-}t\text{-}Bu)_{18}$ cluster. *J Am Chem Soc* 2014;**136**:5000–5.
79. Yang H, Wang Y, Edwards AJ, Yan J, Zheng N. High-yield synthesis and crystal structure of a green Au_{30} cluster co-capped by thiolate and sulfide. *Chem Commun* 2014;**50**:14325–7.
80. Zeng CJ, Qian HF, Li T, Li G, Rosi NL, Yoon B, et al. Total structure and electronic properties of the gold nanocrystal $Au_{36}(SR)_{24}$. *Angew Chem Int Ed* 2012;**51**:13114–8.
81. Wan XK, Yuan SF, Tang Q, Jiang D, Wang QM. Alkynyl-protected Au_{23} nanocluster: a 12-electron system. *Angew Chem Int Ed* 2015;**54**:5977–80.
82. Liu C, Li T, Li G, Nobusada K, Zeng C, Pang G, Rosi NL, Jin R. Observation of body-centered cubic gold nanocluster. *Angew Chem Int Ed* 2015. http://dx.doi.org/10.1002/anie.201502667.
83. Malola S, Lehtovaara L, Knoppe S, Hu K-J, Palmer RE, Buergi T, et al. $Au_{40}(SR)_{24}$ cluster as a chiral dimer of 8-electron superatoms: structure and optical properties. *J Am Chem Soc* 2012;**134**:19560–3.
84. Das A, Li T, Nobusada K, Zeng Q, Rosi NL, Jin R. Total structure and optical properties of a phosphine/thiolate-protected Au_{24} nanocluster. *J Am Chem Soc* 2012;**134**:20286–9.

85. Shichibu Y, Negishi Y, Watanabe T, Chaki NK, Kawaguchi H, Tsukuda T. Biicosahedral gold clusters $[Au_{25}(PPh_3)_{10}(SC_nH_{2n+1})_5Cl_2]^{2+}$ ($n = 2-18$): a stepping stone to cluster-assembled materials. *J Phys Chem C* 2007;**111**:7845–7.
86. Teo BK, Zhang H, Shi XB. Molecular architecture of a novel vertex-sharing biicosahedral cluster $[(p\text{-}Tol_3P)_{10}Au_{13}Ag_{12}Br_8](PF_6)$ containing a staggered-staggered-staggered configuration for the 25-atom metal framework. *Inorg Chem* 1990;**29**:2083–91.
87. Chen J, Zhang QF, Williard PQ, Wang LS. Synthesis and structure determination of a new Au_{20} nanocluster protected by tripodal tetraphosphine ligands. *Inorg Chem* 2014;**53**:3932–4.
88. Wan XK, Yuan SF, Lin ZW, Wang QM. A chiral gold nanocluster Au_{20} protected by tetradentate phosphine ligands. *Angew Chem Int Ed* 2014;**53**:2923–6.
89. Yang H, Wang Y, Huang H, Gell L, Lehtovaara L, Malola S, et al. All-thiol-stabilized Ag_{44} and $Au_{12}Ag_{32}$ nanoparticles with single-crystal structures. *Nat Commun* 2013;**4**:2422.
90. Desireddy A, Conn BE, Guo J, Yoon B, Barnett RN, Monahan BM, et al. Ultrastable silver nanoparticles. *Nature* 2013;**501**:399–402.
91. Yang H, Wang Y, Yan J, Chen X, Zhang X, Häkkinen H, et al. Structural evolution of atomically precise thiolated bimetallic $[Au_{12+n}Cu_{32}(SR)_{30+n}]^{4-}$ ($n = 0, 2, 4, 6$) nanoclusters. *J Am Chem Soc* 2014;**136**:7197–200.
92. Wang Y, Su H, Xu C, Li G, Gell L, Lin S, et al. An intermetallic $Au_{24}Ag_{20}$ superatom nanocluster stabilized by labile ligands. *J Am Chem Soc* 2015;**137**:4324–7.
93. Chen J, Zhang QF, Bonaccorso TA, Williard PG, Wang LS. Controlling gold nanoclusters by diphospine ligands. *J Am Chem Soc* 2014;**136**:92–5.
94. Teo BK, Zhang H, Shi XB. Cluster of clusters: a modular approach to large metal clusters. Structural characterization of a 38-Atom cluster $[(p\text{-}Tol_3P)_{12}Au_{18}Ag_{20}Cl_{14}]$ based on vertex-sharing triicosahedra. *J Am Chem Soc* 1990;**112**:8552–62.
95. Teo BK, Shi X, Zhang H. Pure gold cluster of 1:9:9:1:9:9:1 layered structure: a novel 39-metal-atom cluster $[(Ph_3P)_{14}Au_{39}Cl_6]Cl_2$ with an interstitial gold atom in a hexagonal antiprismatic cage. *J Am Chem Soc* 1992;**114**:2743–5.
96. Dharmaratne AC, Krick T, Dass A. Nanocluster size evolution studied by mass spectrometry in room temperature $Au_{25}(SR)_{18}$ synthesis. *J Am Chem Soc* 2009;**131**:13604–5.
97. Nimmala PR, Yoon B, Whetten RL, Landman U, Dass A. $Au_{67}(SR)_{35}$ nanomolecules: characteristic size-specific optical, electrochemical, structural properties and first-principles theoretical analysis. *J Phys Chem A* 2013;**117**:504–17.
98. Azubel M, Koivisto J, Malola S, Bushnell D, Hura GL, Koh AL, et al. Electron microscopy of gold nanoparticles at atomic resolution. *Science* 2014;**345**:909–12.
99. Schnepf A. $[Ge_9\{Si(SiMe_3)_3\}_3]^-$: a soluble polyhedral Ge_9 cluster stabilized by only three silyl ligands. *Angew Chem Int Ed* 2003;**42**:2624–5.
100. Purath A, Dohmeier C, Ecker A, Koppe R, Krautscheid H, Schnöckel H, et al. Synthesis and structure of a neutral $SiAl_{14}$ cluster. *J Am Chem Soc* 2000;**122**:6955–9.
101. Hartig J, Stosser A, Hauser P, Schnöckel H. A metalloid $[Ga_{23}\{N(SiMe_3)_2\}_{11}]$ cluster: the jellium model put to test. *Angew Chem Int Ed* 2007;**46**:1658–62.
102. Dass A, Nimmala PR, Jupally VR, Kothalawala N. $Au_{103}(SR)_{45}$, $Au_{104}(SR)_{45}$, $Au_{104}(SR)_{46}$ and $Au_{105}(SR)_{46}$ nanoclusters. *Nanoscale* 2013;**5**:12082–5.
103. Negishi Y, Nakazaki T, Malola S, Takano S, Niihori Y, Kurashige W, et al. A critical size for emergence of nonbulk electronic and geometric structures in dodecanethiolate-protected Au clusters. *J Am Chem Soc* 2015;**137**:1206–12.
104. N.F. Zheng group, unpublished.

105. Negishi Y, Sakamoto C, Ohyama T, Tsukuda T. Synthesis and the origin of the stability of thiolate-protected Au_{130} and Au_{187} clusters. *J Phys Chem Lett* 2012;**3**:1624–8.
106. Zeng C, Chen Y, Kirschbaum K, Appavoo K, Sfeir MY, Jin R. Structural patterns at all scales in a nonmetallic chiral $Au_{133}(SR)_{52}$ nanoparticle. *Sci Adv* 2015;**1**:e1500045.
107. Dass A, Theivendran S, Nimmala PR, Kumara C, Reddy Jupally V, Fortunelli A, et al. $Au_{133}(SPh\text{-}t\mathrm{Bu})_{52}$ nanomolecules: X-ray crystallography, optical, electrochemical, and theoretical analysis. *J Am Chem Soc* 2015;**137**:4610–3.
108. Qian HF, Jin R. Controlling nanoparticles with atomic precision: the case of $Au_{144}(SCH_2CH_2Ph)_{60}$. *Nano Lett* 2009;**9**:4083–7.
109. Lopez-Acevedo O, Akola J, Whetten RL, Grönbeck H, Häkkinen H. Structure and bonding in the ubiquitous icosahedral metallic gold cluster $Au_{144}(SR)_{60}$. *J Phys Chem C* 2009;**113**:5035–8.
110. Bahena D, Bhattarai N, Santiago U, Tlahuice A, Ponce A, Bach SBH, et al. STEM electron diffraction and high-resolution images used in the determination of the crystal structure of the $Au_{144}(SR)_{60}$ cluster. *J Phys Chem Lett* 2013;**4**:975–81.
111. Kumara C, Dass A. $(AuAg)_{144}(SR)_{60}$ alloy nanomolecules. *Nanoscale* 2011;**3**:3064–7.
112. Malola S, Häkkinen H. Electronic structure and bonding of icosahedral core-shell gold-silver nanoalloy clusters $Au_{144-x}Ag_x(SR)_{60}$. *J Phys Chem Lett* 2011;**2**:2316–21.
113. Vollet J, Hartig JR, Schnöckel H. $Al_{50}C_{120}H_{180}$: a pseudofullerene shell of 60 carbon atoms and 60 methyl groups protecting a cluster core of 50 aluminum atoms. *Angew Chem Int Ed* 2004;**43**:3186–9.
114. Dass A. Faradaurate nanomolecules: a superstable plasmonic 76.3 kDa cluster. *J Am Chem Soc* 2011;**133**:19259–61.
115. Qian H, Zhu Y, Jin R. Atomically precise gold nanocrystal molecules with surface plasmon resonance. *Proc Natl Acad Sci USA* 2012;**109**:696–700.
116. Malola S, Lehtovaara L, Enkovaara J, Häkkinen H. Birth of the localized surface plasmon resonance in monolayer-protected gold nanoclusters. *ACS Nano* 2013;**7**:10263–70.
117. Knoppe S, Lehtovaara L, Häkkinen H. Electronic structure and optical properties of the intrinsically chiral 16-electron superatom complex $[Au_{20}(PP_3)_4]^{4+}$. *J Phys Chem A* 2014;**118**:4214–21.
118. Clayborne PA, Häkkinen H. The electronic structure of $Ge_9[Si(SiMe_3)_3]_3^-$: a superantiatom complex. *Phys Chem Chem Phys* 2012;**14**:9311–6.
119. Clayborne PA, Lopez-Acevedo O, Whetten RL, Grönbeck H, Häkkinen H. The $Al_{50}Cp^*_{12}$ cluster – a 138-electron closed shell ($L = 6$) superatom. *Eur J Inorg Chem* 2011:2649–52.
120. Ackerson CJ, Powell RD, Hainfeld JF. *Methods Enzym* 2010;**481**:195–230.
121. Dreaden EC, El-Sayed MA. Detecting and destroying cancer cells in more than one way with noble metals and different confinement properties on the nanoscale. *Acc Chem Res* 2012;**45**:1854–65.
122. Marjomäki V, Lahtinen T, Martikainen M, Koivisto J, Malola S, Salorinne K, et al. Site-specific targeting of enterovirus capsid by functionalized monodisperse gold nanoclusters. *Proc Natl Acad Sci USA* 2014;**111**:1277–81.

Chapter 9

Optical Properties and Chirality

Christine M. Aikens
Department of Chemistry, Kansas State University, Manhattan, KS, USA
E-mail: cmaikens@ksu.edu

Chapter Outline

9.1 INTRODUCTION

Ever since the initial discovery of monolayer-protected clusters (MPCs), the optical properties of these systems have been of great interest. In contrast to the single plasmon resonance of spherical gold or silver nanoparticles with diameters over 5 nm, MPCs have structured optical absorption spectra consisting of multiple transitions between energy levels within the nanoparticle (Figure 1).[1] These absorption spectra are characteristic of particular nanoparticle stoichiometries, and can now be used to verify the formation of a given MPC. In this chapter, we will discuss the experimental background of the optical absorption of these systems and then describe our current understanding of these properties. Nonlinear optical properties of these systems will be described briefly. In addition, we will discuss the chirality of these systems

Frontiers of Nanoscience, Vol. 9. http://dx.doi.org/10.1016/B978-0-08-100086-1.00009-9

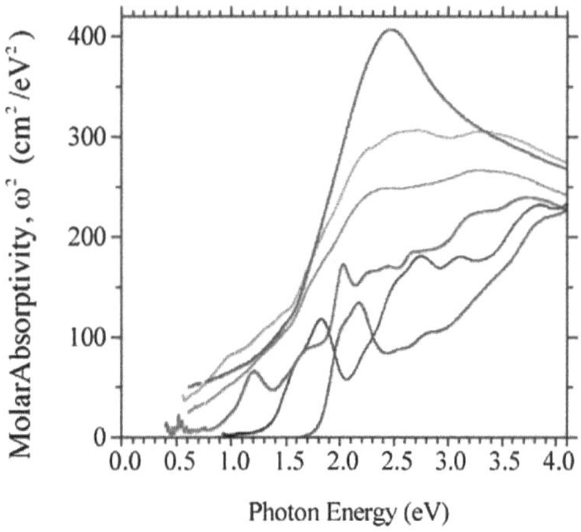

FIGURE 1 Optical absorption spectra of six gold MPCs. $Au_{20}(SR)_{14}$, purple; $Au_{25}(SR)_{18}^{-}$, blue; $Au_{39}(SR)_{23}^{2-}$, green; $Au_{75}(SR)_{42}^{-}$, yellow; $Au_{152}(SR)_{60}$, orange; 66 kDa $Au_n(SR)_m$, red. (see color plate). *Reproduced with permission from Ref. 1. Copyright Springer 2007.*

and their chiroptical properties, including the effects of core chirality and ligand binding.

9.2 BACKGROUND

In 1997, Whetten and coworkers purified ~20 MPC samples with core diameters from 1.4 to 3.2 nm using fractional crystallization.[2] They found that the broad plasmon absorption band centered at 2.4 eV decreased in intensity and broadened with decreasing nanoparticle size. They also noted a strong absorption onset near 1.7 eV that becomes increasingly distinct for the smaller nanoparticle sizes, which can be interpreted as arising from interband (i.e., d → sp) transitions. They found a weak steplike structure in the optical absorption spectra, especially for the smallest clusters, that was especially apparent in derivative spectra. This was attributed to transitions from the essentially continuous d-band levels to the relatively sparse sp-band; this implies that the sparsity of the sp-band increases with decreasing nanoparticle size. In addition, they found that the optical absorption spectra of these systems were essentially independent of the chain length of the alkylthiol ligand passivating the nanoparticle. Whetten and coworkers later found similar results for smaller (~1.1–1.9 nm core diameter) nanocrystals.[3]

Shortly thereafter, Schaaff, Whetten, and coworkers prepared nanoparticles stabilized with glutathione (GSH, N-g-glutamyl-cysteinyl-glycine) and employed gel electrophoresis to separate the MPCs produced in the synthesis.[4,5] These improved separation techniques led to much sharper optical absorption spectra than those previously obtained. They also showed that the absorption spectra of the glutathione-stabilized nanoparticles were essentially identical to those of *n*-hexylthiol-stabilized MPCs both in peak position and intensity of all transitions over the entire spectrum.[5] Separation and mass spectrometry techniques were later improved by Negishi, Tsukuda, and coworkers, who employed polyacrylamide gel electrophoresis to samples of glutathione-stabilized MPCs.[6] They separated nine fractions from their synthesis mixture and measured an optical absorption spectrum displaying significant structure for each. In 2007, Wyrwas et al. examined the evolution of the optical properties of monodisperse nanoparticles of various sizes.[1] They found that the optical absorption spectra change from a spectrum comprised of a series of discrete transitions to a spectrum corresponding to the plasmon resonance of larger nanoparticles as the size of the MPC increases (Figure 1).

Gold nanoparticles protected by ligands other than thiolates have also been found to possess discrete absorption spectra. For example, the well-known undecagold (Au_{11}) cluster used frequently in biological applications and protected by phosphine (and halide) ligands also has a structured optical absorption spectrum.[7,8] Larger phosphine-protected clusters such as greengold (Au_{75}) possess many discrete energy levels, leading to many transitions observable in an optical absorption spectrum.[9]

9.3 OPTICAL PROPERTIES

9.3.1 Fundamental Origin

Our understanding of the optical properties of MPCs increased significantly after X-ray crystal structure determination of several thiolate-stabilized nanoparticle structures: first of $Au_{102}(SR)_{44}$ (Ref. 10) and then of $Au_{25}(SR)_{18}^{-}$.[11,12] These crystal structures conclusively demonstrated that most if not all thiolate-stabilized MPCs could be conceptually divided into a core of metal atoms and a surrounding layer of metal–thiolate oligomers. Aikens, Jin, and coworkers examined the electronic structure of a model $Au_{25}(SH)_{18}^{-}$ nanoparticle using density functional theory (DFT) calculations[12] and found results that largely agreed with the predictions of Whetten and coworkers from a decade before. In these systems, a sparse set of orbitals arising primarily from the 6sp atomic orbitals of gold are located near the Fermi level (Figure 2(a)). These orbitals include the highest occupied molecular orbital (HOMO), lowest unoccupied molecular orbital (LUMO), LUMO+1, etc. The HOMO and LUMO+1 are essentially triply degenerate, the LUMO is doubly degenerate, and the LUMO+2 is singly degenerate. Below these orbitals lies a dense band of orbitals arising from the Au(5d) atomic orbitals and the S(3p) atomic orbitals (initially called the "d-band," but now more appropriately referred to as the "ligand-based d-band" since the gold atoms involved are primarily those in the oligomeric ligand motifs). Again, the degeneracy of these orbitals ranges from 1 to 3. It should be noted that an additional "core d-band" arising from the Au(5d) atomic orbitals of the gold atoms in the core lies lower in energy than the "ligand-based d-band" arising from the oligomeric ligand motifs.

The optical absorption spectrum for the model $Au_{25}(SH)_{18}^-$ MPC was calculated using linear response (LR) time-dependent density functional theory (TDDFT) (Figure 2(b)).[12] The first peak was found to arise primarily from a HOMO to LUMO (sp → sp, or intraband) transition. In contrast, the second peak originates from a combination of transitions: HOMO to LUMO+1

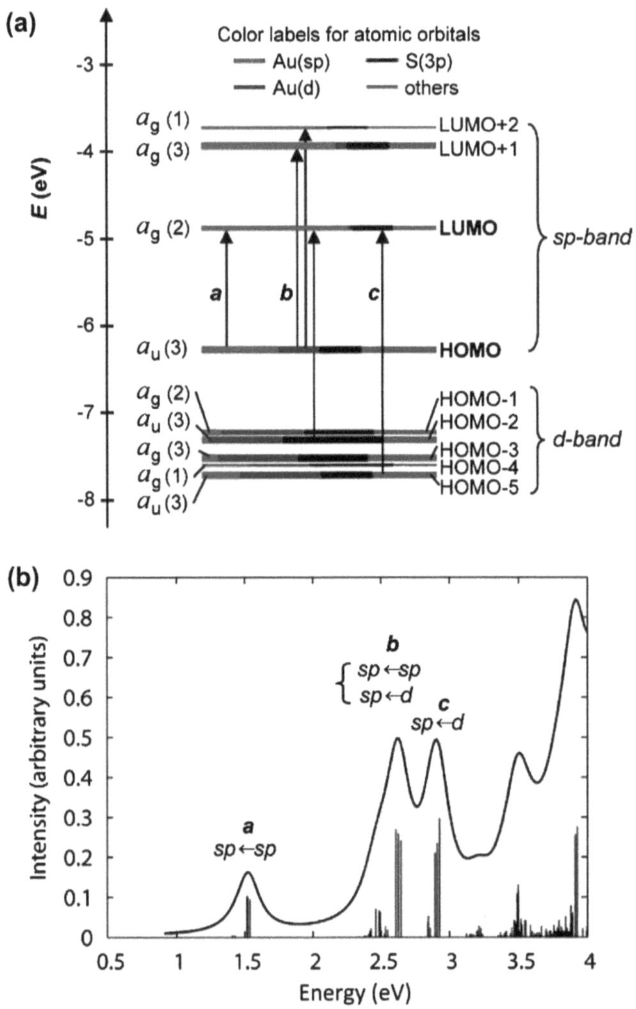

FIGURE 2 (a) Kohn–Sham energy level diagram and (b) Time-dependent density functional theory optical absorption spectrum of $Au_{25}(SH)_{18}^-$. (see color plate). *Figure reproduced with permission from Ref. 12. Copyright American Chemical Society 2008.*

(intraband), HOMO to LUMO+2 (intraband), and HOMO−2 to LUMO (d → sp, or interband). Overall, the positions and intensities of the peaks were found to be in good agreement with experiment, so the origins of the absorption spectrum could be assigned.

As shown by Häkkinen and coworkers using a projected density of states approach, the sp-band orbitals of this nanoparticle have symmetries that closely correspond to P (HOMO) and D (LUMO, LUMO+1) symmetries for a spherical system.[13] This corresponds to the expected S^2P^6 electron configuration for a spherical eight electron system, as discussed in Chapter 8. (The reader is highly encouraged to read Chapter 8 before continuing with this chapter, as the superatomic model is prominently employed in understanding MPC optical properties.) Since the nearly icosahedral core of the $Au_{25}(SR)_{18}^-$ is approximately spherical, these symmetries are expected. Visualization of the Kohn−Sham orbitals for $Au_{25}(SH)_{18}^-$ (Figure 3) demonstrates that the HOMO consists of three nearly degenerate P orbitals that

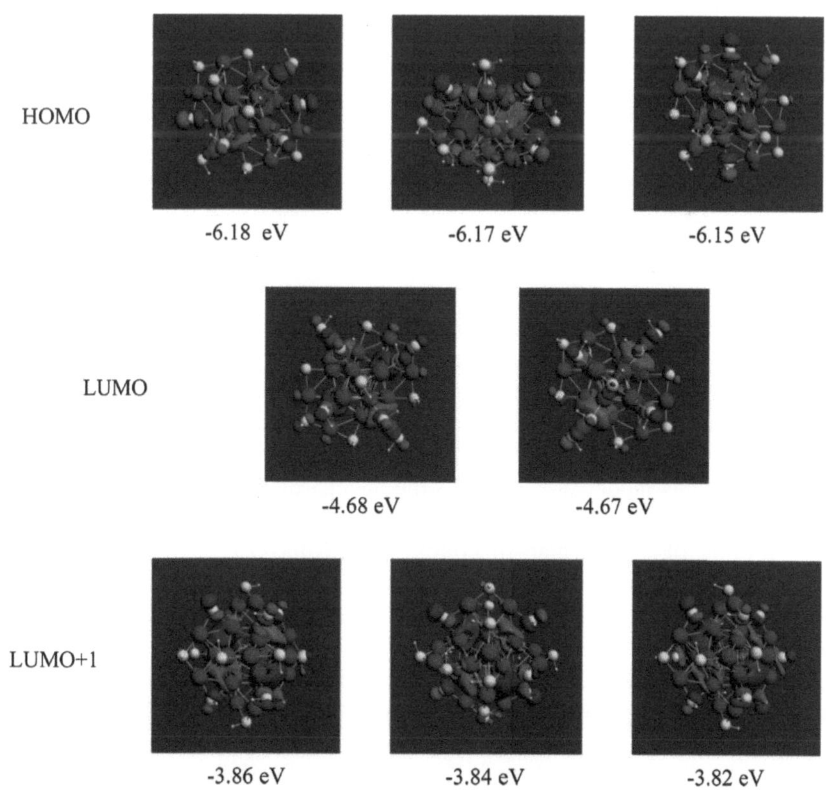

FIGURE 3 Kohn−Sham orbitals and orbital energies of $Au_{25}(SH)_{18}^-$. (see color plate). *Figure reproduced with permission from Ref. 14. Copyright American Chemical Society 2008.*

are delocalized throughout the nanoparticle core, with some contributions from S(3p) atomic orbitals.[14] The LUMO and LUMO+1 have D-like character. The ligand field arising from the six V-shaped RS−Au−SR−Au−SR ligand motifs is responsible for splitting the D orbitals into two sets: a doubly degenerate and a triply degenerate set. The doubly degenerate set is lower in energy because the idealized $Au_{25}S_{18}$ motif has T_h symmetry; consideration of the R-groups lowers the symmetry to C_i. This symmetry lowering has a small effect on the orbital energies in the case of $Au_{25}(SH)_{18}^-$, so the LUMO and LUMO+1 are approximately doubly and triply degenerate, respectively.

Electron density has an important effect on the optical properties of this system. The size of the 13-atom nanoparticle core in $Au_{25}(SR)_{18}^-$ affects the splitting between the two sets of D orbitals. For example, the use of generalized gradient approximation (GGA) functionals (which are well known to predict lattice parameters (bond distances) that are somewhat larger than experiment) in DFT optimizations of the $Au_{25}(SH)_{18}^-$ geometry leads to a core that is expanded relative to the crystal structure (by about 0.07 Å).[15] In contrast, local density approximation (LDA) functionals (which typically slightly underestimate lattice parameters) predict a core in good agreement with experiment. The slight expansion of the core with GGA geometries leads to an increased splitting in the energy levels of the two sets of D orbitals and thus an increased difference in energy between the first two peaks in the absorption spectrum (by 0.15 eV).

Simplified model calculations also provide insights into the fundamental electronic structure of nanoparticle systems, which is reflected in their optical absorption spectra. For example, a charge-perturbed particle-in-a-sphere model was created by Guidez and Aikens to examine the electronic structure of nanoparticles with a spherical core surrounded by oligomeric metal−thiolate units.[16] This model predicts that an increase in the radius of the core while keeping the number of electrons constant leads to a greater splitting between the two sets of D orbitals in the $Au_{25}(SH)_{18}^-$ nanoparticle, in agreement with the geometrical effects discussed above for DFT calculations. In addition, this model shows that as the nanoparticle diameter increases and thus as the number of passivating metal−thiolate oligomeric ligands increases, the splitting of a particular superatomic shell decreases. Thus, the optical absorption spectra are expected to become less split as the nanoparticle size increases, which correlates with the known coalescence to the single plasmon resonance with increasing diameter (Figure 1).

In 2013, Antonello et al. published theoretical spectra of the neutral and cationic $Au_{25}(SR)_{18}$ system.[17] As observed in previous electronic structure studies of the related $Au_{25}(SH)_{18}$ nanoparticle,[18] these authors observed an increased splitting of the three highest occupied P orbitals compared to the anionic system. For the $Au_{25}(SCH_3)_{18}^+$ nanoparticle, they found that the splitting of the P orbitals is even larger than for the neutral. They proposed that the increase in the HOMO-LUMO gap observed from UV–Vis absorption data originates from this splitting.

9.3.2 Accuracy Considerations

As mentioned above, the accuracy of the nanoparticle geometry affects the predicted absorption energies of an MPC. In addition, the level of theory employed in the calculations also affects the absorption spectrum. In this section, we consider the effects of solvents, level of theory employed in the TDDFT calculations, the ligand used, and spin–orbit coupling (SOC).

9.3.2.1 Solvent

Very few studies have examined the impact of solvent on the optical properties of MPCs. In 2009, Aikens described a study of the $Au_{25}(SH)_{18}^-$ nanoparticle in which inclusion of a continuum solvent during the TDDFT calculations had negligible effects on the absorption spectrum.[15] Of the three main peaks observed in the UV–Vis spectrum, the intensity of the third peak arising from interband transitions out of the ligand-based d-band increased slightly as the polarity of the solvent increased. Otherwise, essentially no effects could be noted, which agreed with experimental observations on this system.

In contrast, phosphine- and halide-stabilized nanoparticles have exhibited noticeable effects in their excitation spectra upon addition of a continuum solvent in the calculations. Provorse and Aikens found that the low energy peaks in the absorption spectrum of $Au_2X_2(S-)BINAP$, which is a starting material for the creation of $Au_{11}X_2BINAP_4^+$, blue shift by approximately 50 nm (e.g., from 390 to 340 nm) when a continuum chloroform solvent is employed.[19] In addition, a continuum tetrahydrofuran (THF) solvent was found to blue shift the absorption spectrum of $Au_4(\mu_2-I)_2(PPh_3)_4$ by about 0.3 eV compared to calculations in the vacuum.[20] It is possible that the effect of solvent may depend on the polarity of the molecule and the excitations involved in the relevant region of the absorption spectrum.

9.3.2.2 Level of Theory

The effect of basis set and functional employed on the TDDFT optical absorption spectra have been studied in a few investigations. Aikens found that the predicted excitation spectrum of the $Au_{25}(SH)_{18}^-$ nanoparticle was red shifted by about 0.1 eV when a triple-zeta basis set was used compared to a double-zeta basis set.[15] Since the triple-zeta results were consistently 0.15–0.20 eV lower in energy than experiment,[15] it is possible that using a less flexible basis set such as a double-zeta basis set may appear to provide results in better agreement with experimental values, but this does not provide the "right answer for the right reasons" and should be cautioned against. On the other hand, frozen-core basis sets (in this case, similar to pseudopotential basis sets) provided results very similar to full core basis sets,[15] which suggest that they can be utilized to decrease computational expense without compromising accuracy significantly.

Standard LDA functionals typically provide very poor unoccupied orbital energies, and are thus not commonly used in TDDFT calculations. Drastic underestimation of excitation energies can result. GGA, hybrid, and asymptotically corrected functionals are typically used in TDDFT calculations. For $Au_{25}(SH)_{18}^-$, a comparison of several GGA and asymptotically corrected functionals (LB94, SAOP) showed that the GGA functionals predicted the first excitation peak about 0.1 eV lower in energy than LB94 or SAOP.[15] These energetic underestimations were more pronounced for the second and third absorption peaks. For this example case, LB94 calculations gave essentially the same results as SAOP, but it should be noted that the two functionals can also differ widely; LB94 predicts excitation energies that are 0.15−0.35 eV higher in energy than SAOP values for bare Ag_{20},[21] and 0.5 eV higher in energy for the longitudinal excitation of the Ag_{19}^+ nanorod.[22]

9.3.2.3 Ligands

Early work on the effects of ligands on the optical absorption spectrum examined the replacement of the R groups in $Au_{25}(SR)_{18}^-$ with R = H, CH_3, and CH_2CH_3.[15] Using LB94/TZP.4f, the first absorption peak was found to vary by less than 0.1 eV. The second peak experienced greater changes, with its energy lying about 0.15 eV lower in energy for CH_3 and CH_2CH_3 than for H.

Experimental work by Guo and Murray had shown that *para*-substituted phenyl groups in the $Au_{25}(SR)_{18}^-$ nanoparticle (note that the stoichiometry was originally assigned as $Au_{38}(SR)_{24}$) have similar absorption spectra to R = CH_2CH_2Ph, although the second and third peaks for R = CH_2CH_2Ph actually lie 0.1−0.2 eV higher in energy than the corresponding peaks for R = Ph.[23] In 2010, Aikens examined the $Au_{25}(SPhX)_{18}^-$ (X = H, F, Cl, Br, CH_3, and OCH_3) MPC and predicted that the lowest energy structure has S_6 point group symmetry, with six sets of three π-stacked phenyl groups.[24] The change in symmetry of this system led to a splitting of the superatom P orbitals in which the P_z orbital was predicted to lie 0.12−0.13 eV higher in energy than the P_x and P_y orbitals. (A lower symmetry structure was later suggested as the ground state of this system by Tlahuice-Flores et al., and the P orbital splitting is maintained in this structure.[25]) In addition, the triply degenerate set of D orbitals was split into a set of two degenerate orbitals and a single orbital lying 0.17 eV above the former. The splitting of the P orbitals was found to lead to a double peak structure for the first main peak in the TDDFT absorption spectrum[24] that had not been apparent in the excitation spectra of R = H, CH_3, and CH_2CH_3.[15] Substitution at the para position of the phenyl group did not lead to significant changes in the calculated TDDFT absorption spectra for the first two peaks, but the third peak (which as described above is known to have significant contribution from excitations out of the ligand-based d-band

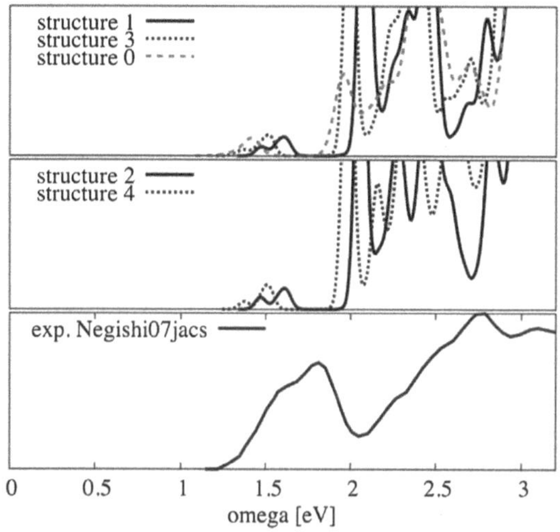

FIGURE 4 Calculated optical absorption of 1: $Au_{25}(SCH_2CH_2Ph)_{18}^-$ MPC (monolayer-protected cluster) extracted from crystal structure; 2: MPC plus TOA+ counterion; 3: MPC optimized in gas phase; 4: MPC plus counterion optimized in gas phase; 0: model MPC with methyl ligands optimized in gas phase. *Figure reproduced with permission from Ref. 26. Copyright American Chemical Society 2010.*

orbitals) shifted by about 0.1 eV. It should be noted that the LB94/DZ calculations employed in Ref. 24 led to underestimations of about 0.4 eV relative to experiment, compared to the 0.15 eV previously reported for $Au_{25}(SH)_{18}^-$. Thus, it is possible that the level of theory may become more important in the TDDFT calculations as ligand size increases.

In 2010, Akola et al. published TDDFT spectra of the full $Au_{25}(SCH_2CH_2Ph)_{18}^-$ MPC, its complex with TOA^+, and the model $Au_{25}(SCH_3)_{18}^-$ system (Figure 4).[26] They determined that the presence of TOA^+ near the anionic nanoparticle led to a red-shifted spectrum compared to the MPC by itself. They found that the absorption spectrum for the system with methyl ligands was somewhat red shifted with respect to the full MPC. They found that the first peak of the absorption spectrum in the full MPC had the characteristic "double-peak" shape with a prominent shoulder that is well known for experimental $Au_{25}(SR)_{18}^-$ compounds, whereas the methyl complex exhibited a single peak. They attributed the change in peak shape to symmetry breaking of the P orbitals induced by the longer ligands.

9.3.2.4 Spin–Orbit Coupling

In 2014, Jiang et al. found that SOC leads to a double peak for the first peak in the TDDFT optical absorption spectrum, even for ligands as small as $R = CH_3$.[27] They ascertained that SOC leads to a 0.20 eV splitting of the P orbitals into a lower energy $1P_{1/2}$ orbital and set of two $1P_{3/2}$ orbitals. Their simulated optical absorption spectrum has a very similar shape to the experimental spectrum, although it is shifted to lower energy by about 0.4 eV.

9.3.3 Effect of Doping on the Optical Absorption Spectrum of $Au_{25}(SR)_{18}^{-}$

9.3.3.1 *Silver Doping*

In 2008, Aikens considered the effects of doping silver atoms into the core and/or metal–thiolate-protecting units in the $Au_{25}(SH)_{18}^{-}$ nanoparticle.[14] In all systems, the electronic structure of an essentially triply degenerate HOMO consisting of P orbitals, an approximately doubly degenerate LUMO consisting of a set of D orbitals, and a nearly triply degenerate LUMO+1 containing the remaining D orbitals was maintained. For the four structures examined (gold core/gold–thiolate ligands, silver/silver–thiolate, silver/gold–thiolate, and gold/silver–thiolate), the optical absorption spectrum was richly structured with many discrete peaks below 4 eV. The splitting between the LUMO and LUMO+1, between the HOMO and LUMO, and between the HOMO and HOMO-1 were found to vary significantly depending on the metal. Systems containing a silver core led to an overall blue shift in the absorption spectrum compared to those with a gold core. The sharpest spectrum arose from the system with the silver core and the gold–thiolate units, whereas the broadest spectrum came from the system with the gold core and silver–thiolate units (Figure 5).

Two years later, Negishi et al. experimentally doped $Au_{25}(SR)_{18}^{-}$ with silver atoms and found that the optical absorption spectrum blue shifted and coalesced with increasing silver content.[28] In 2012, Guidez et al. examined these changes using TDDFT computations.[29] As previously found for $Au_{25}(SCH_3)_{18}^{-}$ (Ref. 30) and larger systems,[31] the lowest energy position for doping of a single silver atom was found to be on the exterior of the core, whereas the highest energy position was doping at the center. In consequence, Guidez et al. examined the optical absorption spectra of isomers containing doping of 1, 2, 4, 6, 8, 10, and 12 silver atoms on the outside of the core. Multiple isomers were found to lie very close (<0.1 eV) in energy, so Boltzmann-averaged spectra were computed for each number of dopant atoms. As previously observed experimentally and predicted theoretically, increasing the number of silver dopants led to a blue shift of the first peak and growth of the second peak, which gave an overall coalescence of the absorption spectrum into a single broad peak.

In 2014, Kumara et al. determined the crystal structure of a silver-doped $Au_{25}(SCH_2CH_2Ph)_{18}^{-}$ MPC and ascertained that about 6.2–6.7 silver atoms on average were present in the structure.[32] Calculations performed on a $Ag_6Au_{19}(SCH_2CH_2Ph)_{18}^{-}$ nanoparticle showed that the P orbitals split upon silver doping, and the highest energy P orbital points toward the silver atoms. This makes sense since the 5s orbitals of silver typically lie higher in energy

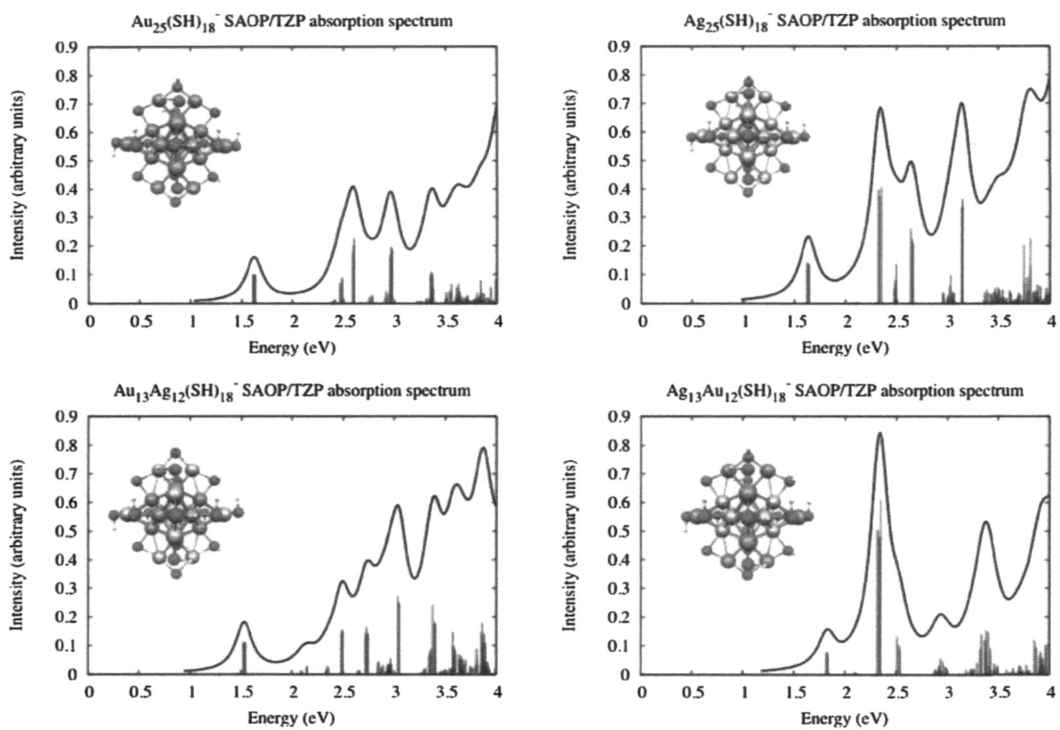

FIGURE 5 TDDFT optical absorption spectra of gold, silver, and mixed gold–silver monolayer-protected clusters. *Figure reproduced with permission from Ref. 14. Copyright American Chemical Society 2008.*

than the 6s orbitals of gold. The authors found that the lowest two LUMOs were D orbitals, but higher orbitals had significant contributions from π^* orbitals on the phenyl groups.

9.3.3.2 Other Metals

The optical absorption of $Au_{25}(SR)_{18}^-$ doped with other metal atoms has also been studied. In 2010, Negishi et al. examined the optical absorption of a neutral $PdAu_{24}(SC_{12}H_{25})_{18}$ system and found two peaks at ~690 and ~620 nm that were consistent with theoretical calculations for Pd in the central site.[33,34] In 2012, Qian et al. examined the Pt-doped MPC and learned that the UV/Vis/NIR spectrum was quite different than the parent $Au_{25}(SCH_2CH_2Ph)_{18}^-$ MPC.[35] They found a distinct peak in the spectrum at 590 nm with a broad peak centered around 1100 nm. They found an optical gap of 0.8 eV, which was much lower than either the pure gold nanoparticle or the Pd-doped species. Like the Pd-doped species, the best match between the calculated TDDFT spectrum and the experimental spectrum was found for doping in the center of the nanoparticle.

Negishi et al. also examined the effects of copper doping on the $Au_{25}(SR)_{18}^-$ nanoparticle.[36] Using mass spectrometry, they found that up to five copper atoms could be doped into the system. The experimental optical absorption spectrum of the $CuAu_{24}(SCH_2CH_2Ph)_{18}$ nanoparticle was found to be shifted slightly to lower energy. The HOMO-LUMO gap was found to reduce by 0.04 eV compared to the pure gold system. DFT calculations predicted that doping on the surface of the cluster was the most stable, followed by doping in the metal−thiolate ligands (by 0.13 eV), and doping in the center of the nanoparticle was found to be the highest energy species (by 0.40 eV). However, the HOMO-LUMO gap was predicted to be the highest for the surface structure (in opposition with experimental findings) and lowest for the centrally doped structure. The shifts in the TDDFT optical absorption spectra reflected the predicted changes in the HOMO-LUMO gap. They suggested that the most energetically stable cluster was not formed under the synthetic conditions employed.

9.3.3.3 Se and Te-Based Ligands

Selenolate ligands have also been used to passivate gold MPCs. These ligands lead to a structure for $Au_{25}(SeR)_{18}^-$ that has a similar framework as $Au_{25}(SR)_{18}^-$.[37] Kurashige et al. determined that the optical absorption spectrum of $Au_{25}(SeC_8H_{17})_{18}^-$ is very similar to that of $Au_{25}(SeC_8H_{17})_{18}^-$ in the 1.2−2.0 eV region, but shifts to lower energy in the 2.5−3.0 eV region. They calculated an optical absorption spectrum for $Au_{25}(SeCH_3)_{18}^-$ that was found to be in reasonable agreement with experiment. Kurashige et al. then examined copper-doped clusters and found that up to nine Cu atoms could dope into the selenolate-stabilized systems.[38] With increasing doping, they found that the first peak in the optical absorption spectrum continuously shifted to lower energy.

Kurashige et al. were also able to synthesize nanoparticles in which TePh ligands were exchanged with thiolates, yielding a stoichiometry of $[Au_{25}(TePh)_n(SC_8H_{17})_{18-n}]^-$.[39] They found that the optical absorption spectrum of this system shifted to lower energy compared to the parent gold–thiolate MPC.

9.3.4 Other Nanoparticle Stoichiometries

9.3.4.1 $Au_{38}(SR)_{24}$

The success of TDDFT calculations for achieving theory–experiment agreement for optical absorption spectra of $Au_{25}(SR)_{18}^-$ has led to their application for many MPCs. One of the earliest nanoparticles of interest for which TDDFT calculations were employed was the $Au_{38}(SR)_{24}$ nanoparticle. In 2008, Pei et al. predicted a structure for this MPC that consisted of a face-fused bi-icosahedral core surrounded by three $Au(SR)_2$ units and six $Au_2(SR)_3$ units.[40] They calculated LR-TDDFT absorption spectra for their lowest energy isomers with $R = CH_3$ and found that the lowest energy isomer was predicted to have four peaks, although the experimental spectrum exhibited three. Two years later, Lopez-Acevedo et al. discovered that a D_3 arrangement of the protecting ligand units led to excellent agreement with X-ray diffraction and circular dichroism (CD) data.[41] They also calculated the optical absorption spectrum of this system and found that, although the spectrum was still predicted to have four main peaks, the intensity of the second was decreased and the intensity of the third was increased compared to the spectrum of Pei et al., so the spectrum agreed slightly better with experiment. The D_3 arrangement of the protecting ligand units was later confirmed by X-ray crystallography.[42]

Like the $Au_{25}(SR)_{18}^-$ nanoparticle, the frontier orbitals of $Au_{38}(SR)_{24}$ near the HOMO-LUMO gap arise primarily from Au(6sp) atomic orbitals and are delocalized throughout the core of the system.[41] Because the $Au_{38}(SR)_{24}$ nanoparticle is elongated, these orbitals do not have spherical symmetry. Instead, they have approximate cylindrical symmetry such as Σ, Π, Δ, etc. Some of these orbitals may be considered to be "dimers" of spherical superatom orbitals within one of the icosahedra in the core, although consideration of cylindrical symmetry does not require this analysis. The strongest peaks in the optical absorption spectrum were found to follow cylindrical selection rules: the strongest longitudinal transitions (i.e., along the long axis of the particle) correspond to excitations between orbitals with the same quantum number but that differ by one in the number of nodes, whereas the strongest transverse transitions (i.e., along the short axis of the particle) arise from excitations between orbitals with the same number of nodes that differ by one in the quantum number. The first three peaks in the optical absorption spectra (up to 1.72 eV) were determined to arise from intraband transitions. The fourth peak (1.89–2.01 eV) was found to arise in large part from transitions out of the ligand-based d-band into the sp-based orbitals. Thus, the interband transitions appear to arise at energies above 1.7 eV, as previously noted by Whetten and coworkers.

9.3.4.2 Large Systems

In 2011, Häkkinen and coworkers performed LR-TDDFT calculations on the $Au_{102}(SCH_3)_{44}$ nanoparticle, which was the largest LR-TDDFT calculation on an MPC to that point.[43] Previous work on the electronic structure of this system had focused on electronic properties such as the HOMO-LUMO gap, but had not calculated the optical absorption spectrum.[44–46] The lowest calculated electronic transition calculated by LR-TDDFT was found at 0.55 eV, in good agreement with the experimental bandgap of 0.5 eV (Figure 6).[43] Above this energy, a dense spectrum of transitions was predicted. Although the calculated excitations were quite congested, four spectral features in the 1.5–3.5 eV region could nonetheless be determined, which were found to be in excellent agreement with baseline-corrected experimental spectra.

In 2012, Koivisto et al. performed a combined theory–experiment investigation on the optical properties of the $Au_{144}(SR)_{60}$ nanoparticle and its silver-doped alloys.[47] They found the experimental optical gap for the $Au_{144}(SCH_2CH_2Ph)_{60}$ MPC to be 0.19 ± 0.01 eV, which was in reasonable agreement with the first LR-TDDFT electronic transition of 0.32 eV for the model $Au_{144}(SH)_{60}$ nanoparticle using a previously predicted[48] structural model for this system (note: the computed transition does not take into account effects of finite temperature that broaden the spectra, so the actual agreement is closer than the raw optical gap numbers imply). As for $Au_{102}(SCH_3)_{44}$, the calculated spectrum is quite dense after the electronic transitions start. Since

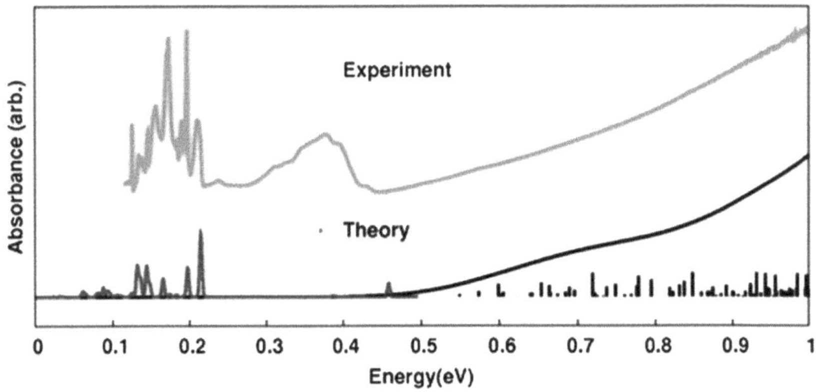

FIGURE 6 Experimental and theoretical absorption spectra of $Au_{102}(SR)_{44}$. *Figure adapted with permission from Ref. 43. Copyright American Chemical Society 2011.*

the sp-band is much more densely spaced for these large systems than for $Au_{25}(SR)_{18}^-$ and $Au_{38}(SR)_{24}$, these broad, complicated spectra are expected.

As the system size increases, interpretation of LR-TDDFT excitations in terms of orbital-to-orbital transitions becomes challenging due to the large number of excited states and the many transitions that contribute to each. In 2013, Häkkinen and coworkers developed a method based on time-dependent density functional perturbation theory (TD-DFPT) for analyzing the character of the optical transitions.[49] They calculate transition contribution maps (TCMs) for particular excitation wavelengths that show bright spots wherever a particular occupied-to-unoccupied orbital transition is important for that excitation. These authors applied this method to $Au_{144}(SH)_{60}$ as well as to a much larger $Au_{314}(SH)_{96}$ nanoparticle. These techniques enabled analysis of the collective nature of the plasmon resonance for systems between 1.5−2 nm in diameter, and they determined that the character of this excitation differs for the two system sizes.

Other researchers have employed a real-time TDDFT (RT-TDDFT) approach for calculating excitations. In this method, rather than diagonalizing a matrix to determine excitation energies as done in LR-TDDFT, the density of the system is propagated in response to a perturbation. The excitation spectrum can be calculated by determining the Fourier transform of the time-dependent dipole moment. RT-TDDFT has the advantage that it can potentially be applied to larger systems than LR-TDDFT since roots of a matrix do not need to be computed, but one must ensure that the simulation is run long enough to achieve convergence. One potential disadvantage of RT-TDDFT is that it does not readily permit an analysis of excitations in terms of orbital-to-orbital transitions.

In 2013, Tlahuice-Flores et al. undertook a combined theory−experiment study of the $Au_{130}(SR)_{50}$ nanoparticle.[50] They suggested a structure with a 75-atom Marks' decahedral core capped by two 15-atom rings that is passivated by 25 $Au(SR)_2$ units. Using RT-TDDFT, they calculated the absorption spectrum for a model MPC with $R = H$ and found moderate agreement with the experimental spectrum. The optical gap of the theoretical spectrum is slightly less than experiment and the two show somewhat different structure between 2.5 and 4.0 eV. The authors attributed the differences to ligand effects, although structural differences and level of theory effects could also play a role.

In 2014, Weissker et al. performed an RT-TDDFT study of $Au_{144}(SCH_3)_{60}$.[51] They used a model of $Au_{144}(SCH_3)_{60}$ with the same connectivity as the original structural model[48] but with ligands arranged to preserve *I* point group symmetry.[52] The RT-TDDFT calculations exhibited sharp oscillations in the absorption spectra, which were observable to some extent in low temperature (77 K) experimental absorption spectra. Overall, the broadening of the spectra that is commonly observed at room temperature due to nuclear vibrations is lessened at low temperature. Both experiment and theory showed that even nanoparticles with 100–300 atoms can exhibit discrete absorption spectra.

9.3.4.3 Gold Nanoparticles with Known Crystal Structures

TDDFT calculations are now routinely employed to examine optical absorption spectra for systems with 20–40 metal atoms. Zeng et al. simulated the UV–Vis spectrum of $Au_{36}(SR)_{24}$ using a model $R = CH_3$ ligand and determined TDDFT excitation spectra in good agreement with experiment (Figure 7).[53] The first peak at 575 nm was assigned to a HOMO to LUMO transition. The second peak around 400 nm was attributed to excitations from the HOMO into higher unoccupied orbitals. In 2014, Nimmala et al. performed TDDFT calculations on $Au_{36}(SR)_{24}$ using the full R = Ph ligand.[54] They found that the first peak at 720 nm arises from the HOMO to LUMO transition and assigned this as a D → D transition; transitions between orbitals with the same spherical symmetry label are forbidden, which explains why this transition is very weak in the experimental spectrum. The first moderately strong peaks in the 575–625 nm region were found to arise from excitations out of the HOMO-9 through HOMO-2 (essentially ligand-based d-band orbitals) into the LUMO, LUMO+1, and LUMO+2 (with D character), so these peaks were assigned as interband transitions. The shoulder at 408 nm in the experimental spectrum (450–475 nm in the TDDFT spectrum) was assigned as interband transitions from the HOMO-9 through HOMO-2 into the lowest set of F orbitals (LUMO+4 to LUMO+6), with a small contribution from the HOMO to LUMO+7 (intraband, with D to F transitions). The strong excitations

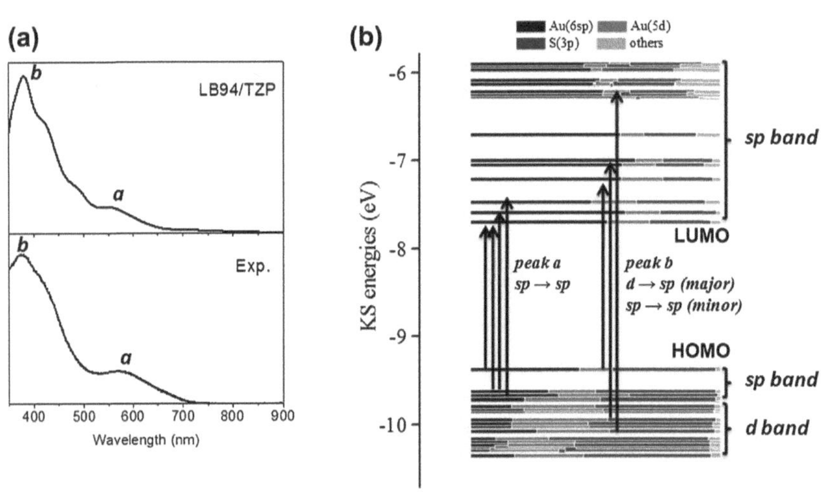

FIGURE 7 (a) Theoretical and experimental optical absorption spectra and (b) Kohn–Sham orbital energy diagram of $Au_{36}(SR)_{24}$. (see color plate). *Figure reproduced with permission from Ref. 53. Copyright American Chemical Society 2013.*

comprising the experimental peak at 373 nm (400–430 nm in the TDDFT spectrum) are dominated by D to F transitions (HOMO to LUMO+8 and LUMO+9), with contributions from interband excitations as well.

In 2013, Knoppe et al. calculated and analyzed TDDFT optical absorption for a $Au_{28}(SCH_3)_{20}$ MPC whose structure was based on a recently reported crystal structure.[55] Both the theoretical and experimental spectra show an absorption onset around 700 nm. In addition, the overall shapes of the spectra were found to be similar. Das et al. employed TDDFT calculations to compute the optical absorption spectrum of a $Au_{23}(SCH_3)_{16}^-$ nanoparticle whose structure was based on their crystal structure with cyclohexanethiol ligands.[56] The overall spectral shapes were quite similar between experiment and theory, with slight differences in the energies (~0.16 eV) that could potentially be attributed to the level of theory or to the use of the methyl ligand. The first peak (529 nm in TDDFT calculations) was determined to arise from an intense HOMO to LUMO transition. A second, weaker HOMO to LUMO+1 transition was found to lie at 507 nm and is also likely a component of the experimental peak centered at 570 nm. Shortly thereafter, Das et al. reported TDDFT spectra of a $Au_{24}(SCH_3)_{20}$ nanoparticle that was closely based on their $Au_{24}(SCH_2Ph{-}^tBu)_{20}$ crystal structure.[57] Again, the spectra were found to have similar shapes (with a ~0.22 eV difference in energy that can again be attributed to the level of theory employed or to the use of a model ligand). The first peak was assigned as a HOMO to LUMO transition with a small contribution from the HOMO-2 to LUMO transition. Because of the differences between the experimental optical absorption spectrum of the $Au_{24}(SCH_2Ph{-}^tBu)_{20}$ MPC and a previously reported[58] spectrum of $Au_{24}(SCH_2CH_2Ph)_{20}$, it is likely that the geometric structure of the gold core and/or the gold–thiolate oligomeric units protecting the core is different between the two nanoparticles.

Another system for which a crystal structure is known is $Au_{30}S(S{-}^tBu)_{18}$.[59] Crasto et al. calculated TDDFT spectra for this system both with and without the μ_3–S atom that was found to be present in the system, and determined that the presence or absence of this atom greatly affects the characteristic 630 nm peak of the full system. Using TCMs, they determined that the strongest contribution to this peak comes from transitions from the HOMO-2 orbital to degenerate HOMO+1 and HOMO+2 orbitals. They attributed the change in the spectrum upon removal of the μ_3–S atom to changes in the length of the core and a reduction in the polarization of the induced density, which led to lower intensity around 630 nm.

9.3.4.4 Gold Nanoparticle Structure Prediction Using Optical Absorption Comparison

As shown above, TDDFT computations have been shown to yield optical absorption spectra in good agreement with experiment when the nanoparticle geometry is accurate. Early structural models of $Au_{38}(SR)_{24}$ (based in part upon theory–experiment agreement of optical absorption spectra) showed the success of this technique by predicting core[40,41] and metal–thiolate ligand[41] arrangements in agreement with independent crystal structure data.[42] In 2009, Pei et al.[60] and Jiang et al.[61] proposed a novel protecting motif $Au_3(SR)_4$ that was longer than those previously observed to passivate the core of $Au_{20}(SR)_{16}$. Pei et al. found several isomers with this structure, and determined that their third isomer (which was predicted to be less than 0.1 eV higher in energy than their lowest energy isomer) yielded a theoretical absorption spectrum in excellent agreement with experiment.

Another MPC of interest to the community is $Au_{44}(SPh)_{28}$. In 2010, Jiang et al. predicted the structure of the dication (which would have 18 delocalized electrons in the core) to be a face-centered cubic Au_{28} core surrounded by eight $Au(SR)_2$ and four $Au_2(SR)_3$ motifs.[62] The computed optical absorption spectrum for this cluster gradually increases from 1 to 4 eV, whereas the experimental spectrum possesses noticeable peaks around 2.2 and 3.2 eV. Thus, this suggested that improved structural models could potentially be determined that would yield closer theory–experiment agreement. Later work by Pei et al. on the neutral $Au_{44}(SPh)_{28}$ MPC proposed a chiral Au_{26} core with eight $Au(SR)_2$ and two $Au_2(SR)_3$ motifs.[63] Their TDDFT calculations showed that this structure would have excitations in the 1.5–2.5 eV region reminiscent of the experimental peak at 2.2 eV (although with a slightly different shape), as well as a strong peak around 3.2 eV (Figure 8). The optical gap was also shown to be ~1.5 eV in good agreement with experiment. In addition, they computed TDDFT spectra for the dianionic version of their system and found that the optical gap was only predicted to be 0.3 eV for this case. Overall, the spectral agreement for the new structure was found to be much better, and Pei et al. suggested that this would be the structure of $Au_{44}(SPh)_{28}$ achieved experimentally.

In 2012, Malola et al. suggested that the $Au_{40}(SR)_{24}$ structure could be visualized as a 26-atom core, consisting of two icosahedra that do not share any atoms, that is passivated by six $Au(SR)_2$ and four $Au_2(SR)_3$ motifs.[64] The experimental optical spectrum for this system is fairly featureless, so several isomers including the two lowest energy chiral isomers yield reasonable agreement with experiment. Jiang later suggested a different structure for $Au_{40}(SR)_{24}$ that is related to $Au_{38}(SR)_{24}$ with 2 extra Au atoms at the "waist" of its 23-atom bi-icosahedral core and passivated by 3 $Au(SR)_2$ and 6 $Au_2(SR)_3$ motifs.[65] Although he could not provide optical absorption spectra at the time, he found that this new structure may be preferred by 0.1–0.5 eV.

Malola et al. pursued the cluster-of-icosahedra approach to examine the optical absorption of systems with up to six icosahedral cores arranged linearly.[66] As the cluster chain length increased, the first peak (corresponding to the longitudinal plasmon resonance) shifted to lower energy and increased in intensity, as expected for nanorod-like structures. TCM maps were determined and showed that this excitation arises mainly from a single transition. For the

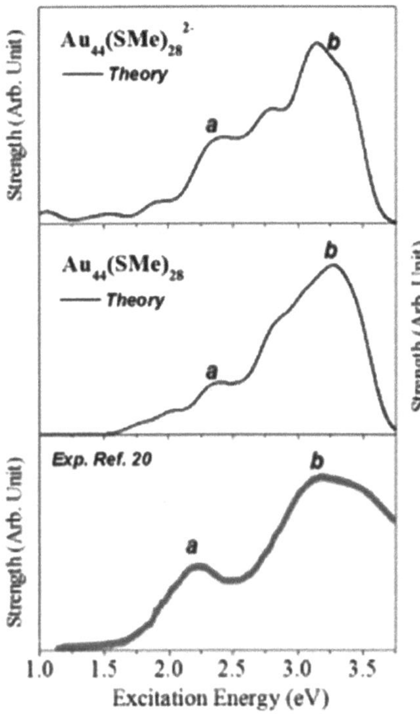

FIGURE 8 Theoretical absorption spectra of (top) $Au_{44}(SCH_3)_{28}^{2-}$ and (middle) neutral $Au_{44}(SCH_3)_{28}$. (bottom) Experimental spectrum of $Au_{44}(SPh)_{28}$ in unknown charge state. *Figure adapted with permission from Ref. 63. Copyright American Chemical Society 2014.*

hexamer, this excitation occurs from the HOMO-1 to the LUMO+3. They also examined the optical absorption spectrum of a model of the $Au_{54}(SH)_{30}$ MPC, which they predicted to have a bi-icosahedral core interacting with a third icosahedron. The TDDFT optical absorption spectrum was found to be more featured than the experimental spectrum, but had three main peaks in good agreement with experiment. As discussed above, an experimental spectrum sometimes includes thermal broadening effects that are not included in a 0 K TDDFT spectrum.

In 2012, Pei et al. proposed a structure for $Au_{24}(SR)_{20}$ involving an elongated Au_8 core passivated by two $Au_3(SR)_4$ and two $Au_5(SR)_6$ motifs that are interlocked.[67] They found that their predicted TDDFT absorption spectra were in reasonable agreement with experimental UV–Vis spectra. They found a first absorption peak around 1.6 eV close to the experimental value. Above 1.6 eV, the theoretical calculations predicted a band of excitations in the 2–2.8 eV range, with a pronounced excitation about 3.1 eV. Although the experimental spectrum is fairly broad in this range, features exist that provide reasonable support for this structure. Since 2012, the system with this stoichiometry has been shown to be quite complicated. A crystal structure of the $Au_{24}(SCH_2Ph{-}^tBu)_{20}$ nanocluster was determined and was found to be passivated by tetramer motifs[57]; however, the selenolate version (i.e., $Au_{24}(SeC_6H_5)_{20}$)[68] was found to have a two trimer and two pentamer passivated structure similar to the one proposed by Pei et al. As discussed in Section 9.3.4.3, the optical absorption spectrum of the $Au_{24}(SCH_2Ph-{}^tBu)_{20}$ MPC was found to be different than that of a previously reported[58] spectrum of $Au_{24}(SCH_2CH_2Ph)_{20}$, so multiple isomers may exist for this stoichiometry.

Shortly thereafter, Tlahuice and Garzón published a structure for the $Au_{18}(SCH_3)_{14}$ MPC possessing an elongated Au_8 core protected by two $Au_2(SR)_3$ and two $Au_3(SR)_4$ protecting units.[69] However, none of their isomers yielded an absorption spectrum in good agreement with experiment; the experimental spectrum for this system has a strong peak with a shoulder around 2.2 eV. They also proposed related structures for a series of nanoclusters created by replacing between one and four of the trimer units of $Au_{20}(SCH_3)_{16}$ with dimer units, to form $Au_{19}(SCH_3)_{15}$, $Au_{18}(SCH_3)_{14}$, $Au_{17}(SCH_3)_{13}$, and $Au_{16}(SCH_3)_{12}$.[70] Moreover, they built geometries for $Au_{13}(SCH_3)_{10}{}^+$, $Au_{14}(SCH_3)_{11}{}^+$, and $Au_{15}(SCH_3)_{12}{}^+$ systems by replacing dimer units of $Au_{12}(SCH_3)_9{}^+$ with trimer units.[70] They calculated TDDFT spectra for each of these systems for future comparison with experiment.

In 2013, Jiang et al. proposed a structure for the smallest known MPC containing "metallic" gold, $Au_{15}(SR)_{13}$, and suggested that its geometry could also be pictured as arising from reduction of a catenane-like species similar to $Au_{24}(SR)_{20}$ suggested by Pei et al. They proposed a geometry with a tetrahedral Au_4 core protected by two trimer units surrounded by a $Au_5(SR)_5$ ring and determined its optical absorption spectrum using TDDFT.[71] This spectrum was found to have many more features than the corresponding experimental spectrum, but the absorption onset of 2.6 eV and the major peak around 3.2 eV agreed well with experiment. Shortly thereafter, Tlahuice-Flores et al.[72] suggested a similar structure for $Au_{15}(SR)_{13}$ with a tetrahedral Au_4 core covered by one tetramer $Au_4(SR)_5$ motif and one heptamer $Au_7(SR)_8$ motif; this structure was very close in energy to the one proposed by Jiang et al. They calculated optical absorption spectra for both structures and predicted absorption onsets below experiment for both. Both structures were found to have a number of peaks in the 2.7–3.5 eV range in agreement with experiment, with the Tlahuice-Flores et al. structure leading to a slightly broader spectrum in better agreement with experiment. However, neither of these two structures yields an optical absorption spectrum that provides conclusive proof for either structure at the current time.

Recently, Yu et al. suggested that the $Au_{22}(SR)_{18}$ MPC has an elongated Au_8 core protected by two $Au_3(SR)_4$ and two $Au_4(SR)_5$ protecting units.[73] The experimental UV–Vis spectrum has two main peaks in the 2.3–2.7 eV range. The TDDFT-simulated spectrum has two peaks at 2.1 and 2.85 eV. In consequence, the theoretical structure yields a spectrum with some similarities to the experimental one, but without the quantitative agreement often seen for the optical absorption comparison.

9.3.4.5 Gold MPCs with Other Ligands

Gold nanoparticles stabilized with phosphine ligands (as well as electron-withdrawing ligands such as thiolates and halides) were some of the earliest systems examined both experimentally and theoretically. For example, the undecagold (Au_{11}) cluster has been examined since the 1960s[74] and is used frequently in biological applications.[75] It is protected by phosphine and either halide or thiolate ligands and has a structured optical absorption spectrum.[7,8] Although the electronic structure of many of these systems has been predicted in depth, calculated absorption spectra have been less common until recently. In 2010, Provorse and Aikens calculated the optical absorption of a particular undecagold compound $Au_{11}(PH_2C_2H_2PH_2)_4X_2^+$ (X = Cl, Br) and found that TDDFT calculations lead to spectra in good agreement with experiment. As expected from earlier predictions by Mingos on various gold–phosphine systems,[76,77] the transitions were found to arise from excitations out of occupied P orbitals into unoccupied D orbitals.

Optical absorption spectra of other gold–phosphine compounds have also been examined using TDDFT calculations. In 2011, Ivanov et al. investigated the optical absorption spectra of $Au_4(PR_3)_4^{2+}$ and $Au_4(\mu_2\text{–}I)_2(PR_3)_4$ compounds.[20] Both systems have two delocalized electrons. They found that for the bare Au_4^{2+} core, the highest occupied orbitals arise from Au(5d) atomic orbitals, with the superatom S orbital from Au(6s) atomic orbitals lying lower in energy. The unoccupied orbitals were found to have P symmetry. For the systems with ligands, the Au(5d) and Au(6s) orbitals mix, so that the unoccupied orbitals do not arise solely from the Au(6s) atomic orbitals. They found that excitation spectra calculated with PH_3 ligands and with $P(CH_3)_3$ ligands did not differ significantly in fundamental origin or shape, although the trimethylphosphine ligands lead to a red shift of 0.25–0.50 eV compared with the simple phosphine. However, the replacement of these ligands with triphenylphosphine changes the nature of the electronic transitions, which now involve excitations into low-lying π^* orbitals. They found that triple-zeta basis sets uniformly lowered the energies of intense transitions by 0.15–0.17 eV, which is similar to the basis set effects on $Au_{25}(SH)_{18}^-$ described in Section 9.3.2.2. These authors also found that the absorption spectra display a great sensitivity to the functional used in the TDDFT calculations, with pure DFT functionals underestimating the energies and functionals involving some degree of Hartree–Fock (i.e., hybrid and long-range corrected) overestimating the energies. Incorporation of a continuum THF solvent was found to systematically blue shift the lower energy transitions by 0.2–0.4 eV, yielding spectra that were in excellent agreement with experiment.

Goel et al. examined the effect of phosphine, amine, and other ligands on the optical absorption spectra of Au_2 and Au_4 clusters.[78] They found that the ligand coordination gave rise to new optically allowed states, which broadened the transitions compared to the bare clusters. They found that many of the optically forbidden states had charge-transfer character, which suggested that long-range corrected functionals were necessary for treating these excited states.

The first theoretical optical absorption study of ligated Au_{20} systems was performed by Aikens and Schatz,[79] who examined the binding of NH_3 and PH_3 to various sites on the tetrahedral gold cluster as models of a recently synthesized[80] tetrahedral $Au_{20}(PPh_3)_8$ complex. They found that amines generated only small changes in the predicted TDDFT absorption spectrum compared to the bare cluster. In contrast, phosphine ligands caused the disappearance of the 2.89 eV in the bare Au_{20} peak and the addition of a strong peak at 3.27 eV. Recently, a different Au_{20}-containing cluster stabilized by phosphine ligands has had its crystal structure determined[81] and its absorption spectrum analyzed. Knoppe et al. investigated the $Au_{20}(PP_3)_4^{4+}$ (PP_3 = tris(2-(diphenylphosphino)ethyl)-phosphine) nanoparticle whose geometry could be described as a 7-atom chiral fanlike structure on top of a 13-atom icosahedral core.[82] The electronic structure of this system was determined to be $1S^2 1P^6 1D^6 2S^2$ in which the splitting of the D orbitals into three occupied and two unoccupied orbitals arises from its nonspherical core. The calculated and experimental optical absorption spectra were found to be in moderate agreement, with excellent agreement between the experimental optical gap of 1.33 eV and the calculated HOMO-LUMO gap of 1.30 eV.

In 2011, Walter et al. presented a $Au_{69}(PR_3)_{20}Cl_{12}^-$ model complex as a possible formulation for the "Schmid Au55" cluster, which was originally given the stoichiometry of $Au_{55}(PPh_3)_{12}Cl_6$.[83] Later that year, Burgess and Keast performed RT-TDDFT calculations on both systems with PH_3 ligands.[84] For all clusters examined, they found an absorption spectrum beginning around 1 eV with an intensity generally increasing through 8 eV. They ascertained that the spectra of the ligated clusters exhibited fewer features than the corresponding bare clusters. Due to the featureless experimental UV–Vis spectrum, neither stoichiometry could be assigned to the experimental system on the basis of optical absorption alone.

Some MPCs can be stabilized by both phosphine and thiolate ligands. In 2007, Shichibu et al. determined the X-ray crystal structure of $[Au_{25}(PPh_3)_{10}(SC_nH_{2n+1})_5Cl_2]^{2+}$ ($n = 2-18$) systems and found that the Au_{25} consists of two icosahedral Au_{13} clusters that share a vertex.[85] They found that this system has a discrete optical absorption spectrum with a low energy peak at 670 nm. Nobusada and Iwasa then calculated the TDDFT spectra for this system and for a related tri-icosahedral structure.[86] They determined the absorption spectrum in good agreement with experiment, which allowed them to assign the first peak as a HOMO to LUMO transition. This transition only arises as a result of the dimeric structure of the nanoparticle. For the tri-icosahedral structure, they predicted peaks at 761 and 1238 nm due to the interaction of the three vertex-sharing Au_{13} icosahedra. In 2013, Iwasa et al.

considered doping of the bi-icosahedral cluster with Pt and Hg atoms at the centers of the two icosahedra.[87] They found that the HOMO-LUMO gap in the predicted TDDFT optical absorption spectrum red shifts by 0.3 eV relative to the pure gold system.

In 2012, Das et al. determined the crystal structure of $Au_{24}(PPh_3)_{10}(SCH_2CH_2Ph)_5X_2^+$ (X = Cl, Br), which was found to be essentially the same as $Au_{25}(PPh_3)_{10}(SC_nH_{2n+1})_5Cl_2^{2+}$ except it was missing the central Au atom.[88] This leads to a slight shortening of the overall length of the nanoparticle by about 0.13 Å. TDDFT calculations on this system with methylthiolate and PH_3 ligands agreed nicely with the experimental spectrum (Figure 9). They found that the first peak in the absorption spectrum is blue shifted compared to $Au_{25}(PPh_3)_{10}(SCH_2CH_2Ph)_5X_2^{2+}$. The authors predicted that this peak arises from a HOMO-1 to LUMO+2 transition instead of a HOMO to LUMO transition. For both the Au_{24} and Au_{25} MPCs, they ascertained that the transitions are caused by interactions between the two Au_{13} icosahedral cores. Similar conclusions were reached by Goh et al., who considered the optical absorption spectrum of the full $Au_{24}(PPh_3)_{10}(SCH_2CH_2Ph)_5Cl_2^+$ compound with 457 atoms.[89]

A few MPCs have been studied that do not contain either thiolate or phosphine ligands. In 2009, Lopez-Acevedo et al. published a combined theoretical–experimental study of recently crystallized carbonyl-protected Au–Fe particles such as $Au_{22}[Fe(CO)_4]_{12}^{6-}$.[90] They found excellent

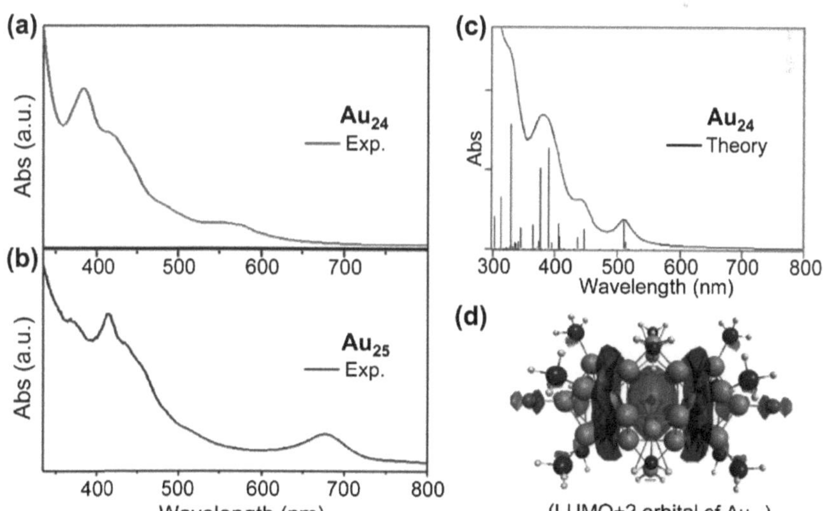

FIGURE 9 Experimental optical absorption spectra of (a) $Au_{24}(PPh_3)_{10}(SCH_2CH_2Ph)_5Cl_2^+$ and (b) $Au_{25}(PPh_3)_{10}(SCH_2CH_2Ph)_5X_2^{2+}$. (c) Theoretical absorption spectrum of $Au_{24}(PH_3)_{10}(SCH_3)_5Cl_2^+$. (d) LUMO+2 orbital. *Figure reproduced with permission from Ref. 88. Copyright American Chemical Society 2012.*

agreement between TDDFT and experimental optical spectra for systems containing Rb or Na counterions and for a charged system with a charge of −4. On the other hand, their computed spectra differed significantly for the −6 charge system, which led the authors to propose that the true charge state in solution is less than the formal charge of −6.

9.3.4.6 Silver–Thiolate MPCs

One of the first silver–thiolate nanoparticles with strong, discrete absorption spectra was synthesized by Bakr et al.[91] It possessed six strong and two weak peaks across the visible and near-UV region. To understand the origin of these discrete peaks, a $Ag_{25}(SH)_{18}^{-}$ model nanoparticle analogous to the recently solved $Au_{25}(SR)_{18}^{-}$ nanoparticle was employed.[14,91] This model MPC possessed several molecular-like peaks whose origin came from delocalized superatom orbitals split by ligand-field splitting in direct analogy with the gold–thiolate MPCs. Although the model provided insights into the origin of the optical properties, the peak locations were not close to the experimental UV–Vis spectrum, which suggested that the silver–thiolate nanoparticle that had been synthesized had a different structure and/or stoichiometry. After a few years, the stoichiometry of this particle was determined to be $Ag_{44}(SR)_{30}^{4-}$.[92] In 2013, two groups solved the X-ray crystal structure of this nanoparticle and found that it consists of an inner icosahedral core of 12 atoms surrounded by a 20-atom decahedral outer core, which is capped with six $Ag_2(SR)_5$ motifs.[93,94] The MPC has 18 delocalized electrons, which suggests that orbitals up to 1D are occupied, while the 2S and 1F orbitals are unoccupied. Indeed, the HOMO was found to be a 1D superatomic orbital, the LUMO to be the 2S orbital, and the LUMO+1 to be a 1F orbital.[93,94] The optical absorption spectrum of $Ag_{44}(SR)_{30}^{4-}$ was calculated by LR-TDDFT and TD-DFPT was employed to analyze the transitions.[93] The two lowest energy transitions were found to arise mainly from 1D → 1F-type transitions. Ligand-based (i.e., $Ag_2(SR)_5$) orbitals contributed to higher energy peaks. Initial TDDFT optical absorption spectra for this cluster achieved very good agreement with experiment for the peak around 650 nm, but only predicted a single peak around 500 nm instead of two in the 450–550 nm range.

In 2013, Yang et al. determined X-ray crystal structures for gold-doped nanoparticles $Au_{12}Ag_{32}(SR)_{30}^{4-}$.[93] They found that the gold atoms are preferentially located in the 12-atom core, whereas silver atoms are present in the outer decahedral core. The absorption spectra for these systems actually display fewer peaks than the pure silver systems, with a single peak in the 490–510 nm region and no sharp peaks (but three weak peaks) at longer wavelengths. Recently, Gell and Häkkinen examined the TDDFT optical absorption spectra for a series of $M_{12}Ag_{32}(SR)_{30}^{4-}$ clusters (M = Au, Ag) and investigated the effects of chemical changes in the metal core, the charge of the cluster, and the nature of the thiolate ligand.[95] They also considered the effects of doping the core with a manganese atom or with a proton. As observed earlier for systems like $Au_{25}(SR)_{18}$, they found that the optical absorption spectrum depended strongly on the charge state of the cluster. Importantly, they found that the higher energy peaks with wavelengths below 641 nm had strong contributions from the π system of the ligands, and methyl ligands were unable to reproduce the features of aromatic ligands. They also hypothesized that their previous calculations using fluorinated phenylthiolate ligands were unable to reproduce the double peak due to charge-transfer considerations, since the nonfluorinated SPh ligands produced a TDDFT spectrum with the double peak structure.

$Ag_{44}(SR)_{30}^{4-}$ is not the only thiolate-stabilized silver MPC that has been characterized optically. In 2009, Cathcart et al. produced chiral silver–thiolate nanoparticles stabilized by ligands such as glutathione, captopril, and cysteine.[96] They employed an etching procedure that yielded a single stoichiometry of MPC with well-defined optical transitions at 660, 490, and 335 nm for the captopril-stabilized system. For glutathione-stabilized systems, the corresponding transitions fall at 635, 480, and 325 nm, and cysteine-stabilized systems display only a single peak at 488 nm. Thus, these MPCs exhibit some ligand dependence on peak locations in their UV–Vis spectra. In 2010, Bigioni and coworkers synthesized and separated a series of silver–glutathione MPCs.[97] They found at least 16 distinct bands and determined optical absorption spectra that displayed one or more discrete peaks (Figure 10). Their band 6 presented an optical spectrum very similar to that of Cathcart et al. Bigioni and coworkers later assigned the stoichiometry of this compound to $Ag_{32}(SG)_{19}$.[98]

In 2013, Bertorelle et al. performed an investigation of $Ag_{31}(SG)_{19}$ and $Ag_{15}(SG)_{11}$ nanoparticles using both experiment and theory.[99] They determined the stoichiometries of these systems via mass spectrometry, and assigned the first one to the band 6 of Bigioni and coworkers since it displayed

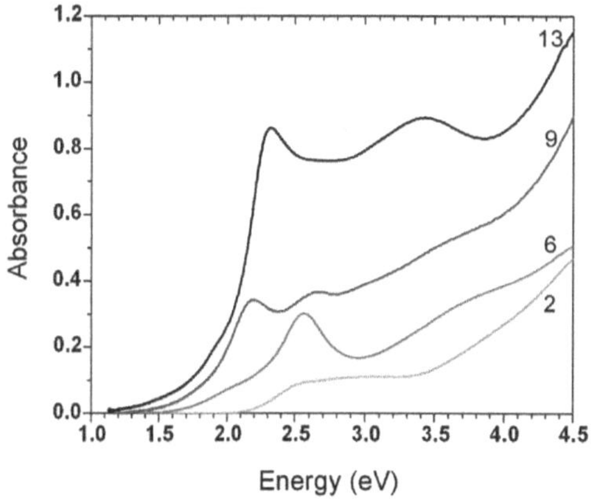

FIGURE 10 Optical absorption spectra of four glutathione-stabilized silver MPCs. *Figure reproduced with permission from Ref. 97. Copyright American Chemical Society 2010.*

only a single strong peak around 490 nm. DFT calculations predicted a structure for $Ag_{31}(SR)_{19}$ with a Ag_{21} core protected by eight SR–Ag–SR units and one SR–Ag–SR–Ag–SR unit. Their TDDFT-computed spectrum for $Ag_{31}(SR)_{19}$ exhibited a strong peak at 510 nm, which was in good agreement with the experimental spectrum. This peak was found to arise from a D to F transition.

9.4 NONLINEAR OPTICAL PROPERTIES

In 2008, Ramakrishna et al. determined that gold MPCs ranging from $Au_{25}(SR)_{18}^-$ up to a system with about 2400 gold atoms were efficient two-photon absorbers.[100] The two-photon absorption (TPA) cross sections of these MPCs were found to be very high. Au_{25} was found to have a cross section of 2700 GM at 1.92 eV. Combined with luminescence in the near-IR region, it was proposed that this system may be useful for multiphoton imaging in biological applications. The larger systems were found to be even more efficient absorbers, leading to potential applications in nanolithography and in optical power limiting.

In 2010, Day et al. calculated the TPA cross section for $Au_{25}(SH)_{18}^-$ using TDDFT calculations. They found that the B3LYP exchange-correlation functional performed the best for the one-photon absorption spectrum. They determined that the high density of states in this system led to strong resonance effects, which led to a wide variation in TPA cross sections. They found that the largest TPA cross sections benefited from resonance with the first excited state. Transitions from the 1P to 1F orbitals (which are forbidden in one-photon absorption) were ascertained to be the strongest in the spectrum.

9.5 CHIRALITY AND CHIROPTICAL PROPERTIES

9.5.1 Fundamental Origin

In 1998, Whetten and coworkers synthesized the $Au_{25}(SR)_{18}^-$ nanoparticle with glutathione (SG) ligands and found that the compound displayed optical activity with strong CD signals (Figure 11).[4] CD spectra arise as a difference in absorption between left circularly polarized light and right circularly polarized light, so the peaks in the spectrum may be either positive or negative. The fine features present in the optical absorption spectrum that suggested a rich spectrum of energy levels in the $Au_{25}(SR)_{18}^-$ nanocluster were even more pronounced in its CD spectrum. For this system, the CD spectrum appeared to be dominated by the interband transitions. Whetten and coworkers later synthesized smaller and larger MPCs with glutathione ligands and likewise found very intense optical activity for these nanoclusters.[5] They suggested several hypotheses that could be used to explain the optical activity, which include (1) the geometry of the metal core is inherently chiral, (2) the ligands adsorb

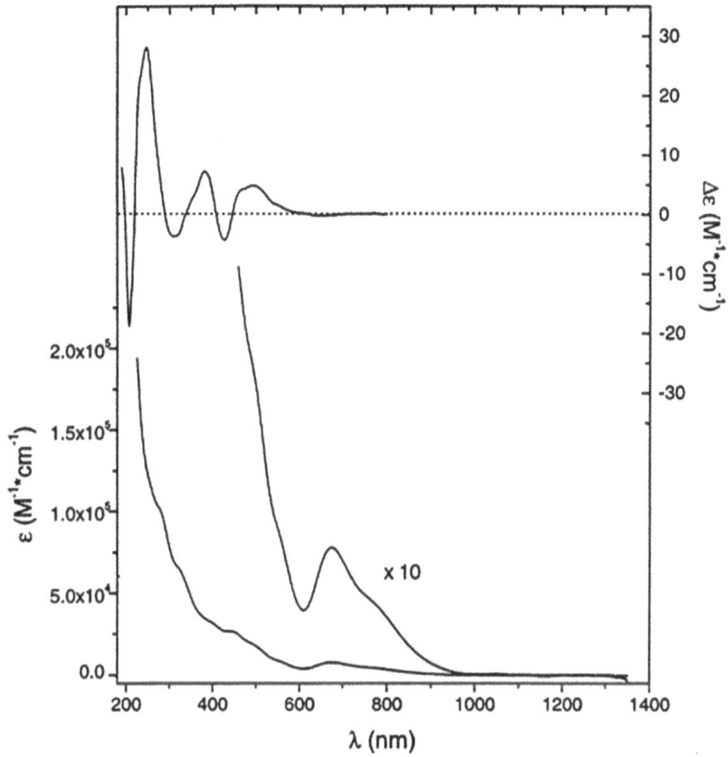

FIGURE 11 Circular dichroism and optical absorption spectra for $Au_{25}(SG)_{18}^{-}$. *Figure reproduced with permission from Ref. 4. Copyright American Chemical Society 1998.*

on the surface of the core in a chiral pattern, or (3) the chiral ligands induce optical activity in the electronic structure of the core even though neither the core itself nor the adsorption pattern is inherently chiral. Other chiral thiolate ligands such as penicillamines[101] have since been shown to also produce strong CD signals in gold MPCs.

Around the same time, Garzón and coworkers determined that the lowest energy structures of neutral bare metal clusters including Au_{38}, Au_{55}, and Au_{75} are disordered (and therefore chiral) rather than the high-symmetry systems often examined.[102–104] This provided support for the first explanation, in which a chiral core is responsible for the strong CD signals observed for the ligand-protected clusters. In contrast, Beratan and coworkers used both TDDFT calculations and a charge-perturbed particle-in-a-box model to examine the CD spectrum of a gold core surrounded by chiral ligands.[105] They found that the presence of chiral ligands was sufficient to induce CD signals in an achiral metal core, which provided support for the third proposed explanation.

In 2006, Tsukuda and coworkers showed that well-defined gold clusters passivated with chiral phosphine ligands could also exhibit strong CD signals.[106] The stoichiometry of the MPC that they examined was determined to be $Au_{11}(BINAP)_4X_2^+$ (BINAP = 2,2′-bis(diphenylphosphino)-1,1′-binaphthyl; X = Cl, Br). A few years later, Provorse and Aikens examined this system theoretically using TDDFT calculations.[19] To predict CD spectra, the transition electric dipole moment as well as the transition magnetic dipole moment must be computed via TDDFT in order to calculate rotatory strengths. For this system, inclusion of a continuum chloroform solvent was found to be critical for accurately predicting the CD spectrum of the Au_2X_2BINAP starting material, so all calculations employed this solvent. They found that the lowest energy structure of $Au_{11}(dpb)_4X_2^+$, where dpb is the model ligand 1,4-diphosphino-1,3-butadiene, has chiral C_2 point group symmetry. The ligands essentially adsorbed on the core in a chiral pattern, which induced C_2 symmetry within the core. The predicted CD spectrum for the model compound was found to agree well with the first two main peaks for the experimental compound (although with a red shift of 0.35 eV). The third peak was found to have the opposite sign as experiment, which could be attributed to the use of the model ligand, since this did not permit excitations into π^* orbitals of the ligands that were found to be important in the 300–350 nm wavelength region. When the ligands were removed from the cluster, the CD spectrum of the bare C_2-symmetric core yielded very weak signals, suggesting that the chirality of the core was not as important of an explanation for strong optical activity as the presence of the chiral field arising from the ligands.

9.5.2 Gold–Thiolate MPC Chirality

In 2007, the seminal X-ray crystal structure determination of the $Au_{102}(p\text{-MBA})_{44}$ (p-MBA = *para*-mercaptobenzoic acid) nanoparticle showed that gold MPCs with achiral ligands could be chiral.[10] The core atoms have approximate D_{5h} symmetry, and the overall chirality of this system arises due to the arrangement of the protecting metal–thiolate oligomeric units. (The symmetry of the core is then distorted slightly due to the presence of the chiral protecting layer.) This chirality has since proven to be a general principle for many thiolate-protected nanoparticles. Although the $Au_{25}(SR)_{18}^-$ nanoparticle is not chiral unless chiral thiolates are employed, other systems such as $Au_{38}(SR)_{24}$ also exhibit an essentially achiral core surrounded in a chiral pattern by metal–thiolate units.[41,42] This has aided structure prediction; for example, the predicted structure of $Au_{144}(SR)_{60}$ for which no crystal structure currently exists has an icosahedral core protected in a chiral pattern by 30 $Au(SR)_2$ units;[48] the overall MPC can potentially have chiral I symmetry depending on the arrangement of the R groups.[52]

9.5.3 Computed CD Spectra for MPCs

The $Au_{38}(SR)_{24}$ nanoparticle was the first thiolate-protected nanoparticle for which calculation of CD spectra helped in the geometrical prediction.[41] The lowest energy chiral D_3 structure of this MPC with R = CH_3 was found to have a CD spectrum with four main peaks below 2.2 eV (Figure 12);

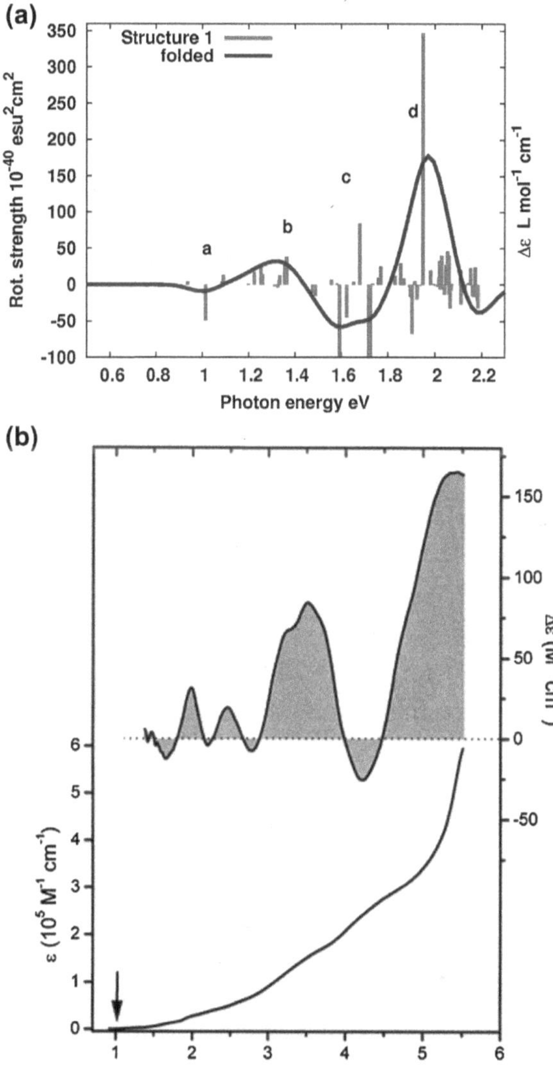

FIGURE 12 (a) TDDFT CD spectrum of $Au_{38}(SCH_3)_{24}$. *(Figure adapted with permission from Ref. 41. Copyright American Chemical Society 2010.)* (b) Experimental CD and absorption spectrum of $Au_{38}(SG)_{24}$. *Figure adapted with permission from Ref. 5. Copyright American Chemical Society 2000.*

the signals for this system were much stronger than corresponding signals for an asymmetric structure based on a C_{3h} arrangement of the protecting units. As discussed in Section 9.3.4.1, the first three peaks arise mainly from transitions within the sp-based orbital manifold. They are relatively weak in the CD spectrum. In contrast, peak d arises primarily from interband transitions; i.e., transitions out of the orbitals arising from Au(5d) and S(3p) contributions on the *chirally arranged* metal–thiolate-protecting units into the sp-based orbitals. This peak is found to be much stronger than peaks a–c in the CD spectrum. Overall, the CD spectrum of the chiral D_3 structure agreed very well with experiment. Although peaks a and b lie essentially below the wavelength range of the experimental apparatus, the moderately strong and negative c peak and the intense and positive d peak exhibited excellent agreement in shape and energy. The agreement enabled assignment of the experimental $Au_{38}(SG)_{24}$ nanoparticle as a clockwise (*C*-) enantiomer.[107,108] Overall, it appears that the chiral arrangement of the metal–thiolate-protecting units affects not only the geometry of an MPC but also its chiroptical properties. Molina et al. then studied the chiroptical properties of an anticlockwise (*A*-) $Au_{38}(SCH_3)_{24}$ nanoparticle that had undergone ligand exchange with an *R*-1-1′-binaphthyl-2,2′-dithiol (*R*-BINAS) ligand.[109] They found that the isomers that were predicted to be more stable by DFT also had CD spectra that were not changed greatly from the spectrum of the parent cluster.

A number of other studies have also been done on the CD spectra of related MPCs. Sánchez-Castillo et al. examined the CD spectra of asymmetric $Au_{25}(SR)_{18}^-$ complexes with achiral ligands and found that these structures provided nonzero CD spectra.[110] They also computed the CD spectrum of a nanoparticle with cysteine ligands and found that the signals were somewhat stronger when a chiral ligand was used.

In 2012, Malola et al. employed TDDFT calculations to determine CD spectra for proposed $Au_{40}(SR)_{24}$ isomers.[64] They found that the isomers with a chiral arrangement of the protecting units yielded stronger CD signals than the ones with a nonchiral Au–S framework. They found that the CD spectrum of their chiral lowest energy isomer was in reasonable agreement with experiment, and suggested that some of the remaining differences could arise due to the orientation of the R-groups.

Tlahuice-Flores and coworkers also utilized CD calculations to provide support for geometrical predictions. In a study on $Au_{18}(SCH_3)_{14}$, they found several isomers with a prolate Au_8 core that produced CD spectra with a negative peak around 2.1 eV and a positive peak around 2.5 eV, in good agreement with experiment.[69] They also predicted CD spectra for characterization of a series of hypothetical compounds around this size range that have not yet been synthesized.[70] In their study of $Au_{15}(SCH_3)_{13}$, they found that the originally predicted isomer for this system did not yield an intense negative peak at 3.5 eV.[72] In contrast, their newly predicted isomer was found to have a strong negative CD peak at 3.5 eV followed by a change to a positive peak around 3.7−3.9 eV. Differences in sign between experiment and theory above 3.9 eV could potentially be attributed to factors such as the use of a model ligand.

Recent crystal structures have also confirmed that many additional MPCs are chiral. During their analysis of the $Au_{28}(SR)_{20}$ crystal structure, Knoppe et al. calculated a CD spectrum for the right-handed *C*-enantiomer of $Au_{28}(SCH_3)_{20}$; based on good agreement with the CD spectrum for the experimental "enantiomer 2," they assigned the clockwise (*C*-) orientation for this enantiomer (Figure 13).[55] In addition, the $Au_{30}S(S{-}^tBu)_{18}$ MPC of Crasto et al. was found to be chiral.[59] The authors computed CD spectra for comparison with future enantioseparation work.

Using the known chiral crystal structures of $Au_{28}(SR)_{20}$ and $Au_{36}(SR)_{24}$, Pei et al. predicted a structure for the slightly larger $Au_{44}(SR)_{28}$ MPC.[63] They predicted a face-centered cubic core surrounded by eight $Au_2(SR)_3$ motifs and two $Au(SR)_2$ motifs, which are arranged in a chiral fashion around the core. As discussed in Section 9.3.4.4, the optical absorption spectrum of this system was found to be in good agreement with experiment. The CD spectrum was also computed by TDDFT and several hundred transitions were found below 4.2 eV. Although this spectrum was not in quite as good agreement as the optical (likely due to its complexity), the number of peaks and overall pattern of sign changes were in reasonable agreement.

In 2014, Knoppe et al. analyzed the CD spectrum of a phosphine-stabilized $Au_{20}(PP_3)_4^{4+}$ nanoparticle.[82] They found that the CD spectra were able to resolve more transitions than the absorption spectra. They learned that although the rotatory strength and anisotropy factors were stronger for the bare Au_{20}^{4+} core than for the bi-icosahedral Au_{23}^{9+} core of $Au_{38}(SR)_{24}$, the situation was reversed when the ligands were added. They attributed this to the effect of the chiral PP_3 ligands not being as strong as for the $Au_2(SR)_3$ units of $Au_{38}(SR)_{24}$.

9.5.4 Silver MPCs

As discussed in Section 9.3.4.6, silver nanoparticles can also display CD signals. In the Cathcart et al. study, the three different chiral ligands led to relatively similar absorption spectra, but vastly different CD spectra.[96] This implies that although the effect of ligands can be small for absorption spectra, this cannot be neglected when considering CD spectra. Other ligands such as penicillamine have been used to synthesize silver MPCs that display strong CD signals.[111,112]

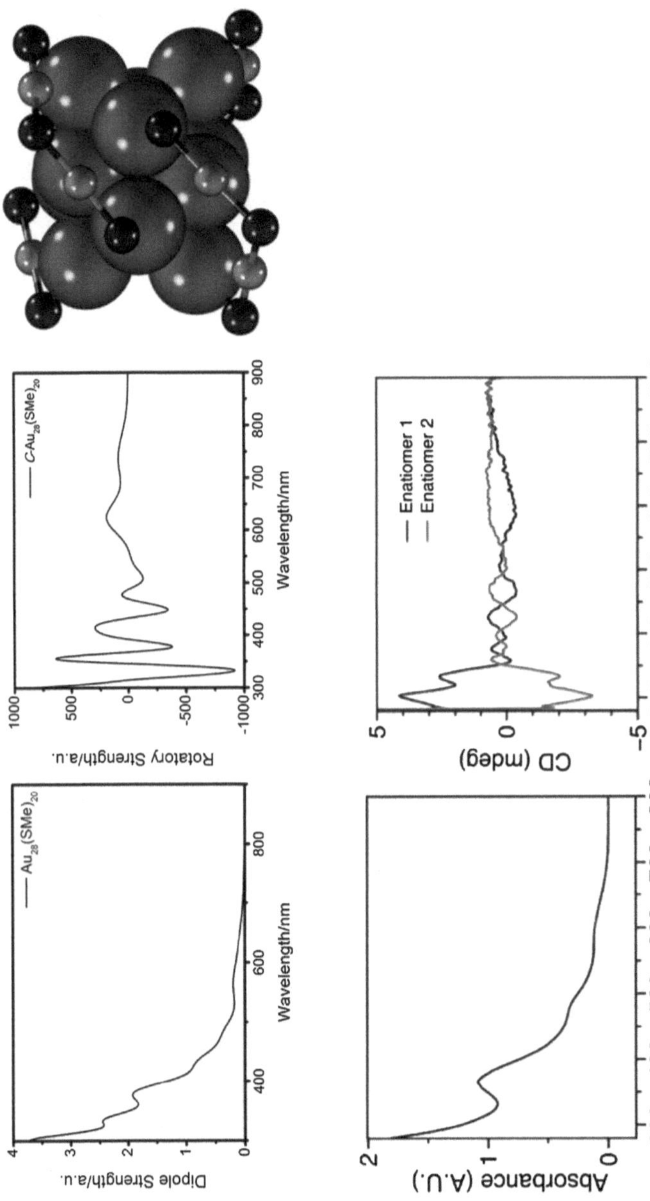

FIGURE 13 Top: Theoretical absorption (left) and circular dichroism (right) spectra for *C*-$Au_{28}(SCH_3)_{20}$. Bottom: Experimental absorption (left) and CD (right) spectra for $Au_{28}(SR)_{20}$. *Figure reproduced with permission from Ref. 55. Copyright American Chemical Society 2013.*

9.6 SUMMARY AND PROSPECTS

The optical absorption and CD spectra of gold and silver MPCs are heavily dependent on their underlying electronic structure. Near the HOMO-LUMO gap, these systems typically consist of a sparse set of delocalized orbitals that arise primarily from the Au(6s) or Ag(5s) atomic orbitals. For systems with a nearly spherical core, these delocalized orbitals have symmetries similar to those of atoms. For systems with other core symmetries such as cylindrical or tetrahedral, the orbitals reflect the corresponding symmetries. These orbitals can be split into several sets due ligand-field splitting or to symmetry-breaking within the core itself. This splitting has important effects on the optical absorption spectra, as multiple peaks result from transitions between these orbitals.

In addition, thiolate-stabilized MPCs have a dense band of orbitals below the sp-band orbitals that arises from S(3p) and Au(5d) or Ag(4d) atomic orbitals from the metal−thiolate oligomeric ligands. Excitations out of these systems are observable in absorption spectra, especially in gold-containing systems.

Systems with phenyl groups directly attached to sulfur or phosphorus often exhibit excitations into π^* ligand orbitals. For these systems, substitution of the aromatic groups with smaller methyl ligands is often not a good approximation for accurate calculation of the absorption spectra. However, in other cases the smaller methyl ligand yields excellent agreement with experiment, so for now this substitution must be analyzed on a case-by-case basis.

Overall, TDDFT calculations can yield absorption spectra in excellent agreement with experiment for systems whose geometries are accurate. It is a useful tool for prediction of unknown structures, as experimental absorption spectra are often available. TDDFT calculations are becoming routinely employed in studies on gold and silver MPCs.

Calculated CD spectra are also useful for interpretation of experimental results as well as for structure prediction. Overall, the accuracy of the CD spectra is not quite as good as the corresponding optical absorption spectra, which appears to be due in part to the many factors that are responsible for the generation of this phenomenon, and the spectra appear to be more sensitive to the choice of ligand than the corresponding optical absorption spectra. Nonetheless, it has become a very useful technique for examining chiral MPCs, of which there are many, and is likely to be routinely utilized in the future.

REFERENCES

1. Wyrwas RB, Alvarez MM, Khoury JT, Price RC, Schaaff TG, Whetten RL. The colours of nanometric gold. *Eur Phys J D* 2007;**43**:91−5.
2. Alvarez MM, Khoury JT, Schaaff TG, Shafigullin MN, Vezmar I, Whetten RL. Optical absorption spectra of nanocrystal gold molecules. *J Phys Chem B* 1997;**101**:3706−12.

3. Schaaff TG, Shafigullin MN, Khoury JT, Vezmar I, Whetten RL, Cullen WG, et al. Isolation of smaller nanocrystal Au molecules: robust quantum effects in optical spectra. *J Phys Chem B* 1997;**101**:7885–91.
4. Schaaff TG, Knight G, Shafigullin MN, Borkman RF, Whetten RL. Isolation and selected properties of a 10.4 kDa gold: glutathione cluster compound. *J Phys Chem B* 1998;**102**:10643–6.
5. Schaaff TG, Whetten RL. Giant gold-glutathione cluster compounds: intense optical activity in metal-based transitions. *J Phys Chem B* 2000;**104**:2630–41.
6. Negishi Y, Nobusada K, Tsukuda T. Glutathione-protected gold clusters revisited: bridging the gap between gold(I)-thiolate complexes and thiolate-protected gold nanocrystals. *J Am Chem Soc* 2005;**127**:5261–70.
7. Bartlett PA, Bauer B, Singer SJ. Synthesis of water-soluble undecagold cluster compounds of potential importance in electron microscopic and other studies of biological systems. *J Am Chem Soc* 1978;**100**:5085–9.
8. Jahn W. Review: chemical aspects of the use of gold clusters in structural biology. *J Struct Biol* 1999;**127**:106–12.
9. Gutiérrez E, Powell RD, Furuya FR, Hainfeld JF, Schaaff TG, Shafigullin MN, et al. Greengold, a giant cluster compound of unusual electronic structure. *Eur Phys J D* 1999;**9**:647–51.
10. Jadzinsky PD, Calero G, Ackerson CJ, Bushnell DA, Kornberg RD. Structure of a thiol monolayer-protected gold nanoparticle at 1.1 Å resolution. *Science* 2007;**318**:430–3.
11. Heaven MW, Dass A, White PS, Holt KM, Murray RW. Crystal structure of the gold nanoparticle $[N(C_8H_{17})_4][Au_{25}(SCH_2CH_2Ph)_{18}]$. *J Am Chem Soc* 2008;**130**:3754–5.
12. Zhu M, Aikens CM, Hollander FJ, Schatz GC, Jin R. Correlating the crystal structure of a thiol-protected Au_{25} cluster and optical properties. *J Am Chem Soc* 2008;**130**:5883–5.
13. Akola J, Walter M, Whetten RL, Häkkinen H, Grönbeck H. On the structure of thiolate-protected Au_{25}. *J Am Chem Soc* 2008;**130**:3756–7.
14. Aikens CM. Origin of discrete optical absorption spectra of $M_{25}(SH)_{18}^-$ nanoparticles (M = Au, Ag). *J Phys Chem C* 2008;**112**:19797–800.
15. Aikens CM. Effects of core distances, solvent, and level of theory on the TDDFT optical absorption spectrum of the thiolate-protected Au_{25} nanoparticle. *J Phys Chem A* 2009;**113**:10811–7.
16. Guidez EB, Aikens CM. Development of a charge-perturbed particle-in-a-sphere model for nanoparticle electron structure. *Phys Chem Chem Phys* 2012;**14**:4287–95.
17. Antonello S, Perera NV, Ruzzi M, Gascón JA, Maran F. Interplay of charge state, lability, and magnetism in the molecule-like $Au_{25}(SR)_{18}$ cluster. *J Am Chem Soc* 2013;**135**: 15585–94.
18. Zhu M, Aikens CM, Hendrich MP, Gupta R, Qian H, Schatz GC, et al. Reversible switching of magnetism in thiolate-protected Au_{25} superatoms. *J Am Chem Soc* 2009;**131**:2490–2.
19. Provorse MR, Aikens CM. Origin of intense chiroptical effects in undecagold subnanometer particles. *J Am Chem Soc* 2010;**132**:1302–10.
20. Ivanov SA, Arachchige I, Aikens CM. Density functional analysis of geometries and electronic structures of gold-phosphine clusters: the case of $Au_4(PR_3)_4^{2+}$ and $Au_4(\mu_2\text{-}I)_2(PR_3)_4$. *J Phys Chem A* 2011;**115**:8017–31.
21. Aikens CM, Li S, Schatz GC. From discrete electronic states to plasmons: TDDFT optical absorption properties of Ag_n ($n = 10, 20, 35, 56, 84, 120$) tetrahedral clusters. *J Phys Chem C* 2008;**112**:11272–9.

22. Johnson HE, Aikens CM. Electronic structure and TDDFT optical absorption spectra of silver nanorods. *J Phys Chem A* 2009;**113**:4445–50.
23. Guo R, Murray RW. Substituent effects on redox potentials and optical gap energies of molecule-like $Au_{38}(SPhX)_{24}$ nanoparticles. *J Am Chem Soc* 2005;**127**:12140–3.
24. Aikens CM. Geometric and electronic structure of $Au_{25}(SPhX)_{18}^-$ (X = H, F, Cl, Br, CH_3, and OCH_3. *J Phys Chem Lett* 2010;**1**:2594–9.
25. Tlahuice-Flores A, Whetten RL, Jose-Yacaman M. Ligand effects on the structure and the electronic optical properties of anionic $Au_{25}(SR)_{18}$ clusters. *J Phys Chem C* 2013;**117**:20867–75.
26. Akola J, Kacprzak KA, Lopez-Acevedo O, Walter M, Grönbeck H, Häkkinen H. Thiolate-protected Au_{25} superatoms as building blocks: dimers and crystals. *J Phys Chem C* 2010;**114**:15986–94.
27. Jiang D-e, Kühn M, Tang Q, Weigend F. Superatomic orbitals under spin-orbit coupling. *J Phys Chem Lett* 2014;**5**:3286–9.
28. Negishi Y, Iwai T, Ide M. Continuous modulation of electronic structure of stable thiolate-protected Au_{25} cluster by Ag doping. *Chem Commun* 2010;**46**:4713–5.
29. Guidez EB, Mäkinen V, Häkkinen H, Aikens CM. Effects of silver doping on the geometric and electronic structure and optical absorption spectra of the $Au_{25-n}Ag_n(SH)_{18}^-$ ($n = 1$, 2, 4, 6, 8, 10, 12) bimetallic nanoclusters. *J Phys Chem C* 2012;**116**:20617–24.
30. Walter M, Moseler M. Ligand-protected gold alloy clusters: doping the superatom. *J Phys Chem C* 2009;**113**:15834–7.
31. Malola S, Häkkinen H. Electronic structure and bonding of icosahedral core-shell gold-silver nanoalloy clusters $Au_{144-x}Ag_x(SR)_{60}$. *J Phys Chem Lett* 2011;**2**:2316–21.
32. Kumara C, Aikens CM, Dass A. X-ray crystal structure and theoretical analysis of $Au_{25-x}Ag_x(SCH_2CH_2Ph)_{18}^-$ alloy. *J Phys Chem Lett* 2014;**5**:461–6.
33. Kacprzak KA, Lehtovaara L, Akola J, Lopez-Acevedo O, Häkkinen H. A density functional investigation of thiolate-protected bimetal $PdAu_{24}(SR)_{18}^z$ clusters: doping the superatom complex. *Phys Chem Chem Phys* 2009;**11**:7123–9.
34. Negishi Y, Kurashige W, Niihori Y, Iwasa T, Nobusada K. Isolation, structure, and stability of a dodecanethiolate-protected Pd_1Au_{24} cluster. *Phys Chem Chem Phys* 2010;**12**:6219–25.
35. Qian H, Jiang D-e, Li G, Gayathri C, Das A, Gil RR, et al. Monoplatinum doping of gold nanoclusters and catalytic application. *J Am Chem Soc* 2012;**134**:16159–62.
36. Negishi Y, Munakata K, Ohgake W, Nobusada K. Effect of copper doping on electronic structure, geometric structure, and stability of thiolate-protected Au_{25} nanoclusters. *J Phys Chem Lett* 2012;**3**:2209–14.
37. Negishi Y, Kurashige W, Kamimura U. Isolation and structural characterization of an octaneselenolate-protected Au_{25} cluster. *Langmuir* 2011;**27**:12289–92.
38. Kurashige W, Munakata K, Nobusada K, Negishi Y. Synthesis of stable Cu_nAu_{25-n} nanoclusters ($n = 1$–9) using selenolate ligands. *Chem Commun* 2013;**49**:5447–9.
39. Kurashige W, Yamazoe S, Yamaguchi M, Nishido K, Nobusada K, Tsukuda T, et al. Au_{25} clusters containing unoxidized tellurolates in the ligand shell. *J Phys Chem Lett* 2014;**5**:2072–6.
40. Pei Y, Gao Y, Zeng XC. Structural prediction of thiolate-protected Au_{38}: a face-fused bi-icosahedral Au core. *J Am Chem Soc* 2008;**130**:7830–2.
41. Lopez-Acevedo O, Tsunoyama H, Tsukuda T, Häkkinen H, Aikens CM. Chirality and electronic structure of the thiolate-protected Au_{38} nanocluster. *J Am Chem Soc* 2010;**132**:8210–8.

42. Qian H, Eckenhoff WT, Zhu Y, Pintauer T, Jin R. Total structure determination of thiolate-protected Au_{38} nanoparticles. *J Am Chem Soc* 2010;**132**:8280–1.
43. Hulkko E, Lopez-Acevedo O, Koivisto J, Levi-Kalisman Y, Kornberg RD, Pettersson M, et al. Electronic and vibrational signatures of the $Au_{102}(p\text{-MBA})_{44}$ cluster. *J Am Chem Soc* 2011;**133**:3752–5.
44. Walter M, Akola J, Lopez-Acevedo O, Jadzinsky PD, Calero G, Ackerson CJ, et al. A unified view of ligand-protected gold clusters as superatom complexes. *Proc Nat Acad Sci U. S. A* 2008;**105**:9157–62.
45. Gao Y, Shao N, Zeng XC. Ab initio study of thiolate-protected Au_{102} nanocluster. *ACS Nano* 2008;**2**:1497–503.
46. Li Y, Galli G, Gygi F. Electronic structure of thiolate-covered gold nanoparticles: $Au_{102}(MBA)_{44}$. *ACS Nano* 2008;**2**:1896–902.
47. Koivisto J, Malola S, Kumara C, Dass A, Häkkinen H, Pettersson M. Experimental and theoretical determination of the optical gap of the $Au_{144}(SC_2H_4Ph)_{60}$ cluster and the $(Au/Ag)_{144}(SC_2H_4Ph)_{60}$ nanoalloys. *J Phys Chem Lett* 2012;**3**:3076–80.
48. Lopez-Acevedo O, Akola J, Whetten RL, Grönbeck H, Häkkinen H. Structure and bonding in the ubiquitous icosahedral metallic gold cluster $Au_{144}(SR)_{60}$. *J Phys Chem C* 2009;**113**:5035–8.
49. Malola S, Lehtovaara L, Enkovaara J, Häkkinen H. Birth of the localized surface plasmon resonance in monolayer-protected gold nanoclusters. *ACS Nano* 2013;**7**:10263–70.
50. Tlahuice-Flores A, Santiago U, Bahena D, Vinogradova E, Conroy CV, Ahuja T, et al. Structure of the thiolated Au_{130} cluster. *J Phys Chem A* 2013;**117**:10470–6.
51. Weissker HC, Escobar HB, Thanthirige VD, Kwak K, Lee D, Ramakrishna G, et al. Information on quantum states pervades the visible spectrum of the ubiquitous $Au_{144}(SR)_{60}$ gold nanocluster. *Nat Commun* 2014;**5**:3785.
52. Bahena D, Bhattarai N, Santiago U, Tlahuice A, Ponce A, Bach SBH, et al. STEM electron diffraction and high-resolution images used in the determination of the crystal structure of the $Au_{144}(SR)_{60}$ cluster. *J Phys Chem Lett* 2013;**4**:975–81.
53. Zeng C, Liu C, Pei Y, Jin R. Thiol ligand-induced transformation of $Au_{38}(SC_2H_4Ph)_{24}$ to $Au_{36}(SPh\text{-}t\text{-}Bu)_{24}$. *ACS Nano* 2013;**7**:6138–45.
54. Nimmala PR, Knoppe S, Jupally VR, Delcamp JH, Aikens CM, Dass A. $Au_{36}(SPh)_{24}$ nanomolecules: X-ray crystal structure, optical spectroscopy, electrochemistry, and theoretical analysis. *J Phys Chem B* 2014;**118**:14157–67.
55. Knoppe S, Malola S, Lehtovaara L, Bürgi T, Häkkinen H. Electronic structure and optical properties of the thiolate-protected $Au_{28}(SMe)_{20}$ cluster. *J Phys Chem A* 2013;**117**: 10526–33.
56. Das A, Li T, Nobusada K, Zeng C, Rosi NL, Jin R. Nonsuperatomic $[Au_{23}(SC_6H_{11})_{16}]^-$ nanocluster featuring bipyramidal Au_{15} kernel and trimeric $Au_3(SR)_4$ motif. *J Am Chem Soc* 2013;**135**:18264–7.
57. Das A, Li T, Li G, Nobusada K, Zeng C, Rosi NL, et al. Crystal structure and electronic properties of a thiolate-protected Au_{24} nanocluster. *Nanoscale* 2014;**6**:6458–62.
58. Zhu M, Qian H, Jin R. Thiolate-protected $Au_{24}(SC_2H_4Ph)_{20}$ nanoclusters: superatoms or not? *J Phys Chem Lett* 2010;**1**:1003–7.
59. Crasto D, Malola S, Brosofsky G, Dass A, Häkkinen H. Single crystal XRD structure and theoretical analysis of the chiral $Au_{30}S(S\text{-t-Bu})_{18}$ cluster. *J Am Chem Soc* 2014; **136**:5000–5.
60. Pei Y, Gao Y, Shao N, Zeng XC. Thiolate-protected $Au_{20}(SR)_{16}$ cluster: prolate Au_8 core with new $[Au_3(SR)_4]$ staple motif. *J Am Chem Soc* 2009;**131**:13619–21.

61. Jiang D-e, Chen W, Whetten RL, Chen Z. What protects the core when the thiolated Au cluster is extremely small? *J Phys Chem C* 2009;**113**:16983–7.
62. Jiang D-e, Walter M, Akola J. On the structure of a thiolated gold cluster: $Au_{44}(SR)_{28}{}^{2-}$. *J Phys Chem C* 2010;**114**:15883–9.
63. Pei Y, Lin S, Su J, Liu C. Structure prediction of $Au_{44}(SR)_{28}$: a chiral superatom cluster. *J Am Chem Soc* 2013;**135**:19060–3.
64. Malola S, Lehtovaara L, Knoppe S, Hu K-J, Palmer RE, Bürgi T, et al. $Au_{40}(SR)_{24}$ cluster as a chiral dimer of 8-electron superatoms: structure and optical properties. *J Am Chem Soc* 2012;**134**:19560–3.
65. Jiang D-e. The expanding universe of thiolated gold nanoclusters and beyond. *Nanoscale* 2013;**5**:7149–60.
66. Malola S, Lehtovaara L, Häkkinen H. A DFT study of linear gold-thiolate superclusters absorbing in the therapeutic NIR window. *J Phys Chem Lett* 2014;**5**:1329–34.
67. Pei Y, Pal R, Liu C, Gao Y, Zhang Z, Zeng XC. Interlocked catenane-like structure predicted in $Au_{24}(SR)_{20}$: implication to structural evolution of thiolated gold clusters from homoleptic gold(I) thiolates to core-stacked nanoparticles. *J Am Chem Soc* 2012; **134**:3015–24.
68. Song Y, Wang S, Zhang J, Kang X, Chen S, Li P, et al. Crystal structure of selenolate-protected $Au_{24}(SeR)_{20}$ nanocluster. *J Am Chem Soc* 2014;**136**:2963–5.
69. Tlahuice A, Garzón IL. On the structure of the $Au_{18}(SR)_{14}$ cluster. *Phys Chem Chem Phys* 2012;**14**:3737–40.
70. Tlahuice A, Garzón IL. Structural, electronic, optical, and chiroptical properties of small thiolated gold clusters: the case of Au_6 and Au_8 cores protected with dimer $[Au_2(SR)_3]$ and trimer $[Au_3(SR)_4)]$ motifs. *Phys Chem Chem Phys* 2012;**14**:7321–9.
71. Jiang D-e, Overbury SH, Dai S. Structure of $Au_{15}(SR)_{13}$ and its implication for the origin of the nucleus in thiolated gold nanoclusters. *J Am Chem Soc* 2013;**135**:8786–9.
72. Tlahuice-Flores A, Jose-Yacamán M, Whetten RL. On the structure of the thiolated Au_{15} cluster. *Phys Chem Chem Phys* 2013;**15**:19557–60.
73. Yu Y, Luo Z, Chevrier DM, Leong DT, Zhang P, Jiang D-e, Xie J. Identification of a highly luminescent $Au_{22}(SG)_{18}$ nanocluster. *J Am Chem Soc* 2014;**136**:1246–9.
74. McPartlin M, Mason R, Malatesta L. Novel cluster complexes of gold(0)-gold(I). *Chem Commun* 1969:334.
75. Safer D, Bolinger L, Leigh Jr JS. Undecagold clusters for site-specific labeling of biological macromolecules: simplified preparation and model applications. *J Inorg Biochem* 1986;**26**:77.
76. Mingos DMP. Molecular-orbital calculations on cluster compounds of gold. *J Chem Soc Dalton Trans* 1976:1163–9.
77. Mingos DMP. Structure and bonding in cluster compounds of gold. *Polyhedron* 1984;**3**:1289–97.
78. Goel S, Velizhanin KA, Piryatinski A, Ivanov SA, Tretiak S. Ligand effects on optical properties of small gold clusters: a TDDFT study. *J Phys Chem C* 2012;**116**:3242–9.
79. Aikens CM, Schatz GC. In: Nagarajan R, Hatton TA, editors. *Nanoparticles: synthesis, stabilization, passivation, and functionalization (ACS symposium series)*. Oxford University Press; 2008. p. 108.
80. Zhang H-F, Stender M, Zhang R, Wang C, Li J, Wang L-S. Toward the solution synthesis of the tetrahedral Au_{20} cluster. *J Phys Chem B* 2004;**108**:12259.
81. Wan X-K, Yuan S-F, Lin Z-W, Wang Q-M. A chiral gold nanocluster Au_{20} protected by tetradentate phosphine ligands. *Angew Chem Int Ed* 2014;**53**:2923–6.

82. Knoppe S, Lehtovaara L, Häkkinen H. Electronic structure and optical properties of the intrinsically chiral 16-electron superatom complex $[Au_{20}(PP_3)_4]^{4+}$. *J Phys Chem A* 2014; **118**:4214–21.
83. Walter M, Moseler M, Whetten RL, Häkkinen H. A 58-electron superatom-complex model for the magic phosphine-protected gold clusters (Schmid-gold, Nanogold®) of 1.4-nm dimension. *Chem Sci* 2011;**2**:1583–7.
84. Burgess RW, Keast VJ. TDDFT study of the optical absorption spectra of bare and coated Au_{55} and Au_{69} clusters. *J Phys Chem C* 2011;**115**:21016–21.
85. Shichibu Y, Negishi Y, Watanabe T, Chaki NK, Kawaguchi H, Tsukuda T. Biicosahedral gold clusters $[Au_{25}(PPh_3)_{10}(SC_nH_{2n+1})_5Cl_2]^{2+}$ ($n = 2–18$): a stepping stone to cluster-assembled materials. *J Phys Chem C* 2007;**111**:7845–7.
86. Nobusada K, Iwasa T. Oligomeric gold clusters with vertex-sharing bi- and triicosahedral structures. *J Phys Chem C* 2007;**111**:14279–82.
87. Iwasa T, Nobusada K, Nakajima A. Electronic and optical properties of vertex-sharing homo- and hetero-biicosahedral gold clusters. *J Phys Chem C* 2013;**117**:24586–91.
88. Das A, Li T, Nobusada K, Zeng Q, Rosi NL, Jin R. Total structure and optical properties of a phosphine/thiolate-protected Au_{24} nanocluster. *J Am Chem Soc* 2012;**134**:20286–9.
89. Goh J-Q, Malola S, Häkkinen H, Akola J. Role of the central gold atom in ligand-protected biicosahedral Au_{24} and Au_{25} clusters. *J Phys Chem C* 2013;**117**:22079–86.
90. Lopez-Acevedo O, Rintala J, Virtanen S, Femoni C, Tiozzo C, Grönbeck H, et al. Characterization of iron-carbonyl-protected gold clusters. *J Am Chem Soc* 2009;**131**:12573–5.
91. Bakr OM, Amendola V, Aikens CM, Wenseleers W, Lee R, Dal Negro L, et al. Silver nanoparticles with broad multi-band linear optical absorption. *Angew Chem Int Ed* 2009; **48**:5921–6.
92. Harkness KM, Tang Y, Dass A, Pan J, Kothalawala N, Reddy VJ, et al. $Ag_{44}(SR)_{30}{}^{4-}$: a silver-thiolate superatom complex. *Nanoscale* 2012;**4**:4269–74.
93. Yang H, Wang Y, Huang H, Gell L, Lehtovaara L, Malola S, et al. All-thiol-stabilized Ag_{44} and $Au_{12}Ag_{32}$ nanoparticles with single-crystal structures. *Nat Commun* 2013;**4**:2422.
94. Desireddy A, Conn BE, Guo J, Yoon B, Barnett RN, Monahan BM, et al. Ultrastable silver nanoparticles. *Nature* 2013;**501**:399–402.
95. Gell L, Häkkinen H. A theoretical analysis of the $M_{12}Ag_{32}(SR)_{40}{}^{4-}$ and $X@M_{12}Ag_{32}(SR)_{30}{}^{4-}$ nanoclusters ($M = Au, Ag$; $X = H, Mn$). *J Phys Chem C* 2015;**119**:10943–8.
96. Cathcart N, Mistry P, Makra C, Pietrobon B, Coombs N, Jelokhani-Niaraki M, et al. Chiral thiol-stabilized silver nanoclusters with well-resolved optical transitions synthesized by a facile etching procedure in aqueous solutions. *Langmuir* 2009;**25**:5840–6.
97. Kumar S, Bolan MD, Bigioni TP. Glutathione-stabilized magic-number silver cluster compounds. *J Am Chem Soc* 2010;**132**:13141–3.
98. Guo J, Kumar S, Bolan M, Desireddy A, Bigioni TP, Griffith WP. Mass spectrometric identification of silver nanoparticles: the case of $Ag_{32}(SG)_{19}$. *Anal Chem* 2012;**84**:5304–8.
99. Bertorelle F, Hamouda R, Rayane D, Broyer M, Antoine R, Dugourd P, et al. Synthesis, characterization and optical properties of low nuclearity liganded silver clusters: $Ag_{31}(SG)_{19}$ and $Ag_{15}(SG)_{11}$. *Nanoscale* 2013;**5**:5637–43.
100. Ramakrishna G, Varnavski O, Kim J, Lee D, Goodson T. Quantum-sized gold clusters as efficient two-photon absorbers. *J Am Chem Soc* 2008;**130**:5032–3.
101. Yao H, Miki K, Nishida N, Sasaki A, Kimura K. Large optical activity of gold nanocluster enantiomers induced by a pair of optically active penicillamines. *J Am Chem Soc* 2005;**127**:15536–43.

102. Garzón IL, Beltrán MR, Posada-Amarillas A, Ordejón P, Artacho E, Sánchez-Portal D, et al. Lowest energy structures of gold nanoclusters. *Phys Rev Lett* 1998;**81**:1600.
103. Michaelian K, Rendón N, Garzón IL. Structure and energetics of Ni, Ag, and Au nanoclusters. *Phys Rev B* 1999;**60**:2000.
104. Garzón IL, Reyes-Nava JA, Rodríguez-Hernández JI, Sigal I, Beltrán MR, Michaelian K. Chirality in bare and passivated gold nanoclusters. *Phys Rev B* 2002;**66**:073403.
105. Goldsmith M-R, George CB, Zuber G, Naaman R, Waldeck DH, Wipf P, et al. The chiroptical signature of achiral metal clusters induced by dissymmetric adsorbates. *Phys Chem Chem Phys* 2006;**8**:63–7.
106. Yanagimoto Y, Negishi Y, Fujihara H, Tsukuda T. Chiroptical activity of BINAP-stabilized undecagold clusters. *J Phys Chem B* 2006;**110**:11611–4.
107. Dolamic I, Knoppe S, Dass A, Bürgi T. First enantioseparation and circular dichroism spectra of Au_{38} clusters protected by achiral ligands. *Nat Commun* 2012;**3**:798.
108. Knoppe S, Bürgi T. Chirality in thiolate-protected gold clusters. *Acc Chem Res* 2014;**47**:1318–26.
109. Molina B, Sanchez-Castillo A, Knoppe S, Garzon IL, Burgi T, Tlahuice-Flores A. Structures and chiroptical properties of the BINAS-monosubstituted $Au_{38}(SCH_3)_{24}$ cluster. *Nanoscale* 2013;**5**:10956–62.
110. Sánchez-Castillo A, Noguez C, Garzón IL. On the origin of the optical activity displayed by chiral-ligand-protected metallic nanoclusters. *J Am Chem Soc* 2010;**132**:1504–5.
111. Nishida N, Yao H, Ueda T, Sasaki A, Kimura K. Synthesis and chiroptical study of D/L-penicillamine-capped silver nanoclusters. *Chem Mater* 2007;**19**:2831–41.
112. Nishida N, Yao H, Kimura K. Chiral functionalization of optically inactive monolayer-protected silver nanoclusters by chiral ligand-exchange reactions. *Langmuir* 2008;**24**: 2759–66.

Chapter 10

Atomically Precise Gold Nanoclusters Catalyzed Chemical Transformations

Yuxiang Chen and Rongchao Jin[1]
Department of Chemistry, Carnegie Mellon University, Pittsburgh, PA, USA
[1]*Corresponding author: E-mail: rongchao@andrew.cmu.edu*

Chapter Outline

10.1 INTRODUCTION

Heterogeneous catalysis based upon metal nanoparticles is of major importance in the chemical industry. Among the precious metals, platinum (Pt) has long been widely studied for many decades, the recent decade has witnessed the "golden rush" in catalytic research.[1–6] Conventional nanocatalysts are composed of polydisperse nanoparticles attached to certain types of supports (such as oxides, porous carbon, etc.), and the polydispersity of nanoparticles poses major challenges in fundamental studies of catalysis.[2] First of all, polydisperse nanoparticles can only give an averaged catalytic reactivity and the precise size dependence cannot be obtained. Second, the surface of the

Frontiers of Nanoscience, Vol. 9. http://dx.doi.org/10.1016/B978-0-08-100086-1.00010-5

nanocatalyst—which is critical important—is hard to be mapped out and thus is poorly defined. Third, the conventional nanocatalysts are prepared via solution-phase methods and various capping agents exist on the particle surfaces, but the exact information is hard to obtain (e.g., what types of capping agents are present and how they are bonded to the particle surfaces). Overall, the polydispersity of nanoparticles and their poorly defined surface structures as well as unknown surface composition preclude atomic/molecular-level insight into the catalytic mechanism, and sometimes it is even difficult to relate the observed catalytic properties to the metal nanoparticles or the support. To overcome these major obstacles, well-defined nanocatalysts are highly desirable, especially when one pursues the fundamental studies of catalysis.

In recent years, *atomically precise* gold nanoparticles (often called nanoclusters) have become available, especially the thiolate-protected gold nanoclusters.[7–9] Such Au nanoclusters are composed of precise numbers of gold atoms (n) and thiolate (-SR) ligands (m), denoted as $Au_n(SR)_m$, with n ranging from ten to a few hundred atoms (equivalent size from subnanometer to 2–3 nm). These nanoparticles are unique in that both the formula and structure are well defined, while conventional nanoparticles can only be represented by an average size and no exact formula can be determined. The attainment of molecularly pure nanoclusters permits the growth of single crystals of nanoclusters for X-ray crystallographic determination of the total structures of nanoclusters, that is, every atom of the nanocluster can be precisely pinned down. Based upon the total structures of nanoclusters, the ultimate atomic/molecular level-understanding of nanocatalysis can be achieved.[2]

Compared to conventional metallic Au nanoparticle catalysts, $Au_n(SR)_m$ nanoclusters possess several distinct features that are of particular interest to catalysis. First of all, metallic gold nanoparticles (2–100 nm) adopt a face-centered cubic (fcc) structure, but $Au_n(SR)_m$ nanoclusters often adopt different types of atom-packing structures[9]; such structures offer an opportunity to probe the structural sensitivity in catalysis. Second, the ultrasmall size not only increases the surface-to-volume ratio, but also induces strong electron-energy quantization in nanoclusters (as opposed to the continuous conduction band in metallic gold nanoparticles)[8]; the quantization effect and the high surface-to-volume effect may exhibit different behavior. Third, some unique structures have been discovered in the crystal structures of nanoclusters, such as the staple-like motifs and volcano-like atomic sites on cluster surfaces—which can act as catalytically active centers in chemical reactions.[2]

In this chapter, we will first briefly discuss some general properties of $Au_n(SR)_m$ nanoclusters relevant to catalysis, and then focus on the catalytic reactions that have been reported to be catalyzed by such nanoclusters. The $Au_n(SR)_m$ nanoclusters are expected to become a promising class of model catalysts, in particular the attainment of X-ray structures of nanoclusters permits precise reactivity–structure correlation. Although much work remains to be carried out, the catalysis research of $Au_n(SR)_m$ nanoclusters hold great promise in revealing the atomic/molecular-level details and significantly advance fundamental understanding of catalytic mechanism.

10.2 OVERVIEW OF $Au_n(SR)_m$ NANOCLUSTERS

10.2.1 Size-Focusing Syntheses and Determination of Atomic Structures

Herein we briefly discuss the "size-focusing" methodology for synthesizing atomically precise $Au_n(SR)_m$ nanoclusters.[10] In this method, an appropriate distribution of size-mixed nanoclusters is first prepared by kinetic control over the reduction reaction of gold precursor (typically Au(I)−SR complexes or polymers) with $NaBH_4$ or other reducing agents, which is the key step for the attainment of single-sized nanoclusters as the final product. Then, these size-mixed nanoclusters are subjected to size-focusing under harsh conditions (e.g., at 80−90 °C and in the presence of excess thiol); under such conditions the unstable nanoclusters are decomposed or converted to the more stable ones, and eventually only the most stable size in the initial size distribution survives the "focusing" process. The initial size distribution should be properly controlled (for example, not too broad—otherwise, two or multiple stable sizes would be resulted after size focusing). A series of size-discrete, robust $Au_n(SR)_m$ nanoclusters have been available through intense research in the past years, such as $Au_{25}(SR)_{18}$,[11] $Au_{36}(SR)_{24}$,[12] $Au_{38}(SR)_{24}$,[13,14] $Au_{102}(SR)_{44}$,[15] $Au_{130}(SR)_{50}$,[16] $Au_{144}(SR)_{60}$,[13,17] $Au_{333}(SR)_{79}$,[18] etc. A more complete account of the $Au_n(SR)_m$ sizes can be found in a recent review.[19] Herein, we choose a few sizes to be used in catalytic tests for a more detailed discussion.

10.2.1.1 The $Au_{25}(SR)_{18}$ Nanocluster

The size-focusing phenomenon was first observed in the synthesis of $Au_{25}(SR)_{18}$ in our early work.[20,21] In this synthesis, the first step involves the spontaneous reduction of Au(III) to Au(I) by thiol (typically 3−5 equivalents per mole of gold), forming Au(I)−SR complexes or polymers (i.e., the intermediates). Subsequently, $NaBH_4$ is added to further reduce Au(I)−SR to Au(0), producing polydisperse nanoclusters protected by thiolate. Over the prolonged aging period, a spontaneous "size-focusing" process was observed, evidenced by the evolution of the optical absorption spectra (Figure 1). Initially, the polydisperse product gave a decay-like spectrum (note: the absorption peaks of individual sizes of nanoclusters were washed out due to polydispersity), but with time, we observed several prominent peaks in the UV−vis spectrum of the crude product, indicating a gradual growth of monodisperse nanoclusters, (Figure 1). Three absorption peaks were observed at 670, 450, and 400 nm, and the final product consists of molecularly pure $Au_{25}(SR)_{18}$ nanoclusters (Figure 2).[21,22]

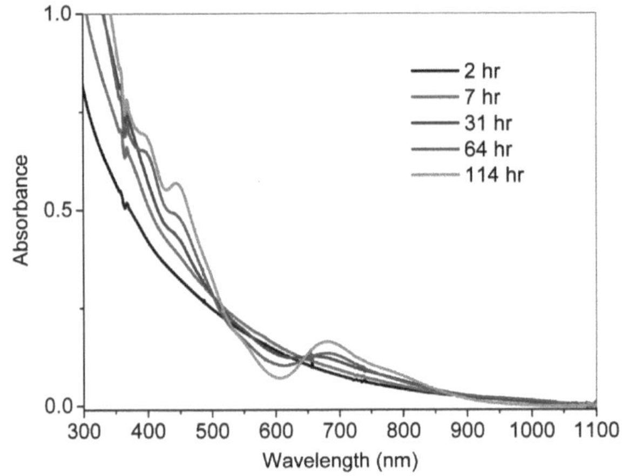

FIGURE 1 UV–vis spectroscopic monitoring of the "size-focusing" process for the growth of $Au_{25}(SR)_{18}$ nanoclusters. *Reproduced from Ref. 21 with permission from The Royal Society of Chemistry.*

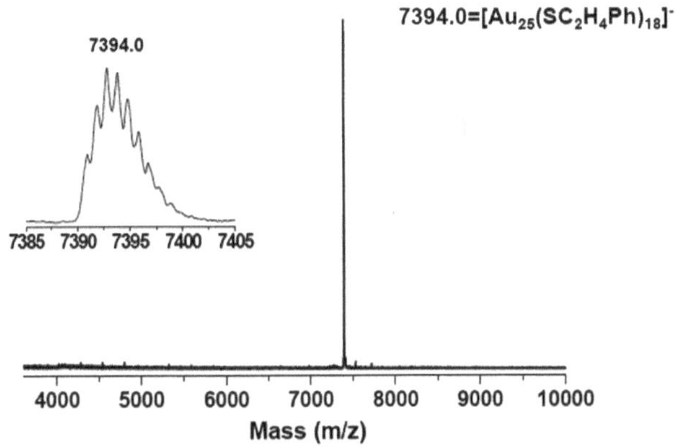

FIGURE 2 Electrospray ionization mass spectrum of $Au_{25}(SC_2H_4Ph)_{18}^-$ nanoclusters. The experimental formula weight is 7394 Da (most abundant ion in the isotope pattern; theoretical value: 7394). *Reproduced with permission from Ref. 22. Copyright 2010 American Chemical Society.*

The case of the $Au_{25}(SR)_{18}$ synthesis illustrates the basic principle of the size-focusing method. This process is based upon the stability property of different-sized nanoclusters.[10] In order to attain atomic monodispersity, we found that it is critical to control the size distribution of the starting $Au_x(SR)_y$ mixture so that only one specific size of nanoclusters survives the size-focusing process. The importance of the initial size distribution for the final molecular purity of nanoclusters is also demonstrated in the case of $Au_{38}(SR)_{24}$ synthesis.[23]

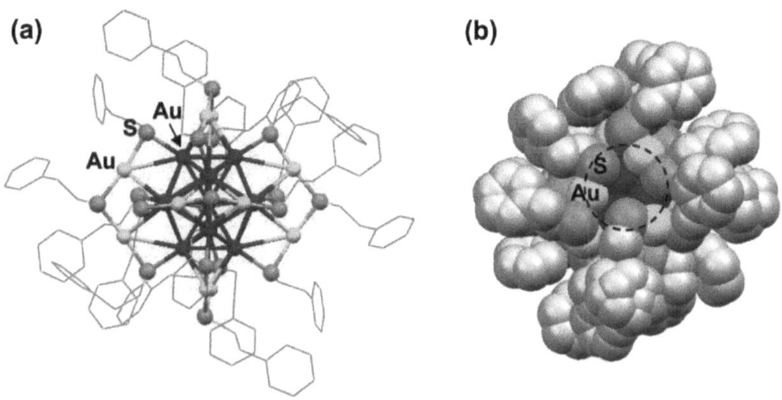

FIGURE 3 X-ray crystal structure of $Au_{25}(SCH_2CH_2Ph)_{18}$ nanoclusters, (a) ball-stick model, (b) space-filling model (magenta: gold atoms of the core, cyan: gold atoms of the surface). (see color plate). *Adapted with permission from Refs 9,11. Copyright 2012/2008 American Chemical Society.*

The molecularly pure nanoclusters permit the growth of single crystals for total structure determination of the nanoclusters. X-ray crystallography reveals that the $Au_{25}(SR)_{18}$ nanocluster comprises an Au_{13} icosahedral core (Figure 3(a), magenta) and an $Au_{12}(SR)_{18}$ shell (Figure 3(a), cyan and yellow),[11,24] consistent with the theoretical structure of $Au_{25}(SCH_3)_{18}$.[25] It is worth pointing out that some of the surface atoms are indeed exposed, such as the triangular Au_3 site as shown in the space-filling model (Figure 3(b)). These sites (resembling "pockets") may become the catalytically active sites since they are readily accessible by reactant molecules in catalytic reactions.

The electronic structure of $Au_{25}(SR)_{18}$ exhibits discrete energy levels caused by the quantum-size effect.[11] The molecular orbitals can be roughly divided into the Au-core orbitals (i.e., primarily contributed by the Au_{13} core) and the shell orbitals (i.e., primarily contributed by the $Au_{12}(SR)_{18}$ part)[11,26]; this core–shell electronic picture is in line with the geometric core–shell structure. Apparently, the core and surface Au atoms possess different oxidation states, i.e., the surface Au atoms are relatively electron-deficient and are formally Au(I), while the core Au atoms are relatively electron-rich. Such a feature has been found to be important in catalytic processes.

10.2.1.2 The $Au_{38}(SR)_{24}$ Nanocluster

The synthesis of $Au_{38}(SR)_{24}$ also followed the size-focusing methodology. The first step was to obtain a crude mixture (in this case, glutathione (SG)-protected $Au_n(SG)_m$ nanoclusters).[23] Then the size-mixed nanoclusters were subjected to a thermal thiol etching process[14,27] (e.g., 80 °C) in a two-phase (water/toluene) system. During the harsh size-focusing process, the starting polydisperse nanoclusters are gradually converted to the most stable product,

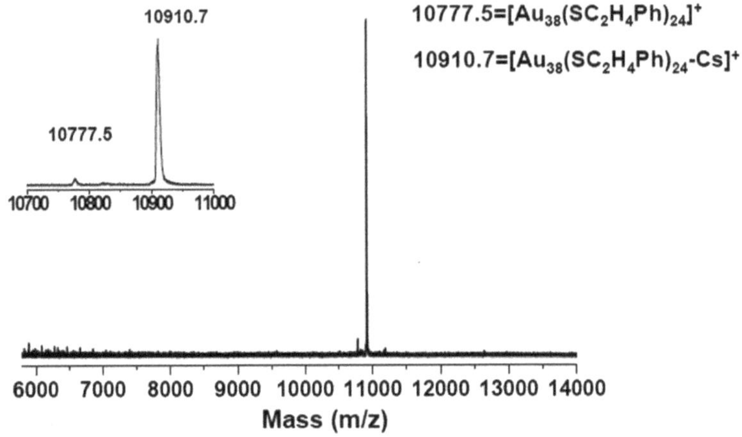

FIGURE 4 Electrospray ionization mass spectrometry spectrum of charge-neutral $Au_{38}(SC_2H_4Ph)_{24}$ nanoclusters. The experimental formula weight of $Au_{38}(SC_2H_4Ph)_{24}$ is 10,777.5 Da (theoretical value: 10,778.05). In mass spectrometric analysis, CsOAc was added to form positively charged $[cluster\text{-}Cs]^+$ adducts. The small peak of $[Au_{38}(SCH_2CH_2Ph)_{24}]^+$ at 10,777.5 was due to induced ionization by Cs^+ salt. *Adapted with permission from Ref. 23. Copyright 2009 American Chemical Society.*

i.e., the $Au_{38}(SC_2H_4Ph)_{24}$ nanocluster, and molecular purity was achieved (Figure 4).[23,28]

Time-dependent optical spectroscopy and matrix-assisted laser desorption ionization coupled with mass spectrometry (MALDI-MS) analyses clearly revealed a size-focusing process.[23] Similar to the case of $Au_{25}(SR)_{18}$ size-focusing synthesis, the starting $Au_x(SC_2H_4Ph)_y$ mixture showed a decaying optical spectrum, indicating a mixture. With reaction going on, several distinct peaks started to emerge, and the final product showed a distinct optical spectrum that is characteristic of $Au_{38}(SC_2H_4Ph)_{24}$ nanoclusters.[13,27] MALDI-MS analysis reveals that, with the increase of reaction time, the relatively large Au nanoclusters ($n > 38$) were decomposed and converted to $Au_{38}(SC_2H_4Ph)_{24}$.[23] After ~40 h, molecularly pure $Au_{38}(SC_2H_4Ph)_{24}$ nanoclusters were obtained. As aforementioned, a key condition to obtain single-sized $Au_{38}(SC_2H_4Ph)_{24}$ nanoclusters is to control the size distribution of the $Au_n(SG)_m$ mixture prior to the size-focusing step. In this case, we found that the solvent played an important role in controlling the size range of the $Au_n(SG)_m$ starting mixture.[23] Acetone was found to produce an appropriate size distribution of $Au_n(SG)_m$ (that permitted high-yielding synthesis of $Au_{38}(SR)_{24}$). Comparison between acetone and methanol (solvents) showed that the acetone-mediated synthesis of $Au_x(SG)_y$ (the starting mixture) had a dominant size range from 8 to 18 kDa, while the methanol system produced a dominant size range below 8 kDa, i.e., smaller than Au_{38}. The acetone-mediated high-yield synthesis of $Au_{38}(SC_2H_4Ph)_{24}$ is attributed to the

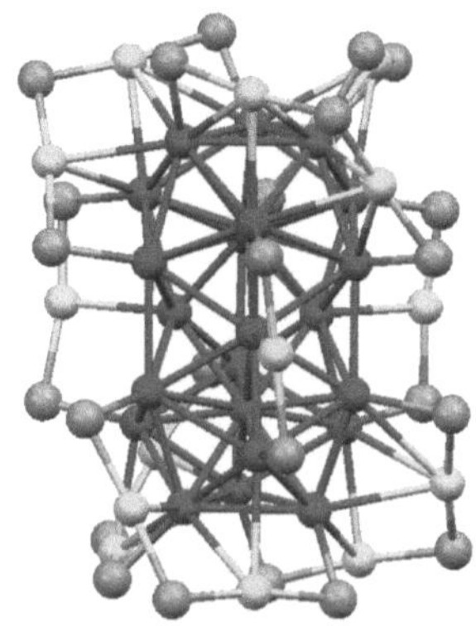

FIGURE 5 The atomic structure of $Au_{38}(SC_2H_4Ph)_{24}$ (magenta (dark gray in print versions): gold atoms of the Au_{23} core, cyan (light gray in print versions): gold atoms of the shell, yellow (gray in print versions): sulfur). All $-C_2H_4Ph$ tails are omitted for clarity. *Reproduced with permission from Ref. 14. Copyright 2010 American Chemical Society.*

down-conversion of those higher-mass $Au_x(SG)_y$ nanoclusters ($38 < x < 100$), while such components were missing in the methanol-mediated synthesis, hence, low yield of Au_{38}. It is worth noting that if the initial size range of $Au_x(SG)_y$ should not be too large, $Au_{144}(SR)_{60}$ nanoclusters could be produced, leading to a mixture of $Au_{38}(SR)_{24}$ and $Au_{144}(SR)_{60}$ nanoclusters.

The atomic structure of the $Au_{38}(SCH_2CH_2Ph)_{24}$ nanocluster comprises a bi-icosahedral Au_{23} core and six dimeric staples $-SR-Au-SR-Au-SR-$ ($Au_2(SR)_3$ for short) as well as three monomeric staples $-SR-Au-SR-$ ($Au(SR)_2$ for short) protecting the Au_{23} core (Figure 5).[14] Apparently, the gold atoms are distributed in two kinds of chemical environments (surface and core), which benefits the catalytic reactions which often involve adsorption and activation of more than one type of reactant molecules.

10.2.1.3 The $Au_{144}(SR)_{60}$ Nanocluster

Both the $Au_{25}(SR)_{18}$ and $Au_{38}(SR)_{24}$ cases explicitly demonstrate the importance of the size distribution of the starting nanoclusters (i.e., prior to the size-focusing step) for the eventual formation of monodisperse nanoclusters. The first step is essentially kinetically controlled, while the second step (i.e., size focusing) is essentially controlled by thermodynamics.[29] To obtain the proper size-distribution of the starting polydisperse $Au_x(SR)_y$ for the synthesis of monodisperse $Au_{144}(SR)_{60}$ nanoclusters,[17] we adjusted the gold salt to thiol ratio (Au/SR) to 1:3 in the first step and as such, the size range of the starting $Au_x(SR)_y$ mixture was controlled to be around 24–36 kDa. In the subsequent size-focusing step, the initial polydispersed nanoclusters were incubated in large amounts of thiol at 80 °C, and after 24 h of size focusing pure $Au_{144}(SC_2H_4Ph)_{60}$ nanoclusters were obtained (Figure 6). Before size focusing, the initial polydispersed nanoclusters showed a decay-like UV–vis spectrum;

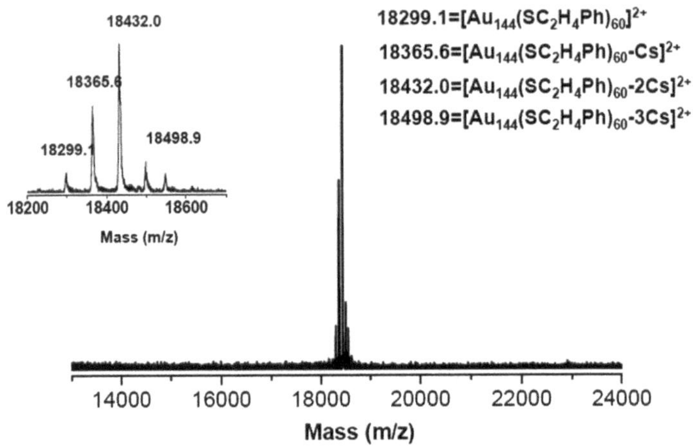

FIGURE 6 Electrospray ionization mass spectrometry (ESI-MS) spectrum of $Au_{144}(SC_2H_4Ph)_{60}$ nanoclusters. CsOAc was added to form positively charged $[\text{cluster-Cs}]^+$ adducts for ESI analysis. The experimental molecular weight is $18{,}299.1 \times 2 = 36{,}598.2$ (theoretical: 36,596.5). *Adapted with permission from Ref. 17. Copyright 2009 American Chemical Society.*

after size focusing, step-like bands were observed at 510 and 700 nm and these absorption bands are characteristic of $Au_{144}(SC_2H_4Ph)_{60}$ nanoclusters.[17] The atomic structure of the $Au_{144}(SC_2H_4Ph)_{60}$ nanocluster has not been solved, as the larger nanoclusters pose major challenges in crystallization.

10.2.2 Thermal Stability

The thermal stability of $Au_n(SR)_m$ nanoclusters is important for catalytic applications as many reactions are run under thermal conditions. Generally, the thiolate-protected gold nanoclusters are thermally robust since they are synthesized by size focusing under harsh conditions. Our motivation is to utilize the ligand-protected nanoclusters as catalysts so that the atomic structures of nanoclusters can be correlated with the catalytic properties, even though the presence of ligands may reduce the catalytic activity. On the other hand, it is also worth comparing ligand-on catalysts with the ligand-off ones. Thus, it is meaningful to find out the temperature at which the surface ligands will be desorbed.

Thermogravimetric analysis (TGA) shows that $Au_n(SR)_m$ (e.g., $Au_{25}(SR)_{24}$, $Au_{38}(SR)_{24}$, and $Au_{144}(SR)_{60}$) starts to lose ligands at ~200 °C and the ligand loss is complete at ~250 °C (Figure 7(a)).[30] The ligand-loss temperature was found to be unaffected by the atmosphere (e.g., N_2, air, O_2, and H_2).[31]

The isothermal stability of nanoclusters was further tested (i.e., maintained at 150 °C for 60 min in air atmosphere). Using $Au_{25}(SR)_{18}$ as a typical example, we found no weight loss (i.e., no weight loss) during the isothermal process

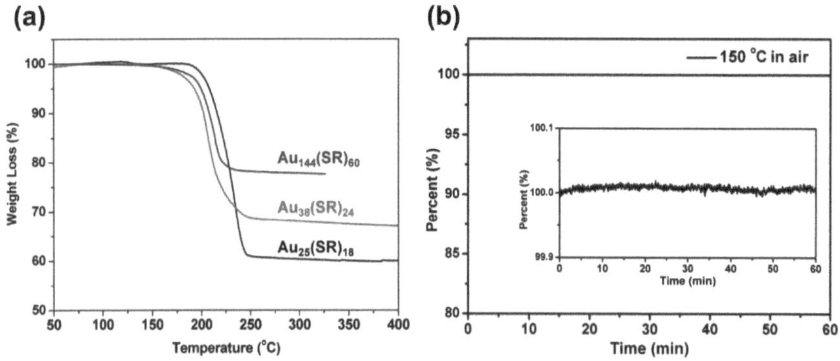

FIGURE 7 (a) Thermogravimetric analysis (TGA) of $Au_n(SR)_m$ nanoclusters (under N_2 atmosphere). (b) Isothermal stability analysis of pure $Au_{25}(SR)_{18}$ nanoclusters (maintained at 150 °C, in air atmosphere for 60 min). No discernable loss of ligands (starting: 100%) was observed. The inset shows a zoom-in of the TGA curve at constant 150 °C. *Reproduced with permission from Ref. 30 (Copyright © 2010 Wiley-VCH Verlag GmbH & Co. KGaA, Weinheim) and Ref. 32 (Copyright © 2013 Elsevier Inc.).*

(Figure 7(b)).[32] Nuclear magnetic resonance (NMR), optical spectroscopy, and mass spectrometry analyses all confirm no change caused by the 150 °C thermal treatment,[33,34] which indicates the high stability of the gold nanoclusters. This is reasonable since the structure of $Au_{25}(SR)_{18}$ is thermodynamically stable according to density functional theory (DFT) analysis, thus, structural isomerization should not occur, unless ligands are lost at higher temperatures.[30]

10.2.3 Reactivity with O_2

Our interest in exploring $Au_n(SR)_m$ for catalysis was inspired by the observation of reversible conversion between $[Au_{25}(SR)_{18}]^0$ and $[Au_{25}(SR)_{18}]^-$.[35,36] The redox properties of nanoclusters are important for those catalytic reactions that involve electron transfer in activating reactants.

The native $Au_{25}(SR)_{18}$ nanoclusters from the synthesis are anionic (i.e., $[Au_{25}(SR)_{18}]^-$, counterion = tetraoctylammonium (TOA^+)).[11] Interestingly, we found that when a solution of $[Au_{25}(SR)_{18}]^-$ was exposed to air, the nanoclusters were gradually converted to charge-neutral $[Au_{25}(SR)_{18}]^0$ clusters,[35] which was first discovered by the color change of the solution (see also the UV–vis spectral changes in Figure 8). Single crystal X-ray crystallography[35] revealed that the product is $[Au_{25}(SCH_2CH_2Ph)_{18}]^0$ and its structure shows the same framework as that of $[Au_{25}(SCH_2CH_2Ph)_{18}]^-$ but without the presence of tetraoctylammonium (i.e., the counterion of the anionic cluster).

The attainment of $[Au_{25}(SCH_2CH_2Ph)_{18}]^0$ and $[Au_{25}(SCH_2CH_2Ph)_{18}]^-$ provides an opportunity for deep understanding of the electronic properties. Further work[36] revealed that the negative charge of $[Au_{25}(SCH_2CH_2Ph)_{18}]^-$ surprisingly resides in the Au_{13} core of the cluster, rather than on the

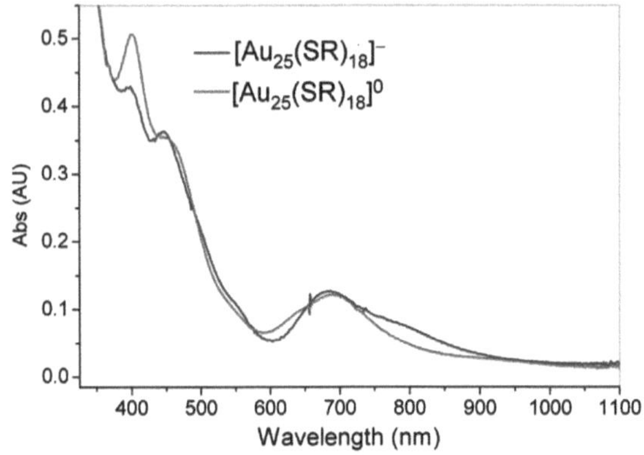

FIGURE 8 Reversible conversion between $[Au_{25}(SR)_{18}]^0$ and $[Au_{25}(SR)_{18}]^-$. *Reproduced with permission from Ref. 35. Copyright 2008 American Chemical Society.*

$Au_{12}(SR)_{18}$ surface or on ligands. This explains why the gold core, instead of the thiolate ligands, was first oxidized (i.e., one-electron loss to yield neutral $[Au_{25}(SCH_2CH_2Ph)_{18}]^0$). Normally, one would expect that O_2 would first oxidize the thiolate ligands on the Au surface, which is indeed the case in bulk gold surfaces functionalized by thiolates. The redox process between $[Au_{25}(SCH_2CH_2Ph)_{18}]^0$ and $[Au_{25}(SCH_2CH_2Ph)_{18}]^-$ is completely reversible.[36] While the interaction of O_2 with $[Au_{25}(SCH_2CH_2Ph)_{18}]^-$ is quite slow at room temperature, it speeds up at higher temperatures, and if using peroxide as the oxidant, the anion-to-neutral conversion is much faster that using O_2. At room temperature, other sizes of nanoclusters such as $Au_{38}(SR)_{24}$, $Au_{144}(SR)_{60}$ were not found to interact with O_2.

10.3 CATALYTIC PROPERTIES OF $Au_n(SR)_m$ NANOCLUSTERS

10.3.1 Catalytic Oxidation

The reactivity of $Au_{25}(SR)_{18}$ nanoclusters with O_2 inspired us to pursue the catalytic application of $Au_{25}(SR)_{18}$ and other $Au_n(SR)_m$ nanoclusters in catalytic oxidation.[37,38] The activation of oxygen (O_2) plays an important role in the gold nanocluster-catalyzed selective oxidation processes. Below we discuss the catalytic properties of nanoclusters in carbon monoxide oxidation, styrene oxidation, and sulfide oxidation.

10.3.1.1 Catalytic Oxidation of Carbon Monoxide

The CO oxidation reaction has been extensively studied in the field of gold catalysis.[1] An interesting observation was the distinct effect of the support; among the common supports, TiO_2 was generally identified to be the best support for nanogold catalysts in the CO oxidation reaction.

Using the ligand-protected $Au_n(SR)_m$ nanoclusters as catalysts for CO oxidation to CO_2, we found some phenomena[31,39] that are distinctly different from the case of bare nanogold catalysts. Starting with the atomically precise, molecularly pure $Au_{25}(SR)_{18}$ nanoclusters, Nie et al.[31] prepared several oxides as the supports for $Au_{25}(SR)_{18}$. Surprisingly, the $Au_{25}(SR)_{18}/TiO_2$ catalyst was found to have no catalytic activity even up to 200 °C (Figure 9(a)), which is in direct contrast with the performance of conventional Au(bare)/TiO_2 catalysts.[1] For other oxide supports, the $Au_{25}(SR)_{18}/Fe_2O_3$ catalyst showed a moderate activity (onset temperature 100 °C), and CeO_2 was found to be the best support (onset temperature 60 °C, 62% conversion at 160 °C). Without any activation, the activities of the $Au_{25}(SR)_{18}/M_xO_y$ catalysts are overall not high compared to the conventional nanogold catalysts.

Interestingly, pretreatment of the $Au_{25}(SR)_{18}/CeO_2$ catalyst in O_2 for 1.5–2 h at 150 °C (denoted as T_{pre}) led to a drastic increase in catalytic activity (onset temperature shifted to room temperature, r.t.), (Figure 9(b)), and 100% CO conversion was reached at 100 °C. Prolonged O_2 pretreatment at

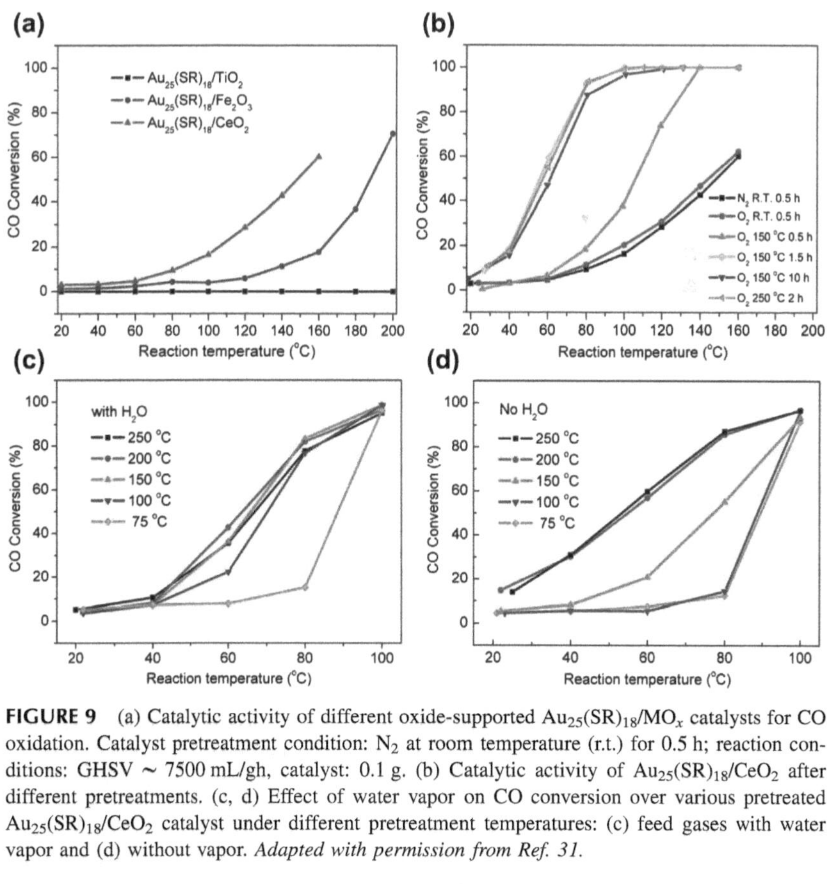

FIGURE 9 (a) Catalytic activity of different oxide-supported $Au_{25}(SR)_{18}/MO_x$ catalysts for CO oxidation. Catalyst pretreatment condition: N_2 at room temperature (r.t.) for 0.5 h; reaction conditions: GHSV ~ 7500 mL/gh, catalyst: 0.1 g. (b) Catalytic activity of $Au_{25}(SR)_{18}/CeO_2$ after different pretreatments. (c, d) Effect of water vapor on CO conversion over various pretreated $Au_{25}(SR)_{18}/CeO_2$ catalyst under different pretreatment temperatures: (c) feed gases with water vapor and (d) without vapor. *Adapted with permission from Ref. 31.*

150 °C (e.g., >2 h) was also tested, but it was found not to further enhance the catalytic activity (Figure 9(b)). In addition, further increasing the T_{pre} to 250 °C (above the thiolate desorption temperature) did not show appreciable improvement in activity. Apparently, O_2 thermal pretreatment had activated the cluster catalyst. In terms of the activation mechanism, one possibility would be the removal of ligands during the thermal pretreatment process, but we did not find evidence of thiolate desorption through in situ mass spectrometry monitoring. Recent work by Wu and coworkers[40] reported partial removal of ligands through X-ray absorption and IR analyses, but there is still a debate on whether the ligands are removed or retained before/during the catalytic process.

It is worth noting that the drastic effect of thermal O_2 pretreatment was not observed in the $Au_{25}(SR)_{18}/TiO_2$ system, nor in the $Au_{25}(SR)_{18}$/oxide catalysts pretreated in N_2 (as opposed to O_2).[31] Theoretical simulations indicated that chemical adsorption of O_2 on ligand-protected $Au_{25}(SR)_{18}$ does not occur[40,41]; if so, the O_2 interaction with $Au_{25}(SR)_{18}$ in solution-phase conversion of the anion to the charge-neutral state should proceed via a physical adsorption pathway, and therefore O_2 activation on the $Au_{25}(SR)_{18}$/oxide catalysts is expected to involve the oxide support.

The presence of water vapor in the feed gases exhibited a promotional effect on the catalyst performance (Figure 9(c) and (d)). For example, in the case of feed-gas-containing vapor, O_2 pretreatment at even lower temperature (e.g., $T_{pre} = 100$ °C) could lead to the same drastic enhancement in activity.[31]

The above results imply that the interface between $Au_{25}(SR)_{18}$ and CeO_2 is critical and the perimeter sites should be the catalytic active site.[31] We speculate that O_2 should first convert to O_2^- by withdrawing an electron from the $Au_{25}(SR)_{18}$ nanocluster, then migrate to the nanocluster/CeO_2 interface and convert to hydroperoxide species on CeO_2; the activated CO should be oxidized at the perimeter sites of the catalyst (Scheme 1). The mechanism of the catalytic CO oxidation by $Au_{25}(SR)_{18}/CeO_2$ catalyst still needs more detailed investigation in further work.

Ma et al.[42] also investigated the CO oxidation reaction using "ligand-off" Au_{25} and Au_{144} nanoclusters as catalysts. The ligands on $Au_{25}(SCH_2CH_2Ph)_{18}$

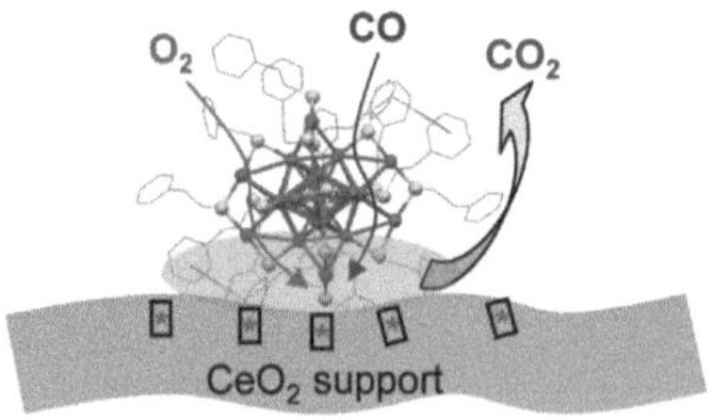

SCHEME 1 Proposed CO oxidation at the perimeter sites of $Au_{25}(SR)_{18}/CeO_2$. *Adapted with permission from Ref. 31.*

and $Au_{144}(SCH_2CH_2Ph)_{60}$ nanoclusters were removed via thermal treatment at 300 °C in air (note: the ligand desorption temperature ~200 °C, Figure 7(a)). No sintering of the Au_{25} and Au_{144} nanoclusters was found, as the ordered heterostructured mesoporous material supports (e.g., CuO-EP-FDU-12 and Co_3O_4-EPFDU-12) can effectively stabilize the "ligand-off" Au_{25} and Au_{144} nanoclusters. The "ligand-off" Au_{25} and Au_{144} clusters were characterized by *Z*-contrast scanning transmission electron microscopy (STEM, where *Z* refers to the atomic number of element) and no size growth was found compared to the initial core size before ligand desorption. The average size of "ligand-off" Au_{144} clusters is determined to be 1.67 ± 0.2 nm which is consistent with the expected size (metal core) of the $Au_{144}(SCH_2CH_2Ph)_{60}$ nanoclusters, Figure 10(a). The "ligand-off" Au_{25} and Au_{144} nanoclusters supported on both CuO-EP-FDU-12 and Co_3O_4-EP-FDU-12 are more active than ligand-on catalysts in CO oxidation at relatively low temperatures (e.g., 56% CO conversion at 20 °C and 100% conversion at 70 °C for the Au_{25} cluster supported on Co_3O_4-EP-FDU-12, Figure 10(b)). Thus, this thermal treatment method can be applied when "ligand-off" gold cluster catalysts are preferred.

Ligand-off Au_{38} clusters supported on oxides were also investigated for CO oxidation.[39] Gaur et al.[43] observed that high activity was not observed until surface thiolate ligands were fully removed by a high-temperature treatment. In contrast, work by Nie et al. showed that a mild pretreatment (e.g., 130 °C) without removing ligands can largely increase the activity.[39] The catalyst after complete removal of thiolate ligands by O_2 pretreatment at 250 °C indeed gave a somewhat lower catalytic activity and lower stability in long-time reaction process.[39] We point out that the oxide support should play a major role and the different results between Nie et al. and Gaur et al. may be caused by the different type of oxide supports used in the tests. The $Au_{38}(SR)$ catalyst also exhibited some effects of water vapor added into the feed gas, similar to the case of $Au_{25}(SR)_{18}$ catalysts.

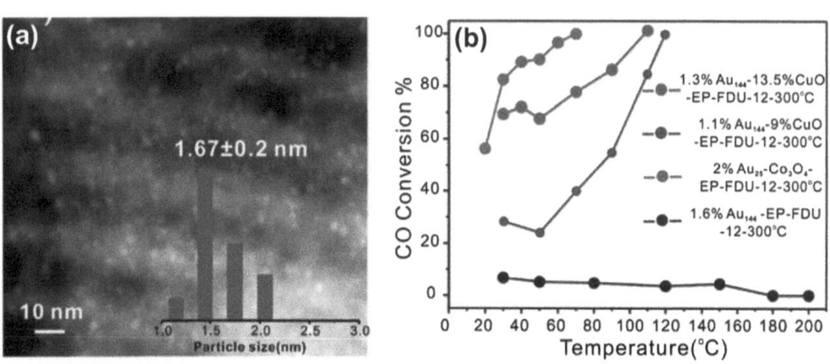

FIGURE 10 (a) Z-contrast scanning transmission electron microscopy image of Au_{144} nanoclusters supported CuO-EP-FDU-12 after calcination at 300 °C. (b) Oxidation of CO as a function of temperature over supported gold cluster catalysts. *Adapted with permission from Ref. 42.*

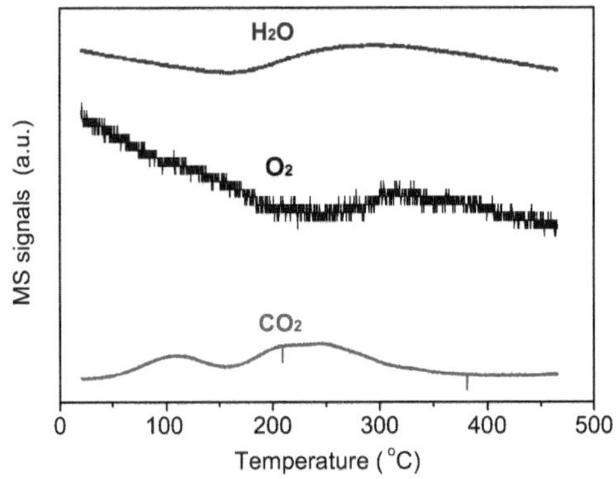

FIGURE 11 Mass spectrometric signals of O_2-TPO profiles of $Au_{38}(SR)_{24}/CeO_2$ catalyst. From top to bottom: H_2O generation (positive peak at >175 °C), O_2 consumption (negative peak between 175 and 310 °C), and CO_2 generation (two peaks at 50–150 °C and 175–320 °C, respectively). *Adapted with permission from Ref. 39.*

To rule out the possibility that O_2 would burn off thiolate ligands at the pretreatment temperatures, Nie et al. performed O_2-TPO experiments over $Au_{38}(SR)_{24}/CeO_2$, in which the signals of O_2 consumption, H_2O and CO_2 generation were monitored by mass spectrometry (Figure 11). A peak of CO_2 generation was observed at ~100 °C, but this is due to the fact that CeO_2 powders capture CO_2 in air, rather than burning of the thiolate ligands on $Au_{38}(SR)_{24}$ clusters since no corresponding H_2O peak was observed at the same temperature, (Figure 11). A distinct O_2 consumption peak occurred at ~230 °C, (Figure 11). The H_2O generation signal (i.e., positive peak) was observed at ~175 °C (onset) and above was attributed to H_2O desorption from the catalyst surface and also possibly due to the H_2O product from O_2 burning of gaseous, desorbed ligands at >175 °C, evidenced by a concurrent CO_2 peak centered at ~230 °C. Overall, the H_2O and CO_2 generation process above 175 °C can be attributed to O_2 burning of desorbed gaseous ligands, consistent with TGA results. However, no burning of ligands was detected during the mild O_2 pretreatment process since the pretreatment temperature is much lower than 175 °C.

10.3.1.2 Selective Oxidation of Styrene

We investigated solution-phase styrene oxidation catalyzed by $Au_n(SR)_m$ nanocluster catalysts (free and SiO_2-supported gold nanoclusters) using O_2 as the oxidant, Scheme 2.[30] The catalytic reaction was run at 80–100 °C for 12–24 h, which gave rise to benzaldehyde as the major product (up to ~70% selectivity) and styrene epoxide (~25% selectivity) and acetophenone (<5% selectivity) using $Au_{25}(SR)_{18}$ nanoclusters.

Using free (i.e., unsupported) Au_{25}, Au_{38}, and Au_{144} nanocluster catalysts, we observed a strong size dependence in the catalytic oxidation reaction of styrene, and the smaller $Au_n(SR)_m$ nanocluster catalyst exhibited much higher

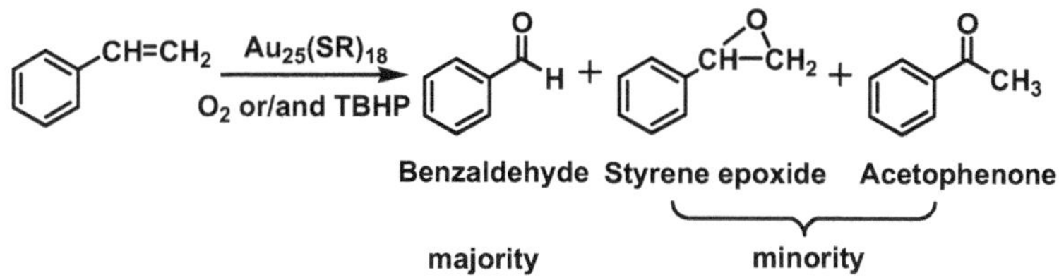

SCHEME 2 Selective oxidation of styrene catalyzed by $Au_{25}(SR)_{18}$ nanocluster catalysts.

catalytic activity.[30] The SiO_2-supported $Au_{25}(SR)_{18}$ catalyst gave rise to comparable performance as that of the free nanocluster catalyst, but the advantage of $Au_{25}(SR)_{18}/SiO_2$ catalyst is its recyclability for reuse in catalytic reaction; no apparent deterioration in activity and selectivity of the reused supported catalyst was observed after multiple cycles.[30]

Furthermore, Zhu et al. compared three oxidant systems using $Au_{25}(SR)_{18}/SiO_2$ catalyst: (1) *tert*-butyl hydroperoxide (TBHP) as the oxidant; (2) TBHP as the initiator and O_2 as the main oxidant; (3) O_2 as the oxidant (without initiator).[30,44] System-A gave a high conversion of styrene (e.g., 86%) with 100% selectivity for benzaldehyde since TBHP is a highly reactive oxidant and easy to activate; system-B gave rise to a much lower activity (25% conversion with 100% selectivity for benzaldehyde) since O_2 is much more difficult to activate than TBHP, and system-C was even lower (18% conversion with 80% selectivity for benzaldehyde). The catalytic results were consistent with the fact that TBHP is much more reactive than O_2. A similar phenomenon on the major effect of the oxidant was also observed in the $Au_{38}(SR)_{24}/SiO_2$ and $Au_{144}(SR)_{60}/SiO_2$ catalysts.[30] The results indicate that the activation of the oxidant is a key step for achieving high conversion of styrene. The results also imply that the active oxygen species in the catalytic cycles may be peroxide-like or hydroperoxide-like species.[30]

Liu et al. used synchrotron radiation-based X-ray absorption fine-structure spectroscopy and ultraviolet photoemission spectroscopy to characterize the $Au_{25}(SCH_2CH_2Ph)_{18}$ catalyst for achieving correlation of the styrene-oxidation catalytic results with the electronic properties of the cluster.[45] Their results revealed that the ligands act as not only stabilizers but also as d-band electron acceptor for Au atoms. They proposed that the d-band unoccupied state population is important for the catalytic activity and selectivity.[45]

Tsukuda and coworkers[46] investigated $Au_{25}(SG)_{18}$/HAP catalyst for selective oxidation of styrene, where HAP refers to hydroxyapatite ($Ca_{10}(PO_4)_6(OH)_2$), Scheme 3. The ligand-off Au_{25}/HAP catalyst was obtained by removing the ligands via 300 °C thermal treatment. The catalyst was explored in styrene oxidation (with anhydrous TBHP as the oxidant) and the reaction yielded styrene oxide with a selectivity of ~92%, in contrast with the results obtained by Zhu et al.[30] in which the benzaldehyde (as opposed to the epoxide) constitutes the major product.

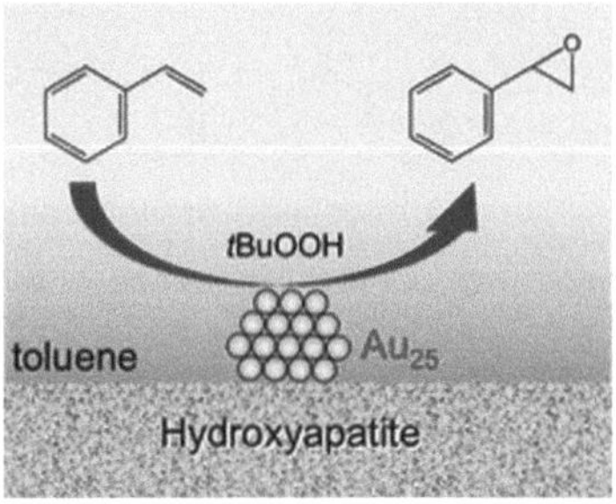

SCHEME 3 Epoxidation of styrene catalyzed by ligands-off Au_{25}/HAP catalyst. *Adapted with permission from Ref. 46.*

10.3.1.3 Catalytic Oxidation of Benzyl Alcohol

Tsukuda and coworkers probed the ligand effect of thiolated nanoclusters in the aerobic oxidation of benzyl alcohol (Scheme 4).[47] They found that the presence of thiolate ligands on $Au_{25}(SC_{12}H_{25})_{18}$ (supported on porous carbon) increases the selectivity for benzaldehyde formation, albeit the activity was reduced. The ligand-off cluster catalyst alters the product selectivity for the benzoic acid product.[47]

Tuel and coworkers[48] deposited $Au_{25}(SPh\text{-}p\text{-}NH_2)_{17}$ clusters into the mesopores of a silica support (SBA-15). Ligand removal by 400 °C calcination led to very small gold nanoparticles (1.8 nm) with a narrow distribution. They found that the ligand-on catalyst could not dissociate O_2, but exceptional activity was observed when radical initiators (e.g., alkyl hydroperoxides) were added.

10.3.1.4 Aerobic Oxidation of Cyclohexane

Tsukuda and coworkers[49] prepared HAP-supported $Au_{10}(SG)_{10}$, $Au_{18}(SG)_{14}$, $Au_{25}(SG)_{18}$, and $Au_{39}(SG)_{24}$ nanoclusters as catalysts (all with 0.2 wt% loading of clusters) for aerobic oxidation of cyclohexane and observed a bell-shaped size effect (Figure 12). Such catalysts were thermally calcined at 300 °C for 2 h in vacuo so that the surface ligands were completely removed. The relatively strong interactions between the HAP support and the clusters prevented size growth of the ligands-off clusters. The catalytic performances of the ligands-off Au_n/HAP catalysts exhibited an optimal size of catalyst at $n \sim 40$ (Au atoms) for the aerobic oxidation of cyclohexane.[49] Cyclohexanol and cyclohexanone were obtained as the primary products with nearly equal yields (~99% total selectivity), with no adipic acid formed in the reaction.

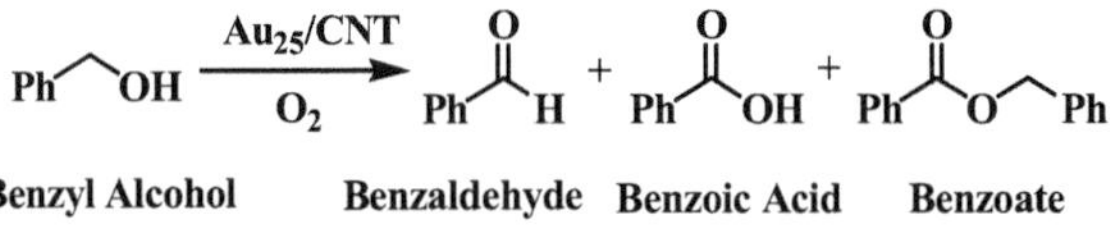

SCHEME 4 Benzyl alcohol oxidation catalyzed by nanoclusters.

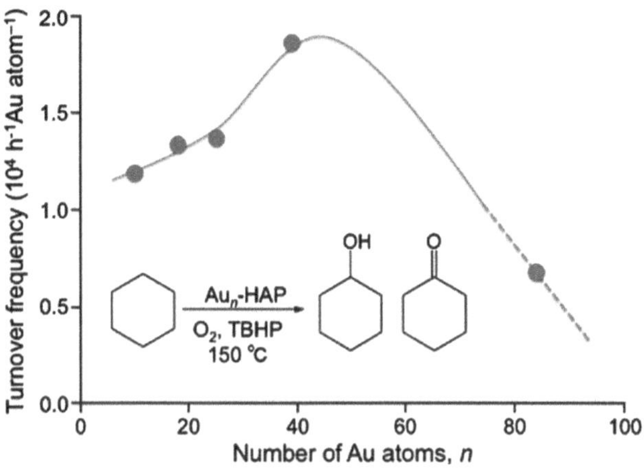

FIGURE 12 Selective oxidation of cyclohexane to cyclohexanol and cyclohexanone with O_2 as the oxidant catalyzed by ligands-off Au_n/HAP (reaction conditions: O_2 pressure: 1 MPa, solvent-free, 150 °C). *Adapted with permission from Ref. 49.*

10.3.1.5 Selective Oxidation of Sulfides

Selective oxidation of sulfides to sulfoxides (Scheme 5) is an important organic transformation and the resulting sulfoxides are versatile intermediates for the preparation of biologically and medicinally important products. The side-reaction of sulfoxidation is the formation of sulfone. It is highly desirable to achieve selective oxidation of sulfides to sulfoxides under mild conditions. In terms of bonding, sulfide has a much less binding affinity toward gold than thiolate, but sulfides can still be chemisorbed onto the surface of gold nanoclusters via the interaction between the sulfur atom (—S—) and the gold surface. Compared to thiolate and sulfide, while sulfoxide (—S(=O)—) can only weakly bind to the gold surface; hence the conversion from sulfide to sulfoxide on the clusters surface should be feasible.

Li et al.[50] investigated the oxide-supported ~1 wt% $Au_{25}(SR)_{18}$ nanocluster catalyst for selective oxidation of sulfides to sulfoxides using iodosylbenzene (PhIO) as the oxidant (Scheme 5). The TiO_2-supported $Au_{25}(SR)_{18}$ nanocluster catalyst gave rise to a high catalytic activity (e.g., ~97% conversion of sulfide with ~92% selectivity for the sulfoxide product). Different types of sulfides have been investigated in order to test the effects of electronic and steric factors in the selective oxidation. The sulfides with electron-rich groups were oxidized with

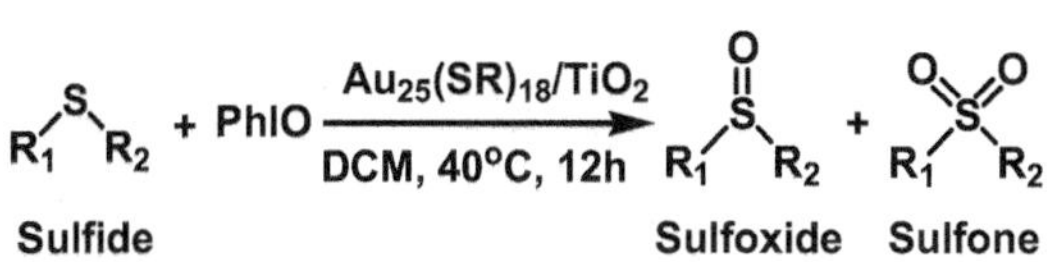

SCHEME 5 Oxidation of sulfides by iodosylbenzene using $Au_{25}(SR)_{18}/TiO_2$ catalyst.

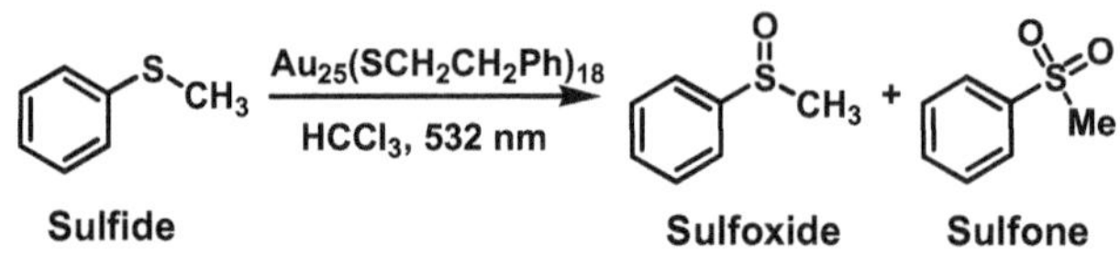

SCHEME 6 Photocatalytic oxidation of sulfides to sulfoxides by $^1O_2$ generated on the photoexcited $Au_{25}(PET)_{18}^-$ catalyst.

higher conversions than those bearing electron-deficient groups, as well as the sulfides with greater steric groups, showed lower conversions. The support effect was also investigated, and the catalyst activity of $Au_{25}(SR)_{18}$/oxide followed the order of $TiO_2 > Fe_2O_3 > CeO_2 > MgO$. The oxide-supported $Au_{25}(SR)_{18}$ catalysts showed excellent recyclability in the sulfoxidation process.[50]

Recent work by Kawasaki et al.[51] reported the nanocluster-induced generation of highly reactive singlet oxygen ($^1O_2$), which can be utilized in catalytic oxidation. The $^1O_2$ was efficiently produced through the direct photosensitization by $Au_{25}(SR)_{18}^-$ nanoclusters (SR = phenylethanethiol, SC_2H_4Ph, or captopril) via visible or near-IR irradiation (e.g., 532, 650, and 808 nm). $^1O_2$ was successfully detected by direct observation of its characteristic emission around 1276 nm as well as by indirect methods of using three different $^1O_2$-selective probes.

Catalytic selective oxidation of sulfides to sulfoxides by $^1O_2$ (Scheme 6) was further demonstrated on the photoexcited $Au_{25}(SC_2H_4Ph)_{18}^-$ nanoclusters. When the reactions were carried out in open vessels in *dark* with O_2 bubbling, almost no sulfoxidation of sulfides was observed. With the assistance of photoexcited $Au_{25}(SR)_{18}^-$ at 532 nm irradiation, an ~100% selective catalytic oxidation of sulfides to sulfoxides was observed. The conversion increased with higher reaction temperatures or longer reaction times. The ~100% selectivity achieved in the sulfoxidation of sulfides was in contrast to the use of other oxidant such as PhIO (87% only). Kawasaki et al.[51] also examined the catalytic oxidation of sulfides to sulfoxides by $^1O_2$ in the presence of $Au_{25}(SR)_{18}^-$ using different wavelengths of photoexcitation (532, 650, and 808 nm). Nearly 100% selectivity was achieved in the sulfoxidation of sulfide, irrespective of the excitation wavelength and the conversion is the largest for photoexcited $Au_{25}(SR)_{18}^-$ upon 650 nm photoexcitation, which is very close to the absorbance at the HOMO-LUMO peak of $Au_{25}(SR)_{18}^-$.

10.3.2 Catalytic Hydrogenation

10.3.2.1 Hydrogenation of Nitrophenol

In addition to catalytic oxidation processes, gold nanoclusters have also been explored in hydrogenation reactions. Generally, gold is not active in activating H_2, but the gold core can store electrons (e.g., from $NaBH_4$) and hence catalyzes hydrogenation. Yamamoto et al.[52] compared the catalytic properties of $Au_{25}(SG)_{18}$ and dimethylformamide (DMF)-stabilized gold nanoclusters (note: not of atomic precision) in the reduction reaction of 4-nitrophenol to

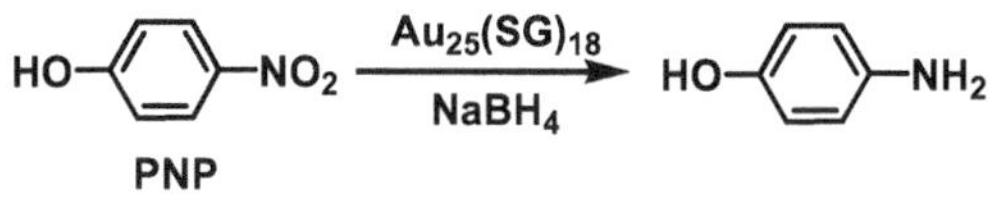

SCHEME 7 Hydrogenation reaction of 4-nitrophenol to 4-aminophenol catalyzed by $Au_{25}(SR)_{18}$ nanocluster catalysts.

4-aminophenol by $NaBH_4$ in water at room temperature, Scheme 7. The reduction process was monitored by UV–vis spectroscopy. The $Au_{25}(SG)_{18}$ nanoclusters exhibited higher catalytic activity than DMF-capped clusters in the nitrophenol reduction; the pseudo-first-order rate constant was $8 \times 10^{-3}\ s^{-1}$ for $Au_{25}(SG)_{18}$ versus $3 \times 10^{-3}\ s^{-1}$ for DMF-capped Au clusters. High catalytic activity of $Au_{25}(SG)_{18}$ catalyst can be achieved at a very low catalyst concentration, even at 1.0 μM. No induction time was observed in the case of $Au_{25}(SG)_{18}$ despite −SG being a strongly binding ligand due to less steric hindrance, while the DMF-stabilized Au clusters exhibited an induction time (~4000 s), which was attributed to the impedance from the surface DMF layer on the clusters when reactants access the cluster surface.[52] For $Au_{25}(SG)_{18}$ nanoclusters, the unique core–shell structure has less steric hindrance and thus renders the catalytic active sites more accessible to reactants.

Shivhare et al.[53] demonstrated intact $Au_{25}(SR)_{18}$ (where, $R = C_6H_{13}$, C_8H_{17}, and $C_{12}H_{25}$) nanoclusters for the reduction of nitrophenol by $NaBH_4$ in mixed THF/water and under N_2 atmosphere. The $Au_{25}(SR)_{18}$ nanoclusters were stable and retained their structural integrity toward $NaBH_4$ treatment, while the larger $Au_{\sim 180}(SC_6H_{13})_{\sim 100}$ nanoclusters were not.

10.3.2.2 Selective Hydrogenation of Aldehydes and Ketones

In terms of chemoselective hydrogenation of different functional groups (e.g., aldehyde group versus nitro group or C=C group), the $Au_n(SR)_m$ nanocluster catalysts were demonstrated to be capable of the selective hydrogenation of α,β-unsaturated aldehydes and ketones to α,β-unsaturated alcohols under mild conditions (60 °C, in mixed solvents (1:1 toluene/acetonitrile)).[37] Under the investigated reaction conditions, the $Au_{25}(SR)_{18}$ nanoclusters were found to preferentially hydrogenate the C=O bond against the C=C bond, and the product (i.e., unsaturated alcohol) was obtained with 76% selectivity, whereas 14% selectivity was observed for unsaturated ketone, and 10% for saturated alcohol.[37] Further exploration of the catalytic reaction led to a nearly complete selectivity for α,β-unsaturated alcohol in a mixed toluene/ethanol (1:1) solvent at lower reaction temperatures (e.g., room temperature), albeit the conversion was not high (<20%).[54] The $Au_{25}(SR)_{18}$ nanocluster catalyst was confirmed to remain intact after the hydrogenation reaction based upon the UV–vis and mass spectrometry analyses.

To enhance the conversion of α,β-unsaturated aldehydes and ketones, oxide-supported nanocluster catalysts were investigated (including $Au_{25}(SR)_{18}/Fe_2O_3$, $Au_{25}(SR)_{18}/TiO_2$, and $Au_{25}(SR)_{18}/SiO_2$), Table 1.[54] Nearly

TABLE 1 Chemoselective Hydrogenation Reaction of α,β-Unsaturated Ketones to α,β-Unsaturated Alcohols Catalyzed by $Au_{25}(SR)_{18}$ Nanocluster Catalysts

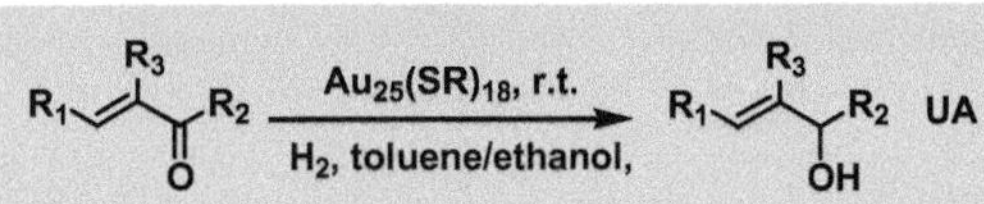

Entry	Substrates	Selectivity for UA	Conversion (%)		
			Au_{25}	Au_{25}/Fe_2O_3	Au_{25}/TiO_2
1	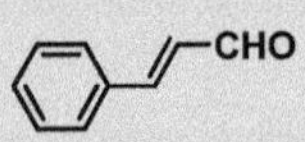	100	38	49	46
2	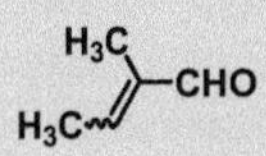	100	43	53	51
3	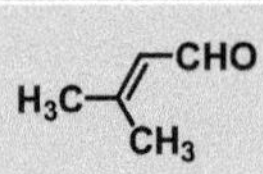	100	44	54	53
4	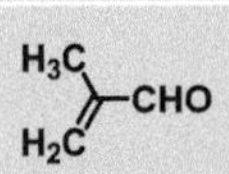	100	39	52	52
5	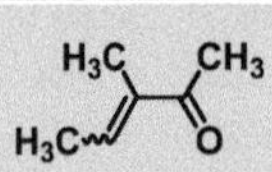	100	29	45	44
6	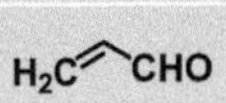	91[a]	46	47	47

[a] *The selectivity for unsaturated alcohol (UA) is 90% (free $Au_{25}(SR)_{18}$ catalyst, Au_{25}), 92% ($Au_{25}(SR)_{18}/Fe_2O_3$ catalyst, Au_{25}/Fe_2O_3), and 91% ($Au_{25}(SR)_{18}/TiO_2$ catalyst, Au_{25}/TiO_2), respectively.*

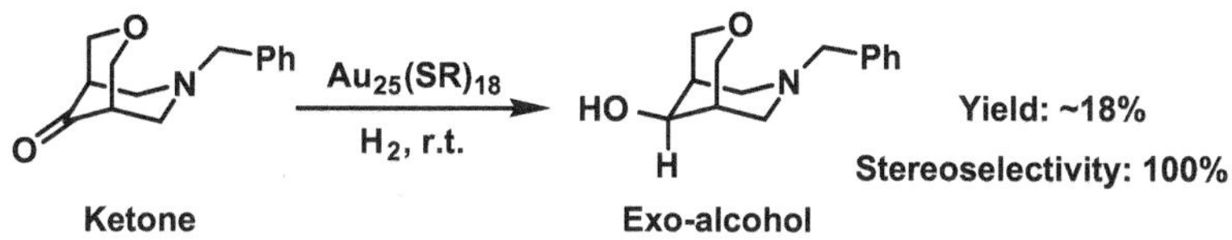

SCHEME 8 Stereoselective hydrogenation reaction of bicyclic ketone to Exo-alcohol catalyzed by $Au_{25}(SR)_{18}$ nanocluster catalysts.

~100% selectivity for α,β-unsaturated alcohols was retained. A range of substrates were tested and all gave rise to high selectivity for the unsaturated alcohol product. It was found that the catalytic activity was largely improved from less than 20% to 40–50%. The catalytic activity of the $Au_{25}(SR)_{18}/Fe_2O_3$ and $Au_{25}(SR)_{18}/TiO_2$ catalysts was almost equivalent (Table 1).

Zhu et al.[55] further expanded the $Au_{25}(SR)_{18}$ nanocluster catalyst for stereoselective hydrogenation of a bicyclic ketone (i.e., 7-(phenylmethyl)-3-oxa-7- azabicyclo [3.3.1] nonan-9-one) to Exo-alcohol (Scheme 8). The stereoselective hydrogenation was run at room temperature in an ethanol/toluene (5:1, V/V)-mixed solvent using H_2 (1 atm) as hydrogen source; ~18% conversion of the bicyclic ketone with 100% selectivity for the Exo-alcohol was obtained; note that the stereoselective product was determined by NMR spectroscopy.[55] The high stereoselectivity (100%) was rationalized to be effected by the atomic spatial restraints and the unique core–shell configuration of the $Au_{25}(SR)_{18}$ nanocluster.

10.3.2.3 Selective Hydrogenation of Nitrobenzaldehyde Derivatives

In recent work, Li et al. synthesized thermally robust $Au_{99}(SPh)_{42}$ nanoclusters and investigated catalytic application in chemoselective hydrogenation of nitrobenzaldehyde derivatives (Scheme 9).[34] The $Au_{99}(SPh)_{42}$ nanocluster is not yet in the metallic state and its HOMO-LUMO gap (or band-gap) was determined to be ~0.7 eV.[34] Ceria-supported $Au_{99}(SPh)_{42}$ was utilized as a catalyst for chemoselective hydrogenation of nitrobenzaldehyde to nitrobenzyl alcohol in water using H_2 gas as the hydrogen source.[34] The catalytic hydrogenation reaction was carried out under mild conditions (80 °C for 12 h). The ~100% selective hydrogenation of the aldehyde group came as a surprise, because conventional nanogold catalysts[56] instead give rise to the reduction product of the nitro group.

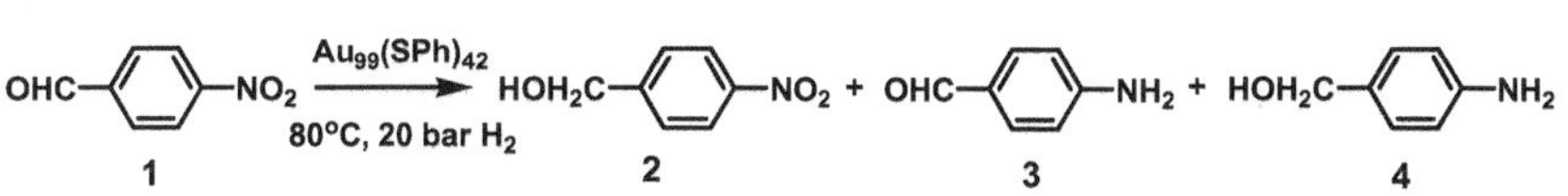

SCHEME 9 Hydrogenation of 4-nitrobenzaldehyde to 4-nitrobenzyl alcohol using the $Au_{99}(SPh)_{42}$ nanocluster as a catalyst.

A range of nitrobenzaldehyde derivatives were tested on the $Au_{99}(SPh)_{42}/CeO_2$ catalyst and high activity and complete selectivity for the reduction of the aldehyde group were obtained (Table 2). From the substrates, the catalytic activity of the $Au_{99}(SPh)_{42}/CeO_2$ catalyst is considerably influenced by the electrophilic aromatic substitution, instead of the electron-rich/deficient side groups. The supported catalysts show excellent recyclability in the reaction.[34]

TABLE 2 Chemoselective Hydrogenation of a Range of Substrates With Nitro Group and Aldehyde Group Using $Au_{99}(SPh)_{42}/CeO_2$ as the Catalyst.[a] Reaction Conditions: Same as Noted in Table 3

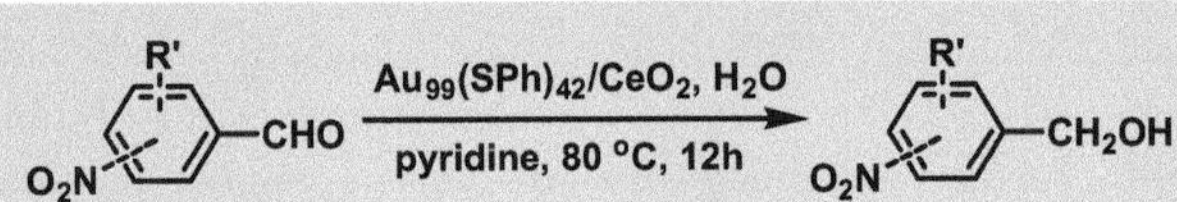

Entry	Substrate	Product	Conv. (%)[b]	Select. (%)[b]
1	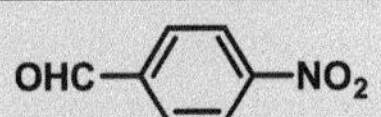	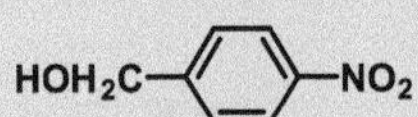	93.1	100
2	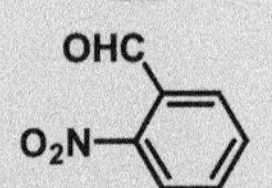	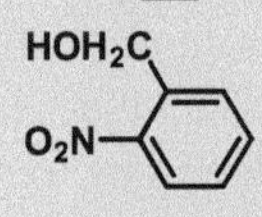	73.2	100
3	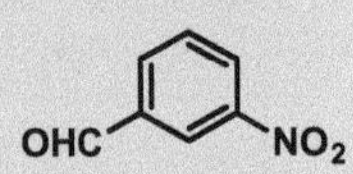	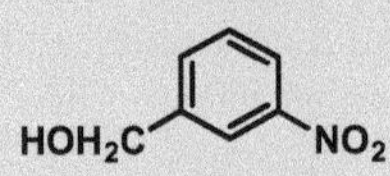	98.5	100
4	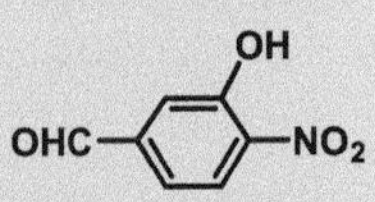	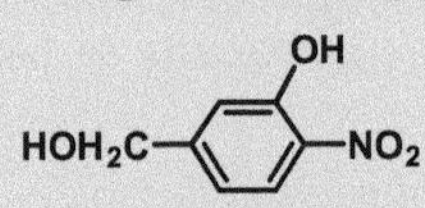	84.1	100
5	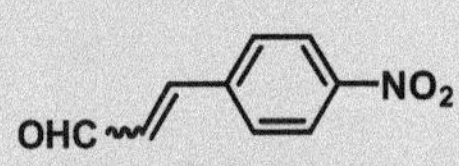	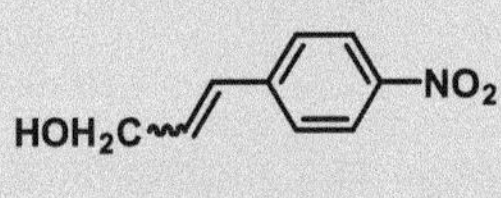	90.0	100
6	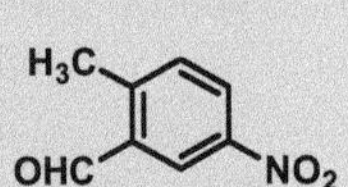	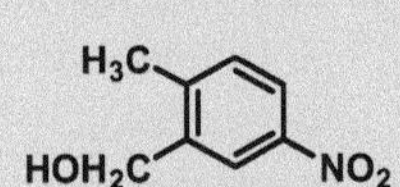	98.9	100
7	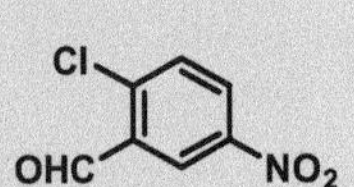	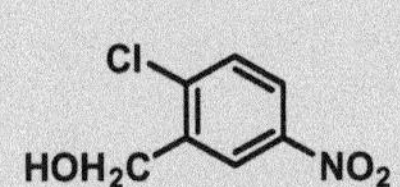	99.5	100

[a]*Reaction conditions: $Au_{99}(SPh)_{42}$/oxide (100 mg, 1 wt% loading) catalysts in 1 mL H_2O, 0.05 mmol 4-nitrobenzaldehyde, 0.1 mmol pyridine, 20 bar H_2, 80°C, 12 h.*
[b]*The conversion (Conv.) of 4-nitrobenzaldehyde and selectivity for 4-nitrobenzyl alcohol were determined by NMR analysis. n.d. = not detected; the conversion and selectivity values are averaged over three measurements (error bar: ~0.5%).*

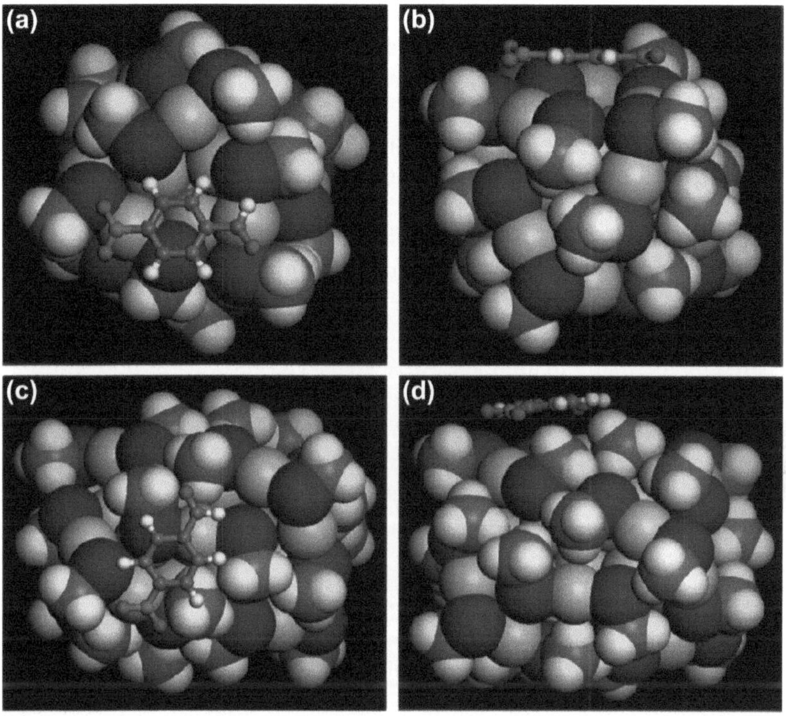

FIGURE 13 The 4-nitrobenzaldehyde molecule adsorbed on $Au_{25}(SCH_3)_{18}^-$ nanocluster in top view (a) and side view (b), and on $Au_{38}(SCH_3)_{24}$ nanocluster in top view (c) and side view (d). Color code: Au, yellow; S, blue; C, gray; H, white; O, red; N, pink. (see color plate). *Reproduced with permission from Ref. 57. Copyright 2014 American Chemical Society.*

We further test the size-dependent catalytic performance of Au_{25}, Au_{36}, Au_{38}, and Au_{99} nanoclusters.[34,57] High activity and selectivity were obtained. Based on the structures of $Au_{25}(SPh)_{18}$ and $Au_{36}(SPh)_{24}$, DFT modeling of the 4-nitrobenzaldehyde adsorption[57] shows that both the $-CHO$ and $-NO_2$ groups are adsorbed on the S–Au–S staples of the gold nanocluster (Figure 13). The adsorption of the 4-nitrobenzaldehyde molecule on the different-sized $Au_n(SR)_m$ nanoclusters is moderately strong and similar in strength. The DFT results suggest that the catalytic activity of the $Au_n(SR)_m$ nanoclusters is primarily determined by the surface areas of the Au nanoclusters, consistent with the observed trend of the conversion of 4-nitrobenzaldehyde versus the cluster size. Overall, experiment combined with theory offers the molecular insight into the hydrogenation of 4-nitrobenzaldehyde and the active site structure on gold nanocluster catalysts.

10.3.2.4 Semihydrogenation of Alkynes

The oxide-supported spherical $Au_{25}(SR)_{18}$ nanocluster was explored for the catalytic semihydrogenation of alkynes to alkenes using H_2 under relatively mild conditions (100 °C, 20 bar H_2).[58] The terminal chain alkynes (e.g., 1-hexyne, 1-octyne, and 1-decyne) and terminal aromatic alkyne (e.g., phenylacetylene) all gave high conversion (>99%) and high selectivity for the alkene product (yield > 99%), Table 3. In contrast, the $Au_{25}(SR)_{18}/TiO_2$ catalyst exhibited no activity for internal alkynes and the conversion for all the tested internal alkynes was below 1% (Table 4, entries 2–4). We proposed a deprotonation activation mechanism for alkynes on the ligand-protected nanocluster catalyst.[58] The terminal alkynes possess a terminal hydrogen (H–C≡C), but internal alkynes have no terminal hydrogen and thus cannot be activated by the $Au_{25}(SR)_{18}$ nanoclusters.

We furthermore tested the *ligand-off* catalyst (Au_{25}/TiO_2),[58] which was prepared by thermally treating the $Au_{25}(SR)_{18}/TiO_2$ powder at 300 °C for 1 h. The bare Au_{25}/TiO_2 catalyst indeed catalyzed the semihydrogenation reactions of internal alkynes including 1-phenyl-1-propyne, diphenylacetylene, and methyl 2-nonynoate (Table 4, entries 6–8), with 52.6–59.7% conversions of internal alkynes and >97% stereoselectivity for *Z*-alkene. Thus, the presence of ligands on Au_{25} inhibits the activation of internal alkynes but not to the terminal alkynes.

TABLE 3 Semihydrogenation of Terminal Alkynes Using $Au_{25}(SR)_{18}/TiO_2$ Catalyst[a]

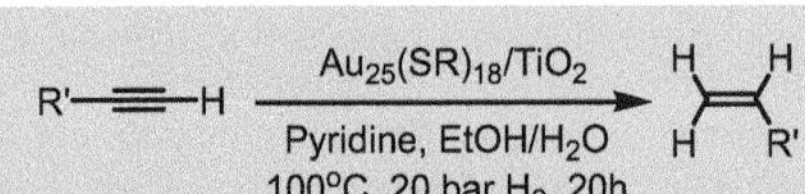

Entry	Alkyne	Alkene	Conv. [%]
1	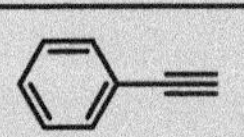	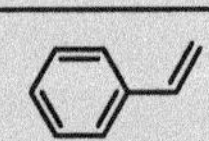	99.2
2	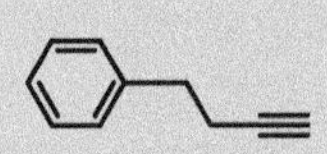	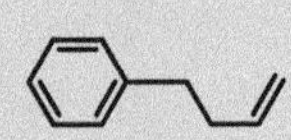	~100
3	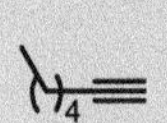		99.5
4	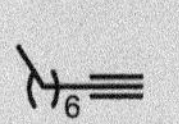		99.6
5	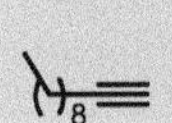	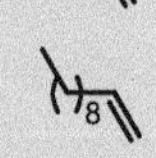	99.4

[a]*Reaction conditions: 100 mg $Au_{25}(SR)_{18}$(1 wt%)/TiO_2 catalyst, 0.1 mmol alkynes, 0.2 mmol pyridine, 1 mL $EtOH/H_2O$ (10:1, V/V), 100 °C, 20 bar H_2, 20 h.*
Adapted with permission from Ref. 58.

TABLE 4 Comparison of Conversion in Semihydrogenation Catalyzed by "Ligand-On" and "Ligand-Off" catalysts[a]

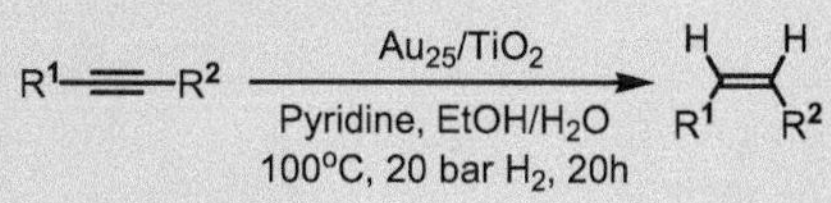

Entry	catalyst	R^1	R^2	Conv.	Sel.[b]
1	Ligand-on Au_{25} sphere	PhC_2H_4	H	>99	–
2		Ph	CH_3	<1	–
3		Ph	Ph	<1	–
4		n-C_6H_{13}	CO_2CH_3	<1	–
5	Ligand-off Au_{25}	PhC_2H_4	H	95.6	–
6		Ph	CH_3	52.8	97
7		Ph	Ph	59.7	>99
8		n-C_6H_{13}	CO_2CH_3	52.6	99

[a]*Reaction conditions: as noted in Table 3.*
[b]*Stereoselectivity (Sel.) for Z-alkene.*
Adapted with permission from Ref. 58.

With respect to the support effect, blank experiments using gold-free TiO_2 gave no conversion, thus the catalytic effect is associated with the Au nanocluster. To further investigate the potential effect of the oxide supports, four different oxides (i.e., CeO_2, SiO_2, TiO_2, and Al_2O_3) were used as supports for the $Au_{25}(SR)_{18}$ nanocluster.[58] No apparent difference in the four types of $Au_{25}(SR)_{18}$/oxide catalysts was observed in the catalytic semihydrogenation reaction of 4-phenyl-1-butyne (i.e., all gave high conversion and selectivity around 99%).

The above results imply that the activation process of the terminal alkynes on $Au_{25}(SR)_{18}$ nanocluster catalyst is different from the case of bare Au_{25} nanocluster and conventional gold nanoparticle catalysts. The structure of the $Au_{25}(SR)_{18}$ nanocluster shows a 13-atom icosahedral Au_{13} core protected by 6 dimeric staple motifs (Figure 14(a)). Two open facets of triangular Au_3 are seen in the structure (Figure 14(b)), with the exposed gold atoms from the three adjacent staples. These sites may constitute the catalytic sites. Previous theoretical simulation indicated the adsorption of phenylacetylene on such Au_3 sites on the surface of $Au_{25}(SR)_{18}$ nanocluster,[32] and the adsorption energy was computed to be −0.40 eV. We proposed that the semihydrogenation reaction involves the following initial adsorption configuration: the phenyl ring

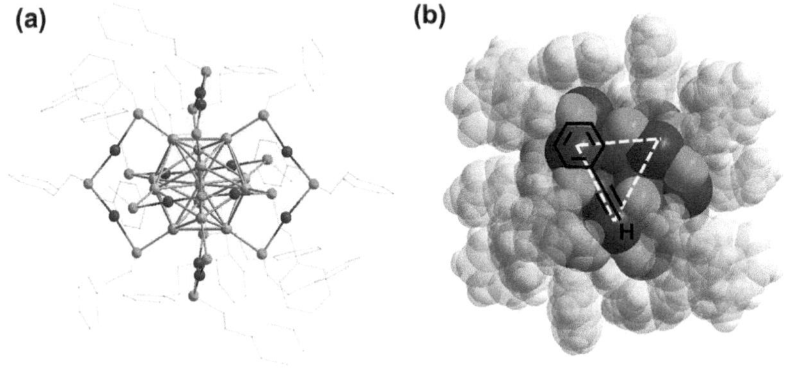

FIGURE 14 (a) The atomic structure of $Au_{25}(SR)_{18}$ nanoclusters. (b) The terminal alkyne molecule adsorption on the surface of the $Au_{25}(SC_2H_4Ph)_{18}$ nanocluster (shown in space-fill mode). Color code: Au, green (dark gray in print versions) and cyan (gray in print versions); S, yellow (light gray in versions). *Adapted with permission from Ref. 58.*

of phenylacetylene faces one of the external Au atoms of the triangular Au_3 site, and the $-C{\equiv}CH$ group of phenylacetylene points to a second Au atom of the Au_3 site (Figure 14(b)). In such a configuration, the third Au atom of the Au_3 site is unoccupied and thus available for possible hydrogen adsorption and activation facilitated by pyridine.

10.3.3 Catalytic Carbon–Carbon Coupling Reactions

Gold nanoclusters also hold great promise for the carbon–carbon coupling reactions, such as Ullmann-type homo-coupling and Sonogashira cross-coupling reactions. The catalytic application in C–C coupling reaction was due to the good capability of gold nanocluster catalysts in activating the C–I bond of iodobenzene and the terminal C–H bond of phenylacetylene. Below we illustrate the C–C coupling reactions.

10.3.3.1 Ullmann-type Homocoupling Reaction

Gold nanoparticles are good catalysts for activating the C–I bond of the iodobenzene in the carbon–carbon homo-coupling process. Li et al.[59] investigated the catalytic activity of the oxide-supported $Au_{25}(SR)_{18}$ catalysts for the Ullmann-type homo-coupling reactions of aryl iodides. Optimization showed the suitable conditions for the homocoupling reaction as follows: DMF as solvent and K_2CO_3 as base, at 130 °C under N_2 atmosphere. Under such conditions, the $Au_{25}(SR)_{18}/CeO_2$ catalyst gave rise to 99.8% conversion of iodobenzene (Table 5). The application of the $Au_{25}(SR)_{18}$-catalyzed homo-coupling reaction was tested with substrates bearing a range of substituents. The electron-rich substrates performed better than electron-deficient substrates in the reaction. The support effect was also investigated, but unlike the case of

TABLE 5 Homocoupling of Aryl Iodides Using Au_{25}/CeO_2 Catalyst

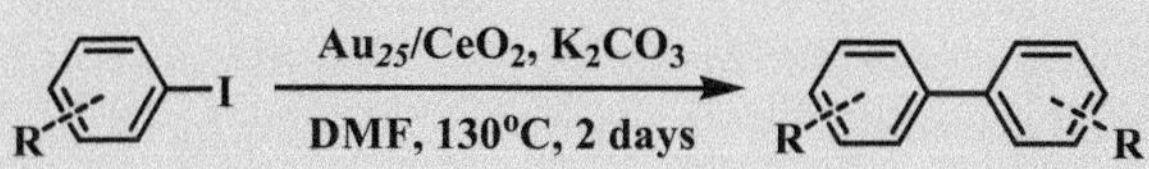

Entry	Substrates	Conversion[%]
1	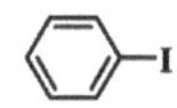	99.8
2	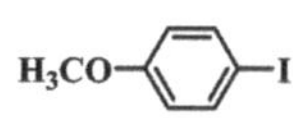	99.5
3	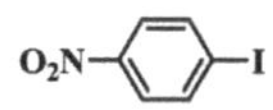	67.5
4	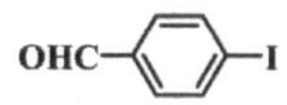	78.2
5	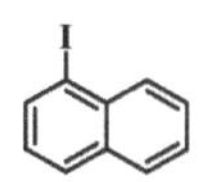	99.7

Adapted from Ref. 59 with permission from The Royal Society of Chemistry.

hydrogenation, no distinct effect of the oxide supports was observed on the investigated supports (including CeO_2, SiO_2, TiO_2, and Al_2O_3).

10.3.3.2 Sonogashira Cross-Coupling Reaction

Li et al.[32] expanded the catalytic application of the oxide-supported $Au_{25}(SR)_{18}$ nanocluster catalysts to the Sonogashira cross-coupling reaction between iodobenzene and phenylacetylene. The supported catalyst was made by impregnation of oxide powders (including CeO_2, TiO_2, MgO, and SiO_2) in a CH_2Cl_2 solution of $Au_{25}(SR)_{18}$ (~1 wt% loading), and then annealed at 150 °C (1 h) in a vacuum oven. The supported gold particle size after the annealing was confirmed by STEM to be identical with that of Au_{25}, and TGA confirmed that the protecting thiolate ligand remained on the surface of gold nanoclusters after the annealing. The $Au_{25}(SR)_{18}$/oxide catalysts after thermal treatment was applied in the Sonogashira cross-coupling reaction of iodobenzene and phenylacetylene.

After optimization of the conditions, the reaction was carried out under N_2 atmosphere at 160 °C using DMF as solvent and K_2CO_3 as base. A high conversion of *p*-iodoanisole (up to 96.1%) with excellent selectivity for cross-coupling product (up to 88.1%) was obtained when it was catalyzed by $Au_{25}(SR)_{18}/CeO_2$ catalyst (Table 6, entry 1). The $Au_{25}(SR)_{18}$ nanocluster

TABLE 6 The Catalytic Performance of $Au_{25}(SR)_{18}$ ($R = CH_2CH_2Ph$) Nanoclusters Supported on Various Oxides as Catalysts for Sonogashira Cross-Coupling Reaction of *p*-Iodoanisole and Phenylacetylene. For the DFT-Simulated Adsorption of Reactants: (a) top View of One of the Two Open Facets of $Au_{25}(SR)_{18}$ Nanoclusters Where Three External Gold Atoms are Exposed; the Other Facet Is on the Back Side. (b) Top View of the Coadsorption of Phenylacetylene and Iodobenzene on the Surface of the $Au_{25}(SR)_{18}$ Nanoclusters. $R' = OCH_3$. Au, Yellow; S, Blue; C, Gray; H, White; I, Green

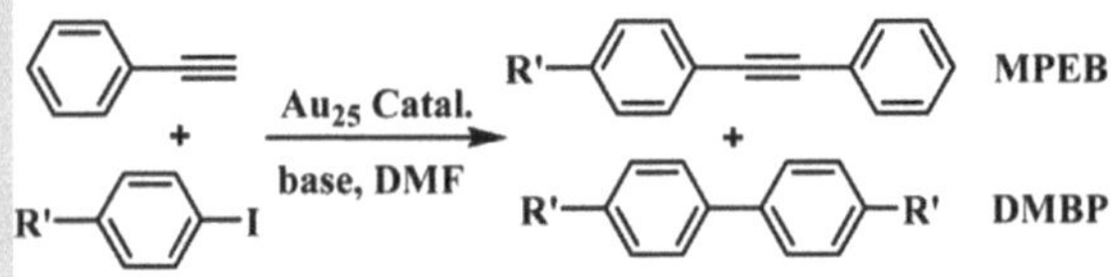

entry	catalyst	conv. (%)	selectivity (%) MPEB	DMBP
1	$Au_{25}(SR)_{18}/CeO_2$	96.1	88.1	11.9
2	$Au_{25}(SR)_{18}/TiO_2$	92.8	82.9	17.1
3	$Au_{25}(SR)_{18}/SiO_2$	90.8	79.3	20.7
4	$Au_{25}(SR)_{18}/MgO$	93.3	80.6	19.4
5	AuNC 2-3 nm (SC_6H_{13})/CeO_2	65.5	57.2	42.8

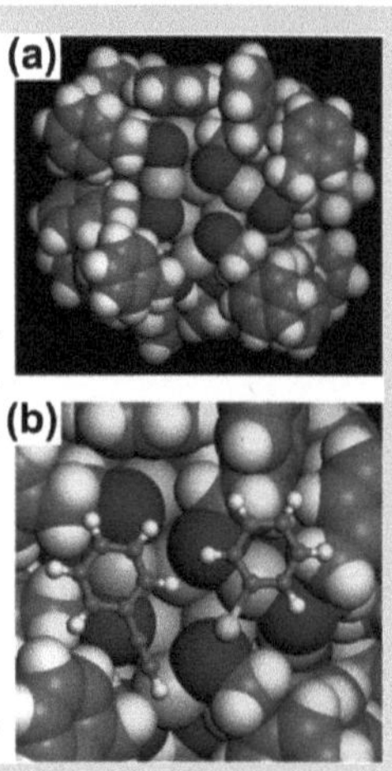

Adapted with permission from Ref. 32.

catalyst performed much better than the larger sized 2–3 nm Au nanoclusters (Table 6, entry 1 vs 5) and CeO_2-supported Au nanoparticles (bare, ~20 nm). Support effects were also investigated, and no distinct effect of the oxide supports was observed (i.e., CeO_2, SiO_2, TiO_2, and MgO). In the recyclability test, the conversion of *p*-iodoanisole exhibited no significant loss, while the selectivity was decreased from 88.1% to 64.5% after 5 cycles. The drop in selectivity should be caused the gradual degradation of nanoclusters in the multiple recycling tests, as larger Au clusters give a much lower selectivity.

DFT calculation of the reactant adsorption shows that both reactants (i.e., iodobenzene and phenylacetylene) prefer to adsorb on the open facet with the phenyl ring facing a surface gold atom.[32] A total adsorption energy reaches −0.90 eV when the two reactants coadsorb on the $Au_{25}(SR)_{18}$ catalyst. The DFT results suggested that the catalytic active sites are associated with the $Au_{25}(SR)_{18}$, which is consistent with the experimental results.

10.3.4 Catalytic Effects of Heteroatom-Doped Nanoclusters

Doping of the atomically precise gold nanoclusters with a second metal (e.g., Pd, Pt) for catalysis has been reported.[60,61] Tsukuda and coworkers compared Au_{25} and Pd_1Au_{24} nanoclusters supported on multiwalled carbon nanotubes for the aerobic oxidation of benzyl alcohol.[60] They showed that single Pd atom doping significantly improved the catalytic performance of the 25-atom nanocluster catalyst. Herein, we illustrate the catalytic activity of the monoplatinum-doped $Pt_1Au_{24}(SR)_{18}$ nanoclusters for the selective oxidation of styrene using $PhI(OAc)_2$ as the oxidant.[61] Various analyses confirmed that the single platinum atom is located in the center of the core (Figure 15), which implies that the shell (i.e., $Au_{12}(SR)_{18}$) of the $Pt_1Au_{24}(SR)_{18}$ nanocluster is the same as that of $Au_{25}(SR)_{18}$. The catalytic reaction was carried out at 70 °C for 10 h in acetonitrile under N_2 atmosphere. Significantly, $Pt_1Au_{24}(SR)_{18}/TiO_2$ catalyst gave much higher catalytic activity (i.e., 90.8% conversion of styrene) and higher selectivity (89.9%) for benzaldehyde than that of $Au_{25}(SR)_{18}/TiO_2$ catalyst (conversion 58.9% and selectivity 54%) under the identical reaction conditions. These results show that the $PtAu_{24}(SR)_{18}/TiO_2$ catalyst has a remarkable enhancement in the catalytic activity/selectivity for styrene oxidation than the undoped $Au_{25}(SR)_{18}/TiO_2$ catalyst.

Wu and coworkers[62] prepared atomically monodisperse 27-metal-atom $Au_{25}Ag_2(SC_2H_4Ph)_{18}$ nanoclusters in high yield (89%) using $Au_{25}(SC_2H_4Ph)_{18}$ as the precursor particle. The two silver atoms in the bimetallic nanocluster were confirmed not to replace the gold atoms but reside on the surface of Au_{25}. These two silver atoms were found to largely enhance the catalytic activity by ~5 times in the hydrolysis of 1,3-diphenylprop-2-ynyl acetate at room temperature in mixed solvents of DMF:H_2O (10:1) and in the presence of K_2CO_3.

10.4 SUMMARY

In this chapter, we have discussed the catalytic application of atomically precise $Au_n(SR)_m$ nanoclusters, including (1) catalytic oxidation (such as selective oxidation of styrene to benzaldehyde or styrene epoxide, selective oxidation of

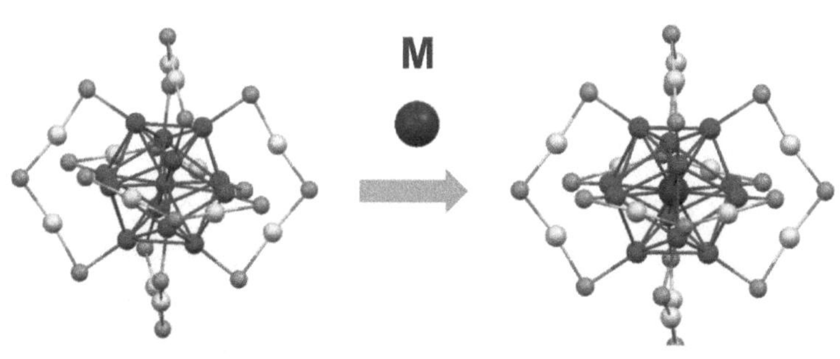

FIGURE 15 Single-atom doping of gold nanocluster (e.g., $Pt_1Au_{24}(SR)_{18}$).

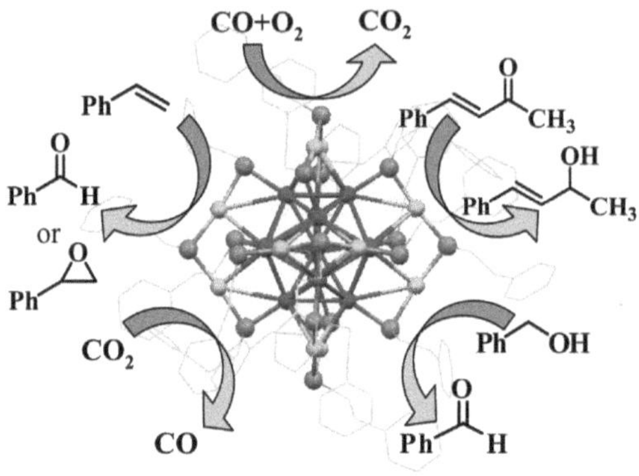

SCHEME 10 Nanocluster-catalyzed chemical reactions. *Reproduced with permission from Ref. 2. Copyright 2013 American Chemical Society.*

sulfides to sulfoxides, selective oxidation of benzyl alcohol to benzaldehyde, and CO oxidation to CO_2), (2) catalytic hydrogenation (such as nitrophenol to aminophenol, and chemoselective hydrogenation), (3) carbon–carbon coupling reactions (such as Ullmann homocoupling of iodobenzene and Sonogashira cross-coupling of iodobenzene and phenylacetylene). In future work, the tunable characteristic of bimetallic nanoclusters[63] is expected to further improve the catalytic reaction conditions toward mild temperatures, and to enhance catalytic activity, selectivity, and stability for specific reaction processes. It is noteworthy that researchers have also explored the catalytic application of such nanoclusters in electrochemical catalysis[64] and photocatalysis.[65–68]

Overall, the demonstrated catalytic reactions indicate that the catalytic activity and selectivity are strongly associated with the unique structure and electronic properties of $Au_n(SR)_m$ nanoclusters (Scheme 10). As a new class of catalysts, $Au_n(SR)_m$ nanoclusters[69-72] will bridge the gap between organometallic complex-based homogeneous catalysis and nanoparticle-based heterogeneous catalysis, and also bridge the gap between the bulk crystal structure model catalysts and real-world catalysts. These atomically defined gold nanoclusters provide exciting opportunities for achieving fundamental understanding of nanocatalysis at the atomic level, including the insight into the size dependence and deep understanding of the molecular activation, catalytic active sites, and catalytic mechanism by correlation with the structures of gold nanoclusters. Future research on atomically precise nanocluster catalysts will contribute to the fundamental catalysis and the new design of highly selective catalysts for specific chemical processes.

ACKNOWLEDGMENT

The work was supported by U.S. Department of Energy-Office of Basic Energy Sciences (Grant DE-FG02-12ER16354).

REFERENCES

1. Haruta M, Tsubota S, Kobayashi T, Kageyama H, Genet MJ, Delmon B. Low-temperature oxidation of CO over gold supported on TiO_2, α-Fe_2O_3, and Co_3O_4. *J Catal* 1993;**144**:175–92.
2. Li G, Jin R. Atomically precise gold nanoclusters as new model catalysts. *Acc Chem Res* 2013;**46**:1749–58.
3. Yamazoe S, Koyasu K, Tsukuda T. Nonscalable oxidation catalysis of gold clusters. *Acc Chem Res* 2014;**47**:816–24.
4. Jin R. The impacts of nanotechnology on catalysis by precious metal nanoparticles. *Nanotechnol Rev* 2012;**1**:31–56.
5. Hutchings GJ, Kiely CJ. Strategies for the synthesis of supported gold palladium nanoparticles with controlled morphology and composition. *Acc Chem Res* 2013;**46**:1759–72.
6. Astruc D, Lu F, Aranzaes JR. Nanoparticles as recyclable catalysts: the frontier between homogeneous and heterogeneous catalysis. *Angew Chem Int Ed* 2005;**44**:7852–72.
7. Negishi Y, Nobusada K, Tsukuda T. Glutathione-protected gold clusters revisited: bridging the gap between gold(I)–thiolate complexes and thiolate-protected gold nanocrystals. *J Am Chem Soc* 2005;**127**:5261–70.
8. Jin R. Quantum sized, thiolate-protected gold nanoclusters. *Nanoscale* 2010;**2**:343–62.
9. Qian H, Zhu M, Wu Z, Jin R. Quantum sized gold nanoclusters with atomic precision. *Acc Chem Res* 2012;**45**:1470–9.
10. Jin R, Qian H, Wu Z, Zhu Y, Zhu M, Mohanty A, et al. Size focusing: a methodology for synthesizing atomically precise gold nanoclusters. *J Phys Chem Lett* 2010;**1**:2903–10.
11. Zhu M, Aikens CM, Hollander FJ, Schatz GC, Jin R. Correlating the crystal structure of a thiol-protected Au_{25} cluster and optical properties. *J Am Chem Soc* 2008;**130**:5883–5.
12. Zeng C, Qian H, Li T, Li G, Rosi NL, Yoon B, et al. Total structure and electronic properties of the gold nanocrystal $Au_{36}(SR)_{24}$. *Angew Chem Int Ed* 2012;**51**:13114–8.
13. Chaki NK, Negishi Y, Tsunoyama H, Shichibu Y, Tsukuda T. Ubiquitous 8 and 29 kDa gold:alkanethiolate cluster compounds: mass-spectrometric determination of molecular formulas and structural implications. *J Am Chem Soc* 2008;**130**:8608–10.
14. Qian H, Eckenhoff WT, Zhu Y, Pintauer T, Jin R. Total structure determination of thiolate-protected Au_{38} nanoparticles. *J Am Chem Soc* 2010;**132**:8280–1.
15. Levi-Kalisman Y, Jadzinsky PD, Kalisman N, Tsunoyama H, Tsukuda T, Bushnell DA, et al. Synthesis and characterization of Au_{102}(p-MBA)$_{44}$ nanoparticles. *J Am Chem Soc* 2011;**133**:2976–82.
16. Negishi Y, Sakamoto C, Ohyama T, Tsukuda T. Synthesis and the origin of the stability of thiolate-protected Au_{130} and Au_{187} clusters. *J Phys Chem Lett* 2012;**3**:1624–8.
17. Qian H, Jin R. Controlling nanoparticles with atomic precision: the case of $Au_{144}(SCH_2CH_2Ph)_{60}$. *Nano Lett* 2009;**9**:4083–7.
18. Qian H, Zhu Y, Jin R. Atomically precise gold nanocrystal molecules with surface plasmon resonance. *Proc Natl Acad Sci USA* 2012;**109**:696–700.
19. Jin R. Atomically precise metal nanoclusters: stable sizes and optical properties. *Nanoscale* 2015;**7**:1549–65.
20. Zhu M, Lanni E, Garg N, Bier ME, Jin R. Kinetically controlled, high-yield synthesis of Au_{25} clusters. *J Am Chem Soc* 2008;**130**:1138–9.
21. Wu Z, Suhan J, Jin R. One-pot synthesis of atomically monodisperse, thiol-functionalized Au_{25} nanoclusters. *J Mater Chem* 2009;**19**:622–6.
22. MacDonald MA, Zhang P, Qian H, Jin R. Site-specific and size-dependent bonding of compositionally precise gold–thiolate nanoparticles from X-ray spectroscopy. *J Phys Chem Lett* 2010;**1**:1821–5.

23. Qian H, Zhu Y, Jin R. Size-focusing synthesis, optical and electrochemical properties of monodisperse $Au_{38}(SC_2H_4Ph)_{24}$ nanoclusters. *ACS Nano* 2009;**3**:3795–803.
24. Heaven MW, Dass A, White PS, Holt KM, Murray RW. Crystal structure of the gold nanoparticle $[N(C_8H_{17})_4][Au_{25}(SCH_2CH_2Ph)_{18}]$. *J Am Chem Soc* 2008;**130**:3754–5.
25. Akola J, Walter M, Whetten RL, Häkkinen H, Grönbeck H. On the structure of thiolate-protected Au_{25}. *J Am Chem Soc* 2008;**130**:3756–7.
26. Aikens CM. Geometric and electronic structure of $Au_{25}(SPhX)_{18}^-$ (X = H, F, Cl, Br, CH_3, and OCH_3). *J Phys Chem Lett* 2010;**1**:2594–9.
27. Qian H, Zhu M, Andersen UN, Jin R. Facile, large scale synthesis of dodecanethiol stabilized Au_{38} clusters. *J Phys Chem A* 2009;**113**:4281–4.
28. Qian H, Liu C, Jin R. Controlled growth of molecularly pure $Au_{25}(SR)_{18}$ and $Au_{38}(SR)_{24}$ nanoclusters from the same polydispersed crude product. *Sci China Chem* 2012;**55**:2359–65.
29. Wu Z, MacDonald MA, Chen J, Zhang P, Jin R. Kinetic control and thermodynamic selection in the synthesis of atomically precise gold nanoclusters. *J Am Chem Soc* 2011;**133**:9670–3.
30. Zhu Y, Qian H, Jin R. An atomic-level strategy for unraveling gold nanocatalysis from the perspective of $Au_n(SR)_m$ nanoclusters. *Chem Eur J* 2010;**16**:11455–62.
31. Nie X, Qian H, Ge Q, Xu H, Jin R. CO oxidation catalyzed by oxide-supported $Au_{25}(SR)_{18}$ nanoclusters and identification of perimeter sites as active centers. *ACS Nano* 2012;**6**:6014–22.
32. Li G, Jiang D-e, Liu C, Yu C, Jin R. Oxide-supported atomically precise gold nanocluster for catalyzing sonogashira cross-coupling. *J Catal* 2013;**306**:177–83.
33. Wu Z, Jin R. Stability of the two Au–S binding modes in $Au_{25}(SG)_{18}$ nanoclusters probed by NMR and optical spectroscopy. *ACS Nano* 2009;**3**:2036–42.
34. Li G, Zeng C, Jin R. Thermally robust $Au_{99}(SPh)_{42}$ nanoclusters for chemoselective hydrogenation of nitrobenzaldehyde derivatives in water. *J Am Chem Soc* 2014;**136**:3673–9.
35. Zhu M, Eckenhoff WT, Pintauer T, Jin R. Conversion of anionic $[Au_{25}(SCH_2CH_2Ph)_{18}]^-$ cluster to charge neutral cluster via air oxidation. *J Phys Chem C* 2008;**112**:14221–4.
36. Zhu M, Aikens CM, Hendrich MP, Gupta R, Qian H, Schatz GC, et al. Reversible switching of magnetism in thiolate-protected Au_{25} superatoms. *J Am Chem Soc* 2009;**131**:2490–2.
37. Zhu Y, Qian H, Zhu M, Jin R. Thiolate-protected au_n nanoclusters as catalysts for selective oxidation and hydrogenation processes. *Adv Mater* 2010;**22**:1915–20.
38. Zhu Y, Qian H, Jin R. Catalysis opportunities of atomically precise gold nanoclusters. *J Mater Chem* 2011;**21**:6793–9.
39. Nie X, Zeng C, Ma X, Qian H, Ge Q, Xu H, et al. CeO_2-supported $Au_{38}(SR)_{24}$ nanocluster catalysts for CO oxidation: a comparison of ligand-on and -Off catalysts. *Nanoscale* 2013;**5**:5912–8.
40. Wu Z, Jiang D-e, Mann AKP, Mullins DR, Qiao Z-A, Allard LF, et al. Thiolate ligands as a double-edged sword for CO oxidation on CeO_2 supported $Au_{25}(SCH_2CH_2Ph)_{18}$ nanoclusters. *J Am Chem Soc* 2014;**136**:6111–22.
41. Lopez-Acevedo O, Kacprzak KA, Akola J, Häkkinen H. Quantum size effects in ambient CO oxidation catalysed by ligand-protected gold clusters. *Nat Chem* 2010;**2**:329–34.
42. Ma G, Binder A, Chi M, Liu C, Jin R, Jiang D-e, et al. Stabilizing gold clusters by heterostructured transition-metal oxide–mesoporous silica supports for enhanced catalytic activities for CO oxidation. *Chem Commun* 2012;**48**:11413–5.
43. Gaur S, Miller JT, Stellwagen D, Sanampudi A, Kumar CSSR, Spivey JJ. Synthesis, characterization, and testing of supported Au catalysts prepared from atomically-tailored $Au_{38}(SC_{12}H_{25})_{24}$ clusters. *Phys Chem Chem Phys* 2012;**14**:1627–34.

44. Zhu Y, Qian H, Das A, Jin R. Comparison of the catalytic properties of 25-atom gold nanospheres and nanorods. *Chin J Catal* 2011;**32**:1145–50.
45. Liu J, Krishna KS, Losovyj YB, Chattopadhyay S, Lozova N, Miller JT, et al. Ligand-stabilized and atomically precise gold nanocluster catalysis: a case study for correlating fundamental electronic properties with catalysis. *Chem Eur J* 2013;**19**:10201–8.
46. Liu Y, Tsunoyama H, Akita T, Tsukuda T. Efficient and selective epoxidation of styrene with TBHP catalyzed by Au_{25} clusters on hydroxyapatite. *Chem Commun* 2010;**46**:550–2.
47. Yoskamtorn T, Yamazoe S, Takahata R, Nishigaki J-i, Thivasasith A, Limtrakul J, et al. Thiolate-mediated selectivity control in aerobic alcohol oxidation by porous carbon-supported Au_{25} clusters. *ACS Catal* 2014;**4**:3696–700.
48. Lavenn C, Demessence A, Tuel A. $Au_{25}(SPh\text{-}pNH_2)_{17}$ nanoclusters deposited on SBA-15 as catalysts for aerobic benzyl alcohol oxidation. *J Catal* 2015;**322**:130–8.
49. Liu Y, Tsunoyama H, Akita T, Xie S, Tsukuda T. Aerobic oxidation of cyclohexane catalyzed by size-controlled Au clusters on hydroxyapatite: size effect in the sub-2 nm Regime. *ACS Catal* 2011;**1**:2–6.
50. Li G, Qian H, Jin R. Gold nanocluster-catalyzed selective oxidation of sulfide to sulfoxide. *Nanoscale* 2012;**4**:6714–7.
51. Kawasaki H, Kumar S, Li G, Zeng C, Kauffman D, Yoshimoto J, et al. Generation of singlet oxygen by photoexcited $Au_{25}(SR)_{18}$ clusters. *Chem Mater* 2014;**26**:2777–88.
52. Yamamoto H, Yano H, Kouchi H, Obora Y, Arakawa R, Kawasaki H. N,N-dimethylformamide-stabilized gold nanoclusters as a catalyst for the reduction of 4-nitrophenol. *Nanoscale* 2012;**4**:4148–54.
53. Shivhare A, Ambrose SJ, Zhang H, Purves RW, Scott RWJ. Stable and recyclable Au_{25} clusters for the reduction of 4-nitrophenol. *Chem Commun* 2013;**49**:276–8.
54. Zhu Y, Qian H, Drake BA, Jin R. Atomically precise $Au_{25}(SR)_{18}$ nanoparticles as catalysts for selective hydrogenation of α, β-unsaturated ketones and aldehydes. *Angew Chem Int Ed* 2010;**49**:1295–8.
55. Zhu Y, Wu Z, Gayathri C, Qian H, Gil RR, Jin R. Exploring stereoselectivity of Au_{25} nanoparticle catalyst for hydrogenation of cyclic ketone. *J Catal* 2010;**271**:155–60.
56. Claus P, Brueckner A, Mohr C, Hofmeister H. Supported gold nanoparticles from quantum dot to mesoscopic size scale: effect of electronic and structural properties on catalytic hydrogenation of conjugated functional groups. *J Am Chem Soc* 2000;**122**:11430–9.
57. Li G, Jiang D-e, Kumar S, Chen Y, Jin R. Size dependence of atomically precise gold nanoclusters in chemoselective hydrogenation and active site structure. *ACS Catal* 2014;**4**:2463–9.
58. Li G, Jin R. Gold nanocluster-catalyzed semihydrogenation: a unique activation pathway for terminal alkynes. *J Am Chem Soc* 2014;**136**:11347–54.
59. Li G, Liu C, Lei Y, Jin R. Au_{25} nanocluster-catalyzed Ullmann-type homocoupling reaction of aryl iodides. *Chem Commun* 2012;**48**:12005–7.
60. Xie S, Tsunoyama H, Kurashige W, Negishi Y, Tsukuda T. Enhancement in aerobic alcohol oxidation catalysis of Au_{25} clusters by single Pd atom doping. *ACS Catal* 2012;**2**:1519–23.
61. Qian H, Jiang D-e, Li G, Gayathri C, Das A, Gil RR, et al. Monoplatinum doping of gold nanoclusters and catalytic application. *J Am Chem Soc* 2012;**134**:16159–62.
62. Yao C, Chen J, Li M-B, Liu L, Yang J, Wu Z. Adding two active silver atoms on Au_{25} nanoparticle. *Nano Lett* 2015;**15**:1281–7.
63. Jin R, Nobusada K. Doping and alloying in atomically precise gold nanoparticles. *Nano Res* 2014;**7**:285–300.

64. Kauffman DR, Alfonso D, Matranga C, Qian H, Jin R. Experimental and computational investigation of Au_{25} clusters and CO_2: a unique interaction and enhanced electrocatalytic activity. *J Am Chem Soc* 2012;**134**:10237–43.
65. Kogo A, Sakai N, Tatsuma T. Photocatalysis of Au_{25}-modified TiO_2 under visible and near infrared light. *Electrochem Commun* 2010;**12**:996–9.
66. Lee M, Amaratunga P, Kim J, Lee D. TiO_2 nanoparticle photocatalysts modified with monolayer-protected gold clusters. *J Phys Chem C* 2010;**114**:18366–71.
67. Lee J, Shim HS, Lee M, Song JK, Lee D. Size-controlled electron transfer and photocatalytic activity of ZnO–Au nanoparticle composites. *J Phys Chem Lett* 2011;**2**:2840–5.
68. Yu C, Li G, Kumar S, Kawasaki H, Jin R. Stable $Au_{25}(SR)_{18}/TiO_2$ composite nanostructure with enhanced visible light photocatalytic activity. *J Phys Chem Lett* 2013;**4**:2847–52.
69. Zeng C, Chen Y, Li G, Jin R. Magic size $Au_{64}(S\text{-}c\text{-}C_6H_{11})_{32}$ nanocluster protected by cyclohexanethiolate. *Chem Mater* 2014;**26**:2635–41.
70. Zeng C, Chen Y, Kirschbaum K, Appavoo K, Sfeir MY, Jin R. Structural patterns at all scales in a nonmetallic chiral $Au_{133}(SR)_{52}$ nanoparticle. *Sci Adv* 2015;**1**:e1500045
71. Chen Y, Zeng C, Kauffman DR, Jin R. Tuning the magic size of atomically precise gold nanoclusters via isomeric methylbenzenethiols. *Nano Lett* 2015;**15**:3603–9.
72. Zeng C, Chen Y, Li G, Jin R. Synthesis of a $Au_{44}(SR)_{28}$ nanocluster: structure prediction and evolution from $Au_{28}(SR)_{20}$, $Au_{36}(SR)_{24}$ to $Au_{44}(SR)_{28}$. *Chem Commun* 2014;**50**:55–7.

Chapter 11

Functionalization and Application

Nirmal Goswami, Jingguo Li and Jianping Xie[1]
Department of Chemical and Biomolecular Engineering, National University of Singapore, Singapore
[1]*Corresponding author: E-mail: chexiej@nus.edu.sg*

Chapter Outline

11.1 INTRODUCTION

Metal nanoclusters (NCs), typically consisting of a few to a hundred metal atoms, represent the missing link between single metal atoms and plasmonic metal nanoparticles (NPs).[1–3] Metal NCs are ultrasmall with particle size below 2 nm, and this size is comparable to the Fermi wavelength of electrons. Particles in this sub-2-nm size regime show unique physical and chemical (or physicochemical) properties, such as strong fluorescence, quantized charging, discrete redox behavior, molecular magnetism, and optical chirality.[4–11] In recent years, the intriguing molecular-like properties of metal NCs have been explored for applications in a wide spectrum of fields ranging from opto-electronics, sensing, catalysis, to biomedicine.[12–23]

One major merit of using metal NCs as functional materials is that one can tailor their physicochemical properties, in a very predictable and controllable manner, to meet the needs of a specific application. For example, the emission

Frontiers of Nanoscience, Vol. 9. http://dx.doi.org/10.1016/B978-0-08-100086-1.00011-7

wavelength of metal NCs can be tuned in the range of visible to near-infrared by simply varying the synthesis conditions, such as using different reducing agents and different reaction temperature or pH.[2,24–26] It is well documented that the physicochemical properties of metal NCs are largely dependent on the NC size, composition, surface, structure, and the surrounding environment; and a subtle change in these parameters may lead to distinctively different NC properties. For example, it was found that doping of only one single Pt atom in a 25-atom Au NC shows remarkable effects on the stability and the catalytic activity of the parental Au_{25} NCs.[27] In recent years, functionalization of metal NCs has attracted increasing interest from the cluster community.[28,29] In particular, the functionalization of metal NCs via a delicately designed and controlled chemical process is considered as an efficient way to enhance and/or enrich the functions of metal NCs, which could largely pave their way toward practical applications.

In this chapter, we focus our discussions on the functionalization and applications of metal NCs. In general, metal NCs possess a core–shell structure (similar as their larger metal NP counterparts), which consists of a metal core and a ligand shell. Both the metal core and the ligand shell could affect the physicochemical properties of the NCs, and they can be readily modified to achieve predesigned properties required by a specific application. The metal core can be modified by some chemical routes. One efficient approach is heteroatom doping, in which one or more heterometallic atoms are selectively doped in the metal core of the NCs. The doping of foreign atoms in the NC core may change their physical and chemical properties. In addition, the heteroatom doping has also been proven as an effective way to impart new physicochemical properties to the parental NCs. The concepts and principles for heteroatom doping, and the effects of doping on the physicochemical properties of the NCs will be discussed in the first part of this chapter.

Similar as the metal core modification, surface modification is another efficient way to functionalize metal NCs and to tailor their physicochemical properties. One good example is the control of the fluorescence properties of metal NCs. It was found that the fluorescence properties of metal NCs are largely determined by the organic ligands on the NC surface.[9] In addition, the organic ligands on the NC surface also dictate the solubility and stability of NCs in solution, which may affect their solution-based applications.[30] Further, the organic ligands on the NC surface may also influence the recognition properties of NCs with other molecules, as well as some molecular-like properties of NCs such as the optical chirality. Therefore, the design and functionalization of the NC surface is crucial for the practical applications of the NCs, and this topic will be discussed in the second part of this chapter.

The as-functionalized metal NCs often feature with the well-defined size, composition, surface, and structure; and they have recently emerged as a new class of functional materials for many applications, which will be discussed in the last part of this chapter In particular, metal NCs often show strong fluorescence and possess good photostability. They are also nontoxic and highly biocompatible. Therefore, the fluorescence properties of metal NCs can be used for many applications. For example, the fluorescence of metal NCs is highly sensitive to some foreign species (or analytes), including heavy metal ions and small biomolecules; such principles have been applied to construct efficient optical sensors for a number of analytes.[18,19] In addition, fluorescent metal NCs are also attractive for biomedical applications owing to their unique physicochemical properties and comparable size with some of the biologically important molecules such as DNA and proteins. For example, fluorescent metal NCs are attractive for bioimaging applications. In addition, the biocompatibility of metal NCs can be further improved by modifying the NC surface with other biocompatible molecules.[23] Some good strategies will be discussed in this chapter. Hopefully the concepts and principles illustrated in this chapter could invite more concerted efforts from the community to develop more efficient strategies for NC functionalization, and to further advance the functionalized metal NCs for practical applications.

11.2 FUNCTIONALIZATION

11.2.1 Metal Core Modification

Modification of the metal core of the NCs could introduce new optical, electronic, or physicochemical properties to metal NCs. Among newly developed strategies of metal core modification, the heteroatom doping is the most attractive as the doping process can be controlled in a good manner by using a delicately designed chemical process. In the last decade, a number of efficient strategies have been developed to synthesize metal NCs with atomic precision, mostly Au NCs protected by thiolate ligands (or thiolated Au NCs for short).[31–34] Such atomically precise Au NCs could be further modified by doping with a certain number of foreign metal atoms, such as Ag, Cu, Pt, and Pd. We highlight some doping strategies in this section as well as discuss the effects of doping on the physicochemical properties of the as-doped metal NCs (Scheme 1).

11.2.1.1 Pd-Doped Au NCs

In 2009, the doping of Pd in atomically precise Au NCs was reported by Murray et al.[35] They have successfully doped one Pd atom to thiolated Au_{25} NCs, leading to the formation of $Au_{24}Pd(SR)_{18}$ NC (here SR denotes a thiolate ligand).[35] In this study, the synthetic protocol for $Au_{24}Pd(SR)_{18}$ was similar as the one used for the preparation of $Au_{25}(SR)_{18}$. The only difference is the starting materials or precursors, where an admixture of the second metal (Pd) and Au was used to prepare $Au_{24}Pd(SR)_{18}$. Another key point illustrated in this study is the location of the Pd atom in $Au_{24}Pd(SR)_{18}$. Since both the size and

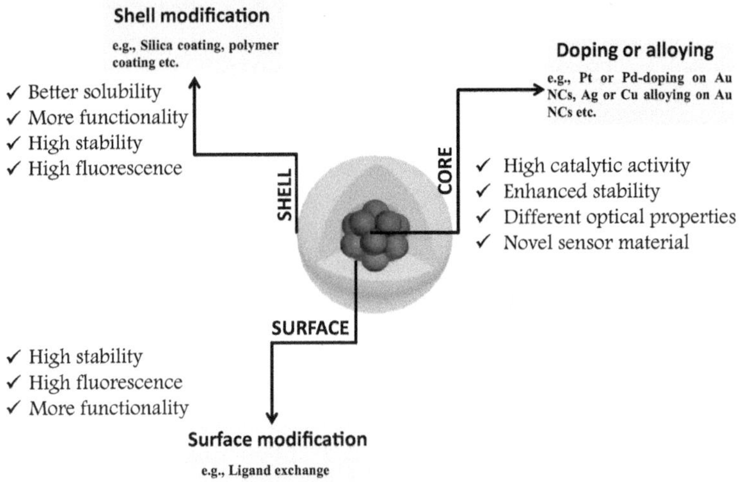

SCHEME 1 Illustration of the various functionalization methods and their effects on the properties of a typical core–shell metal NC.

the surface energy of Pd are different from that of the Au (the calculated atomic radius of Pd is 1.38 Å, which is smaller than that of the Au (1.44 Å); and the surface energy of Pd is 131 meV/Å^2, which is higher than that of the Au (96.8 meV/Å^2)),[36] the Pd atom is most likely located at the center of the metal core rather than on the NC surface. This prediction has been theoretically supported by Häkkinen et al.,[37] Jiang et al.,[38] and Walter et al.[39] In a separate study, Negishi et al.[40] synthesized Pd-doped Au NCs and successfully used high-performance liquid chromatography (HPLC) to obtain high purity $Au_{24}Pd(SR)_{18}$. By employing the density functional theory (DFT) calculations and electrospray ionization mass spectrometry, they predicted that the isolated $[Pd_1Au_{24}(SC_{12}H_{25})_{18}]^0$ has a core–shell structure that can be described as $[Pd_1@Au_{24}(SC_{12}H_{25})_{18}]^0$, in which the central Pd atom is surrounded by a frame of $Au_{24}(SC_{12}H_{25})_{18}$. More recently, Tsukuda et al.[41] examined the occupation site of the Pd dopant within $Pd_1Au_{24}(SC_{12}H_{25})_{18}$ by complementarily probing the local environment of Au and Pd elements using ^{197}Au Mössbauer and Pd K-edge EXAFS spectroscopy. It was found that the doped Pd atom is indeed located at the center of $Pd_1Au_{24}(SC_{12}H_{25})_{18}$.

Similar as thiolated Au_{25} NCs, other magic-sized thiolated Au NCs such as $Au_{38}(SR)_{24}$ and $Au_{144}(SR)_{60}$ can also be doped with Pd atoms; but they may have different doping manners. For example, different from $Au_{25}(SR)_{18}$, $Au_{38}(SR)_{24}$ has two cores and there are two possible doping manners: mono- and bi-doped clusters. For example, Jin et al.[42] observed two Pd-doped NC species ($Pd_1Au_{37}(SR)_{24}$ and $Pd_2Au_{36}(SR)_{24}$) in their samples. Negishi et al.[43] reported the doping of two Pd atoms to $Au_{38}(SR)_{24}$, leading to the formation of

$Pd_2Au_{36}(SR)_{24}$. DFT calculations of $Pd_2Au_{36}(SR)_{24}$ suggest that the parental cluster structure is energetically favored, and compared to $Au_{38}(SR)_{24}$, the only difference is that the two Au atoms in the Au_{13} center are replaced by Pd atoms. For larger-sized thiolated Au NCs such as $Au_{144}(SR)_{60}$, it has been shown that the number of doped Pd atoms could be up to 7.[44] The cluster structures of the Pd-doped Au NCs and their corresponding parental Au NCs are illustrated in Figure 1.

The next question we may ask is what are the new functions that could be offered by the Pd doping. Such new functions often become advantageous for the Pd-doped Au NCs when compared to their parental Au NCs. It was shown that the electrochemical and optical absorption properties of $Au_{25}(SR)_{18}$ are sensitive to the inclusion of Pd atoms.[35] A similar trend was also found in the Pd-doped $Au_{38}(SR)_{24}$ and $Au_{144}(SR)_{60}$. In particular, some of the optical absorption features of the parental Au NCs are diminished when a certain number of Pd atoms are incorporated. Since the optical absorption of metal NCs originates from the metal core, any changes in the optical spectra upon the Pd doping could suggest that the Pd atoms are incorporated into the metal core rather than into the staple motifs. The comparison of the optical absorption spectra (Figure 2) of $Au_{143}Pd_1(SR)_{60}$, $Au_{141}Pd_3(SR)_{60}$, and $Au_{144}(SR)_{60}$ suggests that the Pd-doped $Au_{144-n}Pd_n(SC_2H_4Ph)_{60}$ NCs ($n = 1-7$) feature with a core−shell structure, where the Pd atoms are selectively incorporated into the central Au_{12} core. In another study, it was also found that $Pd_1Au_{24}(SR)_{18}$ shows a higher reactivity toward ligand exchange when compared to $Au_{25}(SR)_{18}$.[45] Negishi et al.[40] also verified the stability of the Pd-doped NCs against degradation in solution and under the laser

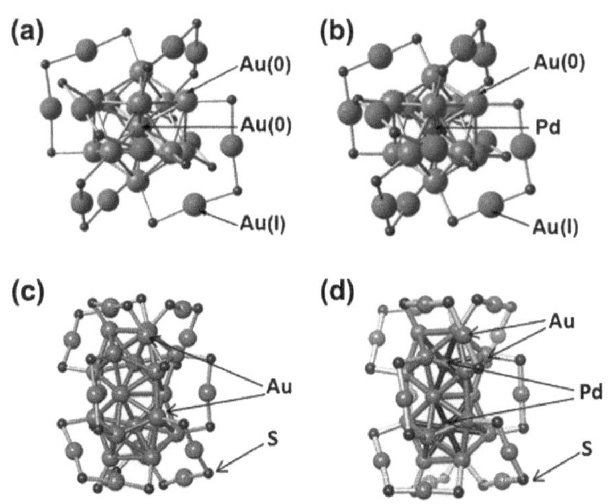

FIGURE 1 Geometrical structure of (a) $Au_{25}(SR)_{18}$, (b) $PdAu_{24}(SR)_{18}$, (c) $Au_{38}(SR)_{24}$, and (d) $Pd_2Au_{36}(SR)_{24}$. *Reproduced with permission from Ref. 41. Copyright (2013) American Chemical Society; and Ref. 43. Copyright (2012) Royal Society of Chemistry.*

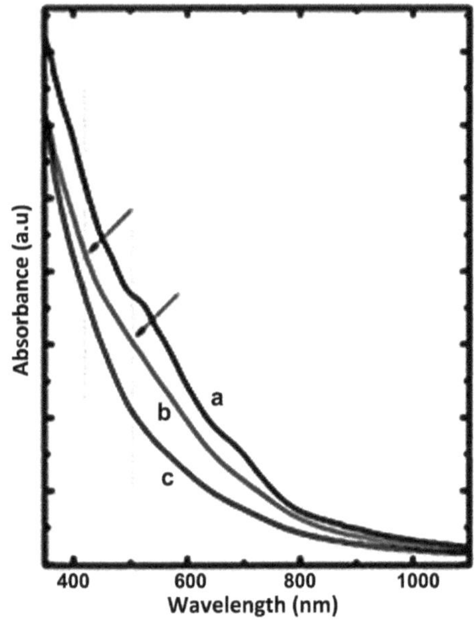

FIGURE 2 Optical absorption spectra of (a) $Au_{144}(SR)_{60}$ in toluene compared to those of (b) $Au_{143}Pd_1(SR)_{60}$, and (c) $Au_{141}Pd_3(SR)_{60}$. *Reproduced with permission from Ref. 44. Copyright (2013) Royal Society of Chemistry.*

dissociation. In both cases, $Pd_1Au_{24}(SR)_{18}$ is more stable than $Au_{25}(SR)_{18}$. A similar trend was also observed in the Pd-doped Au_{38} NCs, where a tetrahydrofuran (THF) solution of a mixture of $Pd_2Au_{36}(SC_2H_4Ph)_{24}$, $Pd_1Au_{37}(SC_2H_4Ph)_{24}$, and $Au_{38}(SC_2H_4Ph)_{24}$ was subjected to a thermal treatment at 60 °C, and $Pd_2Au_{36}(SC_2H_4Ph)_{24}$ showed the highest stability.[42] Similarly, to assess the stability of NCs toward core etching by thiolate ligands, the as-prepared NC mixture was incubated in a neat PhC_2H_4SH solution and heated up to 80 °C. Only $Pd_2Au_{36}(SC_2H_4Ph)_{24}$ was preserved in the solution after 12 h of heating. It should be noted that $Pd_1Au_{37}(SC_2H_4Ph)_{24}$ has a poorer stability compared to $Pd_2Au_{36}(SC_2H_4Ph)_{24}$ and $Au_{38}(SC_2H_4Ph)_{24}$. While the exact reason is still unclear, the asymmetric electronic structure of $Pd_1Au_{37}(SC_2H_4Ph)_{24}$ could be one key factor (in $Pd_1Au_{37}(SC_2H_4Ph)_{24}$ only one center is doped with the Pd atom and the other center is still with the Au). A combined theoretical and experimental research is required to shed more light on the stability fundamentals of Pd-doped Au NCs.

Pd doping could also affect the catalytic properties of the parental Au NCs. For example, Tsukuda et al.[46] reported that the doping of a single Pd atom to Au NCs can significantly improve the catalytic activity of the NCs toward the aerobic oxidation of benzyl alcohol. The modulation of the electronic structures of the NCs by the intracluster electron transfer from Pd to Au is one possible reason for the enhancement of the catalytic activity. More recently,

Bürgi et al.[7] showed that the Pd doping could increase the flexibility of the NC surface. In this study, an achiral thiolate ligand (HSC_2H_4Ph) was used as the protecting ligand for the preparation of Pd-doped Au NCs, leading to the formation of $Pd_2Au_{36}(SR)_{24}$. Using the chiral HPLC, they have successfully separated the enantiomers of $Pd_2Au_{36}(SR)_{24}$, which showed a different circular dichroism (CD) spectrum as compared to that of the parental $Au_{38}(SR)_{24}$, although both spectra featured with comparable anisotropy factors. Moreover, the racemization of the chiral $Pd_2Au_{36}(SR)_{24}$ proceeds at an obviously lower temperature when compared to $Au_{38}(SR)_{24}$, indicating a higher ligand flexibility of the staple motifs of Pd-doped Au NCs. In a separate study, Yao et al.[47] studied the effects of doping on the chiroptical properties of Pd-doped Au NCs protected by a chiral thiolate ligand like glutathione or GSH. This study also suggests that the CD spectrometry is advantageous over the absorption spectrometry for the detection of foreign atoms doping in thiolated Au NCs. As revealed by this study, there is a significant difference in Cotton effects between the Au−Pd NCs and their monometallic NC counterparts. However, the CD profiles of the Pd-doped Au NCs were almost independent on the number of Pd dopants. Probably, the doping of Au NCs with Pd atoms led to the formation of the same $Pd_xAu_{n-x}(SR)_m$, regardless of the feeding ratio of Au to Pd. This has been demonstrated by Murray et al.,[35] where in the Pd doping of the thiolated Au_{25} NCs, the resultant NC species was always $PdAu_{24}(SC_2H_4Ph)_{18}$, regardless of the feeding ratio of Au to Pd during the synthesis.

11.2.1.2 Pt-Doped Au NCs

Since Pt belongs to the same group as Pd in the periodic table, Pt-doped Au NCs is expected to show similar effects as that of Pd. In addition, Pt has a higher surface energy than Au, which implies that Pt would prefer its doping position at the center of the metal core rather than on the NC surface. The Pt-doped Au NCs are considered to be more stable than other metal-doped Au NCs (such as Pd, Cu, or Ag) due to the higher interaction energy between Pt and the surrounding Au_{12} framework. Compared to the successful cases of the Pd-doped Au NCs, syntheses of Pt-doped Au NCs, especially those NCs with atomic precision, are less attempted. In addition, it is also difficult to precisely determine the molecular formula of the Pt-doped Au NCs as Pt and Au has a very small mass difference of ~1.89 Da.

Some successful attempts exist. For example, Jin et al.[27,48] successfully identified $Au_{24}Pt(SC_2H_4Ph)_{18}$ from the parental $Au_{25}(SC_2H_4Ph)_{18}$ (Figure 3(a)). They used a one-phase synthesis method (using THF as the solvent) and obtained a mixture of $Au_{25}(SC_2H_4Ph)_{18}$ and $Au_{24}Pt(SC_2H_4Ph)_{18}$. The Pt-doped $Au_{24}Pt(SC_2H_4Ph)_{18}$ NC was then isolated by the size exclusion chromatography. As expected, the properties of the Pt-doped Au NCs are different from their parental Au NCs. In particular, in the optical absorption spectrum of $Au_{24}Pt(SC_2H_4Ph)_{18}$, a distinct peak in the visible range was located at 590 nm (2.10 eV), with an additional broad peak in the NIR range

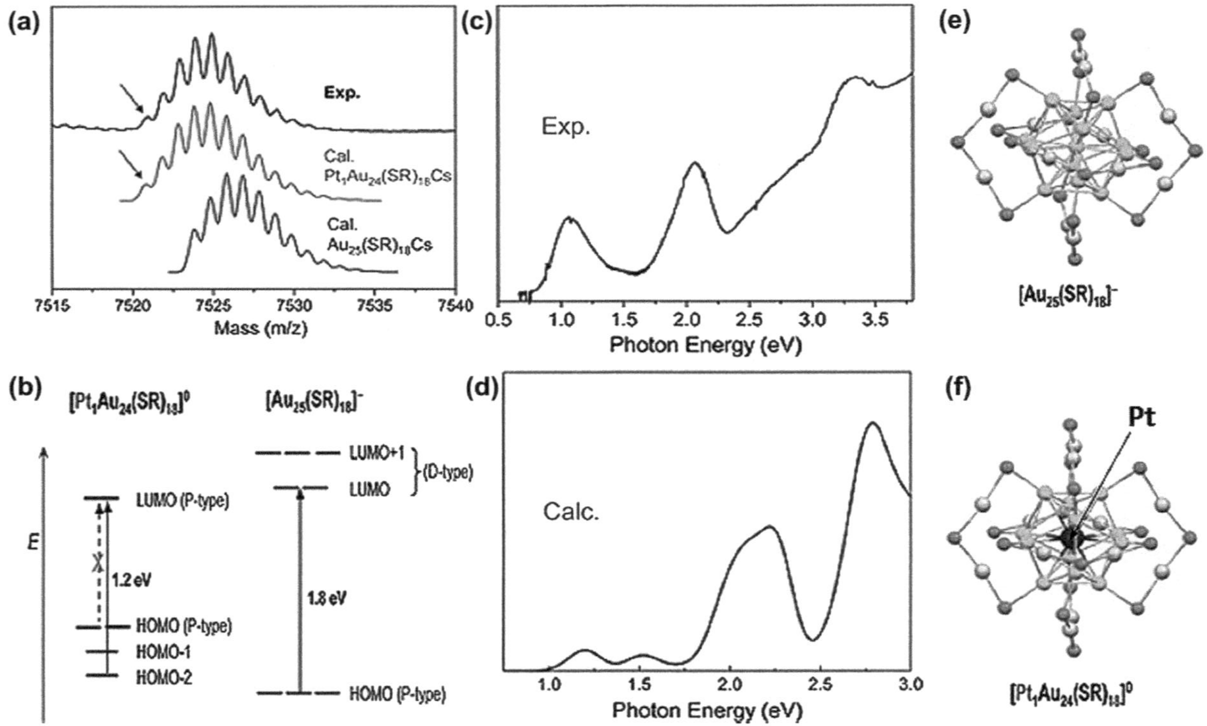

FIGURE 3 (a) Experimental and simulated isotope patterns of $Pt_1Au_{24}(SR)_{18}Cs^+$ and $Au_{25}(SR)_{18}Cs^+$. (b) Electronic energy levels of $Pt_1Au_{24}(SR)_{18}$ and $Au_{25}(SR)_{18}$. (c) Experimental and (d) simulated optical spectrum of $[Pt_1Au_{24}(SR)_{18}]^0$ on the photon energy scales. Cluster structure of (e) $[Au_{25}(SR)_{18}]^-$ and (f) $[Pt_1Au_{24}(SR)_{18}]^0$. *Reproduced with permission from Ref. 27. Copyright (2012) American Chemical Society.*

centered at ~1100 nm (~1.13 eV) (Figure 3(c)). As a result, the solution of $Au_{24}Pt(SC_2H_4Ph)_{18}$ was green, which is in stark contrast to the brown solution of the parental Au NCs. It should be noted that, the NIR absorption was not observed in $Au_{25}(SR)_{18}$ as well as in the Pd-doped $Au_{24}Pd(SC_2H_4Ph)_{18}$. In order to have a deeper understanding of the effects of Pt doping on the optical absorption properties of Au NCs, DFT simulations were performed with an assumption of three possible Pt-doping locations in a model NC, $Au_{24}Pt(SR)_{18}$. It was found that the absorption spectrum of the Pt-doping at the center of the NC is the best match with the experimental data (Figure 3(e) and (f)). In particular, the highest occupied molecular orbital (HOMO) of $Au_{24}Pt(SR)_{18}$ consists of two degenerate 1P orbitals, with the lowest unoccupied molecular orbital (LUMO) of a 1P orbital; whereas in the parental $[Au_{25}(SR)_{18}]^-$, the HOMO consists of a roughly triply degenerate 1P orbital (Figure 3(b)). As a result, the optical absorption spectrum of $[Au_{25}(SR)_{18}]^-$ is simple, exhibiting a dominant peak at 1.8 eV in the 1.0–2.5 eV range, while for $Au_{24}Pt(SR)_{18}$, the HOMO-LUMO transition is forbidden under the dipole selection rule because both the HOMO and LUMO are superatomic 1P orbitals. Therefore the electronic energy levels become more complicated upon the Pt doping, and much effort is required to elucidate the origin of this optical absorption band. Apart from the optical absorption properties, Pt doping could also affect the electronic properties, stability, and catalytic activity of the as-doped Au NCs. Similar to the Pd-doped Au NCs, the Pt-doped Au NCs also showed better stability and enhanced catalytic activity when compared to their parental Au NCs.

11.2.1.3 Ag-Doped Au NCs

Compared to other metal atoms, Ag shares more similar properties with Au. They have a similar atomic radius and a single valence electron. Therefore, Ag is much easier to be doped into Au NCs when compared to Pt or Pd. It is also expected that the electronic structures and the properties of the resulting Ag-doped Au NCs might feature with new functionalities; which could also be controlled by the doping of different amount of Ag atoms to Au NCs.

For example, Negishi et al.[49] first synthesized $Au_{25-x}Ag_x(SC_{12}H_{25})_{18}$ NCs with various Au/Ag mixing ratios. It should be noted that the number of the doped Ag atoms could increase up to 11 when the Au/Ag mixing ratio was 5:20. This suggests that a number of Au atoms have been replaced by Ag atoms, which is in stark contrast with the cases of Pd and Pt doping; in these two systems, only a single Pd or Pt atom could be incorporated into the 25-atom Au NC irrespective of the Au/Pt or Au/Pd mixing ratios. These results also indicate that Ag atoms may not be preferentially located at the center of the NCs. From the theoretical studies, it is evident that regardless of the number of dopant atoms, the cluster structures with the Ag atoms at the surface of the Au_{13} core are the most energetically stable. This prediction has also been experimentally verified, where the single crystal X-ray structural analysis suggests the cluster structure of $Ag_xAu_{25-x}(SC_2H_4Ph)_{18}$ ($x = 4-8$). Recently, a series of water-soluble Ag_xAu_{25-x} NCs protected by mono- and bi-thiolate ligands were successfully synthesized by our group.[50,51] Other atomically precise NCs, such as 38-atom or 144-atom NCs have also been successfully doped with Ag atoms, and it was found that the number of Ag dopants was increased when the size of the NCs increased.[52,53]

While Ag atoms are located at the different position in the as-doped Au NCs as compared to Pt or Pd, the doping effects on the optical absorption properties of the NCs are similar. For example, the electronic structures of the Ag-doped NCs could be continuously varied by increasing the extent of Ag doping. In particular, the increase of Ag number could shift the absorption peak of the NCs to a higher energy, indicating Ag atoms are continuously incorporated into the Au_{13} core. In addition, Ag doping may also influence the fluorescence properties of the NCs. For example, Negishi et al.[49] showed that the Ag-doped Au NCs exhibit fluorescence at a lower emission wavelength when compared to that of $Au_{25}(SC_{12}H_{25})_{18}$. More recently, Zhu et al.[54] demonstrated that when all the 13 Au sites were fully substituted with Ag atoms, the alloyed Au–Ag NCs became highly fluorescent. This suggests that the incorporation of Ag atoms to the Au_{13} core could modulate the electronic structures and the surface states, leading to a remarkable change in their fluorescence properties (Figure 4).

11.2.1.4 Cu-Doped Au NCs

Cu is in the same group as Au in the periodic table, therefore one would expect some intriguing effects upon the doping of Cu at the Au core. The atomic radius of Cu (1.28 Å) is smaller than Au (1.44 Å) and the Au–Cu bond is even stronger than the Au–Au bond. Although quite different from Ag, the Cu-doped Au NCs share some similar features as the Ag-doped Au NCs. For example, more than one Cu atoms can be doped to the 25-atom Au NCs, which means, like Ag, the preferred location of the Cu would be at the surface of the Au_{13} core. However, in case of Cu doping, no specific dependence was found for the dopant amount with respect to the Au to Cu ratio; whereas, Ag showed a strong Au to Ag ratio dependence. Nevertheless, a smaller number of dopants can be incorporated into a specific Au NC when comparing the Cu doping with that of the Ag doping. For example, Jin et al.[55] verified that when 25-atom Au NCs were doped with Ag or Cu, the Ag-doped NCs were more stable than the Cu-doped NCs. More specifically, the Cu-doped Au NCs showed a clear tendency to be dealloyed, resulting in its conversion to $Au_{25}(SR)_{18}$, whereas, the Ag-doped Au NCs showed only small variations in product distribution with time.

Cu doping also affects the optical properties of the as-doped Au NCs. For example, Negishi et al.[56] synthesized the Cu-doped Au NCs, namely $Cu_nAu_{25-n}(SC_2H_4Ph)_{18}$ ($n = 1-5$), via a similar method to that used for

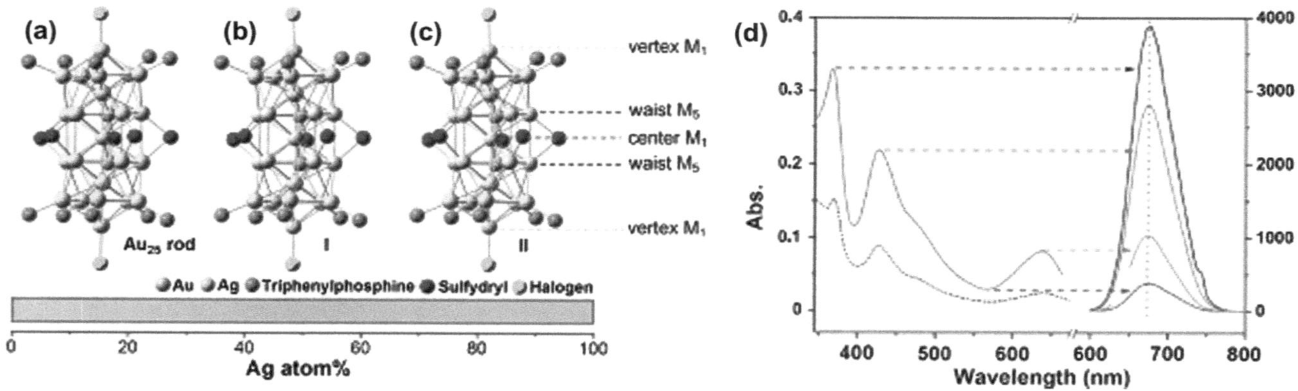

FIGURE 4 Cluster structures of (a) $[Au_{25}(PPh_3)_{10}(SC_2H_4Ph)_5Cl_2]^{2+}$, (b) Au–Ag NCs (type I) obtained from the reaction of polydisperse Au NPs with Ag(I)–thiolate complexes, and (c) Au–Ag NCs (type II) obtained from the reaction of Au_{11} NCs with Ag(I)–thiolate complexes. The general formula of I and II is $[Ag_xAu_{25-x}(PPh_3)_{10}(SC_2H_4Ph)_5Cl_2]^{2+}$ but with different x. (d) Optical absorption, photoexcitation, and photoemission spectra of Au–Ag NCs (type II). *Reproduced with permission from Ref. 54. Copyright (2014) Wiley-VCH.*

the synthesis of the Pd-doped or Ag-doped 25-atom Au NCs. The optical absorption of the $Cu_nAu_{25-n}(SC_2H_4Ph)_{18}$ NCs was different from the Ag-doped Au NCs. In particular, the doping of Cu in Au NCs reduces the NC's HOMO-LUMO gap and continually shifts the emission to longer wavelengths. As evident from the electrochemical measurements of $Cu_nAu_{25-n}(SC_2H_4Ph)_{18}$, the Cu doping led to the reduction of energy levels of the LUMO, resulting in a smaller HOMO-LUMO gap. Another important point revealed in this study is the preferred location of Cu when the NC is doped with a single Cu atom. By taking consideration of three preferred positions in the NCs, i.e., at the center of the icosahedral core, on the surface of icosahedron and in the −SR−Au−SR−Au−SR− staple, Negishi et al.[56] calculated the corresponding optical absorption spectra. The results suggest that the Cu atom in $Cu_1Au_{24}(SR)_{18}$ occupies the central site, although the most energetically favorable structure is the one with the Cu atom on the surface of the icosahedron. Direct experimental data (e.g., single crystal XRD) are required to support this prediction. However, this structure seems reasonable as the atomic size of Cu is smaller than Au, and this suggests that Cu could be more easily trapped at the center. On the other hand, DFT analysis of the $Cu_1Au_{24}(SR)_{18}$ indicates that the geometric structure of the NCs is significantly distorted when Cu is inserted at any of these locations. In contrast, Ag doping only induces a minimal structural distortion, regardless of the doping sites. This theoretical result suggests the different nature of Cu and Ag doping. It is therefore crucial to carefully choose the dopant material to achieve a highly stable Au NC doped with several foreign metals.

Moreover, the amount of dopants as well as the stability of the as-doped NCs can be further varied by replacing the surface thiolate ligands by selenolates. For example, Negishi et al.[57] demonstrated that up to 9 Cu atoms can be doped inside the Au_{25} NCs protected by selenolate ligands; this value is higher than that in its thiolate-protected NC analogs (up to 5 Cu atoms). For large-sized atomically precise NCs, such as 144-atom Au NCs, Dass et al.[58] showed that a maximum of 23 Cu atoms could replace the Au atoms and dope inside the Au NCs. Interestingly, when the number of Cu atoms is above 8, a surface plasmon-like peak is appeared at ~520 nm for the as-doped Au NCs. Although the Cu-doped NCs were less stable compared to those NCs doped with other metals, by applying proper surface chemistry, the properties of these doped NCs could be further improved. The Cu-doped NCs may also offer high catalytic activity, which justifies a future research direction on this topic.

11.2.1.5 Others

While most of the research focus on the functionalization of Au core with other metals, relatively few studies reported the alloying of other metal pairs, such as Cu/Ag,[59,60] Ag/Ni,[61] and Ag/Pt[62] NCs. One good example was presented by Khanna et al.,[61] where $Ag_4Ni_2(DMSA)_4$ NC (DMSA = meso-2,3-dimercaptosuccinic acid) was successfully synthesized. As evidenced from their theoretical study, the NC has an octahedral structure with the Ag atoms occupying the corners of the square plane and the Ni atoms at the apexes. In a follow-up work,[62] the same group reported the synthesis and isolation of $Ag_4Pt_2(DMSA)_4$. It was found that when the synthetic method was modified by replacing the Ni source with a source of its congener Pt, it produced a bimetallic NC with the same stoichiometry as the original synthesis, that is $Ag_4Ni_2(DMSA)_4$. The as-synthesized $Ag_4Pt_2(DMSA)_4$ is water soluble and highly stable in solution. Moreover, $Ag_4Pt_2(DMSA)_4$ has a similar structure as $Ag_4Ni_2(DMSA)_4$. Apart from the fundamental studies, the as-doped NCs with Ni or Pt could be a new class of efficient catalysts due to the presence of highly strained and accessible Ni and Pt atoms, which are well documented as catalytically active elements.

11.2.2 Surface Modification

Proper surface functionalization is one of the most important prerequisites for every possible application, as the surface ligands determine the interactions of functional materials with the environment. In the field of metal NCs, many factors like controlled growth of the NCs during the synthesis, cluster stability in solution, and optical properties, are largely dictated by the nature of the ligands on the NC surface. In addition, for many biological applications such as targeted drug delivery, functional groups on the NC surface could play crucial roles to achieve good performance, especially for the applications require in vivo settings. In this section, we discuss some efficient strategies for the modification and functionalization of metal NC surface, which could further advance them for many practical applications.

In order to attain some new properties or functionalities, the surface ligands of the given NCs could be exchanged by another organic ligand, typically with predesigned functional groups. The ligand exchange is therefore an efficient way to further functionalize metal NCs for practical applications. In a typical ligand exchange process, the choice of the ligand candidate is crucial, which depends on delicate chemistries involving the metal at the core, the solvent that used to host the reaction, and the binding affinity of the ligands to the NC surface. The rule of thumb is that those organic ligands with stronger binding affinities with the NCs may provide NCs better stability, typically in solution. It should be noted that the ligand exchange strategy is well established in the NP field, and the principles and chemistries involved in the exchange reaction are also applicable to metal NCs, which have been recently adopted in the NC community.

11.2.2.1 Ligand Exchange: Phosphine versus Thiolate

In the early stage of NC research, a number of NCs were synthesized by using phosphines as protecting ligands. For example, the phosphine-stabilized undecagold derivatives have been widely studied as labeling reagent for biomolecules in the electron microscopy techniques.[63] However, the applications of the phosphine-stabilized metal NCs face several constraints including the multistep synthetic processes, relatively poor stability in solution, and tedious postsynthetic purification (often time-consuming); such limitations have motivated researchers to search for alternative protecting ligands for metal NCs, and thiol ligands have recently championed out in the NC community. In particular, thiol ligands are good organic molecules that could simultaneously stabilize the materials in solution and functionalize them with predesigned functional groups, typically targeting at a specific application. For example, based on the ligand exchange strategy, Hutchison et al.[64] developed an efficient preparation method that could transform the phosphine-protected Au NPs to a series of thiolate-protected Au NC counterparts. This is a fairly generic method, and it could be used to prepare a diverse family of alkane–thiol-protected Au NCs. In addition, it was found that the time for the exchange reaction was dictated by the type of ligands that were used for the replacement. In particular, the reaction time was usually longer for aromatic thiols, which typically required 12 h or more to achieve a complete ligand exchange. The steric effect between the exchange ligands is one key factor to affect the exchange efficiency and time. The ligand exchange process typically involved three steps (Figure 5), which are (1) a "rapid" replacement of the Au(I)Cl(PPh_3) moieties by −SR ligands, (2) a subsequent "slow" removal of the remaining PPh_3 ligands, and (3) a completion and optimization of the −SR shells.[64] More interestingly, the as-formed NCs after the ligand exchange showed higher stability in solution against heat, aggregation, and decomposition as compared to the pristine NCs. Moreover, the small core size was also preserved during the ligand exchange.

However, some studies also found that the NC size may increase during the ligand exchange. For example, Murray et al.[65] showed that during the ligand exchange with thiols, the phosphine-protected $Au_{55}(PPh_3)_{12}Cl_6$ NC experienced a core size increase to a final thiolated Au_{75} NC. Similarly, Tsukuda et al.[66] reported that when phosphine-protected Au_{11} NCs undergo the ligand exchange with a water-soluble thiol ligand, glutathione or GSH, a thermodynamically stable product—$Au_{25}(SG)_{18}$ was obtained. More recently,

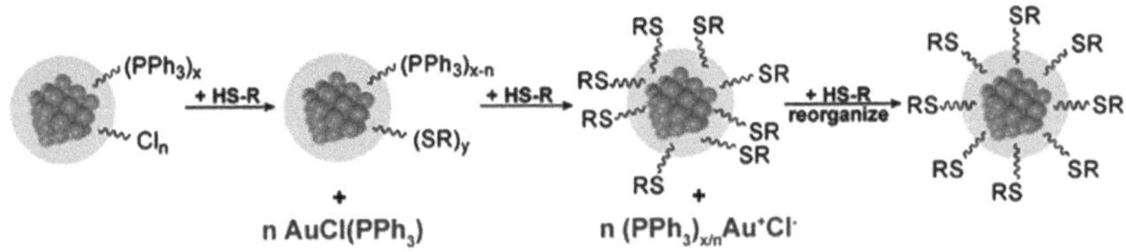

FIGURE 5 A proposed three-stage mechanism for the ligand exchange between phosphine and thiol. *Reproduced with permission from Ref. 64. Copyright (2005) American Chemical Society.*

Hutchison et al.[67] suggested that the starting materials are crucial for the production of NCs with a specific size during the ligand exchange. For example, both $Au_{11}(PPh_3)_7Cl_3$ and $Au_{11}(PPh_3)_8Cl_3$ were used as starting materials, interchangeably; and both were referred to $Au_{11}(PPh_3)_nCl_3$, where the number of core atoms in both NC species is fixed while the number of phosphine ligands is different. The small difference in ligation may lead to a different ligand exchange reactivity. In particular, under the same reaction condition, $Au_{11}(PPh_3)_8Cl_3$ produced thiolated Au NCs with a smaller core size when compared to those NCs generated by using $Au_{11}(PPh_3)_7Cl_3$ as the precursor. Therefore, if the purpose is to preserve a small core size during the ligand exchange, one should employ $Au_{11}(PPh_3)_8Cl_3$ as the precursor and keep the temperature at 50 °C or below. On the other hand, if the purpose is to produce Au_{25} NC, we can start with $Au_{11}(PPh_3)_7Cl_3$ and use a relatively high temperature.

The ligand exchange chemistry can also be applied to synthesize atomically precise NCs from polydisperse Au NPs with core sizes above 2 nm. For example, Jin et al.[68] demonstrated that rod-shaped and spherical Au_{25} NCs can be obtained through a one-phase and two-phase thiol-etching process, respectively, by using the phosphine-protected polydisperse Au NPs (with core size of 1−3.5 nm) as the precursors. The synthetic method mainly involved two steps. In the first step, polydisperse Au NPs were prepared in water/toluene (two phase) using $HAuCl_4$, $NaBH_4$, and triphenylphosphine as the metal precursor, reducing agent, and protecting agent, respectively. In the second step, the as-prepared polydisperse Au NPs were isolated and used as precursors for shape-controlled synthesis of thiolated Au_{25} nanorods and nanospheres by using a one-phase or two-phase thiol-etching protocol, as illustrated in Figure 6. This method could also be extended to other metal NCs.

11.2.2.2 Ligand Exchange: Thiolate versus Thiolate

Thiolate-protected metal NCs possess a strong bond between metal atoms and thiolate ligands, and they are very stable in solution, which have motivated great efforts in the NC community to develop efficient synthetic strategies for thiolate-protected metal NCs or thiolated metal NCs. As a result, a series of atomically precise thiolated NCs have been successfully produced. Some examples are $Au_{25}(SR)_{18}$,[31,33] $Au_{38}(SR)_{24}$,[69] $Au_{67}(SR)_{35}$,[70,71] $Au_{102}(SR)_{44}$,[34] $Au_{144}(SR)_{60}$,[72,73] and those NCs pertaining to Ag such as $Ag_{14}(SC_6H_3F_2)_{12}(PPh_3)_8$,[74] $[Ag_{16}(DPPE)_4(SPhF_2)_{14}$ (where DPPE is 1,2-bis(diphenylphosphino)ethane),[75] and $Ag_{44}(p\text{-MBA})_{30}$ (p-MBA = *para*-mercaptobenzoic acid).[12,76] Most of these NCs are obtained via a size-focusing protocol, which involved the following two steps: (1) preparation of polydisperse NCs via a kinetically controlled process, and (2) size focusing of the polydisperse NCs into a monodisperse (or atomically precise) one.[77] Apart from this protocol, a variety of metal NCs have also

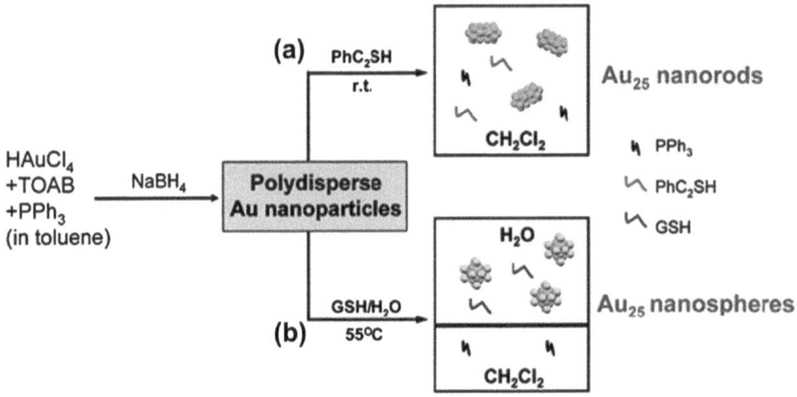

FIGURE 6 Conversion of polydisperse Au NCs into atomically precise Au_{25} nanorods and nanospheres via (a) one-phase and (b) two-phase thiol-etching process, respectively. *Reproduced with permission from Ref. 68. Copyright (2009) American Chemical Society.*

been separated and identified by some high-resolution separation techniques; however, the poor scalability of these separation techniques may further constrain the practical applications of the as-purified NC species; such applications typically require a large amount of samples (e.g., gram scale). As discussed in the last section, the ligand exchange from phosphine-to-thiol is a good way to achieve atomically precise metal NCs protected by thiolate ligands. However, the complete ligand exchange or the molecular purity of the final products could be a key concern for most of the cases.

Ligand exchange between two thiolate ligands is another efficient way to produce high-quality thiolated metal NCs with predesigned ligands. Recent studies suggest that even those stable thiolated Au NCs could undergo ligand exchange with another thiol ligand in solution, which could be readily used to produce metal NCs with different thiol ligands. This thiolate-to-thiolate ligand exchange could be a good platform to achieve predesigned physical and chemical properties of metal NCs such as chirality. For example, the thiolate ligands on Au_{25} NCs could be replaced by other thiolate ligands.[78] Bürgi et al.[6,79,80] have carried out systematic studies on the chiral ligand exchange using different-sized Au NCs. Their data suggest that while Au NCs consisting of several to a few tens of Au atoms only featured with one enantiomer, when they were exposed to free thiols of the opposite configuration, the optical activity in the metal-based electronic transitions could be reversed. In addition, if the incoming thiol has a different structure, the electronic transitions in the metal core were slightly modified. In most cases, the ligand exchange reaction was not complete. Similar to thiolated Au NCs, chirality in other NCs could also be induced. For example, Kimura et al.[81] showed that the chiral functionalization of the optically inactive thiolated Ag NCs could be achieved via a ligand exchange reaction between the racemic penicillamine on the NC surface and the enantiopure D- or L-penicillamine in solution.

However, the question concerning the fate of the NCs upon the ligand exchange still remains. In some cases, the ligand exchange may lead to a size change of the NCs. For example, Jin et al.[32] performed the thiol-to-thiol ligand exchange with 4-tertbutylbenzenelthiol (TBBT) by using a highly stable $Au_{38}(PET)_{24}$ NC (where PET is SCH_2CH_2Ph) as the precursor. It was found that when $Au_{38}(PET)_{24}$ was reacted with TBBT at 80 °C for more than 12 h, it can be transformed into a new NC species, $Au_{36}(TBBT)_{24}$. Interestingly, the structure of the as-synthesized $Au_{36}(TBBT)_{24}$ is distinctively different from that of $Au_{38}(PET)_{24}$, as illustrated in Figure 7. In this case, the starting material is intrinsically chiral, while the product $Au_{36}(TBBT)_{24}$ is achiral. In a follow-up study,[82] the same group shed some light on the reaction pathway of the ligand exchange. The pathway could be roughly divided into four stages: (1) PET to TBBT ligand exchange, (2) TBBT-triggered structural distortion of Au_{38}, (3) disproportionation of Au_{38} to Au_{36} and Au_{40}, and (4) size conversion of Au_{40} to Au_{36}; the details are illustrated in Figure 8. This process gave rise to the formation of Au_{36} NCs at a high yield of ~90% (on the Au atom basis), which is very close to the theoretical yield of ~94% according to the disproportionation mechanism.

11.2.2.3 Ligand Exchange: Thiolate versus Selenolate

To date, the library of metal NCs is full of thiolated NCs due to their superior stability compared to metal NCs protected by other organic ligands. In very recent years, it has been reported that more stable Au NCs could be produced by shifting the ligands of these Au NCs from thiolates to selenolates. Recently, selenolate-protected Au NCs have rapidly emerged as a new class of promising materials in the NC family. Although both sulfur (S) and selenium (Se)

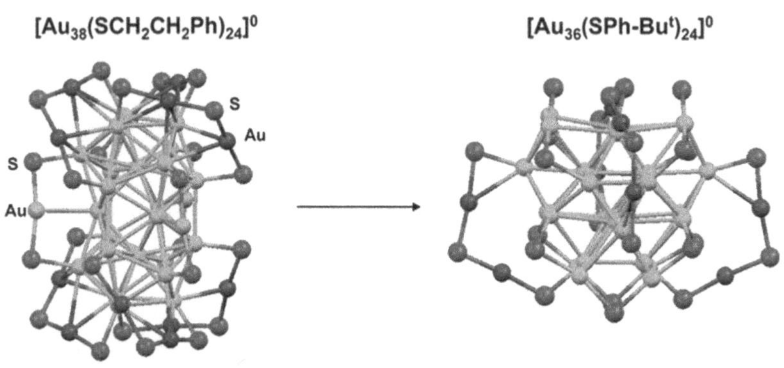

FIGURE 7 Conversion of $Au_{38}(PET)_{24}$ to $Au_{36}(TBBT)_2$ (the carbon tails are not shown for clarity; S atoms, yellow (light gray in print versions); kernel Au atoms, magenta (dark gray in print versions); surface Au atoms, green or cyan (gray in print versions)). *Reproduced with permission from Ref. 82. Copyright (2012) Wiley-VCH.*

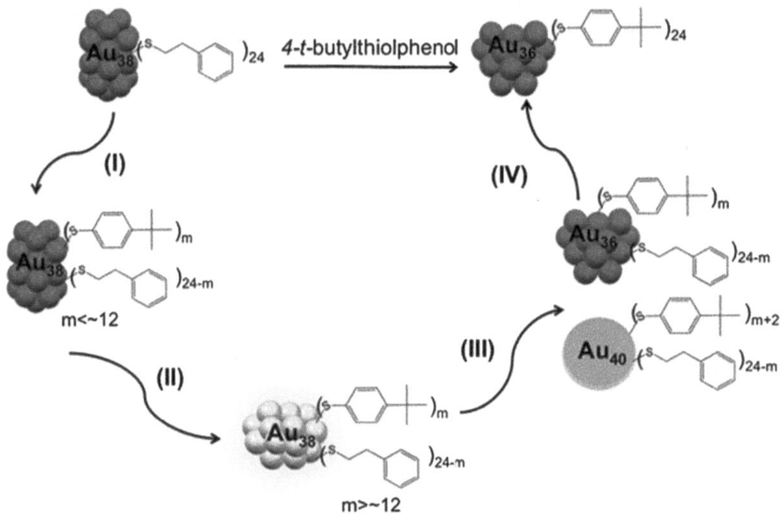

FIGURE 8 Reaction pathway for the conversion of $Au_{38}(PET)_{24}$ to $Au_{36}(TBBT)_{24}$. Stage I: ligand exchange; II: structure distortion; III: disproportionation; and IV: size focusing. *Reproduced with permission from Ref. 82. Copyright (2013) American Chemical Society.*

are in the same group in the periodic table, compared with sulfur, the electronegativity and atomic radius of selenium are closer to those of the Au. As a result, the Au−SeR bond features with more covalent characteristic, and therefore with a higher bonding energy than the Au−SR bond. Thus, it is expected that $Au_n(SeR)_m$ is more stable than $Au_n(SR)_m$. For example, Negishi et al.[83] first reported the direct synthesis of an octaneselenolate-protected Au_{25} NC ($[Au_{25}(SeC_8H_{17})_{18}]^-$). The as-synthesized selenolated Au NC was isolated and characterized by various analytical techniques. The experimental results suggest that $[Au_{25}(SeC_8H_{17})_{18}]^-$ has a similar geometric structure as $[Au_{25}(SC_8H_{17})_{18}]^-$, but features with less charge transfer between the Au atoms and the ligands. This study inspired a new research direction for designing functional metal NCs with different functionalities. The ligand exchange from thiolates to selenolates could affect the physicochemical properties of the NCs. For example, Zhu et al.[84,85] demonstrated an efficient ligand exchange strategy to convert thiolated NCs to selenolated ones. More importantly, the core size of the NCs was preserved during the ligand exchange process, and a complete ligand exchange was achieved. By this efficient strategy, they have successfully obtained $[Au_{25}(SePh)_{18}]^-[(C_8H_{17})_4N]^+$ and $Au_{18}(SePh)_{14}$ from $[Au_{25}(SPh)_{18}]^-[(C_8H_{17})_4N]^+$ and $Au_{18}(SPh)_{14}$, respectively. In addition, it was found that the PhSe-protected Au_{25} NCs were more stable than the $PhCH_2CH_2S$-protected NC counterparts in solution.

In another study, Negishi et al.[30] performed a very detailed investigation on the effects of ligand on the stability of the NCs. They chose

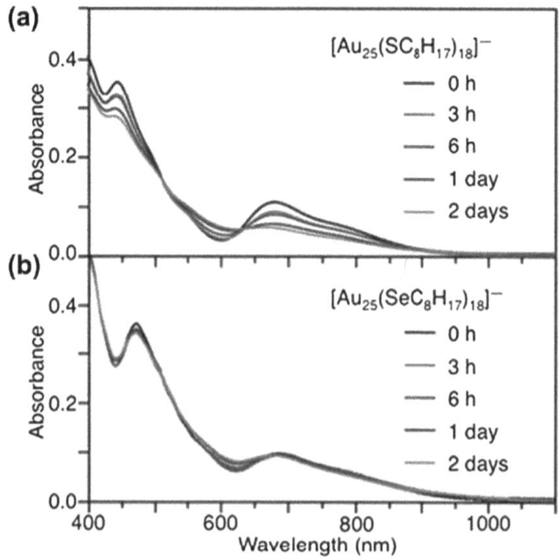

FIGURE 9 Time-course monitoring of the absorption spectra of (a) $[Au_{25}(SC_8H_{17})_{18}]^-$ and (b) $[Au_{25}(SeC_8H_{17})_{18}]^-$ in toluene at 60 °C. *Reproduced with permission from Ref. 30. Copyright (2012) American Chemical Society.*

$[Au_{25}(SC_8H_{17})_{18}]^-$ and $[Au_{25}(SeC_8H_{17})_{18}]^-$ as the model NCs, and compared their stability against degradation in solution, thermal dissolution, and laser fragmentation. In particular, they found that under harsh conditions, $Au_{25}(SR)_{18}$ gradually dissociated with time in solution, while $Au_{25}(SeR)_{18}$ was very stable in solution and no obvious decomposition was observed, as illustrated in Figure 9. The data clearly suggest that, compared to thiolates, selenolates provide a better protection for Au NCs, thus generating NCs with higher stability in solution. This could be due to the stronger Au–Se bond. Similarly, the selenolate-protected 38-atom Au NCs also possessed better stability in solution as compared to its thiolated NC counterparts.[86] Compared to thiolated metal NCs, the successful preparation of selenolate-protected metal NCs, especially those NCs with atomic precision, is less attempted, which calls for more concerted efforts from the NC community to further develop efficient strategies to synthesize high-quality selenolate-protected metal NCs; such metal NCs may find applications in many fields, similar as the applications that have been explored for their thiolated NC counterparts.

11.2.3 Surface Functionalization

As mentioned earlier, ligand-protected metal NCs typically feature with a core–shell structure, where the shell is consisting of organic ligands. Many properties such as solubility, stability, and optical properties are highly dictated by the nature of the ligands on the NC surface. There are a number of efficient strategies for the preparation of atomically precise metal NCs, but many of them are protected by hydrophobic thiolate ligands, which may limit their applications if a good water solubility is required, such as applications in the biological systems. Functionalization of such metal NCs with a well-controlled surface is therefore crucial to further advance the functional NCs for some practical applications. Some concepts and principles are discussed in this section.

11.2.3.1 Silica Coating

Coating NPs with a silica layer is one most widely used method to functionalize the NP surface. Silica coating of metal NPs has several advantages over the organic coating. Apart from the improvement of the solubility and stability of the NPs in water, the silica coating on the metal NP surface may also offer better functionalities interfacing with biomolecules, which could also further improve the biocompatibility of the as-designed materials. Among the newly developed thiolated metal NCs, thiolated Au_{25} NC is the most studied NC species; however, some synthetic protocols for thiolated Au_{25} NCs, especially those protocols to produce NCs with good water solubility, are still far from perfect. In addition, the stability of such Au_{25} NCs in solution is another concern. Some efforts are devoted to mitigating the stability issue of metal NCs in practical applications. Silica coating is one efficient platform to improve the stability of the metal NCs. For example, Pradeep et al.[87] developed an efficient strategy to embed thiolated Au_{25} NCs in a silica layer. In a separate study, Guevel et al.[88] demonstrated that the silica shell could also be used as a protective host to improve the photo- and chemical stability of the fluorescent Au NCs. The high chemical stability of silica-coated metal NCs could further motivate their uses for many biological applications. For example, the silica-coated metal NCs may find applications in bioimaging. Kong et al.[89] developed a straightforward strategy for the preparation of fluorescent Au NCs with silica coating, and have successfully used the as-synthesized nanocomposites for cellular imaging. In another study, Devi et al.[90] reported the synthesis of ultrasmall Au NC arrays that are encapsulated in silica nanospheres, and have applied the materials for fluorescence imaging and catalysis. Similarly, Chen et al.[91] showed that chlorin e6 (Ce6)-conjugated silica-coated Au NCs featured with improved cellular uptake efficiency, which could provide better photodynamic therapeutic efficacy as compared to the free Ce6. While silica has been utilized for coating the NCs for better functionalities, most materials developed so far are limited to Au. The underlying chemistry of the silica coating and the effects of silica coating on the physical and chemical properties of metal NCs may also be applicable to other metal NCs.

11.2.3.2 Polymer Coating

Among the newly developed metal NCs, Au NCs is the most studied due to its inert nature and good biocompatibility. In contrast, there are less attempts on Ag NCs, most likely due to its relatively poorer stability in solution, especially in the presence of oxygen. To date, several atomically precise Ag NCs have been successfully synthesized and characterized; however, some of them are not very stable in solution, especially in a long-term period. Polymer coating could be an efficient way to further improve the stability of metal NCs in solution. For example, polyvinylpyrrolidone (PVP) has the capability to weakly coordinate with metal NCs to improve the stability of the latter.[92] Pradeep et al.[93] reported that the stability of the bovine serum albumin (BSA) protected Ag NCs (Ag_{15}@BSA) in solution was improved by the addition of PVP. In addition, the PVP-protected Au and Pd NCs have been demonstrated as efficient catalysts for the aerobic oxidation of benzylic alcohols in water at an ambient temperature.[94]

11.3 APPLICATIONS

In the past few years, there are extensive research to explore the applications of metal NCs, which typically feature with the well-controlled size, surface, structure, and functionalities; and the applications of metal NCs in optoelectronics, sensing, catalysis, and biomedical applications are only a few of them, as illustrated in Scheme 2. In this section, we highlight some examples, showing that the as-functionalized metal NCs are a new class of promising functional materials for many practical applications.

11.3.1 Sensing

With the rapid development of synthesis strategies, the fluorescence properties of metal NCs are of paramount interest to the NC community. In recent years, fluorescent metal NCs have emerged as a new class of promising fluorescent probes for selective and sensitive detection of many analytes, such as metal ions, anions, and biomolecules. There are some attractive features of fluorescent metal NCs such as good water solubility, large Stokes shifts, low toxicity, and high quantum yields.[18] In this section, we highlight some recent advances in the development of optical sensors by using fluorescent metal NCs as the probes.

11.3.1.1 Metal Ion Sensing

Both the metal core and the ligand shell of the metal NCs can be readily used to construct sensors for metal ion detection. In particular, the core of the metal NCs typically consists of several metal atoms, which feature with a closed shell electronic configuration. As a result, the fluorescent metal NCs may

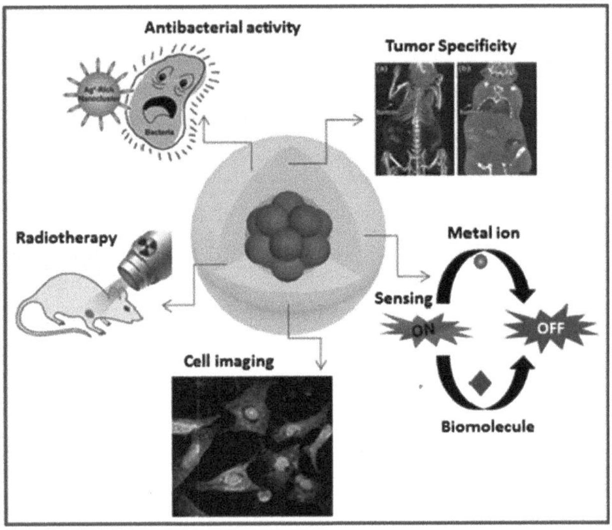

SCHEME 2 Some highlighted applications of functionalized metal NCs. (see color plate).

possess a feature that could specifically recognize some analytes. In addition, the analytes may also interact with the organic shell of the NCs via some specific interactions such as chelation. Based upon these specific interactions, the fluorescence properties of the NCs could be used to detect the analytes, and their intensity change (increase or decrease) can be used to correlate with the concentration of the analytes. We discuss some examples in this section, illustrating the working principles of the NC-based optical sensors for metal ions.

11.3.1.1.1 Based on Fluorescence Quenching

The fluorescence of metal NCs is very sensitive to the environment such as the solvent, temperature, pH, and some foreign species such as heavy metal ions in solution. In particular, the monitoring of heavy metal ions in the environment is of interest to the community due to their hazardous effects on the ecosystem as well as the human health. Hg(II) is one of them. Hg(II) is known to be a toxic and widespread pollutant because it can deleteriously affect the brain, nervous system, and kidney, even at a very low concentration.[95] Therefore, selective detection of Hg(II) in solution is of interest to the environmental and health-care community. It is well documented that the metal center of the NCs is featuring with a closed shell electronic configuration or d^{10} structure, which shows strong affinity with other metal ions with a similar d^{10} electronic configuration.[96] This d^{10}–d^{10} interaction is a metallophilic interaction, which originates from the dispersive forces and it could be further augmented by relativistic effects.[97] The interaction of Hg(II) with Au is well documented in the literature. Metal NCs, particularly Au and Ag NCs, contain a high content of Au(I) and Ag(I) species on their NC surface. Both Au(I) and Ag(I) have a closed shell electronic configuration, i.e., $5d^{10}6S^0$ for Au(I) and $4d^{10}5S^0$ for Ag(I). Thus, the presence of a small amount of Hg(II) in the Au or Ag NC solution could rapidly lead to the formation of strong Hg(II)...Au(I) bond (i.e., $5d^{10}$–$5d^{10}$ interaction), which could quench the fluorescence of the metal NCs. On the basis of this recognition chemistry, we have recently developed a good Hg(II) sensor by using red-emitting Au NCs as the fluorescent probe, as illustrated in Figure 10.[98] The recognition of Hg(II) is based upon the specific Hg(II)...Au(I) interaction, which could efficiently quench the fluorescence of Au NCs in solution. Similarly, several other studies have used the same principle to detect Hg(II).[99,100] In addition, other fluorescent metal NCs could also be used to detect Hg(II). For example, lysozyme-protected Au NCs has been used to detect Hg(II) and $CH_3Hg(I)$ in solution.[101]

The same underlying chemistry i.e., d^{10}–d^{10} interaction, has also been applied to fluorescent Ag NCs to construct sensors for Hg(II). For example, BSA- and lysozyme-stabilized Ag NCs have been utilized for the Hg(II) detection, where the limit of detection (LOD) was found to be 10 nM and 0.6 μM, respectively.[102,103] Apart from the protein-stabilized Ag NCs, oligonucleotide-stabilized Ag NCs were also found as good probes for the selective detection of Hg(II). For example, Wang et al.[104] synthesized fluorescent DNA-stabilized Ag NCs, and have applied them for Hg(II) detection with a LOD of 5 nM. In a separate study, Banerjee et al.[105] found that the

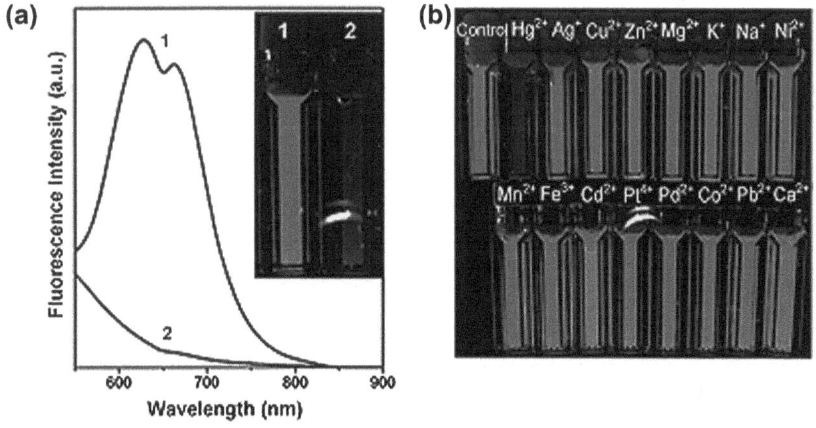

FIGURE 10 (a) Fluorescence spectra ($\lambda_{ex} = 470$ nm) and photographs (inset) of Au_{25}@BSA (20 μm) in aqueous solution under UV light (354 nm) in the absence (1) and presence (2) of Hg(II) ions (50 μm). (b) Photographs (under UV light) of Au_{25}@BSA (20 μm in water) in the presence of various metal ions (where the concentration of the each metal ion is 50 μm). *Reproduced with permission from Ref. 98. Copyright (2010) Royal Society of Chemistry.*

reduced lipoic acid functionalized Ag NCs were capable of sensing Hg(II) by quenching the fluorescence of the Ag NCs in solution. Recently, we reported a unique class of thiolated Ag NCs with high Ag(I) content on the NC surface. The as-synthesized Ag NCs showed strong blue emission in solution, which has been taken to construct an optical sensor for Hg(II) with high selectivity and sensitivity.[20]

Despite a number of studies reported the Hg(II) detection based on the metallophilic interaction, in-depth understandings of the underlying mechanism for the fluorescence quenching through the d^{10}–d^{10} interaction are presently missing. The underlying principles become more complicated as the structures of the protein-stabilized metal NCs are also lacking, and the fluorescence fundamentals of such are not clear; such fundamental issues require more concerted efforts from the community.

There are some recent efforts on addressing these issues. For example, Tang et al.[106] investigated the fluorescence quenching of the metal NCs by mercury ions by using the time-resolved fluorescence and transient absorption techniques, as shown in Figure 11. It was suggested that the lifetime of the metal NCs became shorter with the increase of the Hg(II) concentration, particularly for the slow component. The slow component was assigned to the delayed fluorescence, which suggests that upon excitation, the electron transfer occurs from the triplet state of the NCs to the metal ions. It has also been suggested that Hg(II) ions could interact with the outer semiring of the Au_{25} NCs. Moreover, the fluorescence of the Au_8@BSA and Au_{10}@histidine was not quenched by Hg(II); this suggests that there are no semirings in the Au_8@BSA or Au_{10}@histidine NCs. On the other hand, the lifetimes of the excited states of the Ag NCs are almost identical in the presence of different concentration of Hg(II) ions. In a separate study, Wang et al.[104] suggested that the quenching of the fluorescence of Ag NCs by Hg(II) obeyed a simple static quenching mechanism. They proposed that the Hg(II) ions and oligonucleotide-stabilized Ag NCs may form a kind of nonfluorescent complexes, which could lead to the fluorescence quenching of the NCs. The detailed mechanism of the Hg(II)-induced fluorescence quenching of the NCs is complicated, and more efforts are needed from the community in future studies.

In some cases, Hg(II) ions could interact with the shell or the functional groups on the NC surface. For example, Chang et al.[107] showed that 11-MUDA-protected Au NCs (where 11-MUDA is a carboxyl-terminated thiolate ligand, 11-mercaptoundecanoic acid) are capable of sensing Hg(II) based on the fluorescence quenching via a Hg(II)-induced aggregation mechanism. Under the optimized conditions, the LOD was determined to be 5.0 nM. In another study, Dong et al.[108] developed a label-free method for the "turn-on" detection of Hg(II) using DNA-stabilized Ag NCs as the fluorescent probe. In the presence of Hg(II), the DNA duplex on the NC surface was strengthened by forming a T-Hg(II)-T base pair, which led to the formation of a highly

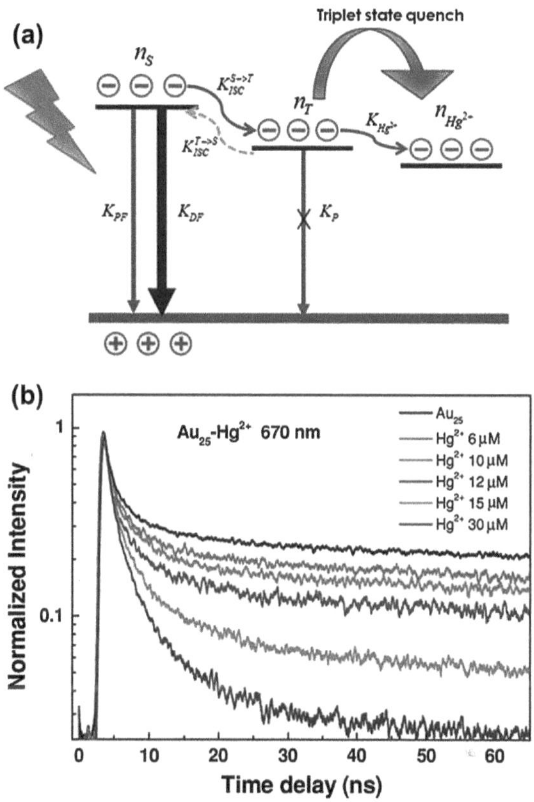

FIGURE 11 (a) Proposed mechanism of triplet electron transfer from Au_{25} NCs to Hg(II) ions (LUMO). (b) Quenching of the delayed fluorescence upon the interaction of Hg(II) with Au_{25}@BSA NCs while the prompt fluorescence is not affected. *Reproduced with permission from Ref. 106. Copyright (2013) Wiley-VCH.*

sequence-dependent fluorescent Ag NC species. The as-designed optical sensor showed good sensitivity with a LOD of 10 nM.

Like Hg(II), Cu(II) is another environmental pollutant. It is also an essential trace element in biological systems. The fluorescence of the metal NCs is sensitive to Cu(II); therefore, fluorescent metal NCs can also be used to construct sensors for Cu(II) detection. For example, Dong et al.[109] developed a simple and sensitive fluorescent sensor for Cu(II) detection based on a specific fluorescence response of the poly(methacrylic acid) (PMAA)-stabilized Ag NCs toward the analytes. Their method was attractive since the response of fluorescent Ag NCs toward Cu(II) was reversible, i.e., the fluorescence of the Ag NCs could be recovered in the presence of a strong chelating ligand such as disodium ethylenediamine tetraacetate (EDTA). In a separate study, Chang et al.[110] developed a simple and convenient method for the detection of Cu(II) using a water-soluble DNA-stabilized Ag NC. The method has good selectivity and sensitivity. In addition, the trace amount of Cu(II) in pond water or in the soil can also be detected, which suggests the practicality of the as-designed optical sensor. Similarly, Pradeep et al.[111] utilized Au_{38}@BSA NC as a "turn-off" sensor for Cu(II) ions. A similar observation was also reported by Sreenivasan et al.,[112] where they used BSA-stabilized Au NCs for the detection of Cu(II) ions in live cells.

11.3.1.1.2 Based on Fluorescence Enhancement

The unique properties of metal NCs could be altered by tuning the surface properties such as the ligand composition and surface charge of the NCs. In addition, compared to the strategy of using fluorescence quenching to report signals, relatively less studies reported the detection of analytes based on fluorescence enhancement. There are some good attempts. For example, Pradeep et al.[111] showed that Ag NPs can enhance the fluorescence of BSA-stabilized Au NCs, which might be caused by altering the radiative decay rate or increasing the absorption of the fluorescent Au NCs. As mentioned earlier, there is a high affinity of Ag(I) with the Au surface as they can form alloyed clusters. The deposition of Ag ions on the surface of Au NCs can be readily used as a recognition chemistry for Ag(I) detection. Although Ag(I) is less toxic compared to other heavy metal ions, a high concentration of Ag(I) ions can still bring harmful side effects to the environment and the human health. Recently, Jin et al.[113] demonstrated that a well-defined NC namely, $Au_{25}(SG)_{18}$ can be used as a good probe to detect Ag ions with a LOD of $\sim$200 nM. This study also sheds some light on the critical factors for the fluorescence enhancement. Among the proposed factors, the oxidation state of $Au_{25}(SG)_{18}$ contributes the most, with the other two possible contributions from the interactions of Au_{25} with Ag(I) and Ag(0). In a separate study, Wu et al.[114] reported that BSA-stabilized Au NCs (Au_{16}@BSA) can be used to construct a fluorescence-enhanced sensor for Ag(I), where the fluorescence of Au NCs was blue shifted and gradually enhanced with the addition of Ag(I) ions. The metal ion-induced fluorescence enhancement has also been applied for the detection of other toxic metal ions.[115]

11.3.1.1.3 Based on FRET

The principle of "fluorescence resonance energy transfer" (FRET) has been used in constructing sensors for heavy metal ions. For example, we have designed a FRET pair using a blue-emitting conjugated oligomer-substituted polyhedral oligomeric silsesquioxane (POSS) and a red-emitting Au NC, and have successfully applied the pair for Hg(II) detection in solution as well as in cells (Figure 12).[116] Due to the spectral overlap between the donor (blue-emitting POSS) emission and the acceptor (red-emitting Au NCs) absorption, a dual-emissive pink fluorescence was observed upon the donor

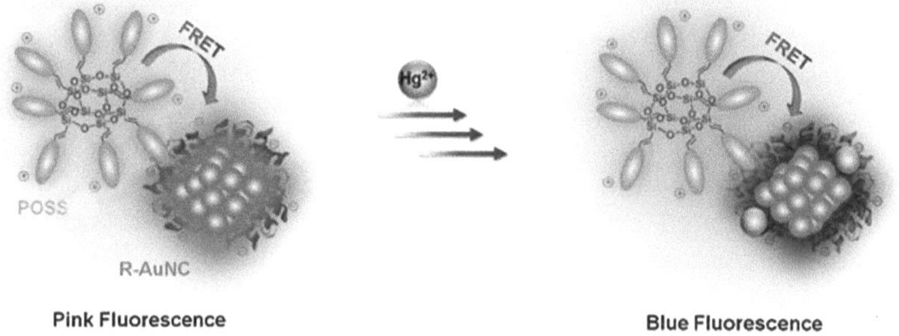

FIGURE 12 Schematic illustration of the visual detection of mercury ions based on a FRET design between the blue-emitting POSS and red-emitting Au NCs. *Reproduced with permission from Ref. 116. Copyright (2011) American Chemical Society.*

excitation. The pink fluorescence specifically changed to blue upon the introduction of Hg(II) ions into the complex solution, as the Hg(II) ions could lead to the quenching of NC fluorescence owing to the strong Hg(II)...Au(I) interaction. Consequently, this allows for a visual detection and a precise quantification of Hg(II) in solution with a LOD of ~0.1 nM. More recently, Chen et al.[117] demonstrated that the fluorescence of Au NCs can be effectively quenched by Ag NPs, originating from the surface plasmon resonance-induced surface energy transfer.

11.3.1.1.4 Based on Scattering

Light scattering belongs to Rayleigh scattering, which can be used to construct optical sensors for metal ions. For example, Zheng et al.[118] used DNA-stabilized Ag NCs as a novel light scattering probe to selectively detect Cu(II) ions. The DNA–Ag–NC conjugates showed weak light scattering signal in the presence of cysteine. However, when Cu(II) ions were added to the DNA–Ag NCs/cys system, the light scattering intensity was dramatically enhanced. It was proposed that the addition of Cu(II) ions in the system led to the formation of particles with larger diameters, which activate the light scattering signal. The LOD was determined to be 0.5 μM, which was below the safe limit of Cu(II) in drinking water (20 μM) set by the U.S. Environmental Protection Agency (EPA).[110]

11.3.1.1.5 Other Metal Ions

Along with Hg(II) and Cu(II), other toxic metal ions such as As(III), Cr(III), and Pb(II) can also be detected by using the fluorescence properties of metal NCs. For example, Banerjee et al.[119] demonstrated that blue-emitting Au NCs can be used to detect As(III) ions in aqueous solution. The fluorescence of MUDA-protected Au NCs could be quenched by several metal ions, however, it will be selectively quenched by Cr(III) ions when EDTA was used as the masking agent for other metal ions. Based on this recognition chemistry, Jin et al.[120] used MUDA-protected Au NCs to construct a "turn-off" fluorescent sensor for Cr(III). Apart from Au NCs, other metal NCs such as alloyed metal NCs and Cu NCs, have also been found as good fluorescent probes. For example, Pal et al.[17] used the blue-emitting Cu NCs to detect Pb(II) in water, where the fluorescence quenching of the Cu NCs was attributed to the NC aggregation induced by Pb(II) ions. Similarly, based on the aggregation-induced fluorescence quenching, Zhang et al.[121] showed that Pb(II) ions in water could be detected by using fluorescent Au−Ag NCs as the probe.

11.3.1.2 Anion Sensing

Like toxic heavy metal ions, some anions are also toxic and they are considered as severe environmental pollutants. The detection of these toxic anions is therefore of interest to the community. Cyanide (CN^-) is one of them, Notably, Lu et al.[122] demonstrated a Au NC-based fluorescent sensor for the detection of cyanide ion in aqueous solution, where the fluorescence of Au NCs was quenched in the presence of cyanide ions due to the etching capability of cyanide. It is known that cyanide can form a very stable $Au(CN)^{2-}$ complex with Au(0) through a strong covalent bond.[123] Based on this, Lu et al.[122] proposed that the fluorescence quenching of Au NCs was attributed to the formation of such gold cyanide complex. Similar to cyanide ion, sulfide ions or some halide ions can also directly react with Au or Ag core due to their high affinity to Au or Ag atoms. Based on this recognition chemistry, Luo et al.[124] developed a Ag NC-based fluorescent and colorimetric sensor for halide ions (Cl^-, Br^-, and I^-), where the fluorescence quenching of the NCs was attributed to the halide-induced oxidative etching and the subsequent aggregation of the NCs. In another study, Cai et al.[125] used fluorescent Au NCs to detect I^- ions, where $Au_{25}(SG)_{18}$ was used as the fluorescent probe. Similarly, Chang et al.[126] reported the use of fluorescent Au−Ag NCs for the detection of sulfide ions.

11.3.1.3 Biosensing

Biomolecules, such as amino acid, peptide, protein, and DNA, play crucial roles in our life. Therefore, detection of these important biomolecules is of interest to the bioanalytical research. Here we introduce several examples on the use of metal NCs for biomolecule detection.

11.3.1.3.1 Detection of Glucose

Glucose is one key physiological biomolecule that could affect a multitude of aspects related to our daily life. For example, glucose has been used as an energy source in most organisms from bacteria to humans. Furthermore, the human health conditions are also dependent on the glucose level in blood, and an abnormal amount of glucose provides crucial information on many diseases such as diabetes and hypoglycemia. Therefore, the development of sensitive and selective sensors for glucose is of importance to biomedical research as well as for food analysis.

It is well known that fluorescent Au NCs can sensitively respond to H_2O_2. Moreover, glucose can be oxidized by oxygen (O_2) in the presence of glucose oxidase (GOD) by the following reaction: $\text{glucose} + O_2 + H_2O \xrightarrow{\text{GOD}} \text{gluconic acid} + H_2O_2$. Therefore, based on the fluorescence quenching of the NCs, the concentration of glucose can be correlated to the amount of the enzymatically generated H_2O_2. Using this principle, Chang et al.[127] developed an assay for H_2O_2 and glucose by using fluorescent Au NCs. It is evident that H_2O_2 can oxidize the 11-MUDA ligands on the NC surface, leading to the formation of an organic disulfide product. As a result, the fluorescence of the NCs was quenched. Further combining the fluorescent Au NCs with GOD enables a sensitive determination of glucose. The LOD for glucose was 1 μM; this value is comparable to those obtained from the electrochemical methods.[128] Similarly, Dong et al.[129] used BSA-stabilized Au NCs for the detection of glucose, where the LOD was 5 μM. There are also other examples for glucose detection by using fluorescent metal NCs.[130]

11.3.1.3.2 Detection of Biothiols

Thiol-containing biomolecules, such as cysteine, homocysteine, and glutathione, are biologically important analytes; they can also be determined based on the fluorescence response of the metal NCs. For example, Dong et al.[131] observed that the fluorescence of PMAA-stabilized Ag NCs was effectively quenched by cysteine. By virtue of this specific response, a sensitive and selective detection method was developed for cysteine. It was proposed that cysteine could directly interact with Ag NCs via the strong Ag–S interaction, followed by the oxidation of silver by oxygen. Similarly, Wang et al.[132] used DNA-stabilized Ag NCs to detect cysteine, homocysteine, and glutathione. It was found that, in the presence of biothiols, a nonfluorescent coordination complex was formed via the strong Ag–S interaction. The fluorescence quenching was attributed to the static quenching, which was further supported by the shift of the corresponding absorption spectra as well as the unchanged fluorescence lifetimes. In another study, Qu et al.[133] used DNA-stabilized Ag NCs to construct a fluorescent sensor for biothiols.

More recently, we developed a simple method to detect cysteine with high selectivity and sensitivity by using highly fluorescent Ag NCs protected by the interference biothiol, glutathione (GSH).[134] In this study, both the physicochemical properties (e.g., the unique Ag surface chemistry and strong fluorescence) of the NC core and the physical properties (e.g., steric hindrance) of the organic shell of the GSH–Ag NCs were used for cysteine detection.

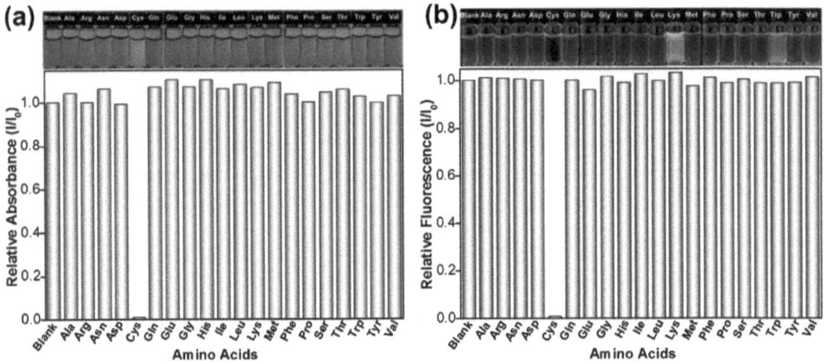

FIGURE 13 Top panel: digital photos under (a) visible light and (b) UV light; and bottom panel: relative optical intensity (I/I_0) of (a) absorption at 489 nm and (b) fluorescence at 647 nm of aqueous solutions of GSH−Ag NCs (10 μM) in the presence of 50 μM of different amino acids. (see color plate). *Reproduced with permission from Ref. 134. Copyright (2012) American Chemical Society.*

Owing to the strong Ag−S interaction, the fluorescent GSH−Ag NCs showed superior selectivity for cysteine over other amino acids (without thiol groups) (Figure 13). Our study also exemplifies the utilization of the organic shell of the NCs to further improve the sensor performance.

Similar to Ag NCs, fluorescent Au NCs can also be used to detect thiols. For example, Tseng et al.[135] used blue-emitting Au_8 NCs for the detection of glutathione (GSH), where the addition of a small amount of GSH led to the fluorescence quenching of the NCs due to the core etching. The method was sensitive and has been applied to detect GSH in a blood drop. In another study, based on the fluorescence enhancement strategy, Liu et al.[136] reported a cysteine sensor by using BSA-stabilized Au NCs as the probe. They suggest that cysteine might reduce the surface defects of the Au NCs, and subsequently enhance their fluorescence.

11.3.1.3.3 Detection of Adenosine Triphosphate (ATP)

Dong et al.[137] used fluorescent Ag NCs as probes to detect ATP. In this sensor system, without the ATP, the ATPS1/ATPS2 was at its free state and only few base pairs existed, leading to a weak fluorescence in solution. The introduction of ATP significantly enhanced the fluorescence of the NCs in solution. In a follow-up study, the same group designed a simple label-free sensor for ATP detection by using the dsDNA-stabilized Cu NCs as the probe.[138] It was known that dsDNA can serve as a template for the synthesis of fluorescent Cu NCs whereas ssDNA could not. The introduction of ATP could trigger the structure switching of the reporter DNA, leading to the formation of an aptamer−ATP complex, which destructed the double helix structure and quenched the fluorescence of Cu NCs. In a separate study, Willner et al.[139] developed a hybrid system consisting of nucleic acid-functionalized Ag NCs and graphene oxide (GO), which can be used for the detection of ATP. Similarly, Chen et al.[140] used another hybrid system namely, AuNC@GSH-Fe^{3+}, for the detection of ATP. It was found that Fe^{3+} ion quenched the fluorescence of AuNC@GSH. However, the fluorescence can be restored by those phosphate-containing molecules.

11.3.1.3.4 Detection of Proteins

Development of functionalization methods, such as conjugation of the receptor molecules on the NC surface, can facilitate the use of functional metal NCs for protein detection. For example, Chang et al.[141] developed a FRET-based competitive homogeneous fluorescence quenching assay for the detection of proteins by using a breast cancer marker protein, PDGF AA (platelet-derived growth factor AA), and conjugating them with a fluorescent Au NC. The as-designed composite serves as a donor while a thiol derivative PDGF-binding aptamer-conjugated Au NP serves as an acceptor. Through the interaction between anti-PDGF AA aptamer and PDGF, the fluorescence of Au NCs was quenched by Au NPs via the resonance energy transfer. The addition of free PDGF can, however, restore the fluorescence of PDGF-modified NCs through the competitive binding. In another study, the same group used water-soluble fluorescent Au NCs to construct fluorescent sensor for Concanavalin A (Con A) and *Escherichia coli*.[142] Based on the aggregation-induced fluorescence quenching principle, they showed that the mannose-protected Au NCs were capable of sensing Con A with high sensitivity (LOD = 75 pM). On the other hand, the incubation of the probes with *E. coli* revealed that the mannose-protected Au NCs could bind to the bacteria, yielding intensely fluorescent cell clusters. From the linear relationship between the fluorescence of the Au NCs and the concentration of *E. coli*, the LOD was determined to be 7.20×10^5 cells/mL.

In another study, Leblanc et al.[143] developed an immunoassay platform to detect human IgG antigen based on the fluorescence quenching mechanism by using Poly(amidoamine) or (PAMAM)-stabilized Au NCs as the probe. The PAMAM-stabilized Au NCs were first coupled with the goat-derived polyclonal antihuman IgG. The introduction of human IgG to the antihuman IgG/PAMAM−Au NC complexes led to the fluorescence quenching of the Au NCs. Similarly, Chang et al.[144] used protein A-modified Au NCs as a fluorescent probe for the detection of human immunoglobulin G (hIgG) in a homogeneous solution.

Trypsin is another biomolecule which can quench the fluorescence of Au NCs. For example, Xu et al.[145] reported that BSA-stabilized Au NCs can be used to detect trypsin, where the quenching mechanism is based on the trypsin digestion of the BSA in BSA-stabilized Au NCs and a subsequent change of the Au NC surface. Another protein, namely thrombin, has also been detected using the strong recognition chemistry of DNA aptamers. Martinez et al.[146] synthesized cytosine-rich DNA aptamer stabilized Ag NCs, which were

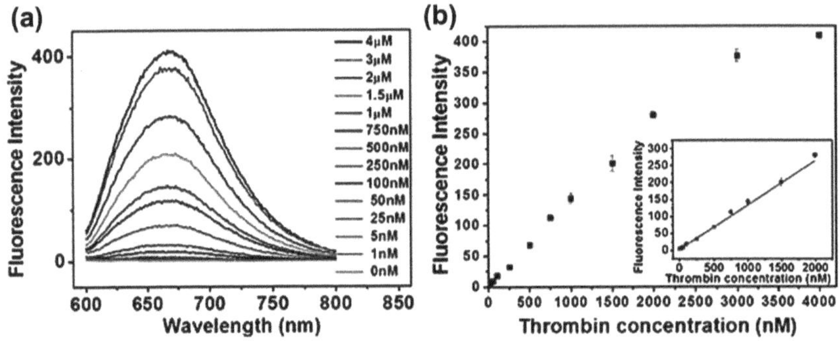

FIGURE 14 (a) Fluorescence spectra from the assay for human α-thrombin (0–4 μM). (b) The fluorescence intensity increases with the concentration of human α-thrombin. *Reproduced with permission from Ref. 147. Copyright (2012) American Chemical Society.*

further conjugated with an antithrombin aptamer sequence to specifically recognize thrombin. Similarly, Zhu et al.[147] synthesized fluorescent Ag NCs by using a guanine-rich DNA sequence. For the detection of human α-thrombin, two aptamers (Apt15 and Apt29) were used, where the first aptamer was linked with an NC nucleation sequence and the second one was linked through a complementary sequence to the G-rich overhang. The binding of the two aptamers to the target proteins could bring the G-rich overhang to a close proximity to Ag NCs. As a result, the fluorescence of the Ag NCs was enhanced (Figure 14).

11.3.1.3.5 Detection of DNA

The ability to detect DNA is critically important in biotechnology. Fluorescent metal NCs hold great promise as good optical probes for the detection of DNA/RNA. A number of studies have been reported by Wang et al.[148,149] For example, by taking the advantage of highly sequence-dependent generation of fluorescent DNA–Ag NCs, the fluorescent Ag NCs can specifically identify a single nucleotide mutation in the sickle cell anemia gene mutation (Figure 15). The same group also found that the fluorescence of DNA-stabilized Cu NCs was sensitive to the base type located in the major grooves.[149] Based on this, fluorescent Cu NCs have been used as a good probe to detect the DNA hybridization. On the basis of a "turn-off" model through the photo-induced electron transfer between the DNA–Ag NCs and G-quadruplex/hemin complex, a Ag NC-based molecular beacon was also designed, which can be used to detect biomolecules such as DNA and ATP.[150] In another study, a "turn-on" model for the detection of guanine-rich DNA sequences was developed by Werner et al.[151] They found that the fluorescence of DNA–Ag NCs can be enhanced up to 500-fold when the NCs were placed in proximity to guanine-rich DNA sequences. By virtue of this specific response, they designed a simple "light-up" probe, called NanoCluster

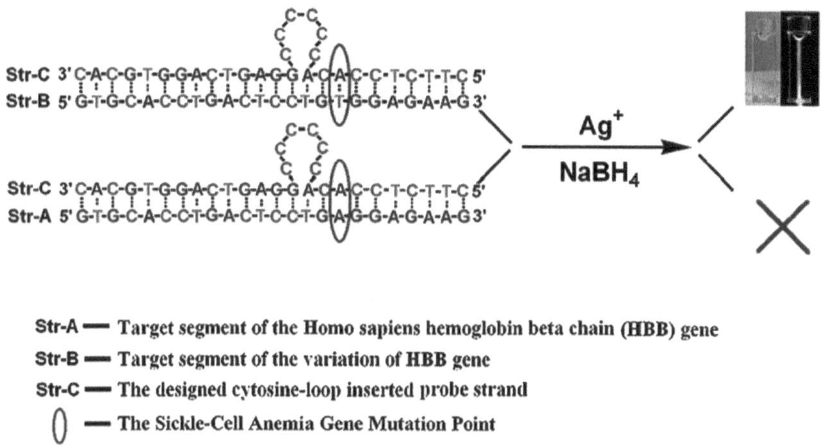

FIGURE 15 Schematic illustration of the use of two different DNA duplexes with inserted cytosine loops as synthetic scaffolds to produce fluorescent Ag NCs for the determination of the sickle cell anemia gene mutation. *Reproduced with permission from Ref. 148. Copyright (2009) American Chemical Society.*

Beacon, which consists of two short linear DNA probes (an NC probe and a G-rich probe). When they are brought into proximity through the hybridization with target DNA, the weak fluorescence of DNA–Ag NCs was significantly enhanced. Similarly, Chang et al.[152] synthesized oligonucleotide-stabilized Ag NCs and used them as fluorescent probes to detect single nucleotide polymorphisms.

11.3.2 Bioimaging

Akin to the NC-based sensor development, research on the NC-based fluorescence imaging has also made a rapid progress in the past few years. Fluorescent metal NCs have attractive features of small sizes, bright and tunable fluorescence, large Stokes shift, low toxicity, good biocompatibility, and good photo- and chemical stability. In some cases, the fluorescent metal NCs are prepared with biomolecules such as peptides, proteins, and DNA, which could further improve their biocompatibility in the biological systems. We provide a brief overview of the metal NC-based fluorescence imaging in this section.

11.3.2.1 Case Study with Fluorescent Au NCs

Among the existing fluorescent metal NCs, Au NCs have been the most successfully employed as fluorescent labels because of their high stability and well-developed synthetic techniques. Although the fluorescence of metal NCs is the most important properties for imaging, functionalization with a particular bioprobe is also demonstrated to be useful as it may further facilitate the uptake of functional NCs by specific cells. For example, Yang et al.[153] exploited both the ligand exchange and the shell functionalization strategies to prepare a new class of Au NCs as fluorescent probes for nuclear targeting and intracellular imaging. At first, 11-mercaptoundecanoic acid (MUDA)-functionalized Au NCs were prepared from the PAMAM-stabilized Au NCs via the ligand exchange. Following by that a specific peptide named SV40 nuclear localization signal (NLS) was anchored to the NC surface. From the confocal microscope images, it can be seen that the HeLa cells treated by MUDA-Au NCs did not show fluorescence, which means that the MUDA-Au NCs were not able to efficiently enter the cells due to their negative surface charge. On the other hand, an intensive blue emission (panel (c and e) of Figure 16) was observed from NLS-MUDA-Au NCs, indicative of the internalization of the MUDA-Au NCs associated with NLS by HeLa cells. It should be noted that this study used a common 1-Ethyl-3-(3-dimethylaminopropyl)carbodiimide or (EDC) coupling chemistry, which has also been used to functionalize dihydrolipoic acid (DHLA)-protected Au NCs with biologically relevant molecules such as PEG, BSA, avidin, and streptavidin.[154] Due to the specific avidin–biotin interaction, only streptavidin-conjugated AuNC@DHLA can be used for the staining of the biotin-containing cells, human hepatoma cells ($HepG_2$), with high intensity. In addition, like quantum dots, the unconjugated AuNC@DHLA can only be nonspecifically uptaken by the cells, as observed in the endothelial cells after ~5 h of incubation. By virtue of this specific streptavidin–biotin interaction, Pradeep et al.[155] also applied fluorescent Au_{23} NCs for the imaging of human hepatoma (HepG2) cell lines.

Folic acid (FA), which can be readily conjugated to the ligand shell of the NCs, is another efficient bioprobe for targeting, which has been widely used to help the NCs entering the cells. For example, Pradeep et al.[156] reported the imaging of the oral carcinoma KB cells using FA-conjugated fluorescent Au_{25} NCs. The KB cells have overexpressed folate receptors, therefore, FA-conjugated Au NCs could be significantly internalized by the cells. The characteristic NIR emission of the Au NCs was also used to image the cells. In addition, Pradeep et al.[111] also conjugated FA to BSA-Au NCs and used it for folate-targeted imaging. Recently, Zhao et al.[157] developed a core–shell multifunctional nanocarrier for targeted anticancer drug delivery, where an FA-conjugated amphiphilic hyperbranched block copolymer was used as a functional shell on the NC surface. Because of the FA-mediated cellular uptake, the nanocarriers exhibited high anticancer activity.

Moreover, some monoclonal antibodies such as herceptin, also show good targeting effects. The herceptin-modified BSA-Au NCs exhibited good targeting of HER2 breast cancer cells with overexpressed Erb2.[158] In some cases, it was also observed that the surface charge of the NCs plays a crucial role for intracellular targeting and tracking. For example, Lin et al.[159] demonstrated that when NCs with different surface charges were introduced into the culture

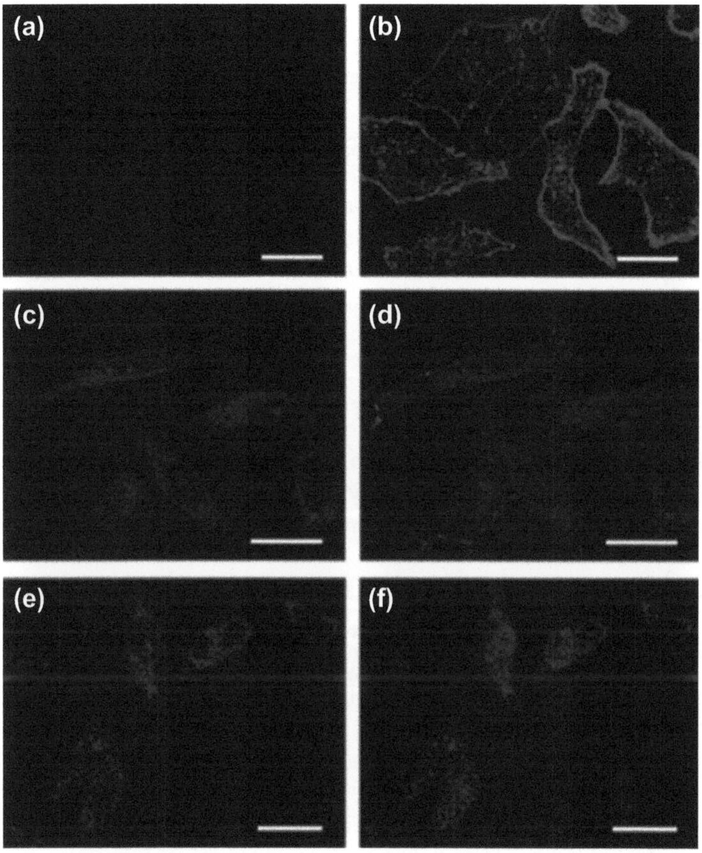

FIGURE 16 Confocal microscope images of intracellular delivery of the Au NCs. HeLa cells were treated with MUDA-Au NCs (a and b) and NLS-MUDA-Au NCs (c to f) for 1.5 h. The left panel shows the one-color image. The right panel shows the two-color colocalization image of the HeLa cells incubated with NLS-MUDA-Au NCs were counter-stained with a specific membrane dye (WGA–Alexa 594) and a nuclear dye (SYTO 59). Scale bar: 25 μm. (see color plate). *Reproduced with permission from Ref. 153. Copyright (2008) Royal Society of Chemistry.*

media of HeLa cells, those NCs carrying a neutral surface charge cannot penetrate the cell membrane and therefore could not be uptaken by the cells. However, the positively charged NCs can penetrate the cell membrane. In addition, several reports are available in the literature on the investigation of fluorescent Au NCs with different ligands for in vitro imaging. Interested readers may refer this to some recent review articles.[14,23,160,161]

11.3.2.2 Case Study with Fluorescent Ag NCs

One good example on the Ag NC-based imaging was demonstrated by Baskakov et al.[162] They prepared an organic–inorganic hybrid NCs composed of Ag and thioflavin T, and demonstrated that the hybrid NCs can be used to image amyloid fibrils. Dickson et al.[163–165] performed several systematic studies on the fluorescent Ag NC-based cell imaging. For example, they used the fluorescent Ag NCs synthesized via an intracellular protocol for the imaging of live cells.[163] In a follow-up study, they also reported the protein- and antibody-modified Ag NCs to stain the NIH 3T3 cells and monitored their subsequent internalization inside the cells.[164]

It is known that aptamers have a high targeting capability because of their specificity and strong binding affinity to their targets. Based on this strategy, Huang et al.[166] prepared yellow-emitting Ag NCs by using DNA. The sequence of DNA was designed to contain two fragments; one is the DNA aptamer fragment that can selectively bind to the target, and the other is a fragment rich with cytosine that can improve the fluorescence of the as-synthesized Ag NCs. It has been shown that the Apt-Ag NCs probe can be used for the detection of PrP^C and to image the SK-N-SH cells. Similarly, Wang et al.[167] reported an efficient way to prepare fluorescent Ag NC-aptamer assemblies, which could be used to target cancer cells.

11.3.2.3 Case Study with Fluorescent Cu NCs

In contrast to Au- and Ag NCs-based cellular imaging, studies on fluorescent Cu NCs are relatively less due to the complicated synthetic procedure of Cu NCs. There are some successful attempts. For example, Zhang et al.[168] prepared red-emitting BSA-stabilized Cu NCs and used them for cellular imaging. When CAL-27 cells were incubated with the BSA-Cu NCs for 24 h, an intense red emission was seen inside the cells, which suggests an efficient uptake of the NCs by the cells. In addition, the confocal laser fluorescence microscopy analysis suggests that the fluorescence was distributed not only in the cytoplasm but also in the nucleus. This suggests that the BSA-Cu NCs could be used as fluorescent probes for nucleus imaging. Similarly, Feng et al.[169] synthesized fluorescent Cu NCs by using a peptide as the template. They also used the as-synthesized Cu NCs as a fluorescent probe for the imaging of HeLa cells. More recently, Cao et al.[170] used tannic acid (TA)-protected Cu NCs (TA-Cu NCs) to image A549 cells. All these examples suggest that fluorescent Cu NCs could also be promising for bioimaging applications.

11.3.2.4 Case Study with Fluorescent Pt NCs

Fluorescent Pt NCs can also be used for bioimaging. For example, Inouye et al.[171] reported the synthesis and bioimaging application of a blue-emitting Pt NC. In another study, the same group reported a green-emitting Pt NC for the imaging of human epithelial carcinoma HeLa cells.[172] Note that, in the HeLa cells, the chemokine receptors are highly expressed. Therefore, they conjugated the Pt NCs with anti-chemokine receptor (anti-CXCR4-Ab) by using Protein A as the adapter protein. From the confocal laser fluorescence microscopic images, a green emission was observed on the cellular membrane. However, no fluorescence was seen in the control sample. More recently, Wang et al.[173] demonstrated that the GSH-protected Pt NCs could be readily utilized for the effective imaging of cancer cells (i.e., HeLa and HepG2 cells).

11.3.3 Antimicrobial Activity

From the ancient times, silver is known for its capability for the treatment of burns, wounds, and several bacterial infections. Recently, nanotechnology is coming up as promising tool to fight against bacteria and viruses. The development of alternative antimicrobial agents is becoming more important due to the emergence of antibiotic-resistant bacteria and the limitations of some existing antibiotics. In particular, Ag NPs have emerged as a new class of promising antimicrobial agents as they showed good antimicrobial efficacy against bacteria, viruses, and other eukaryotic microorganisms.[174] Therefore, in the past two decades, extensive research has been performed on the development of Ag-NP-based antibacterial agents. It has been demonstrated that the antibacterial efficacy of Ag NPs is largely shape and size dependent. The mechanism of the dissociation of Ag^+ ions from the NPs was recently proposed. However, a direct use of Ag^+ ions and their complexes as antimicrobial agents is not advisable as it may cause some adverse cytotoxic effects.

Therefore, the use of ultrasmall Ag NCs could be a potential solution as the Ag NCs have smaller sizes compared to the NPs. Their surfaces can also be assembled with a high concentration of Ag^+ ions. For example, we recently investigated the antimicrobial activity of the GSH-protected Ag NCs, which showed superior efficiency against *Pseudomonas aeruginosa*.[24] In a follow-up study,[175] we also found that the antimicrobial activity of the Ag NCs are highly dependent on the oxidation states of Ag species on the NC surface. For example, we found that the Ag^+-rich NCs showed higher efficacy when compared to the Ag^0-rich NCs, although they are stabilized by the same GSH ligand (Figure 17). In another study, Chen et al.[176] demonstrated that the lysozyme-protected Au NCs can be used to target notorious pathogenic bacteria, including pan-drug resistant Acinetobacter baumannii (PDRAB), Vancomycin-Resistant Enterococci (VRE), *Escherichia coli*, and *Staphylococcus aureus*.

11.3.4 Cancer Radiotherapy

In cancer radiotherapy, high energy radiation such as X-rays, gamma rays, electron beams, or protons are used to kill or damage cancer cells and stop them from growing and multiplying. Some good radiosensitizer materials are carbon ($Z = 6$), iodine ($Z = 53$), gadolinium ($Z = 64$), gold ($Z = 79$), and platinum ($Z = 80$). Among which, the high Z materials are the most attractive

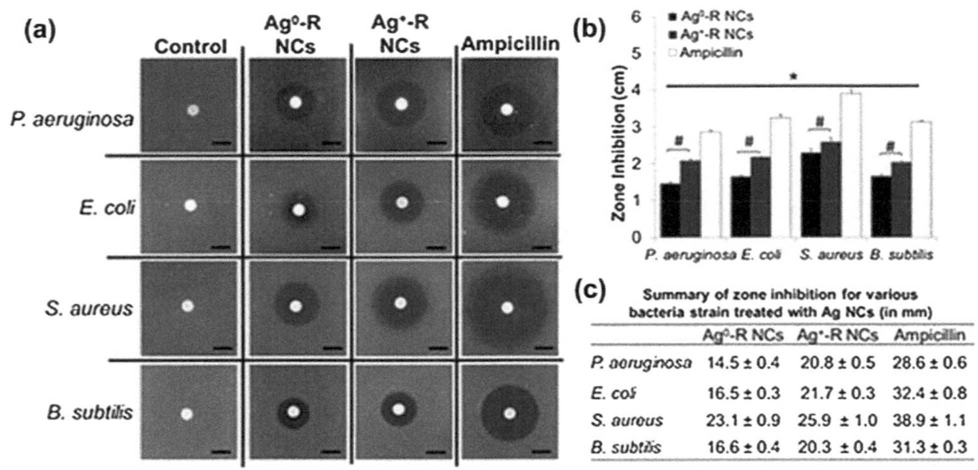

FIGURE 17 (a) Agar diffusion assay suggesting that the GSH-Ag^{+}-R NC-treated samples produce wider zone inhibition (b and c). *Reproduced with permission from Ref. 175. Copyright (2014) Springer.*

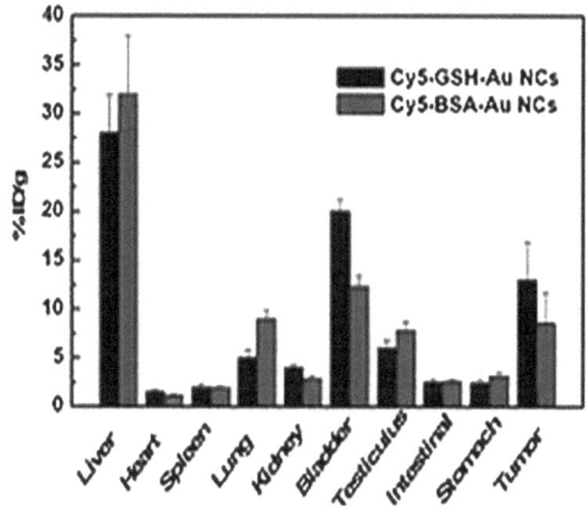

FIGURE 18 Biodistribution of GSH-Au_{25} NCs and BSA-Au_{25} NCs at 24 h after injection. *Reproduced with permission from Ref. 180. Copyright (2014) Wiley-VCH.*

due to their strong absorptions and high efficiency in generating the secondary radiation and electrons. In this regard, there is a growing interest to develop Au and Pt NP-based radiosensitizers for cancer radiotherapy.[177,178] In addition, it has been recently shown that smaller-sized Au NPs could feature with higher efficiency in terms of tumor deposition and radiation enhancement. Therefore, the biomolecule-functionalized Au NCs have been suggested as good radiosensitizer candidates as the several Au atoms inside the Au NCs may further improve the therapeutic efficiency by providing a locally high Au concentration in tumor (often with synergistic effects). We have recently investigated the sensitizing effect of the GSH-protected 25-atom Au NCs. It was found that the GSH-protected Au_{25} NCs were able to preferentially accumulate in tumors via an enhanced permeability and retention (EPR) effect and strongly enhanced the cancer radiotherapy (Figure 18).[179,180] On the other hand, the BSA-protected Au_{25} NCs mainly be accumulated in the liver and spleen, and can be used as the agent for liver cancer therapy.[181] In another study, Chen et al.[182] demonstrated that the conjugates of BSA-stabilized Au_{25} NCs and FA can be used for early tumor diagnosis and targeted therapy.

11.4 SUMMARY AND PROSPECTS

In this chapter, we discussed the recent advances on the functionalization and applications of metal NCs. It is evident that upon functionalization, metal NCs can achieve improved stability in solution, additional (typically predesigned) physicochemical properties, and enhanced catalytic properties. Thus, it is expected that in future more functionalization strategies will be developed as it could further enrich the library of functional NCs. There are some other interesting topics. For example, doping of metal NCs with some rare earth metal ions such as Eu^{3+}, Er^{3+}, or Gd^{3+} may also enrich or enhance the functionalities of the NCs. In addition, the fluorescent metal NCs may also be functionalized with a magnetic material to achieve dual functions in one NC species.

Fluorescent metal NCs with well-controlled functionalities have recently emerged as a new class of promising functional materials for many practical applications, such as sensing, bioimaging, antimicrobial agents, and cancer therapy. In the NC-based optical sensors, the selectivity and sensitivity of the NC probes are largely dictated by the properties of the metal core or the ligand shell, which may provide opportunities for chemists to further tailor these attributes to cater the sensor development for a specific analyte. However, the working principles and some fundamental issues related to NC-based sensor systems are presently lacking, which require more concerted efforts from the community. For cancer therapy, some metal NCs, such as Au and Pt NCs, have been proven to be promising as radiosensitizers, but this field is still in its infancy and more systematic studies are needed. We hope the concepts and principles discussed in this chapter may invite more concerted efforts from the community on functionalization and applications of metal NCs, which may finally pave the way of functional metal NCs toward practical applications.

ACKNOWLEDGMENT

This work is financially supported by the Ministry of Education, Singapore, under the Grant of R-279-000-409-112.

REFERENCES

1. Jin R. Quantum sized, thiolate-protected gold nanoclusters. *Nanoscale* 2010;**2**:343–62.
2. Zheng J, Nicovich PR, Dickson RM. Highly fluorescent noble-metal quantum dots. *Annu Rev Phys Chem* 2007;**58**:409–31.
3. Lu Y, Chen W. Sub-nanometre sized metal clusters: from synthetic challenges to the unique property discoveries. *Chem Soc Rev* 2012;**41**:3594–623.
4. Qian H, Zhu M, Wu Z, Jin R. Quantum sized gold nanoclusters with atomic precision. *Acc Chem Res* 2012;**45**:1470–9.
5. Li G, Jin R. Atomically precise gold nanoclusters as new model catalysts. *Acc Chem Res* 2013;**46**:1749–58.
6. Gautier C, Bürgi T. Chiral inversion of gold nanoparticles. *J Am Chem Soc* 2008;**130**:7077–84.
7. Barrabés N, Zhang B, Bürgi T. Racemization of chiral $Pd_2Au_{36}(SC_2H_4Ph)_{24}$: doping increases the flexibility of the cluster surface. *J Am Chem Soc* 2014;**136**:14361–4.
8. Lopez-Acevedo O, Tsunoyama H, Tsukuda T, Häkkinen H, Aikens CM. Chirality and electronic structure of the thiolate-protected Au_{38} nanocluster. *J Am Chem Soc* 2010;**132**:8210–8.
9. Wu Z, Jin R. On the ligand's role in the fluorescence of gold nanoclusters. *Nano Lett* 2010;**10**:2568–73.

10. Guo R, Murray RW. Substituent effects on redox potentials and optical gap energies of molecule-like $Au_{38}(SPhX)_{24}$ nanoparticles. *J Am Chem Soc* 2005;**127**:12140–3.
11. McCoy RS, Choi S, Collins G, Ackerson BJ, Ackerson CJ. Superatom paramagnetism enables gold nanocluster heating in applied radiofrequency fields. *ACS Nano* 2013;**7**:2610–6.
12. Desireddy A, Conn BE, Guo J, Yoon B, Barnett RN, Monahan BM, et al. Ultrastable silver nanoparticles. *Nature* 2013;**501**:399–402.
13. Luo Z, Zheng K, Xie J. Engineering ultrasmall water-soluble gold and silver nanoclusters for biomedical applications. *Chem Commun* 2014;**50**:5143–55.
14. Tan X, Jin R. Ultrasmall metal nanoclusters for bio-related applications. *WIREs Nanomed Nanobiotechnol* 2013;**5**:569–81.
15. Kong Y, Chen J, Gao F, Brydson R, Johnson B, Heath G, et al. Near-infrared fluorescent ribonuclease-a-encapsulated gold nanoclusters: preparation, characterization, cancer targeting and imaging. *Nanoscale* 2013;**5**:1009–17.
16. Shiang YC, Huang CC, Chen WY, Chen PC, Chang HT. Fluorescent gold and silver nanoclusters for the analysis of biopolymers and cell imaging. *J Mater Chem* 2012;**22**:12972–82.
17. Goswami N, Giri A, Bootharaju MS, Xavier PL, Pradeep T, Pal SK. Copper quantum clusters in protein matrix: potential sensor of Pb^{2+} ion. *Anal Chem* 2011;**83**:9676–80.
18. Yuan X, Luo Z, Yu Y, Yao Q, Xie J. Luminescent noble metal nanoclusters as an emerging optical probe for sensor development. *Chem Asian J* 2013;**8**:858–71.
19. Zhang L, Wang E. Metal nanoclusters: new fluorescent probes for sensors and bioimaging. *Nano Today* 2014;**9**:132–57.
20. Yuan X, Yeow TJ, Zhang Q, Lee JY, Xie J. Highly luminescent Ag^+ nanoclusters for Hg^{2+} ion detection. *Nanoscale* 2012;**4**:1968–71.
21. Mathew A, Pradeep T. Noble metal clusters: applications in energy, environment, and biology. *Part Part Syst Charact* 2014;**31**:1017–53.
22. Yamazoe S, Koyasu K, Tsukuda T. Nonscalable oxidation catalysis of gold clusters. *Acc Chem Res* 2013;**47**:816–24.
23. Goswami N, Zheng K, Xie J. Bio-NCs – the marriage of ultrasmall metal nanoclusters with biomolecules. *Nanoscale* 2014;**6**:13328–47.
24. Yuan X, Setyawati MI, Tan A, Ong CN, Leong DT, Xie J. Highly luminescent silver nanoclusters with tunable emissions: cyclic reduction-decomposition synthesis and antimicrobial properties. *NPG Asia Mater* 2013;**5**:e39.
25. Zheng J, Zhang C, Dickson RM. Highly fluorescent, water-soluble, size-tunable gold quantum dots. *Phys Rev Lett* 2004;**93**:077402.
26. Díez I, Pusa M, Kulmala S, Jiang H, Walther A, Goldmann A, et al. Color tunability and electrochemiluminescence of silver nanoclusters. *Angew Chem Int Ed* 2009;**48**:2122–5.
27. Qian H, Jiang D-e, Li G, Gayathri C, Das A, Gil RR, et al. Monoplatinum doping of gold nanoclusters and catalytic application. *J Am Chem Soc* 2012;**134**:16159–62.
28. Negishi Y, Kurashige W, Niihori Y, Nobusada K. Toward the creation of stable, functionalized metal clusters. *Phys Chem Chem Phys* 2013;**15**:18736–51.
29. Kurashige W, Niihori Y, Sharma S, Negishi Y. Recent progress in the functionalization methods of thiolate-protected gold clusters. *J Phys Chem Lett* 2014;**5**:4134–42.
30. Kurashige W, Yamaguchi M, Nobusada K, Negishi Y. Ligand-induced stability of gold nanoclusters: thiolate versus selenolate. *J Phys Chem Lett* 2012;**3**:2649–52.
31. Zhu M, Aikens CM, Hollander FJ, Schatz GC, Jin R. Correlating the crystal structure of a thiol-protected Au_{25} cluster and optical properties. *J Am Chem Soc* 2008;**130**:5883–5.

32. Zeng C, Qian H, Li T, Li G, Rosi NL, Yoon B, et al. Total structure and electronic properties of the gold nanocrystal $Au_{36}(SR)_{24}$. *Angew Chem Int Ed* 2012;**51**:13114–8.
33. Heaven MW, Dass A, White PS, Holt KM, Murray RW. Crystal structure of the gold nanoparticle $[N(C_8H_{17})_4][Au_{25}(SCH_2CH_2Ph)_{18}]$. *J Am Chem Soc* 2008;**130**:3754–5.
34. Jadzinsky PD, Calero G, Ackerson CJ, Bushnell DA, Kornberg RD. Structure of a thiol monolayer-protected gold nanoparticle at 1.1 Å resolution. *Science* 2007;**318**:430–3.
35. Fields-Zinna CA, Crowe MC, Dass A, Weaver JEF, Murray RW. Mass spectrometry of small bimetal monolayer-protected clusters. *Langmuir* 2009;**25**:7704–10.
36. Ferrando R, Jellinek J, Johnston RL. Nanoalloys: from theory to applications of alloy clusters and nanoparticles. *Chem Rev* 2008;**108**:845–910.
37. Kacprzak KA, Lehtovaara L, Akola J, Lopez-Acevedo O, Häkkinen H. A density functional investigation of thiolate-protected bimetal $PdAu_{24}(SR)_{18}(z)$ clusters: doping the superatom complex. *Phys Chem Chem Phys* 2009;**11**:7123–9.
38. Jiang D-e, Dai S. From superatomic $Au_{25}(SR)_{18}^{-}$ to superatomic $M@Au_{24}(SR)_{18}^{q}$ core–shell clusters. *Inorg Chem* 2009;**48**:2720–2.
39. Walter M, Moseler M. Ligand-protected gold alloy clusters: doping the superatom. *J Phys Chem C* 2009;**113**:15834–7.
40. Negishi Y, Kurashige W, Niihori Y, Iwasa T, Nobusada K. Isolation, structure, and stability of a dodecanethiolate-protected Pd_1Au_{24} cluster. *Phys Chem Chem Phys* 2010;**12**: 6219–25.
41. Negishi Y, Kurashige W, Kobayashi Y, Yamazoe S, Kojima N, Seto M, et al. Formation of a $Pd@Au_{12}$ superatomic core in $Au_{24}Pd_1(SC_{12}H_{25})_{18}$ probed by ^{197}Au Mössbauer and Pd K-edge EXAFS spectroscopy. *J Phys Chem Lett* 2013;**4**:3579–83.
42. Qian HF, Ellen B, Zhu Y, Jin RC. Doping 25-atom and 38-atom gold nanoclusters with palladium. *Acta Phys Chim Sin* 2011;**27**:513–9.
43. Negishi Y, Igarashi K, Munakata K, Ohgake W, Nobusada K. Palladium doping of magic gold cluster $Au_{38}(SC_2H_4Ph)_{24}$: formation of $Pd_2Au_{36}(SC_2H_4Ph)_{24}$ with higher stability than $Au_{38}(SC_2H_4Ph)_{24}$. *Chem Commun* 2012;**48**:660–2.
44. Kothalawala N, Kumara C, Ferrando R, Dass A. $Au_{144-x}Pd_x(SR)_{60}$ nanomolecules. *Chem Commun* 2013;**49**:10850–2.
45. Niihori Y, Kurashige W, Matsuzaki M, Negishi Y. Remarkable enhancement in ligand-exchange reactivity of thiolate-protected Au_{25} nanoclusters by single Pd atom doping. *Nanoscale* 2013;**5**:508–12.
46. Xie S, Tsunoyama H, Kurashige W, Negishi Y, Tsukuda T. Enhancement in aerobic alcohol oxidation catalysis of Au_{25} clusters by single Pd atom doping. *ACS Catal* 2012;**2**:1519–23.
47. Yao H, Kobayashi R. Chiral monolayer-protected Au–Pd bimetallic nanoclusters: effect of palladium doping on their chiroptical responses. *J Colloid Interface Sci* 2014;**419**:1–8.
48. Christensen SL, MacDonald MA, Chatt A, Zhang P, Qian H, Jin R. Dopant location, local structure, and electronic properties of $Au_{24}Pt(SR)_{18}$ nanoclusters. *J Phys Chem C* 2012;**116**:26932–7.
49. Negishi Y, Iwai T, Ide M. Continuous modulation of electronic structure of stable thiolate-protected Au_{25} cluster by Ag doping. *Chem Commun* 2010;**46**:4713–5.
50. Dou X, Yuan X, Yao Q, Luo Z, Zheng K, Xie J. Facile synthesis of water-soluble $Au_{25-x}Ag_x$ nanoclusters protected by mono- and bi-thiolate ligands. *Chem Commun* 2014;**50**:7459–62.
51. Dou X, Yuan X, Yu Y, Luo Z, Yao Q, Leong DT, et al. Lighting up thiolated Au@Ag nanoclusters via aggregation-induced emission. *Nanoscale* 2014;**6**:157–61.
52. Kumara C, Dass A. AuAg alloy nanomolecules with 38 metal atoms. *Nanoscale* 2012;**4**:4084–6.

53. Kumara C, Dass A. $(AuAg)_{144}(SR)_{60}$ Alloy nanomolecules. *Nanoscale* 2011;**3**:3064–7.
54. Wang S, Meng X, Das A, Li T, Song Y, Cao T, et al. A 200-fold quantum yield boost in the photoluminescence of silver-doped Ag_xAu_{25-x} nanoclusters: the 13th silver atom matters. *Angew Chem Int Ed* 2014;**53**:2376–80.
55. Gottlieb E, Qian H, Jin R. Atomic-level alloying and de-alloying in doped gold nanoparticles. *Chem Eur J* 2013;**19**:4238–43.
56. Negishi Y, Munakata K, Ohgake W, Nobusada K. Effect of copper doping on electronic structure, geometric structure, and stability of thiolate-protected Au_{25} nanoclusters. *J Phys Chem Lett* 2012;**3**:2209–14.
57. Kurashige W, Munakata K, Nobusada K, Negishi Y. Synthesis of stable Cu_nAu_{25-n} nanoclusters ($n = 1-9$) using selenolate ligands. *Chem Commun* 2013;**49**:5447–9.
58. Dharmaratne AC, Dass A. $Au_{144-x}Cu_x(SC_6H_{13})_{60}$ nanomolecules: effect of Cu incorporation on composition and plasmon-like peak emergence in optical spectra. *Chem Commun* 2014;**50**:1722–4.
59. Lan GY, Chen WY, Chang HT. Characterization and application to the detection of single-stranded DNA binding protein of fluorescent DNA-templated Copper/Silver nanoclusters. *Analyst* 2011;**136**:3623–8.
60. Su YT, Lan GY, Chen WY, Chang HT. Detection of copper ions through recovery of the fluorescence of DNA-templated copper/silver nanoclusters in the presence of mercaptopropionic acid. *Anal Chem* 2010;**82**:8566–72.
61. Biltek SR, Mandal S, Sen A, Reber AC, Pedicini AF, Khanna SN. Synthesis and structural characterization of an atom-precise bimetallic nanocluster, $Ag_4Ni_2(DMSA)_4$. *J Am Chem Soc* 2012;**135**:26–9.
62. Biltek SR, Sen A, Pedicini AF, Reber AC, Khanna SN. Isolation and structural characterization of a silver–platinum nanocluster, $Ag_4Pt_2(DMSA)_4$. *J Phys Chem A* 2014;**118**:8314–9.
63. Hainfeld JF. Gold cluster-labeled antibodies. *Nature* 1988;**333**:281–2.
64. Woehrle GH, Brown LO, Hutchison JE. Thiol-functionalized, 1.5-nm gold nanoparticles through ligand exchange reactions: scope and mechanism of ligand exchange. *J Am Chem Soc* 2005;**127**:2172–83.
65. Balasubramanian R, Guo R, Mills AJ, Murray RW. Reaction of $Au_{55}(PPh_3)_{12}Cl_6$ with thiols yields thiolate monolayer protected Au_{75} clusters. *J Am Chem Soc* 2005;**127**:8126–32.
66. Shichibu Y, Negishi Y, Tsukuda T, Teranishi T. Large-scale synthesis of thiolated Au_{25} clusters via ligand exchange reactions of phosphine-stabilized Au_{11} clusters. *J Am Chem Soc* 2005;**127**:13464–5.
67. McKenzie LC, Zaikova TO, Hutchison JE. Structurally similar triphenylphosphine-stabilized undecagolds, $Au_{11}(PPh_3)_7Cl_3$ and $[Au_{11}(PPh_3)_8Cl_2]Cl$, exhibit distinct ligand exchange pathways with glutathione. *J Am Chem Soc* 2014;**136**:13426–35.
68. Qian H, Zhu M, Lanni E, Zhu Y, Bier ME, Jin R. Conversion of polydisperse Au nanoparticles into monodisperse Au_{25} nanorods and nanospheres. *J Phys Chem C* 2009;**113**:17599–603.
69. Qian H, Eckenhoff WT, Zhu Y, Pintauer T, Jin R. Total structure determination of thiolate-protected Au_{38} nanoparticles. *J Am Chem Soc* 2010;**132**:8280–1.
70. Azubel M, Koivisto J, Malola S, Bushnell D, Hura GL, Koh AL, et al. Electron microscopy of gold nanoparticles at atomic resolution. *Science* 2014;**345**:909–12.
71. Nimmala PR, Yoon B, Whetten RL, Landman U, Dass A. $Au_{67}(SR)_{35}$ nanomolecules: characteristic size-specific optical, electrochemical, structural properties and first-principles theoretical analysis. *J Phys Chem A* 2013;**117**:504–17.

72. Qian H, Jin R. Ambient synthesis of $Au_{144}(SR)_{60}$ nanoclusters in methanol. *Chem Mater* 2011;**23**:2209–17.
73. Qian H, Jin R. Controlling nanoparticles with atomic precision: the case of $Au_{144}(SCH_2CH_2Ph)_{60}$. *Nano Lett* 2009;**9**:4083–7.
74. Yang H, Lei J, Wu B, Wang Y, Zhou M, Xia A, et al. Crystal structure of a luminescent thiolated Ag nanocluster with an octahedral Ag_{64}^{+} core. *Chem Commun* 2013;**49**: 300–2.
75. Yang H, Wang Y, Zheng N. Stabilizing subnanometer Ag(0) nanoclusters by thiolate and diphosphine ligands and their crystal structures. *Nanoscale* 2013;**5**:2674–7.
76. Yang H, Wang Y, Huang H, Gell L, Lehtovaara L, Malola S, et al. All-thiol-stabilized Ag_{44} and $Au_{12}Ag_{32}$ nanoparticles with single-crystal structures. *Nat Commun* 2013;**4**:2422.
77. Jin RC, Qian HF, Wu ZK, Zhu Y, Zhu MZ, Mohanty A, et al. Size focusing: a methodology for synthesizing atomically precise gold nanoclusters. *J Phys Chem Lett* 2010;**1**:2903–10.
78. Dass A, Holt K, Parker JF, Feldberg SW, Murray RW. Mass spectrometrically detected statistical aspects of ligand populations in mixed monolayer $Au_{25}L_{18}$ nanoparticles. *J Phys Chem C* 2008;**112**:20276–83.
79. Beqa L, Deschamps D, Perrio S, Gaumont AC, Knoppe S, Bürgi T. Ligand exchange reaction on $Au_{38}(SR)_{24}$, separation of $Au_{38}(SR)_{23}(SR')_1$ regioisomers, and migration of thiolates. *J Phys Chem C* 2013;**117**:21619–25.
80. Knoppe S, Dharmaratne AC, Schreiner E, Dass A, Bürgi T. Ligand exchange reactions on Au_{38} and Au_{40} clusters: a combined circular dichroism and mass spectrometry study. *J Am Chem Soc* 2010;**132**:16783–9.
81. Nishida N, Yao H, Kimura K. Chiral functionalization of optically inactive monolayer-protected silver nanoclusters by chiral ligand-exchange reactions. *Langmuir* 2008; **24**:2759–66.
82. Zeng C, Liu C, Pei Y, Jin R. Thiol ligand-induced transformation of $Au_{38}(SC_2H_4Ph)_{24}$ to $Au_{36}(SPh\text{-}t\text{-}Bu)_{24}$. *ACS Nano* 2013;**7**:6138–45.
83. Negishi Y, Kurashige W, Kamimura U. Isolation and structural characterization of an octaneselenolate-protected Au_{25} cluster. *Langmuir* 2011;**27**:12289–92.
84. Xu Q, Wang S, Liu Z, Xu G, Meng X, Zhu M. Synthesis of selenolate-protected $Au_{18}(SeC_6H_5)_{14}$ nanoclusters. *Nanoscale* 2013;**5**:1176–82.
85. Meng X, Xu Q, Wang S, Zhu M. Ligand-exchange synthesis of selenophenolate-capped Au_{25} nanoclusters. *Nanoscale* 2012;**4**:4161–5.
86. Kurashige W, Yamazoe S, Kanehira K, Tsukuda T, Negishi Y. Selenolate-protected Au_{38} nanoclusters: isolation and structural characterization. *J Phys Chem Lett* 2013;**4**:3181–5.
87. Habeeb Muhammed MA, Pradeep T. $Au_{25}@SiO_2$: quantum clusters of gold embedded in silica. *Small* 2011;**7**:204–8.
88. Le Guevel X, Hotzer B, Jung G, Schneider M. NIR-emitting fluorescent gold nanoclusters doped in silica nanoparticles. *J Mater Chem* 2011;**21**:2974–81.
89. Zhang Z, Zhang P, Guo K, Liang G, Chen H, Liu B, et al. Facile synthesis of fluorescent $Au@SiO_2$ nanocomposites for application in cellular imaging. *Talanta* 2011;**85**:2695–9.
90. Samanta A, Dhar BB, Devi RN. Ultrasmall gold cluster arrays encapsulated in silica nanospheres: applications in fluorescence imaging and catalysis. *J Phys Chem C* 2011;**116**:1748–54.
91. Huang P, Lin J, Wang S, Zhou Z, Li Z, Wang Z, et al. Photosensitizer-conjugated silica-coated gold nanoclusters for fluorescence imaging-guided photodynamic therapy. *Biomaterials* 2013;**34**:4643–54.

92. Ott LS, Hornstein BJ, Finke RG. A test of the transition-metal nanocluster formation and stabilization ability of the most common polymeric stabilizer, poly(vinylpyrrolidone), as well as four other polymeric protectants. *Langmuir* 2006;**22**:9357–67.
93. Mathew A, Sajanlal PR, Pradeep T. A fifteen atom silver cluster confined in bovine serum albumin. *J Mater Chem* 2011;**21**:11205–12.
94. Tsunoyama H, Sakurai H, Negishi Y, Tsukuda T. Size-specific catalytic activity of polymer-stabilized gold nanoclusters for aerobic alcohol oxidation in water. *J Am Chem Soc* 2005;**127**:9374–5.
95. Holmes P, James KAF, Levy LS. Is low-level environmental mercury exposure of concern to human health? *Sci Total Environ* 2009;**408**:171–82.
96. Pyykkö P. Theoretical chemistry of gold. *Angew Chem Int Ed* 2004;**43**:4412–56.
97. Sculfort S, Braunstein P. Intramolecular d^{10}–d^{10} interactions in heterometallic clusters of the transition metals. *Chem Soc Rev* 2011;**40**:2741–60.
98. Xie J, Zheng Y, Ying JY. Highly selective and ultrasensitive detection of Hg^{2+} based on fluorescence quenching of Au nanoclusters by Hg^{2+}–Au^{+} interactions. *Chem Commun* 2010;**46**:961–3.
99. Hofmann CM, Essner JB, Baker GA, Baker SN. Protein-templated gold nanoclusters sequestered within sol–gel thin films for the selective and ratiometric luminescence recognition of Hg^{2+}. *Nanoscale* 2014;**6**:5425–31.
100. Liu P, Shang L, Li H, Cui Y, Qin Y, Wu Y, et al. Synthesis of fluorescent α-chymotrypsin a-functionalized gold nanoclusters and their application to blot-based technology for Hg^{2+} detection. *RSC Adv* 2014;**4**:31536–43.
101. Lin YH, Tseng WL. Ultrasensitive sensing of Hg^{2+} and CH_3Hg^{+} based on the fluorescence quenching of lysozyme type VI-stabilized gold nanoclusters. *Anal Chem* 2010;**82**:9194–200.
102. Guo C, Irudayaraj J. Fluorescent Ag clusters via a protein-directed approach as a Hg(II) ion sensor. *Anal Chem* 2011;**83**:2883–9.
103. Zhou T, Huang Y, Li W, Cai Z, Luo F, Yang CJ, et al. Facile synthesis of red-emitting lysozyme-stabilized Ag nanoclusters. *Nanoscale* 2012;**4**:5312–5.
104. Guo W, Yuan J, Wang E. Oligonucleotide-stabilized Ag nanoclusters as novel fluorescence probes for the highly selective and sensitive detection of the Hg^{2+} ion. *Chem Commun* 2009:3395–7.
105. Adhikari B, Banerjee A. Facile synthesis of water-soluble fluorescent silver nanoclusters and Hg^{II} sensing. *Chem Mater* 2010;**22**:4364–71.
106. Yu P, Wen X, Toh YR, Huang J, Tang J. Metallophilic bond-induced quenching of delayed fluorescence in Au_{25}@BSA nanoclusters. *Part Part Syst Charact* 2013;**30**:467–72.
107. Huang CC, Yang Z, Lee KH, Chang HT. Synthesis of highly fluorescent gold nanoparticles for sensing mercury(II). *Angew Chem Int Ed* 2007;**46**:6824–8.
108. Deng L, Zhou Z, Li J, Li T, Dong S. Fluorescent silver nanoclusters in hybridized DNA duplexes for the turn-on detection of Hg^{2+} ions. *Chem Commun* 2011;**47**:11065–7.
109. Shang L, Dong S. Silver nanocluster-based fluorescent sensors for sensitive detection of Cu(II). *J Mater Chem* 2008;**18**:4636–40.
110. Lan GY, Huang CC, Chang HT. Silver nanoclusters as fluorescent probes for selective and sensitive detection of copper ions. *Chem Commun* 2010;**46**:1257–9.
111. Habeeb Muhammed MA, Verma PK, Pal SK, Retnakumari A, Koyakutty M, Nair S, et al. Luminescent quantum clusters of gold in bulk by albumin-induced core etching of nanoparticles: metal ion sensing, metal-enhanced luminescence, and biolabeling. *Chem Eur J* 2010;**16**:10103–12.

112. Durgadas CV, Sharma CP, Sreenivasan K. Fluorescent gold clusters as nanosensors for copper ions in live cells. *Analyst* 2011;**136**:933–40.
113. Wu Z, Wang M, Yang J, Zheng X, Cai W, Meng G, et al. Well-defined nanoclusters as fluorescent nanosensors: a case study on $Au_{25}(SG)_{18}$. *Small* 2012;**8**:2028–35.
114. Yue Y, Liu TY, Li HW, Liu Z, Wu Y. Microwave-assisted synthesis of BSA-protected small gold nanoclusters and their fluorescence-enhanced sensing of silver(I) ions. *Nanoscale* 2012;**4**:2251–4.
115. Kawasaki H, Hamaguchi K, Osaka I, Arakawa R. Ph-dependent synthesis of pepsin-mediated gold nanoclusters with blue green and red fluorescent emission. *Adv Funct Mater* 2011;**21**:3508–15.
116. Pu KY, Luo Z, Li K, Xie J, Liu B. Energy transfer between conjugated–oligoelectrolyte-substituted POSS and gold nanocluster for multicolor intracellular detection of mercury ion. *J Phys Chem C* 2011;**115**:13069–75.
117. Chen CW, Wang CH, Wei CM, Hsieh CY, Chen YT, Chen YF, et al. Highly sensitive emission sensor based on surface plasmon enhanced energy transfer between gold nanoclusters and silver nanoparticles. *J Phys Chem C* 2009;**114**:799–802.
118. Liu G, Feng DQ, Chen T, Li D, Zheng W. DNA-templated formation of silver nanoclusters as a novel light-scattering sensor for label-free copper ions detection. *J Mater Chem* 2012;**22**:20885–8.
119. Roy S, Palui G, Banerjee A. The as-prepared gold cluster-based fluorescent sensor for the selective detection of As^{III} ions in aqueous solution. *Nanoscale* 2012;**4**:2734–40.
120. Sun J, Zhang J, Jin Y. 11-Mercaptoundecanoic acid directed one-pot synthesis of water-soluble fluorescent gold nanoclusters and their use as probes for sensitive and selective detection of Cr^{3+} and Cr^{6+}. *J Mater Chem C* 2013;**1**:138–43.
121. Wang C, Cheng H, Sun Y, Xu Z, Lin H, Lin Q, et al. Nanoclusters prepared from a silver/gold alloy as a fluorescent probe for selective and sensitive determination of lead(II). *Microchim Acta* 2014:1–7.
122. Liu Y, Ai K, Cheng X, Huo L, Lu L. Gold-nanocluster-based fluorescent sensors for highly sensitive and selective detection of cyanide in water. *Adv Funct Mater* 2010;**20**:951–6.
123. Wang XB, Wang YL, Yang J, Xing XP, Li J, Wang LS. Evidence of significant covalent bonding in $Au(CN)_2^-$. *J Am Chem Soc* 2009;**131**:16368–70.
124. Qu F, Li NB, Luo HQ. Polyethyleneimine-templated Ag nanoclusters: a new fluorescent and colorimetric platform for sensitive and selective sensing halide ions and high disturbance-tolerant recognitions of iodide and bromide in coexistence with chloride under condition of high ionic strength. *Anal Chem* 2012;**84**:10373–9.
125. Wang M, Wu Z, Yang J, Wang G, Wang H, Cai W. $Au_{25}(SG)_{18}$ as a fluorescent iodide sensor. *Nanoscale* 2012;**4**:4087–90.
126. Chen WY, Lan GY, Chang HT. Use of fluorescent DNA-templated gold/silver nanoclusters for the detection of sulfide ions. *Anal Chem* 2011;**83**:9450–5.
127. Shiang YC, Huang CC, Chang HT. Gold nanodot-based luminescent sensor for the detection of hydrogen peroxide and glucose. *Chem Commun* 2009:3437–9.
128. Xu JJ, Feng JJ, Zhong X, Chen HY. Low-potential detection of glucose with a biosensor based on the immobilization of glucose oxidase on polymer/manganese oxide layered nanocomposite. *Electroanalysis* 2008;**20**:507–12.
129. Jin L, Shang L, Guo S, Fang Y, Wen D, Wang L, et al. Biomolecule-stabilized Au nanoclusters as a fluorescence probe for sensitive detection of glucose. *Biosens Bioelectron* 2011;**26**:1965–9.

130. Hussain AMP, Sarangi SN, Kesarwani JA, Sahu SN. Au-nanocluster emission based glucose sensing. *Biosens Bioelectron* 2011;**29**:60–5.
131. Shang L, Dong S. Sensitive detection of cysteine based on fluorescent silver clusters. *Biosens Bioelectron* 2009;**24**:1569–73.
132. Han B, Wang E. Oligonucleotide-stabilized fluorescent silver nanoclusters for sensitive detection of biothiols in biological fluids. *Biosens Bioelectron* 2011;**26**:2585–9.
133. Huang Z, Pu F, Lin Y, Ren J, Qu X. Modulating DNA-templated silver nanoclusters for fluorescence turn-on detection of thiol compounds. *Chem Commun* 2011;**47**:3487–9.
134. Yuan X, Tay Y, Dou X, Luo Z, Leong DT, Xie J. Glutathione-protected silver nanoclusters as cysteine-selective fluorometric and colorimetric probe. *Anal Chem* 2012;**85**:1913–9.
135. Chen TH, Tseng WL. (Lysozyme type VI)-stabilized Au_8 clusters: synthesis mechanism and application for sensing of glutathione in a single drop of blood. *Small* 2012;**8**:1912–9.
136. Cui ML, Liu JM, Wang XX, Lin LP, Jiao L, Zhang LH, et al. Selective determination of cysteine using BSA-stabilized gold nanoclusters with red emission. *Analyst* 2012; **137**:5346–51.
137. Zhou Z, Du Y, Dong S. DNA–Ag nanoclusters as fluorescence probe for turn-on aptamer sensor of small molecules. *Biosens Bioelectron* 2011;**28**:33–7.
138. Zhou Z, Du Y, Dong S. Double-strand DNA-templated formation of copper nanoparticles as fluorescent probe for label-free aptamer sensor. *Anal Chem* 2011;**83**:5122–7.
139. Liu X, Wang F, Aizen R, Yehezkeli O, Willner I. Graphene oxide/nucleic-acid-stabilized silver nanoclusters: functional hybrid materials for optical aptamer sensing and multiplexed analysis of pathogenic DNAs. *J Am Chem Soc* 2013;**135**:11832–9.
140. Li PH, Lin JY, Chen CT, Ciou WR, Chan PH, Luo L, et al. Using gold nanoclusters as selective luminescent probes for phosphate-containing metabolites. *Anal Chem* 2012;**84**:5484–8.
141. Huang CC, Chiang CK, Lin ZH, Lee KH, Chang HT. Bioconjugated gold nanodots and nanoparticles for protein assays based on photoluminescence quenching. *Anal Chem* 2008;**80**:1497–504.
142. Huang CC, Chen CT, Shiang YC, Lin ZH, Chang HT. Synthesis of fluorescent carbohydrate-protected Au nanodots for detection of concanavalin a and *Escherichia coli*. *Anal Chem* 2009;**81**:875–82.
143. Triulzi RC, Micic M, Giordani S, Serry M, Chiou WA, Leblanc RM. Immunoasssay based on the antibody-conjugated PAMAM-dendrimer-gold quantum dot complex. *Chem Commun* 2006:5068–70.
144. Shiang YC, Lin CA, Huang CC, Chang HT. Protein a-conjugated luminescent gold nanodots as a label-free assay for immunoglobulin g in plasma. *Analyst* 2011;**136**:1177–82.
145. Hu L, Han S, Parveen S, Yuan Y, Zhang L, Xu G. Highly sensitive fluorescent detection of trypsin based on BSA-stabilized gold nanoclusters. *Biosens Bioelectron* 2012;**32**:297–9.
146. Sharma J, Yeh HC, Yoo H, Werner JH, Martinez JS. Silver nanocluster aptamers: in situ generation of intrinsically fluorescent recognition ligands for protein detection. *Chem Commun* 2011;**47**:2294–6.
147. Li J, Zhong X, Zhang H, Le XC, Zhu JJ. Binding-induced fluorescence turn-on assay using aptamer-functionalized silver nanocluster DNA probes. *Anal Chem* 2012;**84**:5170–4.
148. Guo W, Yuan J, Dong Q, Wang E. Highly sequence-dependent formation of fluorescent silver nanoclusters in hybridized DNA duplexes for single nucleotide mutation identification. *J Am Chem Soc* 2009;**132**:932–4.
149. Jia X, Li J, Han L, Ren J, Yang X, Wang E. DNA-hosted copper nanoclusters for fluorescent identification of single nucleotide polymorphisms. *ACS Nano* 2012;**6**:3311–7.

150. Zhang L, Zhu J, Guo S, Li T, Li J, Wang E. Photoinduced electron transfer of DNA/Ag nanoclusters modulated by g-quadruplex/hemin complex for the construction of versatile biosensors. *J Am Chem Soc* 2013;**135**:2403–6.
151. Yeh HC, Sharma J, Han JJ, Martinez JS, Werner JH. A DNA–silver nanocluster probe that fluoresces upon hybridization. *Nano Lett* 2010;**10**:3106–10.
152. Lan GY, Chen WY, Chang HT. One-pot synthesis of fluorescent oligonucleotide Ag nanoclusters for specific and sensitive detection of DNA. *Biosens Bioelectron* 2011;**26**:2431–5.
153. Lin SY, Chen NT, Sum SP, Lo LW, Yang CS. Ligand exchanged photoluminescent gold quantum dots functionalized with leading peptides for nuclear targeting and intracellular imaging. *Chem Commun* 2008:4762–4.
154. Lin CAJ, Yang TY, Lee CH, Huang SH, Sperling RA, Zanella M, et al. Synthesis, characterization, and bioconjugation of fluorescent gold nanoclusters toward biological labeling applications. *ACS Nano* 2009;**3**:395–401.
155. Muhammed MAH, Verma PK, Pal SK, Kumar RCA, Paul S, Omkumar RV, et al. Bright, NIR-emitting Au_{23} from Au_{25}: characterization and applications including biolabeling. *Chem Eur J* 2009;**15**:10110–20.
156. Retnakumari A, Setua S, Menon D, Ravindran P, Muhammed H, Pradeep T, et al. Molecular-receptor-specific, non-toxic, near-infrared-emitting Au cluster-protein nanoconjugates for targeted cancer imaging. *Nanotechnology* 2010;**21**:055103.
157. Chen T, Xu S, Zhao T, Zhu L, Wei D, Li Y, et al. Gold nanocluster-conjugated amphiphilic block copolymer for tumor-targeted drug delivery. *ACS Appl Mater Interfaces* 2012;**4**:5766–74.
158. Wang Y, Chen J, Irudayaraj J. Nuclear targeting dynamics of gold nanoclusters for enhanced therapy of $HER2^+$ breast cancer. *ACS Nano* 2011;**5**:9718–25.
159. Jao YC, Chen MK, Lin SY. Enhanced quantum yield of dendrimer-entrapped gold nanodots by a specific ion-pair association and microwave irradiation for bioimaging. *Chem Commun* 2010;**46**:2626–8.
160. Shang L, Dong S, Nienhaus GU. Ultra-small fluorescent metal nanoclusters: synthesis and biological applications. *Nano Today* 2011;**6**:401–18.
161. Shang L, Nienhaus GU. Gold nanoclusters as novel optical probes for in vitro and in vivo fluorescence imaging. *Biophys Rev* 2012;**4**:313–22.
162. Makarava N, Parfenov A, Baskakov IV. Water-soluble hybrid nanoclusters with extra bright and photostable emissions: a new tool for biological imaging. *Biophys J* 2005;**89**:572–80.
163. Yu J, Patel SA, Dickson RM. In vitro and intracellular production of peptide-encapsulated fluorescent silver nanoclusters. *Angew Chem Int Ed* 2007;**119**:2074–6.
164. Yu J, Choi S, Richards CI, Antoku Y, Dickson RM. Live cell surface labeling with fluorescent Ag nanocluster conjugates†. *Photochem Photobiol* 2008;**84**:1435–9.
165. Choi S, Yu J, Patel SA, Tzeng YL, Dickson RM. Tailoring silver nanodots for intracellular staining. *Photochem Photobiol Sci* 2011;**10**:109–15.
166. Zhou YW, Li CM, Liu Y, Huang CZ. Effective detection and cell imaging of prion protein with new prepared targetable yellow-emission silver nanoclusters. *Analyst* 2013;**138**:873–8.
167. Yin J, He X, Wang K, Qing Z, Wu X, Shi H, et al. One-step engineering of silver nanoclusters-aptamer assemblies as luminescent labels to target tumor cells. *Nanoscale* 2012;**4**:110–2.
168. Wang C, Wang C, Xu L, Cheng H, Lin Q, Zhang C. Protein-directed synthesis of pH-responsive red fluorescent copper nanoclusters and their applications in cellular imaging and catalysis. *Nanoscale* 2014;**6**:1775–81.

169. Huang H, Li H, Wang AJ, Zhong SX, Fang KM, Feng JJ. Green synthesis of peptide-templated fluorescent copper nanoclusters for temperature sensing and cellular imaging. *Analyst* 2014;**139**:6536–41.
170. Cao H, Chen Z, Zheng H, Huang Y. Copper nanoclusters as a highly sensitive and selective fluorescence sensor for ferric ions in serum and living cells by imaging. *Biosens Bioelectron* 2014;**62**:189–95.
171. Tanaka SI, Miyazaki J, Tiwari DK, Jin T, Inouye Y. Fluorescent platinum nanoclusters: synthesis, purification, characterization, and application to bioimaging. *Angew Chem Int Ed* 2011;**50**:431–5.
172. Tanaka S, Aoki K, Muratsugu A, Ishitobi H, Jin T, Inouye Y. Synthesis of green-emitting Pt_8 nanoclusters for biomedical imaging by pre-equilibrated Pt/PAMAM (G4-OH) and mild reduction. *Opt Mater Express* 2013;**3**:157–65.
173. Chen D, Gao S, Ge W, Li Q, Jiang H, Wang X. One-step rapid synthesis of fluorescent platinum nanoclusters for cellular imaging and photothermal treatment. *RSC Adv* 2014;**4**:40141–5.
174. Rai M, Yadav A, Gade A. Silver nanoparticles as a new generation of antimicrobials. *Biotechnol Adv* 2009;**27**:76–83.
175. Yuan X, Setyawati M, Leong D, Xie J. Ultrasmall Ag^{+}-rich nanoclusters as highly efficient nanoreservoirs for bacterial killing. *Nano Res* 2014;**7**:301–7.
176. Chen WY, Lin JY, Chen WJ, Luo L, Wei-Guang Diau E, Chen YC. Functional gold nanoclusters as antimicrobial agents for antibiotic-resistant bacteria. *Nanomedicine* 2010;**5**:755–64.
177. Zhang XD, Wu D, Shen X, Chen J, Sun YM, Liu PX, et al. Size-dependent radiosensitization of PEG-coated gold nanoparticles for cancer radiation therapy. *Biomaterials* 2012;**33**:6408–19.
178. Erika P, Samuel L, Hynd R, Noriko U, Katsumi K, Yoshiya F, et al. Platinum nanoparticles: a promising material for future cancer therapy? *Nanotechnology* 2010;**21**:085103.
179. Zhang XD, Luo Z, Chen J, Shen X, Song S, Sun Y, et al. Ultrasmall $Au_{10-12}(SG)_{10-12}$ nanomolecules for high tumor specificity and cancer radiotherapy. *Adv Mater* 2014;**26**:4565–8.
180. Zhang XD, Chen J, Luo Z, Wu D, Shen X, Song SS, et al. Enhanced tumor accumulation of sub-2 nm gold nanoclusters for cancer radiation therapy. *Adv Healthc Mater* 2014;**3**:133–41.
181. Zhang XD, Wu D, Shen X, Liu PX, Fan FY, Fan SJ. In vivo renal clearance, biodistribution, toxicity of gold nanoclusters. *Biomaterials* 2012;**33**:4628–38.
182. Chen H, Li S, Li B, Ren X, Li S, Mahounga DM, et al. Folate-modified gold nanoclusters as near-infrared fluorescent probes for tumor imaging and therapy. *Nanoscale* 2012;**4**:6050–64.

Index

Note: Page numbers followed by "f" and "t" indicates figures and tables respectively.

D

E

F

M

N

U

V

W

X

Z

Kolibri 360 GmbH
Ferdinand-Jühlke-Straße 7
99095 Erfurt, Deutschland
produktsicherheit@kolibri360.de